Gas Treating with Chemical Solvents

GAS TREATING WITH CHEMICAL SOLVENTS

GIANNI ASTARITA
University of Naples and University of Delaware

DAVID W. SAVAGE
Exxon Research and Engineering Company

ATTILIO BISIO
Exxon Research and Engineering Company

A Wiley-Interscience Publication

JOHN WILEY & SONS

New York Chichester Brisbane Toronto Singapore

Library of Congress Cataloging in Publication Data:
Astarita, Giovanni.
Gas treating with chemical solvents.

Bibliography: p.
Includes index.
1. Gases—Purification. 2. Solvents. I. Savage, David W. II. Bisio, Attilio. III. Title.
TP242.A87 1982 665.7 82-11016
ISBN 0-471-05768-1

Printed in the United States of America

10 9 8 7 6 5 4 3 2 1

To Nerina, Katherine Jane, and Rosemary

Foreword

The removal of acidic impurities from industrial gases is an essential step in many of the processes used in the natural gas, petroleum, and chemical industries. Companies within these industries have devoted countless hours to developing improved methods for the purification of gases. In addition, industry associations have been working for many years to meet the educational needs of their personnel.

In the natural gas industry, the American Gas Association, the Gas Producers' Association, and the Gas Conditioning Conferences of the University of Oklahoma have worked vigorously to educate engineers, operating personnel, and students. The data and procedures used by field and plant engineers to determine operating and design parameters have been assembled and made widely available to individuals throughout the world. However, it has become increasingly clear that there is a growing gap between applicable theoretical developments and actual industrial practice.

This volume is a first attempt to fill that gap. Undoubtedly, it will be widely used within industries where "gas treating" is an essential part of their business. Interest in gas treating is increasing both from the need to reduce energy consumption and because of the development of alternative energy sources. One should not consider this volume an end, but a new beginning, in industry's education of its technical personnel.

In addition, one should consider that there are only a few books available that present an adequate blend of theory and practice. Also, there is need for current industrial practice to be reflected in both undergraduate and graduate courses, as well as in industrial and academic research. It is hoped that this book will be of use in specialized graduate courses in mass transfer, thermodynamics of multiphase systems, and gas-liquid reactions. Researchers interested in an aspect of potential importance to gas treating should find the book of assistance in putting their thoughts and programs into perspective.

PETER J. LUCCHESI

Vice President
Corporate Research
Exxon Research and Engineering Company
Linden, New Jersey
September 1982

Preface

The subject of this book is the rational analysis of gas treating processes where liquids are used to absorb acid gases from gaseous streams. We have tried to cover all the relevant issues of the subject, with particular emphasis on aspects unique to the use of chemically reactive solvents. By chemical solvents we mean liquids containing components that react chemically with the acid gases after they dissolve in the liquid phase. We have treated in considerably less detail those aspects common to purely physical acid gas absorption processes. We presume that our readers have a working knowledge of classical gas absorption and mass transfer processes.

The book consists of four parts: industrial gas treating, theory, applications, and industrial problems.

The first part, industrial gas treating, consists of Chapter 1, where an overall discussion of all parts of the subject matter is given in a descriptive form. Also in Chapter 1, a few general concepts that enter again and again in subsequent parts, in different forms and for different reasons, are introduced in a basically intuitive way.

The second part, theory, deveops the *methodology* for the rational analysis and mathematical modeling of gas treating processes. The attempt here is primarily one of classification of categories of problems and of the respective techniques of solution. This part consists of two subdivisions; the first dealing with the *equilibrium* (or, with a possibly misleading but commonly used term, thermodynamic) behavior of chemical gas treating systems, and the second one dealing with their *kinetic*, or rate, behavior. The thermodynamic part is given in Chapter 2, whereas Chapters 3-8 are devoted to the kinetic behavior. The reason for such a large fraction of the book being devoted to the kinetic behavior is related to the fact that the chemical nature of the solvents used gives rise to peculiar kinds of phenomena mainly in this area. These phenomena have to do with the way in which the diffusion of dissolved acid gases in the liquid phase is coupled intimately with chemical reaction.

The third part, applications, applies the theoretical machinery to actual gas treating processes of common industrial use. This part, consisting of Chapters 9-12, is dedicated to the analysis of what we regard as the most important actual industrial processes of gas treating. These are carbon dioxide removal, hydrogen sulfide and sulfur contaminants removal, simultaneous and selective removal of mixed acid gases, and sulfur dioxide removal. In this part we move from the general which characterizes the theoretical part, to the specific which is required when analysis of a given industrial process is needed. The theoretical analysis is applied to systems characterized by specific values of the relevant parameters (e.g., reaction rate constants); experimental confirmation of the theoretical analysis is stressed. We hope that the reader, at the end of Chapter 12, will be convinced that, given any gas treating system, it is possible to follow a logical path of analysis that will lead ultimately to a fundamentally based understanding of the process and to the ability to construct a mathematical model of it capable of quantitative predictions of satisfactory accuracy. Such an understanding is essential to optimum economic design and operation of acid gas removal processes.

The last three chapters cover industrial problems as they are encountered in actual practice: the problems of gathering, efficiently and economically, the basic required data (Chapter 13); the problems related to the selection of the most appropriate process for any given situation (Chapter 14); and finally the problems of scaleup, design, and optimization of processes once they have been selected and once enough physico-chemical information is available (Chapter 15). It is perhaps obvious that, the better the understanding of the phenomena involved as developed up to and including Chapter 12, the more efficient will be the performing of the tasks discussed in the last three chapters and the more economical will be the resulting gas treating processes.

It is perhaps useful to suggest to the reader a "road map" to the book. One possible itinerary is, of course, going from the first to the last page without skipping anything. However, some industrial practitioners may find the theoretical part (Chapters 2-8) developed in a way that is too formal, and an alternate itinerary may be as follows. The reader may start from whatever specific process he or she may be interested in. Reading of the relevant chapter will introduce the theoretical concepts, the full understanding of which may require study of some of the theoretical material in previous chapters. In particular, as far as the theory of the rate behavior is concerned, Chapter 3 gives all the essential concepts, whereas Chapters 4-8 present a more detailed analysis of rate behavior.

One of the decisions that authors of a book need to make concerns what policy to follow in regard to literature references. We have decided to restrict references to a minimum. Modern, computer-aided literature searches have made books an obsolete source for such searches. Therefore, we have restricted our references to only two types: to those works from which we extracted directly some key concept or some specific information, and to those works of a general nature that we actually suggest the reader should consult. Of course, the large majority of the material in the book is not original, but we ourselves often do not know where in the literature we originally found the seed of concepts which we discuss. Some of the material in the book is original, or has been published only recently by ourselves; such material has not undergone the test of prolonged exposure in the open literature, and the reader should be alerted to this. Chapters 2, 7, and 8 contain most of the material of this type.

We wish to thank several people who have helped in a variety of ways in producing this book. K. B. Bischoff, J. J. Carberry, J. E. Myers, and J. Wei offered comments at the initial stage when we were deciding whether to write the book at all, and have made useful suggestions on the general layout. J. P. Crilly, B. Eisenberg, F. Gioia, R. R. Johnson, S. V. Joshi, C. J. Kim, G. T. Rochelle, J. A. Sarsten, G. Sartori, and G. R. Say have either read parts of the first draft and commented on it or generally influenced our thinking. Parts of the first draft were also used as class notes for a graduate course given at the Chemical Engineering Department of the University of Delaware in the fall of 1979 and many of the students taking the course made useful suggestions for improvement. One of us (GA) has received a grant from the National Research Council of Italy, Grant #79.02810.11, in partial support of this work.

Finally, we wish to acknowledge support from the Management of Exxon Research and Engineering Company. Although the authors have drawn on their own experiences with Exxon Corporation and its affiliated companies, this book is not an exposition of Exxon practices or operations in any particular aspect of gas treating. This book represents only the opinions and conclusions of the authors who are solely responsible for its content.

GIANNI ASTARITA

Newark, Delaware

DAVID W. SAVAGE

Linden, New Jersey

ATTILIO BISIO

Linden, New Jersey

September 1982

Acknowledgments

The following copyright holders have granted permission for their figures and tables to be reprinted: Academic Press and J. Cone (Figure 13.3.1), American Chemical Society (Figures 10.3.1, 11.1.7, and 13.3.2), American Chemical Society and R. W. Parrish (Figure 9.5.1), American Institute of Chemical Engineers and L. Crabs (Figures 9.5.9 and 15.7.1), American Institute of Chemical Engineers and D. J. Kubek (Figure 14.3.2), American Institute of Chemical Engineers and M. S. Kuh (Figure 15.6.2), American Institute of Chemical Engineers and R. W. Parrish (Figure 15.8.2), American Institute of Chemical Engineers and K. J. Stokes (Figures 14.3.1 and 15.7.3), Bureau of Mines, Department of Interior (Figure 2.4.1), Chemical Engineering (Figures 1.3.5 and 15.6.5, Tables 15.6.1, 15.6.2, and 15.6.3), Chemical Engineering Journal (Figure 15.3.3), Chemical Engineering Progress (Figures 1.3.6, 15.5.6, 15.6.3, and 15.6.6 and Table 15.8.1), Chemical Engineering Science (Figures 5.4.1, 10.3.2, 10.3.3, and 13.4.10 and Example 4.1), Marcel Dekker (Figures 1.3.1, 1.8.1, 9.1.1, 9.1.2, 9.1.3, 9.4.5, 9.5.8, 9.5.10, and 15.6.4), Deutschen Gesellschaft fur Mineral Olwissenschaft und Kohlechemie (Figures 10.1.1 and 10.1.2), Gas Conditioning Conference and B. Eisenberg (Figures 2.6.2 and 2.6.3), Gulf Publishing Company (Figures 1.3.3, 1.3.9, 1.10.2, and 14.4.1 and Tables 1.3.3 and 1.3.5), Hydrocarbon Processing (Figures 1.3.8, 1.8.2, 11.1.3, 12.1.4, 12.1.5, and 15.7.2 and Table 11.1.1), Institute of Chemical Engineers (Figures 12.1.2, 15.2.2, and 15.2.3), Japan Petroleum Institute and G. G. Frye (Figure 10.1.5), Journal of Air Pollution (Figure 11.1.6), Journal of Chemical Engineering of Japan (Figure 15.3.4), Linde Reports on Science and Technology (Figure 11.1.4), McGraw-Hill Book Company (Figures 15.1.2, 15.1.3, and 15.1.4), Norton Company, Chemical Process Products Division (Figure 15.3.1), Oil and Gas Journal (Figures 1.3.3, 9.1.4, 10.1.4, 11.1.1, 11.1.2, 11.1.8, and 11.1.9 and Table 1.3.2), The Petroleum Engineer (Figure 1.10.4), and the Tennessee Valley Authority and D. K. Fleming (Figures 14.2.1, 14.2.2, and 14.2.3).

Contents

PART ONE: INDUSTRIAL GAS TREATING

PART TWO: THEORY

PART FOUR: INDUSTRIAL PROBLEMS

List of Frequently Used Symbols

A, A'	The volatile component(s)
a, a'	Concentrations of A, A' in liquid
$\bar{a}$	Value of a that would equilibrate local composition
$\tilde{a}$	Interfacial area per unit volume
B	A chemical base
B_j	The jth non-volatile component
b_j	Concentration of B_j
$\hat{b}_j$	Value of b_j normalized to molarity, $\hat{b}_j = b_j/m$
C_c	Cyclic capacity
D	Diffusivity of A
D_j	Diffusivity of B_j
$\tilde{D}_j$	Equivalent diffusivity of ion B_j
E	Rate promotion factor
f	Fugacity
$g(\ )$	Function defined by Eq. 7.2.2
g_0	Value of g corresponding to the bulk-liquid composition
G	Inert gas molar flow rate
$\tilde{G}$	Inert gas molar superficial velocity
H, H'	Henry's Law constants for A, A' in reactive solvent
H^0	Value of H in pure solvent
$\Delta\mathrm{H}^{\mathrm{v}}$	Enthalpy of vaporization
$\Delta\mathrm{H}^{\mathrm{Diss}}$	Enthalpy of dissolution
I	Enhancement factor
I_F	Value of I in fast reaction regime
I_i	Ionic strength
I_∞	Value of I in instantaneous reaction regime
I_∞^s	Value of I_∞ in low driving force asymptote

k	Reaction rate constant
k_G	Gas phase mass transfer coefficient
k_L	Chemical mass transfer coefficient in liquid phase
k_L^0	Physical mass transfer coefficient in liquid phase
K	Thermodynamic equilibrium constant
L	Solution volumetric flow rate
$\tilde{L}$	Solution superficial velocity
M	Number of independent chemical reactions
m	Solution molarity
N, N'	Rate of absorption of A, A' per unit interface area (negative in desorption)
N^0	Number of transfer units
P	Defined by Eq. 15.3.13
p, p'	Partial pressures of A, A'
$p^*, p^{*\prime}$	Equilibrium values of p, p'
p^0	Vapor pressure of pure component
q_j	Order of reaction with respect to B_j
R	Gas constant
r	Rate of reaction
r_0	Rate of reaction at interface per unit area
$\overrightarrow{r}$	Rate of forward reaction
$\overleftarrow{r}$	Rate of reverse reaction
S_T	Thermodynamic selectivity, defined by Eq. 2.7.1
T	Temperature
t_D	Diffusion time
t_r	Reaction time
v	Partial molar volume
x	Distance from interface
Y	Gas phase mol fraction
y, y'	Fractional chemical saturation
$\bar{y}$	Total saturation, $\bar{y} = \alpha/m$
α	Total concentration of A, physically dissolved plus chemically combined
α_j	Concentration ratio, $\alpha_j = \alpha_0/b_{j0}$
β_j	Defined by Eq. 7.3.17
γ_A	Activity coefficient
δ	Film thickness
η	Effectiveness, defined by Eq. 3.7.3
λ	Distance from interface to reaction plane
λ_j	Coefficients for molarity equation, defined in Eq. 2.3.4
μ_j	Number of moles of A chemically combined in B_j

ν_j	Stoichiometric coefficient of B_j
ν_{jk}	Stoichiometric coefficient of B_j in kth reaction
ξ	Extent of reaction, defined by Eq. 2.2.4
$\overline{\xi}$	Defined by Eq. 6.2.12
ξ_j	Defined by Eq. 3.5.2
π	Total pressure
$\overline{\rho}$	Molar density of liquid
ϕ	Ratio of diffusion and reaction times

Suffixes

A	of A
a	activity based
B	in the feed gas (bottom of countercurrent absorber)
$_0$	in the bulk (subscript)
^{0}	pure component (superscript)
I or II	phase designation
i	at the interface
f	fugacity based (mass transfer coefficient)
j	of the jth component
g	gas phase
l	liquid phase
k	of the kth reaction
$*$	at equilibrium
ϵ	which equilibrates the bulk
max	maximum
min	minimum
Diss	dissolution
T	in the treated gas (top of countercurrent absorber)
∞	at infinite dilution

Gas Treating with Chemical Solvents

PART ONE: INDUSTRIAL GAS TREATING

1 Practice and Problems of Industrial Gas Treating

A chemical engineer is a professional man experienced in the design, construction, and operation of plants in which materials undergo chemical or physical change.

BRITISH INSTITUTION OF CHEMICAL ENGINEERS, INAUGURAL MEETING, 1922.

We have come to the conclusion that a chemical engineer as such does not in reality exist.

A. DUCKHAM, PRESIDENTIAL ADDRESS, BRITISH INSTITUTION OF CHEMICAL ENGINEERS, 1924.

This first chapter is meant to introduce the reader, in an informative way, to the industrial practice of gas treating and to the unsolved problems in the area. Basic concepts of classical chemical engineering that are required for the understanding of subsequent chapters are also reviewed briefly.

1.1. GENERAL BACKGROUND

Gas treating is the term used to describe the separation from gases of acidic impurities such as acid gases (CO_2, H_2S, SO_2), organic sulfur compounds, and certain other impurities. The main impurities encountered are listed in Table 1.1.1. Acid gas removal generally refers to removal of CO_2 and H_2S. Removal of SO_2 from gases is sometimes treated as a separate technology—flue gas desulfurization—although the separation principles and concepts are the same for both types of impurities.

In acid gas removal, the most common impurities are CO_2 and H_2S which often occur in quite large concentrations (5-50%) in industrial processes such as hydrogen manufacture, ammonia production, and natural gas purification. Organic sulfur contaminants tend to occur at much lower concentrations (below 0.5%). When

TABLE 1.1.1. Impurities Encountered in Gas Treating

Type of impurity	Acid gas removal	Flue gas desulfurization
Acid gases		
	Carbon dioxide (CO_2)	Sulfur dioxide (SO_2)
	Hydrogen sulfide (H_2S)	
Organic sulfur compounds		
	Carbonyl sulfide (COS)	
	Carbon disulfide (CS_2)	
	Thiophene	
	Mercaptans (RSH)[a]	
	Organic sulfides (RSSR, RSR)	
Other impurities		
	H_2O	H_2O
	HCN	CO_2
	NH_3	O_2
	Hydrocarbons	
	Particulates	Particulates
	SO_2	NO_x
	SO_3	
	Tar	

[a]R is used to designate an alkyl group.

present, the other impurities such as HCN, NH_3, and SO_3 tend to occur at very low levels but they can often be quite troublesome to remove. The removal, along with CO_2 and H_2S, of trace sulfur and nitrogen compounds will increase in importance in the future as new technologies such as synthetic fuels are introduced.

In flue gas desulfurization, the principal impurity, SO_2, occurs at a relatively low concentration (say, 0.25%), depending on the sulfur content of the fuel (coal or oil) that is being burned.

Some major industrial processes that need gas treating are listed in Table 1.1.2. The table lists for each process the acid gases to treating and the cleanup targets for each impurity present. Some processes (e.g., ammonia manufacture from natural gas), require removal of just one acidic impurity, CO_2. Others such as natural gas purification often require the simultaneous removal of several impurities such as CO_2, H_2S, and COS. In petroleum desulfurization by catalytic treatment under hydrogen, the sulfur is released as H_2S. In this situation, H_2S is the sole acidic impurity to be removed from the recycle H_2.

The cleanup target, the allowable extent of the impurity in the treated gas to meet product specifications, varies markedly from process to process and with the nature of the acidic impurity. In general, the cleanup target for H_2S is particularly

TABLE 1.1.2. Major Industrial Processes that Need Gas Treating

Process	Acid gases to treating[a]	Common cleanup targets (% acid gas)
Hydrogen manufacture	CO_2	$<0.1\%$ CO_2
Petroleum desulfurization	$CO_2 + H_2S + COS$	10 ppm H_2S
Coal liquefaction		
Chemicals		
Ammonia manufacture	CO_2	<16 ppm $CO_2 + CO$
(H_2/N_2 mixture)	$CO_2 + H_2S + COS$	0.01 ppm H_2S
Natural gas purification		
Pipeline gas	H_2S, CO_2, COS, RSH, etc.	<4 ppm H_2S; $<1\%$ CO_2
LNG feedstock		1–2 ppm H_2S; <50 ppm CO_2
Syn gas for chemicals (H_2/CO)		<500 ppm CO_2; <0.01 ppm H_2S
Coal gasification		
SNG (high Btu gas)	CO_2, H_2S, COS	500 ppm CO_2; 0.01 ppm H_2S
Intermediate Btu gas		
Low Btu gas		100 ppm H_2S
Oil desulfurization	H_2S	100 ppm H_2S
Refinery fuel gas treating	H_2S, CO_2, COS	100 ppm H_2S
Ethylene manufacture (steam cracker gas treating)	H_2S, CO_2	~1 ppm H_2S, 1 ppm CO_2
Flue gas desulfurization	SO_2	90% removal
Utilities (electric)		
Refineries, etc.		

[a]Typical composition of feed gases to treating are given in Tables 1.3.1, 1.3.2, 1.3.3, and 1.3.4.

TABLE 1.1.3. Basis for Separation of Acid Gases from Gaseous Streams

Component	Mol. wt.	B.Pt. (°C)	M.Pt. (°C)	Dipole moment, debye	Acidity (pK)	Chemical reaction
CO_2	44	-78.5	-5	0.0	6.37 (H_2CO_3)	Acid–Base (A–B)
H_2S	34	-60	-83	0.97	7.04	A–B; oxidation
SO_2	64	-10	-76	1.63	1.82 (H_2SO_3)	A–B
COS	60	-50	-138	0.72	—	A–B; hydrolysis
CS_2				0.0	—	A–B; hydrolysis
CH_3SH	48	6	-121	1.26	—	A–B; hydrolysis
C_4H_4S				0.56	—	—
H_2	2	-253	-257	0.0	—	—
CH_4	16	-162	-183	0.0	—	—
CO	28	-191	-207	0.11	—	—
C_2H_6	30	-89	-183	0.0	—	—
C_3H_8	44	-42	-188	0.0	—	—

stringent, e.g., 4 ppm by volume for pipeline gas at 1000 psi and 0.1 ppm for chemical applications such as ammonia synthesis. A stringent target is necessary for H_2S to avoid catalyst poisoning, corrosion, and environmental release. The CO_2 specification may be relatively loose for natural gas (<1%), but for ammonia and LNG manufacture, CO_2 must be reduced to 10 ppm and 100 ppm, respectively. The CO_2 cleanup targets are set to avoid catalyst poisoning or limit the amount of diluent CO_2 in the treated gas. Cleanup target plays an important role in the selection of a gas treating process.

Distillation, which is often the most economical approach for separating fluid mixtures, is not generally attractive compared to absorption into a liquid agent for acid gas separation from gaseous streams. The parent components of gaseous streams containing acidic impurities are usually light gases including hydrogen, methane, ethane, propane, carbon monoxide, and so on. Boiling points of these constituents and typical acid gas impurities are listed in Table 1.1.3; these are so low that cryogenic distillation would be needed.

Selective absorption of the acidic impurities from the parent gas into a liquid solvent may be based on physical absorption (based on a difference in solubility) or on absorption with reaction with a chemical base. The chemical approach takes advantage of the relative acidity of the impurities. In the case of removal of CO_2, H_2S, SO_2, and COS, acid-base chemistry is the basis for the majority of industrially important separations. Other chemical approaches which are applicable for certain acid gases include oxidation in the case of H_2S and hydrolysis in the case of COS and other sulfur contaminants.

1.2. CATEGORIES OF GAS TREATING PROCESSES

The main categories of gas treating processes are shown in Table 1.2.1. Absorption into a liquid agent, either physical absorption or absorption into a solution of a chemical base, is the most commonly used approach. This type of process is used

TABLE 1.2.1. Categories of Gas Treating Processes

Absorption into a liquid	Bulk removal
Physical—polar organic solvent	Bulk removal
Chemical—solution of a base	Bulk removal
Cryogenic separation	Bulk removal
Adsorption onto a solid	Bulk removal; Final or trace purification
Zeolites, activated carbon, clays	Bulk removal; Final or trace purification
Chemical conversion to another compound	Final or trace purification
Gas phase reaction, e.g., methanation	Final or trace purification
Reaction with solid, e.g., Fe_2O_3, ZnO	Final or trace purification
Reaction with slurried solids (SO_2)	Final or trace purification
Chemical oxidation (H_2S)	Final or trace purification
Hydrolysis (COS)	Final or trace purification

TABLE 1.2.2. Commercial Use of Different Types of Solvent Gas Treating Processes[a]

Solvent	Number of installations
Acid gas removal	>1000
Aqueous alkanolamine	
MEA	
DEA	
DGA	
DIPA	
Promoted hot potassium carbonate	>740
Organic promoters	
Inorganic promoters	
Organic solvent–alkanolamine	>130
Sulfolane/DIPA	
MeOH/MEA/DEA	
Aqueous solution of potassium salt of amino acids	~100
Physical (organic) solvents	73
Propylene carbonate	
Polyethylene glycol dialkyl ether	
N-methylpyrrolidone	
Chilled methanol	
Flue gas desulfurization	~200
Lime/limestone slurries in water	
Sodium sulfite	

[a]See Table 1.2.3 for chemical structures of alkanolamines.

almost exclusively for bulk removal of acidic contaminants. By bulk removal we mean removal of the majority of the acid gas present in high concentration down to a level such as 0.1-2.0% in the treated gas.

In certain circumstances, other types of gas treating processes such as adsorption (pressure swing) and cryogenic methods may be attractive for bulk acid gas removal. The relative attractiveness of pressure swing adsorption (PSA) compared to absorption into a liquid agent is generally highest in small plants. Cryogenic methods represent only a very small fraction of commercial applications. An improved cryogenic scheme for bulk CO_2 removal at high pressure has been described recently.

For removal of the last traces of acidic impurities from gases, adsorption onto a solid and chemical conversion to another compound (e.g., methanation) are used extensively.

The cumulative commercial use of different types of chemical and physical solvents for acid gas removal is shown in Table 1.2.2. Based on number of installations, the most widely used chemical solvents are aqueous alkanolamines (aminoalcohols) and promoted hot potassium carbonate processes. These are followed by organic solvent solutions of aminoalcohols and aqueous solutions of potassium salts of amino acids. Compared to over 2000 installations using chemical solvents, less

TABLE 1.2.3 Common Alkaline Reagents used in Chemical Solvents[a]

Monoethanolamine (MEA)	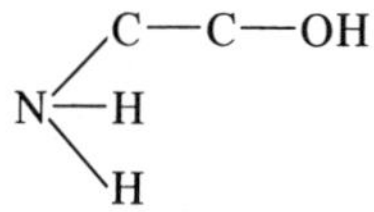
Diethanolamine (DEA)	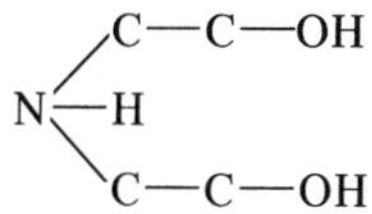
Diisopropanolamine (DIPA)	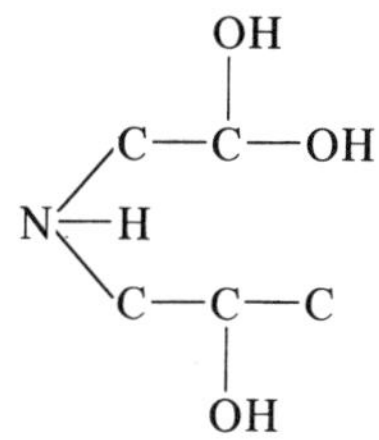
β, β′ Hydroxyaminoethylether (DGA)	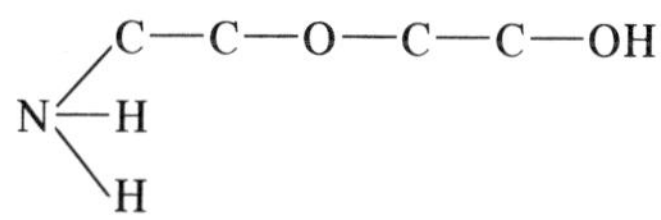
Potassium carbonate (with promoters)	K_2CO_3
Potassium glycinate	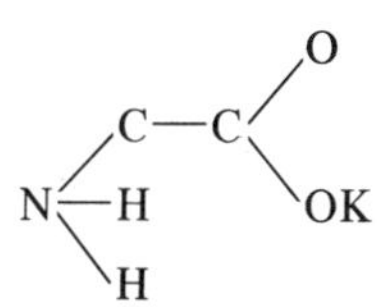
Caustic	NaOH

[a]Trade names for the major chemical solvent gas treating processes are given in Chapter 14 (Table 14.1.1).

than 100 physical solvent processes are in place. Common physical solvents include organic solvents such as alcohols, glycols, propylene carbonate, and pyrrolidones.

There are indications that physical solvent gas treating processes may assume increasing importance in the future as energy costs continue to rise (see also the discussion in Sections 1.5 and 1.8). Chemical solvents are expected to remain pre-eminent for low-pressure gas treating (< about 50 psi) and for higher pressure treating applications where thermal regeneration of the scrubbing solvent represents an attractive utilization of low-level waste heat.

Flue gas desulfurization is an infant technology compared to acid gas removal. However, there are over 50 installations already in the United States and a larger number in the Far East.

The most commonly used chemical solvents (usually aqueous or organic solvent solutions of organic nitrogen bases) are listed in Table 1.2.3. The second major

TABLE 1.2.4. CO_2, H_2S, and Propane Solubilities and Some Physical Properties of Organic (Physical) Solvents

Solvent[a]	Skeletal formula[b]	CO_2[c] solubility	H_2S[c] solubility	C_3H_8[c] solubility	B.Pt. (°C)	Viscosity (CPS)
Water	H—O—H	0.8	2.5	0.05	100	1.0
Propylene carbonate	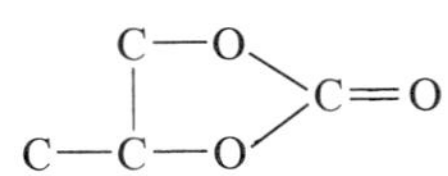	3.5	13.2	1.8, 3.5[d]	240	2.4
Dimethyl ether of polyethylene glycol	C—O—C—C—(O—C—C)$_n$—O—C	3.6	26.0	4.7	276	3.6
Methanol at 263K	C—O—H	8.0	41.0	—	—	—
at 243K		15.0	—	—	—	—
N-Methyl-Pyrolidone	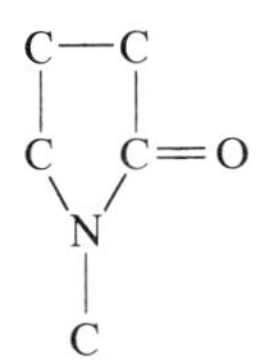	3.9	49.0	3.8	201	1.7

Tetrahydrothiophene-1,1-dioxide +3% water (sulfolane)	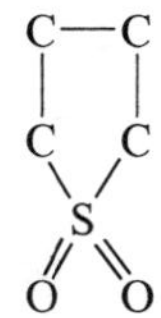	2.9	15.9	1.1	285	11.5
Methyl cyanoacetate	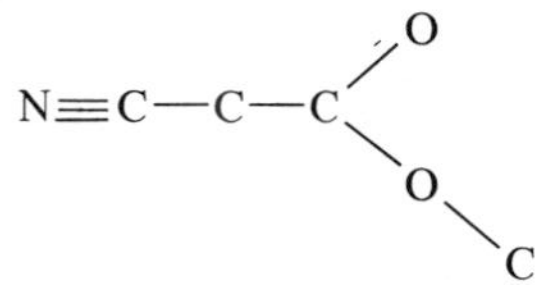	3.2	10.7	1.3	202	1.8
N-methyl-ε-Caprolactam	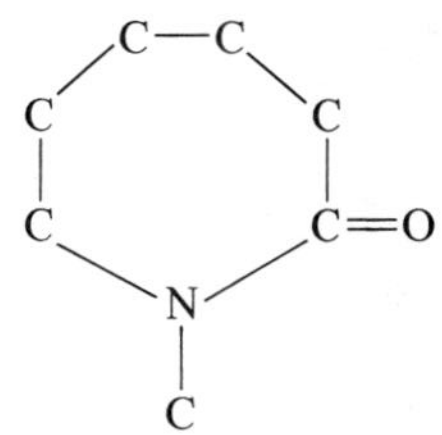	3.0	37.8	3.8	183	6.0

[a]Trademarks of common physical solvent processes are given in Chapter 14 (Table 14.1.1).
[b]In skeletal formulas, all H atoms except those associated with hydroxyl groups have been omitted.
[c]Units of solubility are cc gas/cc liquid/atm gas pressure. Solubilities refer to pure gases at 70–80°F.
[d]Two different values are reported in the literature.

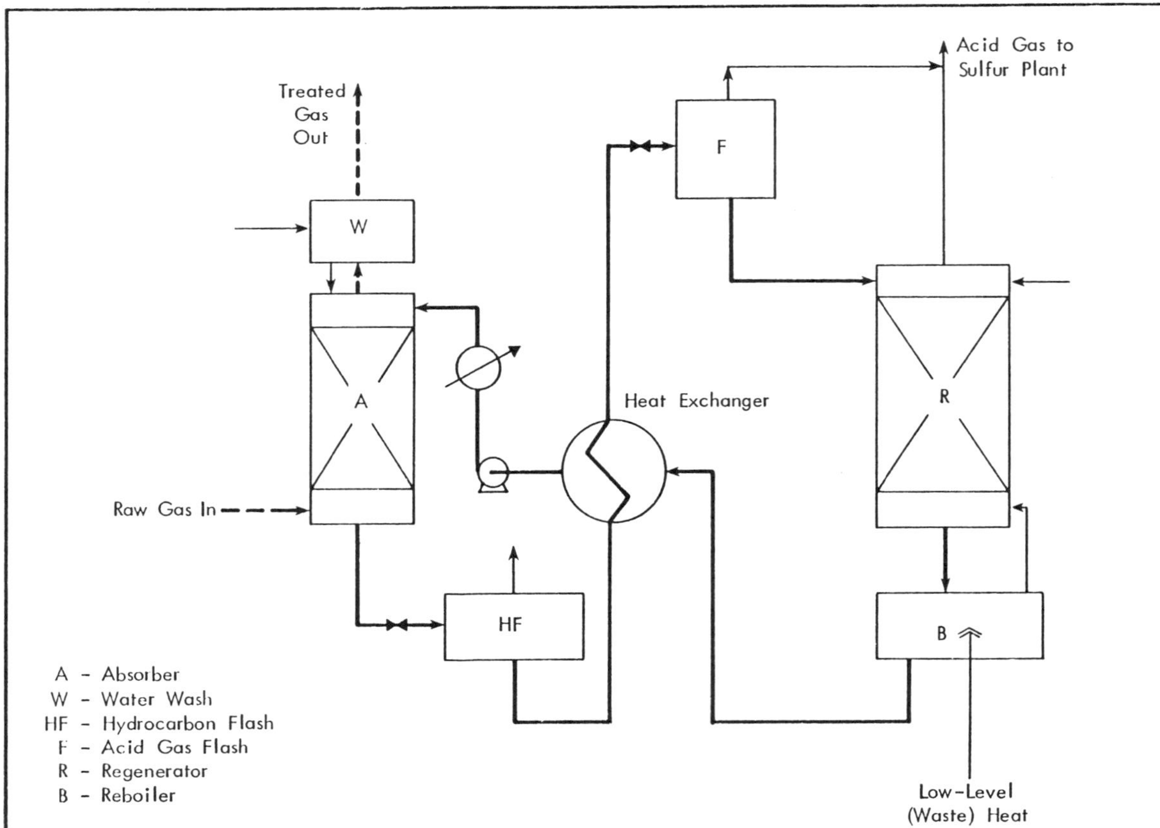

FIGURE 1.2.1. Idealized solvent absorption acid gas removal process.

class of solvents is based on aqueous potassium carbonate solutions with inorganic or organic additives called promoters. Aqueous sodium hydroxide (caustic) finds use as a final purification treatment.

Skeletal formulas and selected physical properties of the more important physical solvents are given in Table 1.2.4. Absorption in water is mainly of historical interest. The organic solvents that have seen most use include propylene carbonate, the dimethyl ether of polyethylene glycol, chilled methanol and *n*-methyl pyrrolidone. Tetrahydrothiophene-1,1-dioxide (sulfolane), and diethylene glycol are used mainly in conjunction with aminoalcohols.

Acid gas removal processes based on solvent absorption all use variations on a basic process flowsheet which is shown in idealized form in Figure 1.2.1. The raw gas is contacted countercurrently with regenerated (lean) solvent in an absorber tower which usually operates at low temperature (313-373K) and high pressure (100-1000 psig). Treated gas leaves overhead from the absorber, sometimes after water wash to prevent carryover of solvent components. The loaded (rich) solvent leaving from the bottom of the absorber is reduced in pressure, sometimes in stages, to flash off dissolved hydrocarbons and a portion of the acid gases. The rich solvent may also be heated against the returning hot regenerated lean solvent in a rich/lean heat exchanger. The regenerator is usually a low-pressure countercurrent tower in which the remaining dissolved acid gases are stripped out with steam generated by reboiling the lean solution. In practice, the hot rich solution is usually allowed to flash into the top of the regenerator tower (i.e., no separate flash drum is provided).

No gas treating process is a stand-alone operation. Every gas treating unit is integrated into a larger process. Common integration features include utilization of waste heat available in the raw gas or from an associated process unit (e.g., sulfur plant or steam reformer) to reboil the stripped solvent. Mass integration is also important. When H_2S is present in the acid gas, the separated acid gas must go directly to a sulfur plant (e.g., a Claus process) to minimize emissions of H_2S to the atmosphere. In CO_2 removal, particularly in connection with ammonia manufacture, the separated CO_2 may become the feedstock for making urea, or it may be recovered as a pure product (e.g., for carbonation of beverages or for injection into the ground to enhance recovery of oil and gas). The use of CO_2 for enhanced recovery is a relatively new technology that is expected to experience significant growth.

1.3. INDUSTRIAL GAS TREATING EXAMPLES

In this section we discuss four industrial gas treating examples chosen to illustrate the broad scope of gas treating problems and the variety of treating approaches that are available.

CO_2 Removal in Ammonia Manufacture

A simplified flowsheet of a high capacity, single-train plant for making ammonia by steam reforming of natural gas is shown in Figure 1.3.1. After sulfur removal,

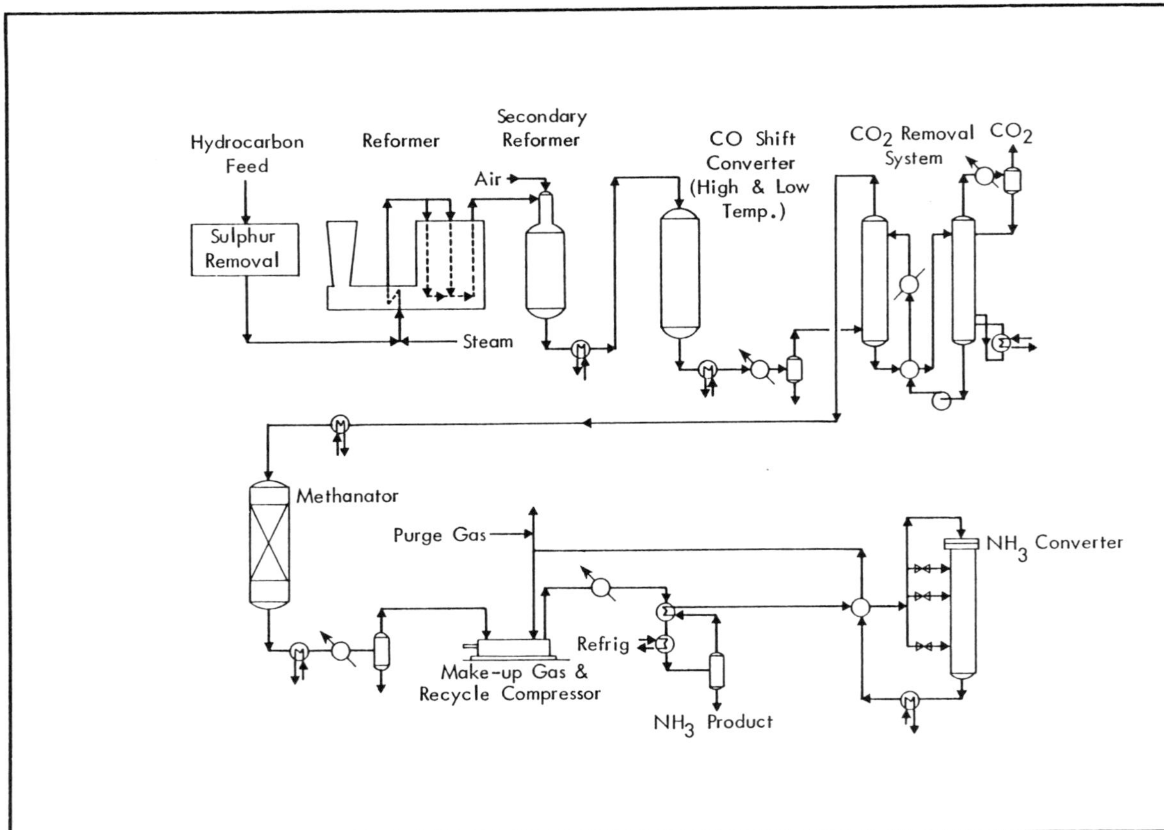

FIGURE 1.3.1. Flowsheet for high capacity, single-train ammonia plant.

TABLE 1.3.1. **Inlet Gas Treating in Ammonia Manufacture by Steam Reforming of Natural Gas**

Component	Mole% (dry basis)
H_2	58.3
N_2	18.3
CO_2	20.4
CO	1.0
H_2O	–
CH_4	1.7
Ar	0.3

the natural gas (largely methane) is catalytically reformed in the presence of steam to generate raw H_2, CO, CO_2, CH_4 synthesis gas. The appropriate amount of nitrogen is added by injecting air in a secondary reformer. Carbon monoxide (CO) is removed by shift conversion, generating additional H_2 and CO_2. The composition of the 345 psig raw inlet gas to the CO_2 removal section is given in Table 1.3.1.

Carbon dioxide is removed in two steps; the first step, using a chemical solvent, reduces CO_2 to about 0.1% by volume. The second step, a catalytic gas-phase methanation reaction, reduces total carbon oxides (CO_2 + CO) to less than 10 ppm, as required to protect the ammonia synthesis catalyst. This sequential combination of steps is usually more economical than the use of a chemical solvent alone to achieve $<$10 ppm CO_2 in the treated gas. Furthermore, a chemical solvent will not reduce CO to very low levels.

A representative ammonia plant CO_2 removal system is shown schematically in Figure 1.3.2. The chemical solvent is an organic-promoted hot potassium carbonate solution. The shift effluent (gas mixture of $H_2/N_2/CO_2/H_2O$) after giving up much of its sensible/latent heat to reboil the scrubbing solution, enters the base of the countercurrent absorber at about 383K. The absorber is of the split-flow type; partially regenerated (semi-lean) solution is fed to the top of the lower bulk-removal section of the absorber, while cooled, deeply regenerated (lean) solution is fed to the top of the upper cleanup section. Loaded (rich) solution leaves the bottom of the absorber at about 394K, the increase in temperature across the absorber reflecting the exothermic heat of CO_2 absorption. No rich–lean exchanger is needed, owing to the high absorber temperature of hot carbonate scrubbing. Rich solution is flashed directly into the top of the regenerator tower. The majority of the solution (semi-lean) is drawn off at the midpoint of the two-stage regenerator and pumped back to the absorber midpoint. A minor portion of the circulating solution is regenerated deeply by stripping with steam in the lower lean section of the regenerator. From the reboiler, lean solution is pumped to the top of the absorber. The split-flow design requires less energy for solution regeneration than a single stage configuration (see Figure 9.5.10).

In big ammonia plants, the towers are large absorber-reactor vessels typically

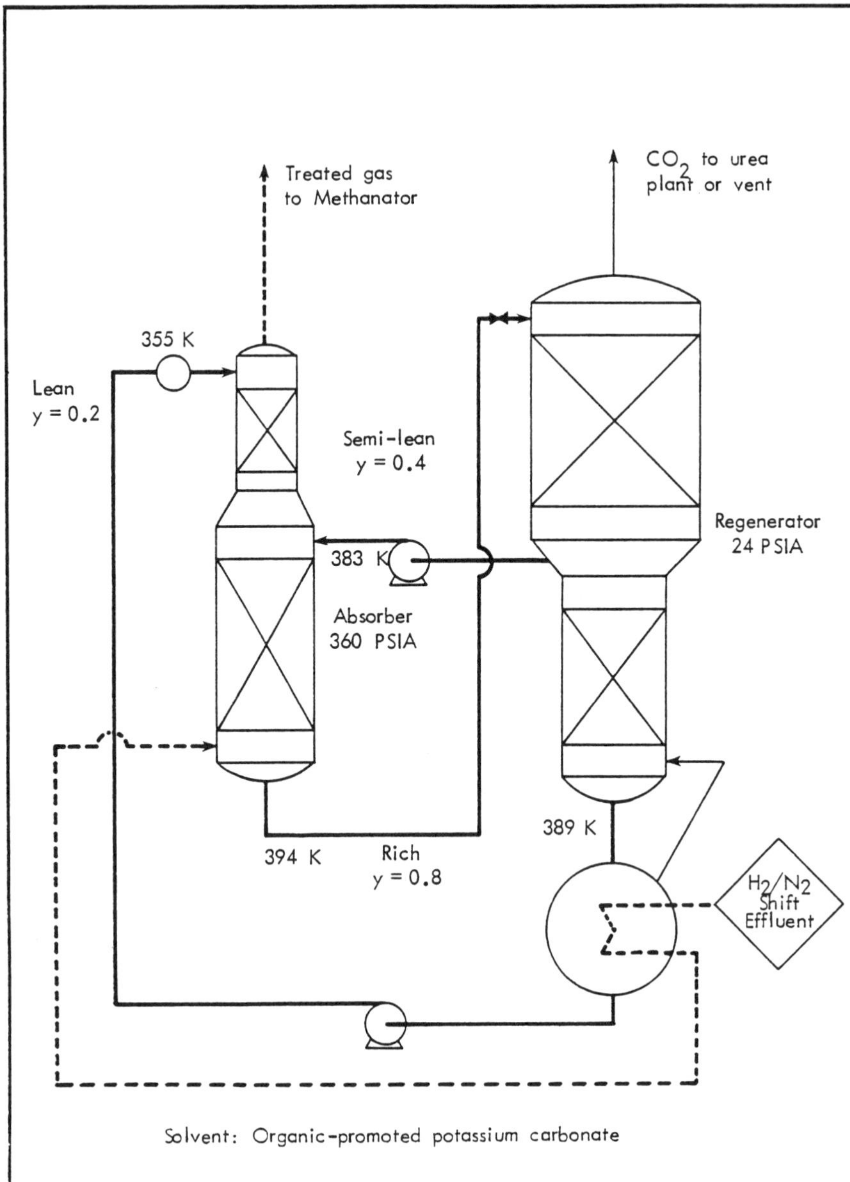

FIGURE 1.3.2. Ammonia plant carbon dioxide removal system using chemical absorption.

15 ft in diameter and over 100 ft high, containing either packing or trays for efficient gas-liquid contacting.

The loading of CO_2 in the aqueous potassium carbonate solvent is usually expressed in terms of the fractional saturation of the chemical solvent. Fractional saturation y is the fraction of potassium carbonate converted chemically to bicarbonate in the overall reaction:

$$CO_2 + K_2CO_3 + H_2O \rightleftharpoons 2KHCO_3 \tag{1.3.1}$$

Fractional saturation is given by

$$y = \frac{[HCO_3^-]}{[K^+]} \tag{1.3.2}$$

where $[HCO_3^-]$ is the molar concentration of bicarbonate ion, and $[K^+]$ is the potassium ion concentration. In principle, y can vary between zero and unity if no promoters are present. In practice, the y of the lean solution is typically 0.1-0.25, whereas that of the rich solution is typically 0.7-0.85.

Simultaneous Removal of H_2S, CO_2, and COS in Natural Gas Treating

Natural gas treating plants are usually situated within a few miles of the wellhead and may serve different gas fields and many individual wells. Many new gas fields are in remote locations and harsh environments. A representative gas gathering system and processing plant is shown in Figure 1.3.3. Natural gas treating systems

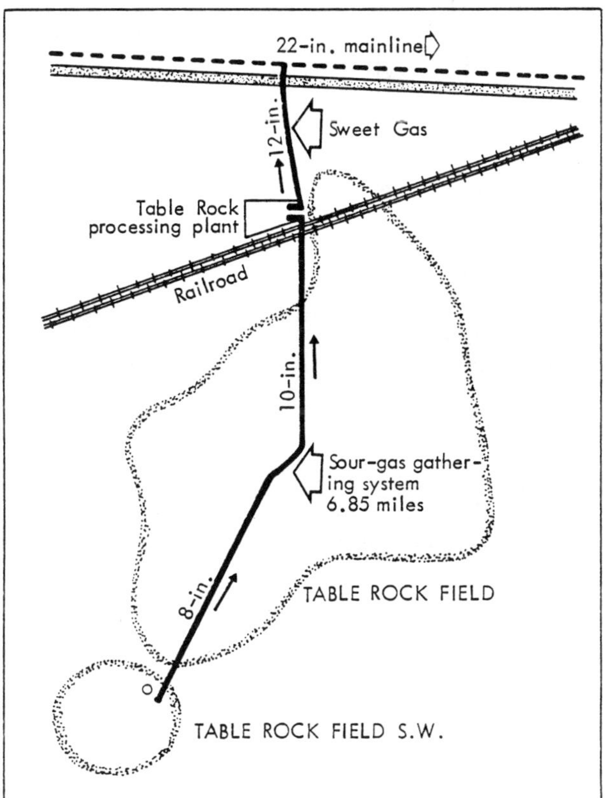

FIGURE 1.3.3. Gathering system and processing plant for sour natural gas field.

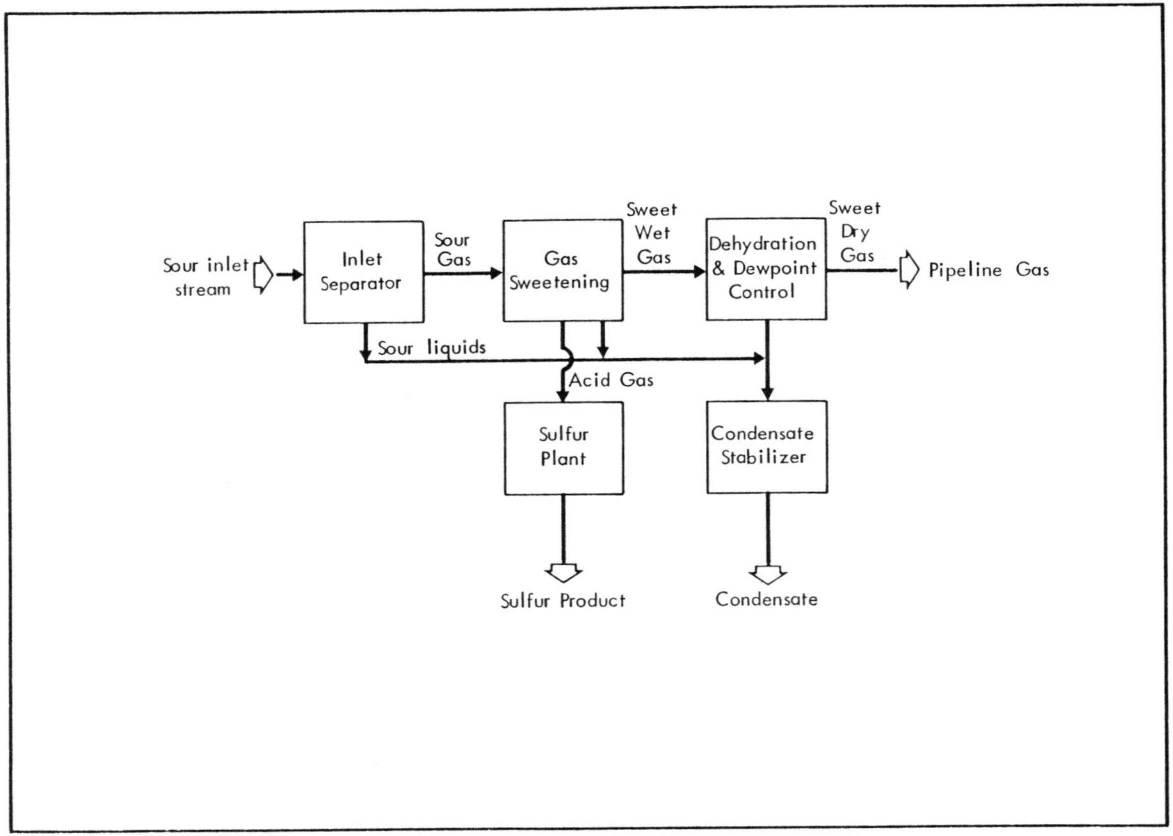

FIGURE 1.3.4. Sour natural gas processing plant.

TABLE 1.3.2. **Representative Sour Natural Gas Compositions**

Component	Wyoming (mole%)	Florida (mole%)
H_2	0.28	–
N_2	4.20	1.38
C_1	71.15	35.76
C_2	2.01	9.78
C_3	0.49	5.63
i-C_4	0.07	1.49
n-C_4	0.23	3.68
C_5^+	0.25	31.58
CO_2	17.56	2.15
H_2S	3.76	8.55
COS	trace	–

must have considerable flexibility to handle fluctuating raw gas rates and the possibility of changes in raw gas composition with time.

A sour natural gas process plant is shown in Figure 1.3.4. The sour inlet stream, which is usually a two-phase mixture, goes first to a separator to separate sour gas from raw hydrocarbon liquids. The sour gas is sweetened (treated) and finally dehydrated to meet pipeline specifications. Acid gas (CO_2 + H_2S + COS) is sent to an integrated sulfur plant for sulfur recovery as elemental sulfur.

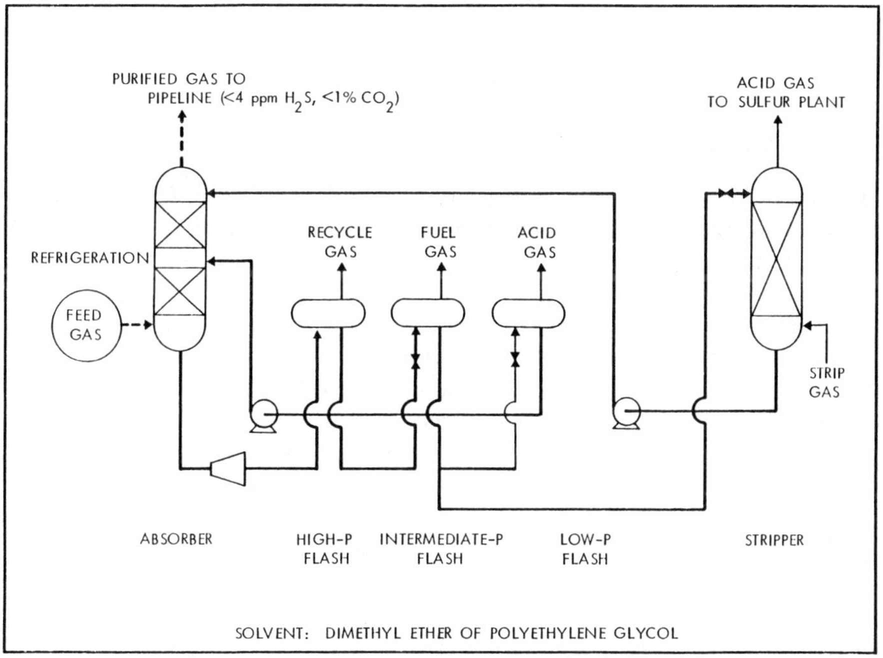

FIGURE 1.3.5. Acid gas separation flowsheet using physical absorption.

Sour gas composition can vary over a wide range. Two examples from recently installed natural gas treating plants are given in Table 1.3.2.

In the case of both of the above gases, the H_2S-to-CO_2 ratio is high enough for processing the separated acid gas to sulfur in a Claus plant. (A split-flow type of Claus plant may have to be used with Wyoming gas owing to the rather low H_2S-to-CO_2 ratio.) A sweetening process that can simultaneously remove the acidic impurities is therefore suitable.

Figure 1.3.5 shows an acid gas separation flowsheet using physical absorption in the dimethyl ether of polyethylene glycol. The sour gas is scrubbed countercurrently in a refrigerated (273K) absorber, giving pipeline gas of <4 ppm H_2S, <15 ppm total sulfur and $<1\%$ CO_2. Mild refrigeration is used to improve the capacity of the solvent for CO_2. The rich solvent is regenerated by flashing and stripping, usually with an extraneous strip gas (e.g., N_2). The flashes are carried out at successively decreasing pressures. The first flashes (high and intermediate pressure) serve to separate dissolved hydrocarbons that are either recycled to the feed gas (after recompression) or used as fuel gas. The third low-pressure flash releases a portion of the acid gas that goes into the sulfur plant along with stripper acid gas from the stripper overhead. A feature of larger modern physical solvent scrubbers is the use of hydraulic turbines on the high-pressure rich solvent to drive the solvent recycle pumps.

In physical solvent processes, loading of acid gas is best expressed as mole fraction or mole ratio, although many data are reported as a Bunsen coefficient, volume of acid gas per volume of solution (V/V). Acid gas loadings, particularly of CO_2, are quite low ($\sim$0.05 or less) unless the partial pressure of CO_2 in the feed gas is very high. With these low capacities, large amounts of physical solvent must be circulated.

SO_2 Removal from Flue Gases

The use of high-sulfur coal as a fuel for generating electricity or for firing process furnaces requires the installation of air quality control systems to prevent SO_2 emissions to the environment. Figure 1.3.6 shows a typical control system for a coal-fired electricity generating station.

With coal of very low sulfur content, the flue gas is treated for particulate removal in electrostatic precipitators and dispersed through a stack. With sulfurous coal, the system must be modified by addition of fans, SO_2 absorbers, and a reheater to provide phase buoyancy. The absorbers are tied in to reactors, a thickener, and waste disposal plants. Typical flue gas compositions are given in Table 1.3.3.

A simplified flow diagram for removal of SO_2 from combustion gases using limestone scrubbing is shown in Figure 1.3.7. The combustion product gases from the precipitator are scrubbed with powdered limestone ($CaCO_3$) slurried in water to effect removal of at least 90% of the SO_2 in the gas. Absorbed SO_2 reacts to form calcium sulfite and sulfate which is disposed of as a sludge after dewatering. The overall reaction leading to sulfate formation may be represented as:

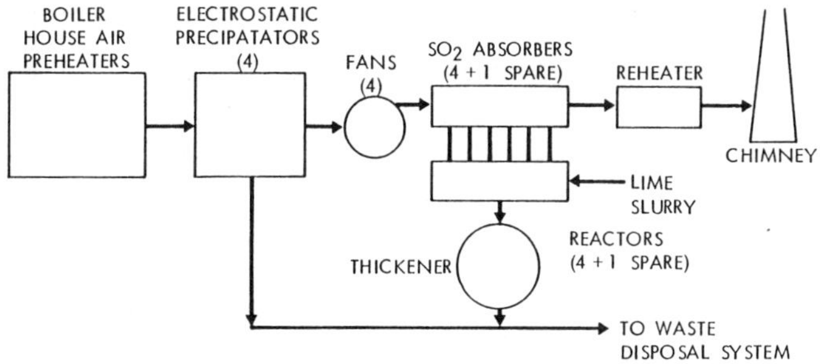

FIGURE 1.3.6. Air quality control system for coal-fired electricity generating station.

$$CaCO_{3(s)} + SO_{2(g)} + 2H_2O + 0.5\ O_2 \longrightarrow CaSO_4 \cdot 2H_2O_{(s)} + CO_{2(g)} \quad (1.3.3)$$

This application differs from the ones previously discussed in that the separating agent is a slurry of reactive solid rather than a homogeneous solution. This introduces operational problems of slurry flow and scaling in the absorbers.

TABLE 1.3.3. Typical Flue Gas Compositions

	Mole%	
Component	Oil-Fired[a]	Coal-Fired[b]
SO_2	0.245	0.245
SO_3	0.005	0.005
O_2	1.0	3.0
CO_2	13	15
H_2O	10	7
N_2	~76	~75

[a]4% sulfur fuel
[b]3% sulfur fuel

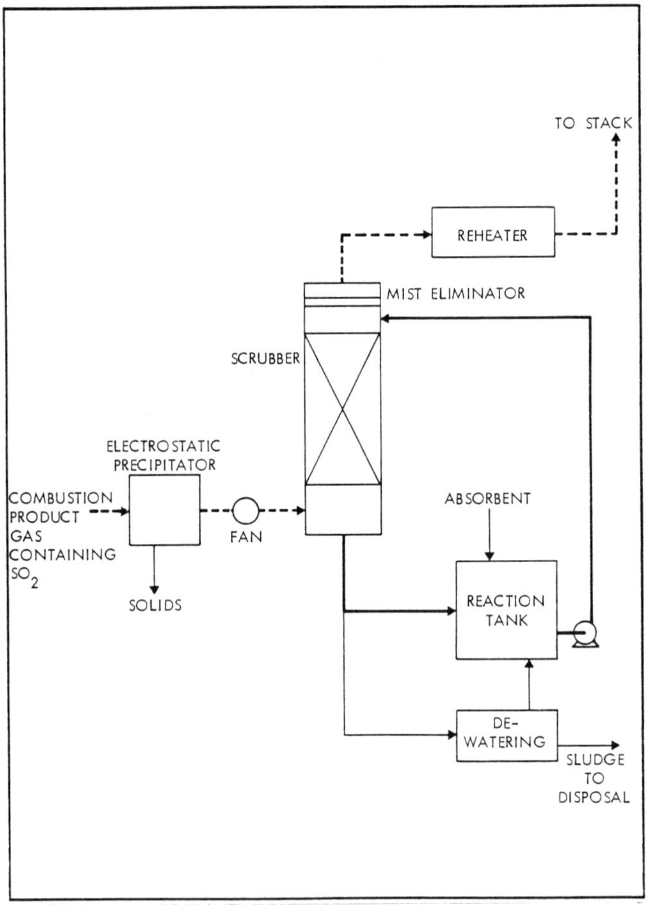

FIGURE 1.3.7. Simplified flow diagram for removal of sulfur dioxide from combustion gases using limestone slurry.

Selective H_2S Removal from Gases that Contain H_2S and CO_2

The manufacture of fuel gases and chemical synthesis gas from high-sulfur coal or high-sulfur heavy oil introduces new gas treating problems that did not arise in connection with traditional gasification and reforming processes. Gasification of high-sulfur carbonaceous substances generates a gas mixture containing acidic impurities CO_2, H_2S, and COS with H_2S-to-CO_2 ratios of from 1-to-10 to 1-to-30. Conventional treating of these gases with alkaline reagents leads to simultaneous removal of all acid gas constituents. Owing to the low concentration of H_2S in the separated acid gas, the latter is not a suitable feed for a Claus plant for sulfur recovery. To provide a suitable Claus plant feed gas, the gas mixture must be treated to selectively remove the H_2S (and the COS if possible), leaving as much as possible of the CO_2 in the parent gas mixture. It should also be mentioned that any CO_2 removed

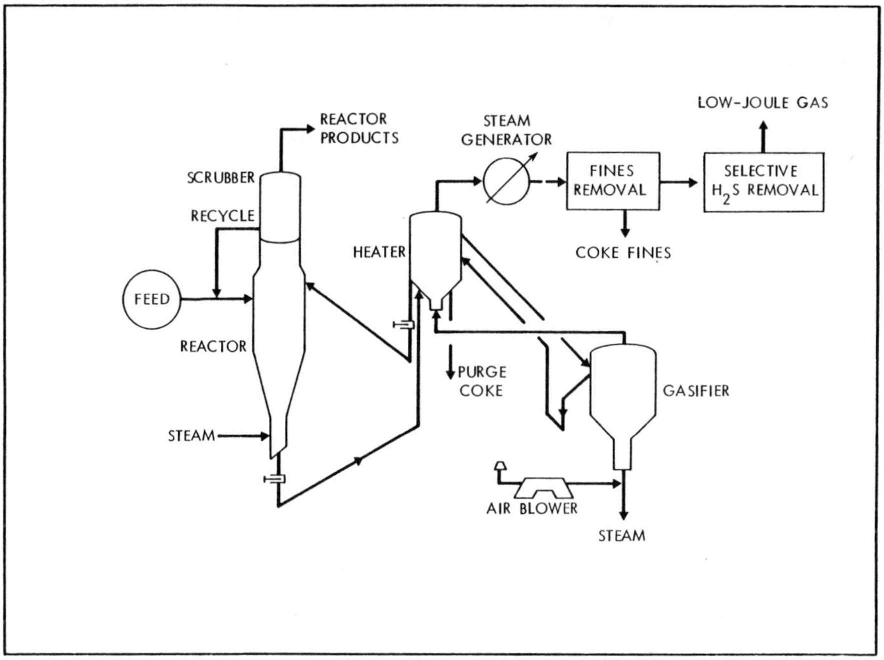

FIGURE 1.3.8. Simplified flowsheet of FLEXICOKING process.

increases the cost of the gas treating operation by increasing both the circulation rate of the solvent and the heat load to the regenerator reboiler.

An example of an application requiring selective H_2S removal is the manufacture of low heating value fuel gas from the residual portion of crude petroleum. Figure 1.3.8 shows a simplified flow plan of a coking process (FLEXICOKING) for converting the high-boiling residual portion of petroleum to low-boiling products and a fuel gas (~150 Btu) suitable for firing furnaces.

In FLEXICOKING, hot residuum feed is injected into the coker reactor where it is cracked thermally to a full range of vaporized products plus a coke product which is deposited on the circulating fluidized coke particles. The fluidized coke particles circulate between the reactor and a heater vessel where they pick up the heat required to vaporize the feed and supply the endothermic heat of cracking. Another stream of fluidized coke particles recycles between the heater and the gasifier vessel where the total coke produced is gasified at 1173K with steam and air. The hot gas passes through a steam generator and into a water scrubber where fine coke particles are removed. Finally, the gas is treated to remove H_2S and possibly COS. CO_2 removal is not normally required or economically justified.

The low heating value feed gas has the composition shown in Table 1.3.4.

Two approaches to selective H_2S removal can be considered. One involves chemical oxidation of H_2S to elemental sulfur. The other employs the concept of kinetic selectivity to selectively remove H_2S from CO_2 taking advantage of the faster rate of absorption of H_2S in solutions of tertiary and other amines. Although H_2S

TABLE 1.3.4. Flexicoker Low Heating Value Gas to Treating

Component	Mole%
CO	17.9
N_2	50.0
CH_4	1.4
H_2	18.2
CO_2	11.4
H_2S	1.0
COS	0.02
NH_3	0.06

and CO_2 have similar acidities, the rates of reaction with tertiary amines differ markedly owing to the different chemical reaction mechanisms involved. H_2S, being a Bronsted (proton) acid, reacts extremely rapidly by the following proton-transfer reaction with tertiary amine in solution:

$$H_2S_{(aq)} + R_3N \rightleftharpoons R_3NH^+ + HS^- \tag{1.3.4}$$

On the other hand, CO_2, which is a Lewis acid, must first react with water by a slow chemical reaction before it is transformed to the bicarbonate ion. The mechanism of the CO_2 reaction is believed to be:

$$CO_{2\,(aq)} + OH^- \overset{\text{slow}}{\rightleftharpoons} HCO_3^- \tag{1.3.5}$$

$$H_2O + R_3N \overset{\text{fast}}{\rightleftharpoons} R_3NH^+ + OH^- \tag{1.3.6}$$

There is some evidence that reaction 1.3.5 may be catalyzed by the amine, so that its rate is larger than can be accounted for by the existing OH^- concentration.

By employing a short controlled contact time between gas and tertiary amine solution, it is possible to effect selective H_2S removal combined with good cleanup and reasonable solution capacity for H_2S. The flowsheet is entirely conventional (see Figure 1.2.1) except for the short absorber needed for low contact time.

Ammonia and primary and secondary amines may also be used for kinetically selective H_2S removal in short-contact-time absorbers. These reagents are not as selective as tertiary amines owing to a faster rate of CO_2 absorption caused by the formation of carbamate ion (see Section 1.9):

$$CO_2 + 2R_2NH \rightleftharpoons R_2NH_2^+ + R_2NCOO^- \tag{1.3.7}$$

Nevertheless, kinetic selectivity may be sufficiently high to make the use of ammonia and non-tertiary amines attractive in certain circumstances. Figure 1.3.9 shows a schematic of a spray tower for selective H_2S absorption from coke-oven gas

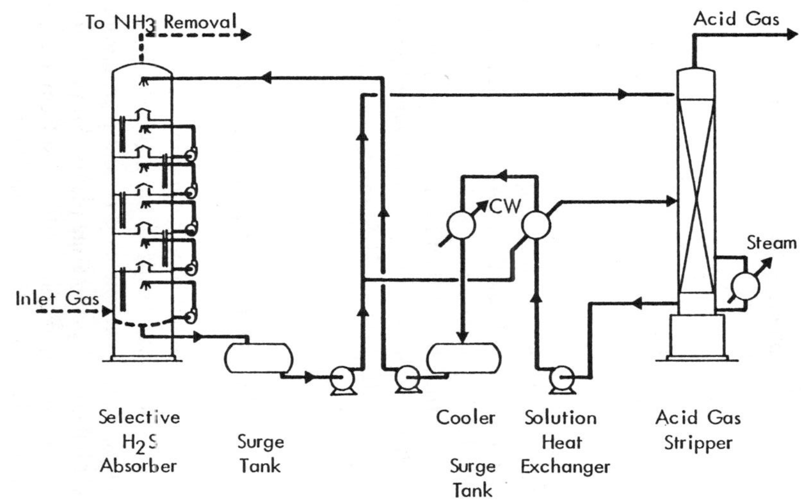

FIGURE 1.3.9. Spray tower for selective hydrogen sulfide absorption from coke-oven gas using ammonia solution.

TABLE 1.3.5. Typical Non-Hydrocarbon Impurities in Coke-Oven Gas

Impurity (type)	Mole%
H_2S	0.3–3.0
CO_2	1.5–2.0
CS_2	0.02
COS	0.01
Thiophene	0.01
Mercaptans	0.003
NH_3	1.1
HCN	0.1–0.25
Pyridines	0.004
Nitric oxide	0.0001

using ammonia solution. Ammonia is used as the selective reagent because it is a constituent of coke-oven gas. Typical non-hydrocarbon impurities in coke-oven gas are given in Table 1.3.5.

A spray tower is an attractive contacting device for short-contact-time treating of dirty gases as it has low pressure drop and provides high gas-phase mass transfer needed to maximize the rate of H_2S absorption, which is largely gas-phase controlled.

1.4. ECONOMIC FACTORS IN GAS TREATING

Solvent circulation rate is the single most important factor in the economics of gas treating with chemical solvents. Solvent circulation rate, through its influence on the size of pumps, lines, heat exchangers, and regeneration tower, has a large influence on the capital cost of gas treating plants. Solvent rate also has a major influence on the energy (steam) requirement for solvent regeneration because the reboiler heat duty is associated directly with liquid rate.

Rates of mass transfer, in both absorption and desorption, affect the tower heights. The primary influence of rate on investment is in the absorber which is usually a high-pressure vessel. The rate of mass transfer in desorption may influence the steam required for stripping. In non-equilibrium or kinetically selective processes, the rates of mass transfer are extremely important as they set the selectivity for the separation of two acid gases.

Other factors that play an important role in gas treating economics include solution corrosivity, which determines materials of construction particularly in the flash and regenerator because of high temperature and acidity; solvent cost, especially as it relates to makeup requirements; and process reliability, which determines the service factor.

Recognition of the important influence of solvent capacity (and hence solvent circulation rate) on gas treating costs has come only recently. Table 1.4.1 shows

TABLE 1.4.1. Investment and Fuel Costs for Natural Gas Sweetening with Aqueous Diethanolamine[a]

	Traditional process	High concentration		
DEA conc. in water, molar	2.5	3.5	3.5	3.5
Acid gas loading, SCF/gal	3.0	4.4	5.8[b]	7.0[c]
GPM	1736	1190	892	744
Relative investment	1.0	0.76	0.62	0.56
Relative annual fuel cost	1.0	0.63	0.42	0.33

[a]Basis: Sour gas rate, 5×10^7 SCFD; total pressure, 533 psia; H_2S in sour gas, 7.5%; CO_2 in sour gas, 7.5%; and treated gas spec., $\frac{1}{4}$ grain H_2S/100 CF.
[b]High load DEA process.
[c]Société Nationale des Pétroles d'Aquitaine DEA process.

investment and fuel costs for natural gas sweetening with the aqueous diethanolamine process as a function of solvent working capacity for CO_2 and H_2S. In the table, cases are shown for various solvent circulation rates. Solvent rate is reduced by increasing working capacity. This is achieved by increasing DEA concentration in solution and allowing the acid gas "rich" loading in solution to rise above the traditional level. In the most favorable case, solvent circulation rate is reduced by over 50% relative to the traditional process, leading to reduction of investment by nearly 50% and reduction of fuel cost by over 65%. Traditionally, capacity was limited by corrosion which was avoided by holding amine concentration and acid gas loadings down. New understanding and corrosion inhibitor technology allow these traditional limits to be lifted.

1.5. CYCLIC CAPACITY CONCEPT

The loading of chemical solvents for gas absorption is determined in the limit by considerations of vapor-liquid equilibrium. At any temperature, for a given solvent, there is a definite relationship between the loading of acid gas in the solvent y and the equilibrium vapor pressure of the acid gas p^* over the liquid. This vapor-liquid equilibrium (VLE) relationship can be expressed mathematically by an equation of the form

$$p^* = f(T, y) \tag{1.5.1}$$

Often, VLE data are available as a plot of p^* against y at constant T. VLE curves for CO_2 at 313K in a physical solvent and a chemical solvent are shown in Figure 1.5.1. Because many decades of pressure are usually covered in a gas absorption-desorption cycle, the pressure scale is logarithmic. The loading scale is linear and is expressed in mole fraction units for a physical solvent and as fractional saturation for a chemical solvent.

For the physical solvent (propylene carbonate), acid gas (CO_2) solubility (load-

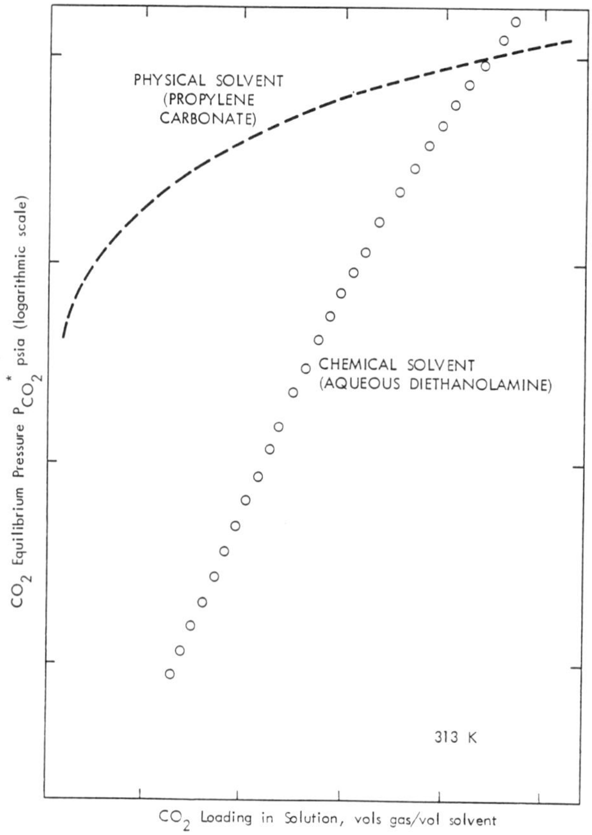

FIGURE 1.5.1. Vapor–liquid equilibrium curves.

ing) is small at low pressures, but it becomes quite large at high pressures (e.g., 200 psi). The chemical solvent (aqueous DEA), on the other hand, exhibits quite high loadings at low CO_2 pressures but quickly becomes chemically saturated at moderate pressures. Thus, when removing acid gases at low partial pressures in the feed gas, chemical solvents will generally be preferred, whereas for high partial pressures, physical solvents may have advantages over chemical solvents.

It should be borne in mind that the comparison in Figure 1.5.1 is between a physical solvent other than water and a chemical solvent that is an aqueous solution. In the chemical solvent, once the acid gas loading has reached the point of chemical saturation, additional absorption of acid gas can only take place by physical absorption in the solvent, which is water and exhibits a comparatively low physical solubility for acid gases. The chemical solvent curve in Figure 1.5.1 eventually bends to the right and takes the shape of the physical solvent curve, but only at unrealistically high acid gas partial pressures. An interesting possibility is the use, as chemical solvents, of non-aqueous solutions, where loading at low acid gas pressures will be governed by chemical saturation and at high pressure by the comparatively high physical solubility in the non-aqueous solvent. Non-aqueous solutions of chemical bases have been used in industry and will probably be used more in the future; these are discussed in some detail in Chapter 14.

The important parameter in practice is the cyclic capacity C_C of the solvent. The cyclic capacity is the maximum total number of moles of acid gas (component A) that can be extracted from the gas stream by a unit volume of the solvent. It is evident that C_C is given by (see Figure 1.5.2)

$$C_C = (y_{\max} - y_{\min})\, m \tag{1.5.2}$$

where $y_{\max}$ is the maximum or "rich" loading of the solvent leaving the absorber, $y_{\min}$ is the minimum or "lean" loading of the solvent entering the absorber (leaving the reboiler), and m is the molarity, i.e., the concentration of reactive species in the liquid solution.

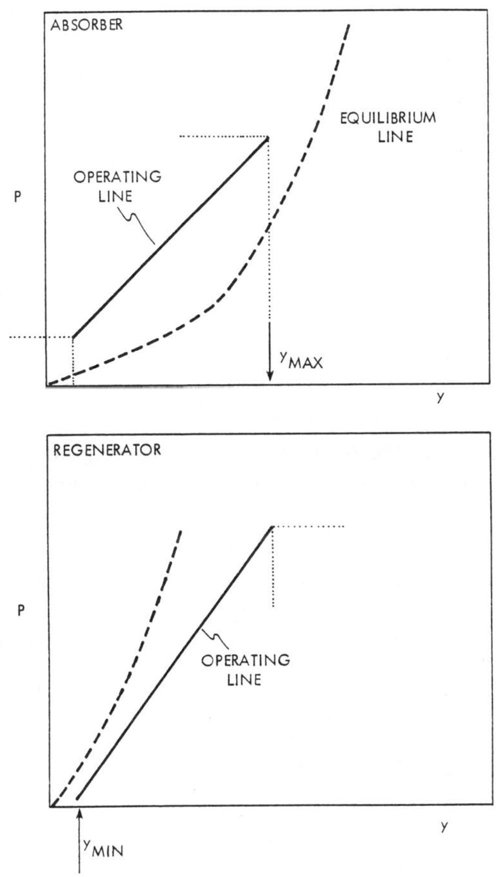

FIGURE 1.5.2. Cyclic capacity concept for a countercurrent absorber–regenerator process using a chemical solvent.

The importance of the concept of cyclic capacity can easily be understood from the following argument. Given a certain gas treating load Φ_T (i.e., the number of moles per unit time of component A to be removed from the gas stream), the volumetric liquid circulation rate L is subject to the following obvious condition:

$$L > \frac{\Phi_T}{C_C} \tag{1.5.3}$$

Equation 1.5.3 shows that a cyclic capacity as high as possible is desirable.

The values of y_{max} and y_{min} for a given operation are essentially determined by the VLE behavior of the system. y_{max} is determined by the condition that, at the gas inlet section of the absorber (absorber bottom), a finite driving force for absorption needs to be available. In the limit where this driving force is negligibly small, one may write

$$p_B = p^*(y_{max}, T_B) \tag{1.5.4}$$

where p_B is the partial pressure of A in the feed gas and T_B is the temperature at the bottom of the absorber.

The value of y_{min} is determined by the condition that at the top of the absorber there should be a finite driving base for acid gas absorption to achieve the desired degree of cleanup of acid gas. In the limit of zero driving force at the absorber top, one may write

$$p_T = p^*(y_{min}, T_T) \tag{1.5.5}$$

where p_T is the partial pressure of A in the treated gas and T_T is the temperature at the top of the absorber.

Thus knowing p_B (from total pressure and composition of the feed gas), p_T (from total pressure and cleanup target of the treated gas), and the absorber top and bottom temperatures, the thermodynamic cyclic capacity can be estimated from VLE curves as shown in Figure 1.5.2.

The actual cyclic capacity of a solvent can never be larger than the thermodynamic cyclic capacity calculated above. The actual capacity will, in fact, be smaller owing to the need to provide finite driving forces for mass transfer and other limitations such as maximum rich solution loading allowed to avoid corrosion or precipitation.

1.6. RATE OF MASS TRANSFER

Mass transfer between two phases can, in general, be described as follows. Figure 1.6.1 is a qualitative profile of fugacity distribution for a component transferring from Phase I to Phase II. The equation for the flux N, moles transferred per unit time, and unit interface area can be written as

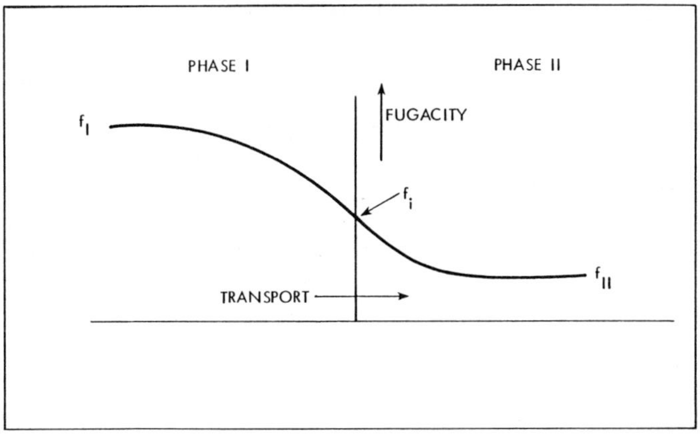

FIGURE 1.6.1. Rate concept for physical mass transfer.

$$N = k_I^f(f_I - f_i) = k_{II}^f(f_i - f_{II}) \tag{1.6.1}$$

where k_I^f, k_{II}^f are fugacity-based mass transfer coefficients.

Several comments are important with regard to Eq. 1.6.1. First of all, it is a *definition* of the mass transfer coefficients; actual calculation of the values of the latter from first principles can seldom be performed inasmuch as a detailed knowledge of the fluid mechanics in both phases would be required. However, in the classical theory, one can prove that under comparatively mild conditions, the mass transfer coefficients are independent of the driving forces (e.g., k_{II}^f is independent of $f_i - f_{II}$). This, in turn, implies that the flux N is a *linear* function of the driving force.

In the case of chemical solvents, this linearity *does not* hold and the mass transfer coefficient in the phase where the reactions take place is, in general, a highly nonlinear function of the driving force. This problem will be discussed in detail in Chapters 3–8.

The crucial assumption in Eq. 1.6.1 is that fugacity is continuous across the interface; i.e., f_i has the same value on both sides of the interface, or, equivalently, *physical equilibrium prevails at the interface.*

The case of interest in gas treating is where the two phases involved are a gas and a liquid. For this case, it is usual to express the driving forces in terms of partial pressures for the gas phase and in terms of concentrations for the liquid phase. We here restrict attention to the case where no chemical reactions take place (i.e., to the case where a physical solvent is being used).

Let A be the component that is being transferred. Let a_0 be the concentration of A in the bulk of the liquid phase, and let p be its partial pressure in the bulk of the gas phase. For the sake of simplicity, we restrict attention to the case where the equilibrium behavior of the system is linear, say, Henry's Law applies. In that case, the partial pressure p^* that equilibrates the bulk-liquid concentration a_0 is given by

$$p^* = \mathrm{H}a_0 \tag{1.6.2}$$

where H is Henry's constant. If in the actual system $p > p^*$, absorption (i.e., transfer of A from the gas to the liquid) will take place; if $p < p^*$, desorption will take place.

Consider, in particular, the case of absorption. The volatile component A is transferred from the bulk of the gas to the gas–liquid interface under the influence of a driving force $p - p_i$, where p_i represents the partial pressure in the gas phase at the interface. The first part of Eq. 1.6.1 will be written as

$$N = k_G(p - p_i) \tag{1.6.3}$$

where k_G is the partial-pressure-based gas-phase mass transfer coefficient. The second part of Eq. 1.6.1 is written as

$$N = k_L^0(a_i - a_0) \tag{1.6.4}$$

where k_L^0 is the concentration-based liquid-phase mass transfer coefficient and a_i is the interface concentration of A in the liquid.

Equations 1.6.3 and 1.6.4, even if the values of k_G and k_L^0 are known, do not allow the calculation of the flux N unless a relationship between p_i and a_i is known. The assumption that fugacity is continuous across an interface reduces, when Henry's Law applies, to

$$p_i = \mathrm{H}a_i \tag{1.6.5}$$

Combination of Eqs. 1.6.3–1.6.5 yields

$$N = \frac{p/\mathrm{H} - a_0}{1/k_L^0 + 1/\mathrm{H}k_G} \tag{1.6.6}$$

Equation 1.6.6 shows that there are two possible limiting behaviors in the system considered, namely:

1. *Gas Phase Control.* Suppose that the following condition is fulfilled:

$$\frac{k_L^0}{\mathrm{H}k_G} \gg 1 \tag{1.6.7}$$

In this case, Eq. 1.6.6 can be approximated by

$$N = k_G(p - \mathrm{H}a_0) \tag{1.6.8}$$

Condition 1.6.7 implies that mass transfer in the liquid phase is much easier than in the gas phase, so that only a negligible fraction of the overall driving force $P/\mathrm{H} - a_0$

is "used" in the liquid phase. Consequently, a_i is almost equal to a_0, and Eq. 1.6.8 follows from 1.6.3 and 1.6.5.

2. *Liquid-Phase Control.* Conversely, suppose that the following condition is fulfilled:

$$\frac{k_L^0}{\mathrm{H}k_G} \ll 1 \tag{1.6.9}$$

In this case, Eq. 1.6.6 can be approximated by

$$N = k_L^0 \left(\frac{p}{\mathrm{H}} - a_0\right) \tag{1.6.10}$$

and the overall driving force is almost entirely "used" in the liquid phase, so that p_i is almost equal to p, and Eq. 1.6.10 follows from 1.6.4 and 1.6.5.

Notice that liquid-phase control is more likely to occur the larger the value of Henry's constant H (i.e., the less soluble the absorbing component is). Gas treating operations are, of course, difficult when the components to be absorbed out of the gas are slightly soluble, so that the case where Eq. 1.6.9 holds true deserves special attention.

All equations given in this section apply also to desorption, provided one agrees to measure the rate of desorption by a negative number.

1.7. GAS SOLUBILITY IN LIQUIDS

In the dissolution of gases in liquids, entropy and energy considerations act in different directions. As molecules are more disordered in a gas than in solution, the drive to achieve greater disorder tends to inhibit dissolution of gas in a liquid. On the other hand, the drive to reduce energy generally favors dissolution, inasmuch as solution of a gas in a liquid is usually an exothermic process. Assuming that two gases have the same entropy of solution, the one with the largest heat of solution (ΔH) will tend to be most soluble. Raising the temperature is expected to decrease gas solubility. These rather simple expectations are borne out by experimental data for acid gas and light hydrocarbon gas absorption in methanol (Figure 1.7.1), methyl-2-pyrrolidone, propylene carbonate, and the dialkyl ether or polyethylene glycol. In Figure 1.7.1, the slope of the curves is a measure of the heat of solution, since Clapeyron's Law applies (see also the discussion at the end of this section).

The general condition of equilibrium for a component A present in two different phases is that the fugacity of the component should be the same in both phases. In the special case of a gaseous and a liquid phase, the condition is

$$f_A^g = f_A^\ell \tag{1.7.1}$$

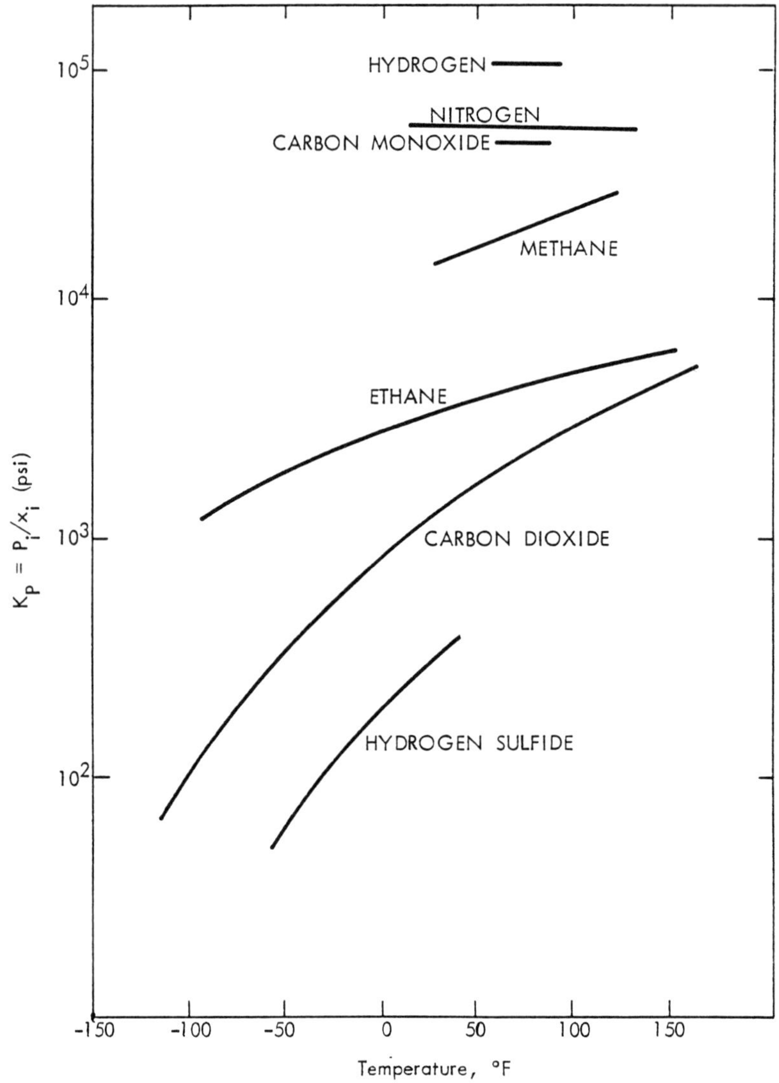

FIGURE 1.7.1. Solubility of gases in methanol.

The fugacity in the gas phase f_A^g is given by

$$f_A^g = \nu_A p^* \tag{1.7.2}$$

where ν_A is the fugacity coefficient. The value of ν_A depends on temperature and pressure but is independent of the composition of the gas phase (gas mixtures are ideal solutions, even if they do not behave as ideal gases).

The fugacity in the liquid phase f_A^ℓ is given by

$$f_A^\ell = x_A \gamma_A p_A^0 \tag{1.7.3}$$

where p_A^0 is the vapor pressure of the pure component, x_A is the mole fraction, and γ_A is the activity coefficient. (When the temperature is above the critical temperature of A, p_A^0 needs to be estimated by extrapolation.)

The value of γ_A depends on temperature and pressure (weakly) and on the composition of the liquid phase (strongly). As $x \to 0$, the value of γ_A approaches a finite limit $\gamma_{\infty A}$, which is called the activity coefficient at infinite dilution. For sufficiently small mole fractions x_A, one can, therefore, use the following approximation (which is of order zero in x_A):

$$\gamma_A = \gamma_{\infty A} \tag{1.7.4}$$

(i.e., the value of γ_A may be regarded as independent of x_A). The value $\gamma_{\infty A}$, however, still depends on the liquid phase composition, say, on the concentrations of the *other* components in the liquid phase.

If x_A is sufficiently small to justify Eq. 1.7.4, the concentration of A in the liquid phase is given by

$$a = x_A \bar{\rho} \tag{1.7.5}$$

where $\bar{\rho}$ is the molar density of the liquid phase. Combination of Eqs. 1.7.1–1.7.5 yields

$$p^* = \frac{\gamma_{\infty A} p_A^0}{\bar{\rho} \nu_A} a \tag{1.7.6}$$

Comparison of Eqs. 1.6.2 and 1.7.6 shows that Henry's constant is given by

$$\mathrm{H} = \frac{\gamma_{\infty A} p_A^0}{\bar{\rho} \nu_A} \tag{1.7.7}$$

This shows that H depends on:

Temperature, strongly, since p_A^0 is a strong increasing function of temperature

Pressure, weakly, since all quantities on the right-hand side of Eq. 1.7.7 are either independent or weak functions of pressure* (ν_A will be discussed later)

Liquid phase composition, insofar as $\gamma_{\infty A}$ depends on it

Furthermore, Eq. 1.7.7 is based on the assumption that $x_A \ll 1$, or equivalently that $a \ll \bar{\rho}$. In aqueous solutions, $\bar{\rho}$ is of the order of 50 moles/liter and, therefore,

*The pressure dependence of H can be expressed as

$$\frac{\partial \ln \mathrm{H}}{\partial \ln \Pi} = \frac{\bar{v}^{\infty} \Pi}{RT} \tag{1.7.8}$$

where Π is the pressure and $\bar{v}^{\infty}$ is the partial molar volume of A in the liquid at infinite dilution. The right-hand side of Eq. 1.7.8 is always very small, of the order of 10^{-3}, since $\bar{v}^{\infty}$ is of the order of a liquid specific volume and RT/Π is of the order of a gas specific volume.

Eq. 1.7.4 is justified up to values of a of the order of 0.5 moles/liter; values of interest in gas treating are nearly always well below that; therefore, Eq. 1.7.4 is justified.

As far as ν_A is concerned, its value is approximately unity at sufficiently low pressures (by definition, $\nu_A \rightarrow 1$ when the pressure approaches zero). As a rough rule of thumb, ν_A may be taken as unity whenever the total pressure does not exceed five atmospheres. At higher pressures, corrections for non-ideality of the gas phase should be considered.

When the liquid phase is a physical solvent, the value of H can be determined experimentally by measuring the gas solubility. Values of H for H_2S, CO_2, and COS, relative to water are reported in Figure 1.7.2. Table 1.7.1 gives expressions for the accurate estimation of H for H_2S, CO_2, COS, SO_2, and light hydrocarbon gases in water. Table 1.7.2 gives expressions for H for H_2S, CO_2, and COS in four polar organic solvents.

An approximate equation for the effect of temperature on the solubility of a gas dissolving in a liquid is obtained by considering the thermodynamic behavior of a one-component two-phase system. The Clapeyron equation for vaporization

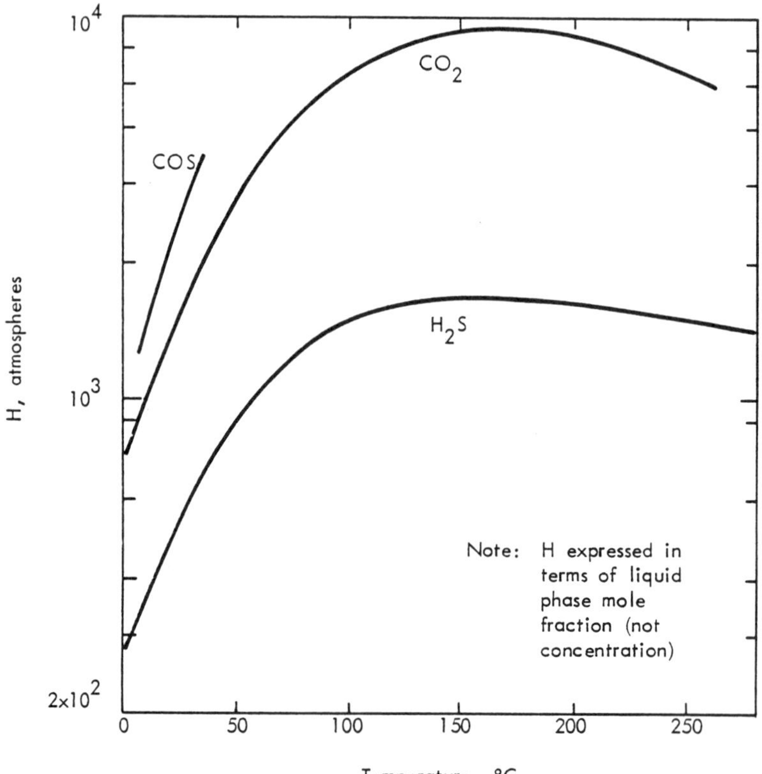

FIGURE 1.7.2. Henry's Law constants for H_2S, CO_2, and COS in water.

TABLE 1.7.1. Solubilities of Acid Gases and Some Light Hydrocarbon Gases in Water[a]

Gas	Temperature range (°K)	A (cal K^{-1} mol^{-1})	B (cal mol^{-1})	C (cal K^{-1} mol^{-1})	D (cal K^{-2} mol^{-1})
H_2S	273–333	-297.158	16347.7	40.2024	0.00257153
CO_2	273–353	-317.658	17371.2	43.0607	-0.00219107
COS	273–303	-439.589	23896.1	60.3429	–
SO_2	283–386	-29.8850	5709.15	0.601884	–
Methane	275–353	-365.183	18106.7	49.7554	-0.000285033
Ethane	275–353	-533.392	26565.0	74.6240	-0.004573130
Propane	273–347	-628.866	31638.4	88.0808	–

[a]The table gives the coefficients in the equation $R\ln x_2 = A + B/T + C \ln T + DT$, where: R = gas constant, T = °K, x_2 = mole fraction of solute, and ln = natural logarithm.

of a pure liquid is

$$\frac{dp^0}{dT} = \frac{\Delta H^v}{T\Delta V^v} \tag{1.7.9}$$

where ΔH^v and ΔV^v are the enthalpy and volume changes associated with the phase transition.

At low pressure the equation can be simplified by making the assumptions that: (1) the vapor phase is ideal, and (2) $V^L \ll V^G$ which is given by RT/p. With these assumptions, the Clapeyron equation becomes

$$\Delta H^V = -R \frac{d(\ln p^0)}{d(1/T)} \tag{1.7.10}$$

which is known as the Clapeyron–Clausius equation.

The thermodynamic argument leading to Eq. 1.7.9 can be duplicated for the case of a gas dissolving into a liquid, leading, under the same approximations, to the following equation:

$$\Delta H^{\text{Diss}} = -R \frac{d \ln p^*}{d(1/T)} \tag{1.7.11}$$

If the gas is only moderately soluble (say, if Henry's Law applies), ΔH^{Diss} is the infinite dilution value of the heat of solution (i.e., it is independent of the concentration a of the gas in the liquid phase). Equation 1.7.11 is a powerful tool for estimating heat effects associated with gas absorption (i.e., for estimating ΔH^{Diss} by differentiating p^* versus T data). Furthermore, often ΔH^{Diss} is not very different from ΔH^V and, therefore, an estimate can be obtained simply from vapor pressure data of the pure component.

TABLE 1.7.2. Solubilities of Acid Gases in Polar Organic Solvents[a]

Solvent	Gas	Temperature range (°K)	D_1	D_2	D_3	D_4
Methanol	H_2S	222.59–279.68	-3397.628	-1.180	76929.474	613.507
	CO_2	192.85–345.65	-95.381	-0.034	619.929	19.125
Propylene Carbonate	H_2S	263.15–373.15	363.035	0.088	-11973.508	-60.692
	CO_2	263.15–373.15	-153.742	-0.056	2124.752	29.423
N-Methyl-Pyrrolidone	H_2S	283.15–333.15	-5497.082	-1.555	145346.333	960.973
		344.26–449.82	-413.436	-0.111	9245.006	73.242
		283.15–449.82	-325.303	-0.089	6675.050	58.170
	CO_2	283.15–333.15	-2350.771	-0.659	62245.557	411.198
		344.26–477.59	-195.025	-0.038	5840.518	33.573
		283.15–477.59	204.578	0.0428	-7735.222	-32.842
	COS	283.15–322.04	3096.739	0.877	-84494.871	-539.069
		344.26–477.59	52.272	0.0068	-3405.706	-6.957
		283.15–477.59	-23.568	-0.0081	-790.176	5.608
Polyethyleneglycol-Dipropylether	H_2S	328.15–413.15	478.368	0.113	-16501.839	-79.989
		273.15–323.15	-3516.25	-1.071	87704.298	621.717
		273.15–413.15	-165.776	-0.0435	2611.478	30.007
	CO_2	273.15–318.15	-7132.428	-2.162	181108.658	1258.857
		323.15–393.15	-5671.782	-1.332	172115.462	964.532
		373.15–393.15	445.630	0.119	-13856.91	-75.774

[a]The table gives the coefficients in the equation $\ln H = D_1 + D_2 T + D_3/T + D_4 \ln T$, where H = Henry's Law constant (= p/x with p in atm and x mole fraction), T = °K, and ln = natural logarithm.

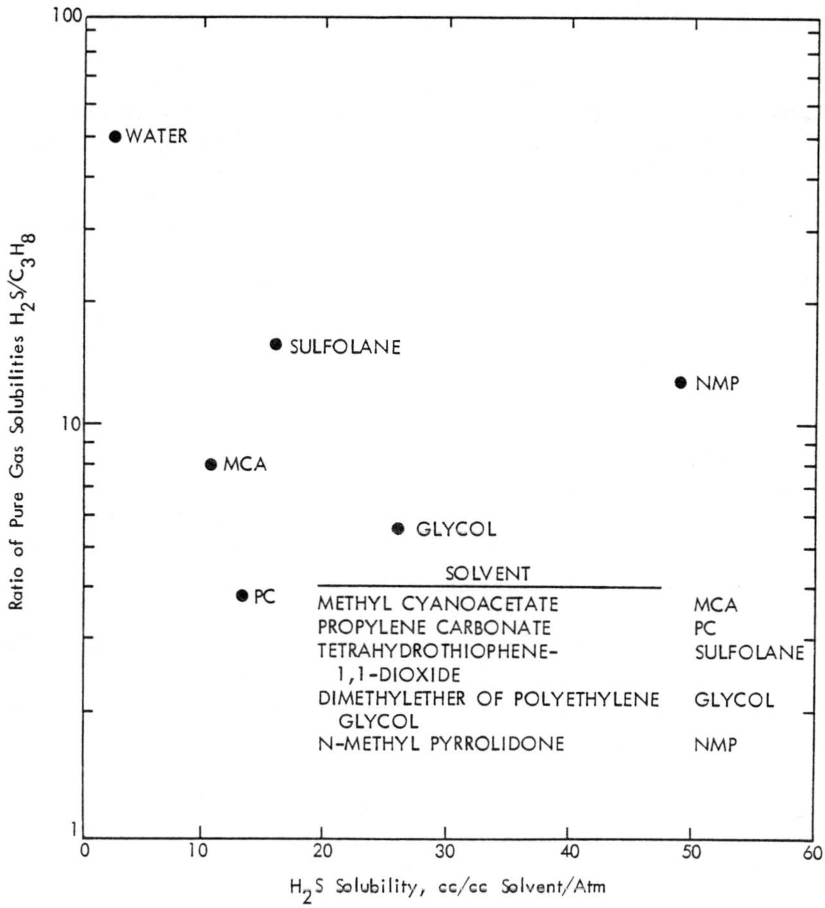

FIGURE 1.7.3. Relative propane solubility in some physical solvents.

The solubility of light hydrocarbons in solvents is an important consideration, as it is generally desirable to minimize hydrocarbon pickup relative to acid gas absorption. Hydrocarbons picked up by the solvent represent loss of product gas. In addition, they may contaminate the acid gases from the regenerator and cause problems in the Claus sulfur plant. Few, if any, data for mixed acid gas-hydrocarbon solubilities are available. Figure 1.7.3 plots the ratio of pure gas solubilities H_2S/propane (log scale) versus H_2S solubility for a variety of physical solvents. Among the organic (physical) solvents, Sulfolane and *n*-methyl pyrrolidone exhibit relatively low solubility for propane.

1.8. CHEMICAL AND PHYSICAL SOLUBILITY

The principal differences between physical and chemical solvents can be summarized as in Table 1.8.1. These differences imply that physical solvents tend to be

TABLE 1.8.1. Principal Differences Between Physical and Chemical Solvents

Physical	Chemical
Capacity (solubility) proportional to acid gas partial pressure in feed gas	Capacity not highly sensitive to acid gas partial pressure
Low heat of absorption	High heat of absorption
Desorb by flashing (sometimes)	Need low-level steam for desorption
Difficult to remove acid gas completely (need many stages)	Easily reduces acid gas to low level in treated gas

preferred when the partial pressure of acid gas in the feed is high, whereas chemical solvents are preferred when feed acid gas partial pressure is low. Both types of solvents are capable of reducing acid gas to low levels in the treated gas, but chemical solvents can do this with many fewer theoretical contacting stages owing to the different shape of the vapor-liquid equilibrium curve, as shown in Figure 1.8.1. Physical solvents require much taller towers to achieve a high degree of acid gas cleanup. In general, physical solvent processes require higher capital investments than chemical solvent processes owing to the taller towers, the need for refrigeration, and the use of rotating machinery.

The expectation that physical solvents tend to be favored when treating gases with high partial pressures of acid gas is confirmed by a recent study of various chemical and physical absorbents for acid gas removal in coal gasification plants. The results of the study are summarized in Figure 1.8.2.

With chemical solvents, it is generally desirable to use as high a concentration (molarity) as possible of the chemical base. However, there may be several limitations on the molarity of chemical solvents in practice. Increase in molarity generally (but not always) leads to:

Increase in the equilibrium vapor pressure p^* (this point will be discussed in detail in Chapter 2).

Increase in hydrocarbon solubility.

Increase in corrosion.

Increase in viscosity.

Increase of solvent temperature rise.

Danger of precipitation occurring at rich conditions.

Historically, the principal limitation to molarity in practice has been set by corrosion considerations. However, the recent introduction of corrosion inhibitors has allowed this limitation to be lifted in certain circumstances.

The practical range of molarity is usually 2–6 gmol/liter. Thus industrial chemical solvents are solutions of high ionic strength and their behavior can be expected to be highly non-ideal.

The principal advantages of chemical solvents may be listed as:

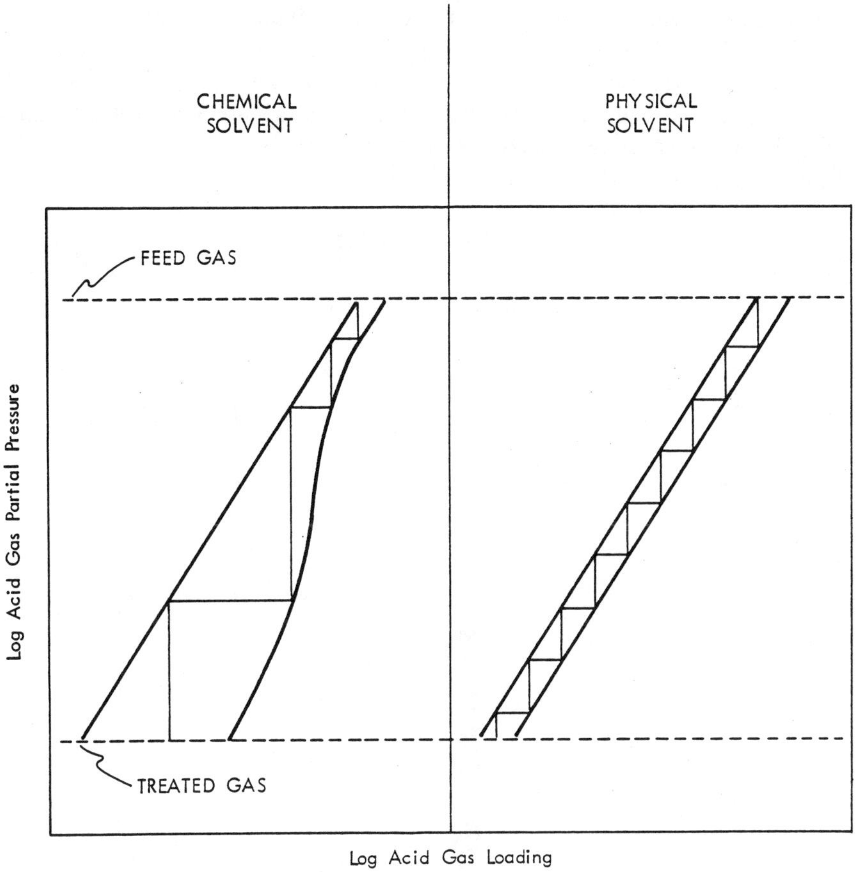

FIGURE 1.8.1. Absorber theoretical stage requirements.

Increased capacity, especially when removing acid gases having low partial pressure.

High degree of removal (cleanup) of acid gas.

Absorption mass transfer coefficient increased (if the reaction is fast enough to occur appreciably near the interface).

Desorption mass transfer coefficient increased.

Potential for selectively absorbing one reacting gas over another based on a difference in rates of absorption (this type of non-equilibrium selectivity is called kinetic selectivity).

Against these considerable advantages certain disadvantages must be considered, including the cost of chemical bases, high heat of absorption, corrosion of carbon steel, side reactions (base degradation), and possible environmental concerns.

The thermodynamics of physical and chemical solubility of gases in liquids can be represented by the various cases depicted in Figure 1.8.3. Case A is the simple one which was discussed in the preceding section. However, acid gases are weak electrolytes, and Case B should be analyzed in some detail. We do this by considering the specific example of SO_2 solubility in water.

Upon dissolution of SO_2 in water, the following reactions may occur:

$$SO_2 + H_2O = H_2SO_3 \tag{1.8.1}$$

$$H_2SO_3 = HSO_3^- + H^+ \tag{1.8.2}$$

Should only reaction 1.8.1 take place, the system would be indistinguishable from one where no reaction at all takes place. In fact, let K_1 be the equilibrium con-

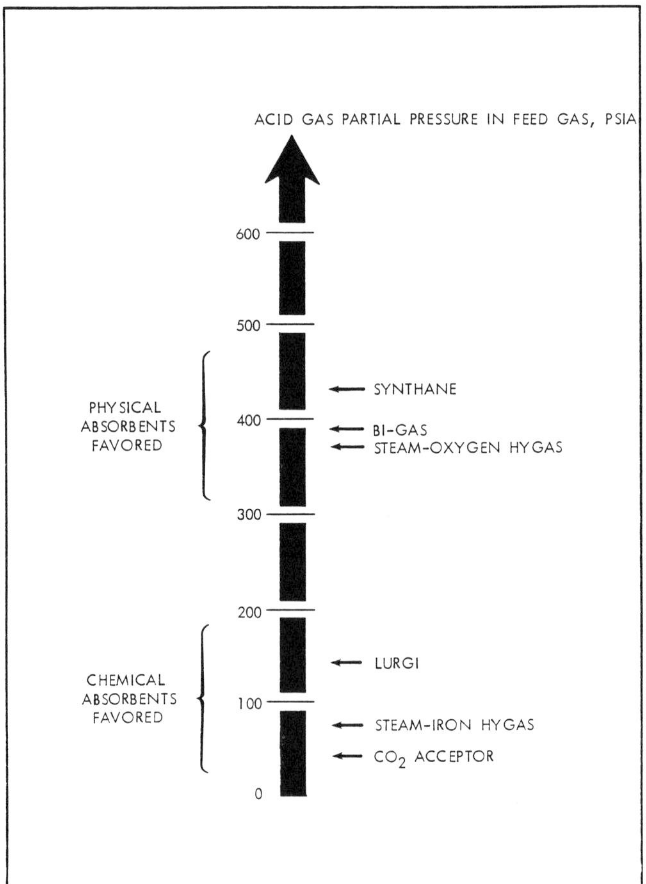

FIGURE 1.8.2. Place of chemical and physical absorbents in coal gasification plants.

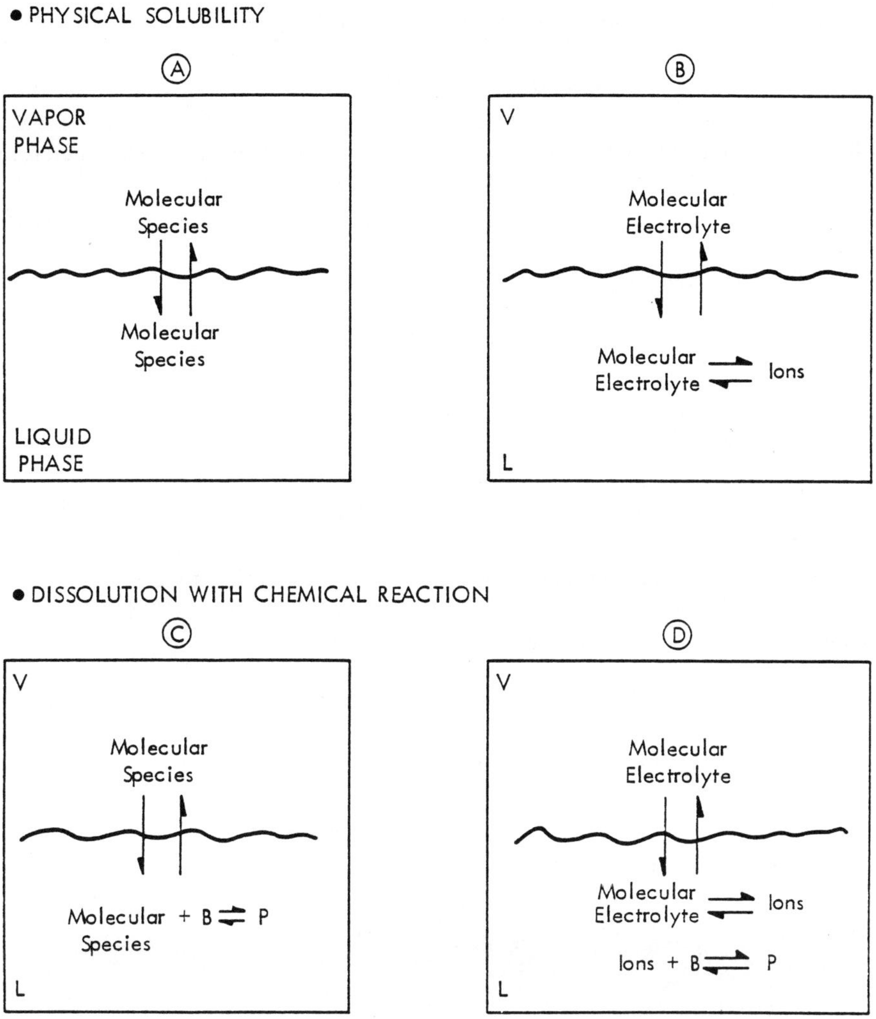

FIGURE 1.8.3. Physical and chemical solubility of gases in liquids.

stant of reaction 1.8.1:

$$K_1 = \frac{[H_2SO_3]}{[SO_2]\,[H_2O]} \tag{1.8.3}$$

Inasmuch as the liquid phase is a dilute solution, $[H_2O]$ is a constant. It follows that

$$[H_2SO_3] = (K_1[H_2O])\,[SO_2] \tag{1.8.4}$$

where the quantity in round brackets is a constant.

Let α be the total concentration of SO_2 in the liquid, say,

$$\alpha = [SO_2] + [H_2SO_3] + [HSO_3^-] \tag{1.8.5}$$

Should only reaction 1.8.1 take place, one would have $[HSO_3^-] = 0$ and, therefore,

$$\alpha = [SO_2]\,(1 + K_1[H_2O]) \tag{1.8.6}$$

From Eq. 1.6.2, one obtains

$$p^* = H_{SO_2}[SO_2] = H\alpha \tag{1.8.7}$$

where the "apparent" Henry's constant H is

$$H = \frac{H_{SO_2}}{1 + K_1[H_2O]} \tag{1.8.8}$$

Since the only measurable quantities are α and p^*, it is useless to keep track of the distinction between physically dissolved SO_2 and sulfurous acid H_2SO_3. Consequently, we write reactions 1.8.1 and 1.8.2 in the form

$$SO_2 + H_2O = HSO_3^- + H^+ \tag{1.8.9}$$

with the understanding that by $[SO_2]$ one means really $[SO_2] + [H_2SO_3]$.

Reaction 1.8.9 does make the VLE behavior different from the one relative to Case A. In fact, let K be the equilibrium constant for reaction 1.8.9 (i.e., the first dissociation constant for sulfurous acid):

$$K = \frac{[HSO_3^-]\,[H^+]}{[SO_2]} \tag{1.8.10}$$

In view of the stoichiometry of reaction 1.8.9, one has, if no other ions are in the liquid,

$$[HSO_3^-] = [H^+] = \sqrt{K[SO_2]} \tag{1.8.11}$$

The total concentration is given by

$$\alpha = [SO_2] + [HSO_3^-] = \frac{p^*}{H} + \sqrt{\frac{Kp^*}{H}} \tag{1.8.12}$$

which shows that the relationship between α and p^* is not linear. Figure 1.8.4 is a plot of SO_2 solubility data which exhibit the kind of behavior predicted by Eq. 1.8.12.

Differentiation of Eq. 1.8.12 yields

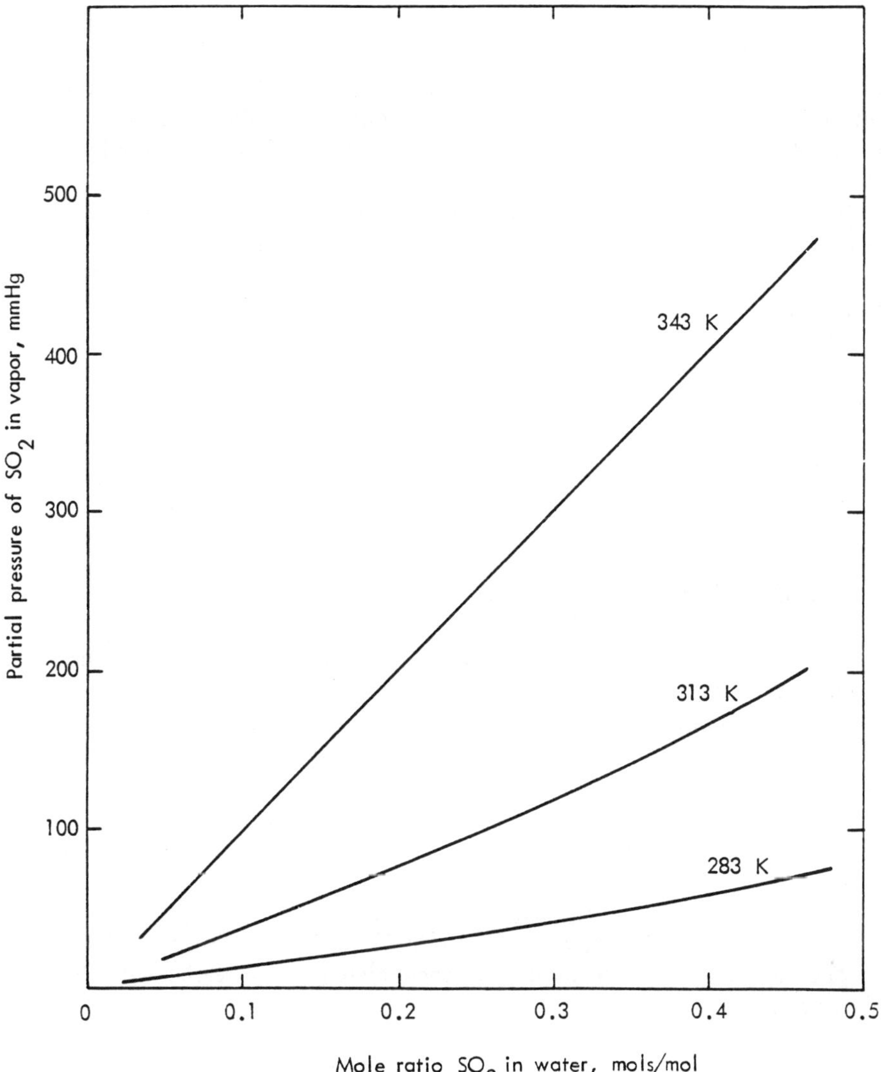

FIGURE 1.8.4. Solubility of SO_2 in water.

$$\frac{d\alpha}{dp^*} = \frac{1}{H} + \frac{1}{2}\sqrt{\frac{K}{Hp^*}} \tag{1.8.13}$$

which shows that, as $p^* \to \infty$, the slope of the curves in Figure 1.8.4 approaches the (constant) value H. This allows the extraction of the Henry's Law constant from curved equilibrium lines.

Turning attention back to Eq. 1.8.12, the non-linear term is seen to become negligible when

$$p^* \gg KH \tag{1.8.14}$$

For SO_2 in water, the right-hand side of Eq. 1.8.14 is of the order of 10^{-2} atm and, therefore, curvature of the equilibrium line is important in the industrially relevant range of partial pressures (at 50°C, for SO_2, $K = 1.25 \times 10^{-2}$ gmol/liter, and $H = 2$ atm liter/gmol).

In the case of CO_2 and H_2S, the value of H is appreciably larger, say, of the order of 100 atm liter/gmol. However, CO_2 and H_2S are much weaker acids than SO_2; say, the value of K is of the order of 10^{-7} gmol/liter. Therefore, the right-hand side of Eq. 1.8.14 is of the order of 10^{-5} atm, which means that the curvature of the equilibrium curve would be observable only at unrealistically low partial pressures. For CO_2 and H_2S in water, the behavior is of the type observed in Case A.

When the liquid phase contains a reactive solute B, cases C and D in Figure 1.8.3 should be considered. This leads to rather complex equilibrium behavior. The equilibrium behavior for chemical solvents will be discussed in Chapter 2.

1.9. CHEMISTRY OF ACID GASES

Chemical bonding in the most common acid gases, H_2S, CO_2, and SO_2, is shown in Figure 1.9.1. H_2S is actually an acid (also called a proton donor Bronsted acid), whereas CO_2 and SO_2 are anhydrides (also called electron acceptor Lewis acids) which need to hydrate to become acids. H_2S and CO_2 have comparable acidities whereas SO_2 is a much stronger acid.

The chemistry of H_2S in aqueous solutions is relatively straightforward. It reacts with a base B by a proton transfer reaction to form the hydrosulfide ion:

$$H_2S_{(aq)} + B \rightleftharpoons BH^+ + HS^- \tag{1.9.1}$$

The most important chemical bases encountered in gas treating with chemical solvents include:

water 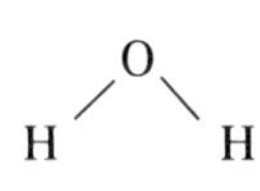

hydroxyl ion 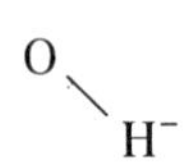

ammonia, amines 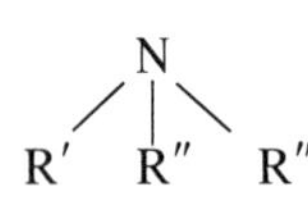

The second dissociation to form the sulfide (S^{2-}) ion can usually be neglected because the equilibrium constant for the reaction

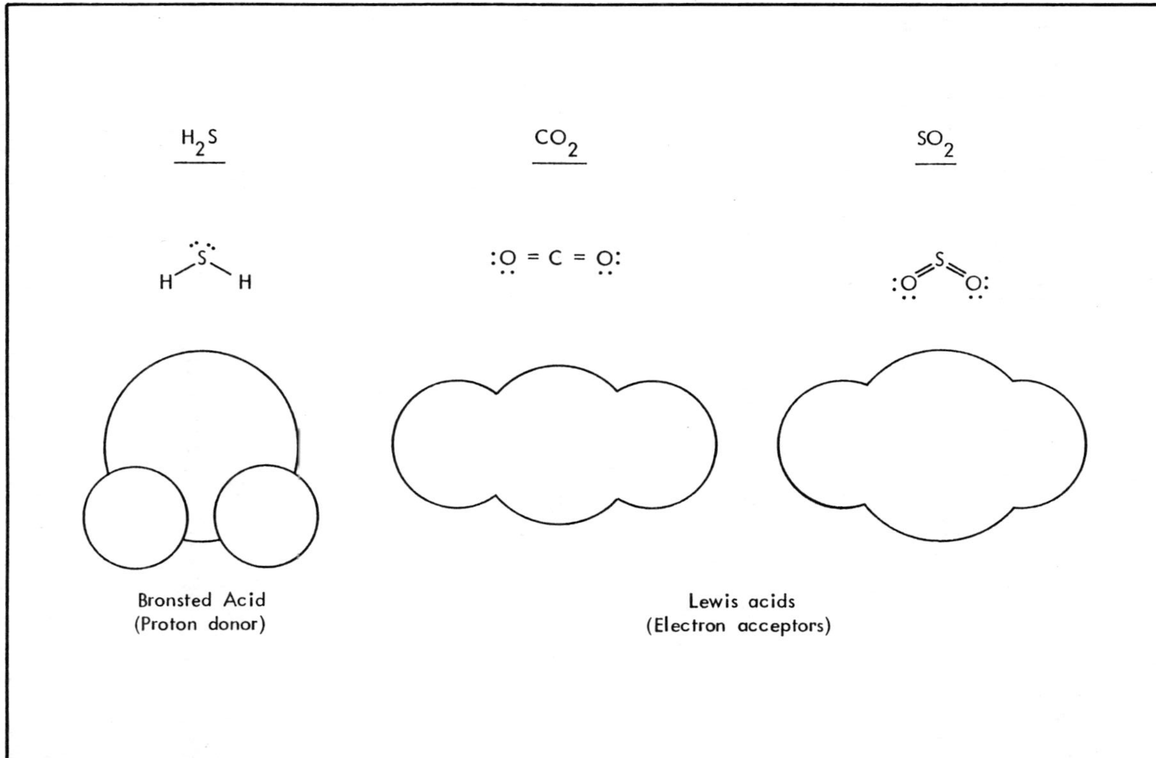

FIGURE 1.9.1. Electronic structures and molecular shapes of H_2S, CO_2, and SO_2.

$$HS^- \underset{}{\overset{H_2O}{\rightleftharpoons}} H^+ + S^{2-} \tag{1.9.2}$$

is about 10^{-14}.

All proton transfer reactions are extremely fast and may be regarded as occurring instantaneously as compared to the time-scale for diffusion processes in solution.

Another class of H_2S removal processes is based upon the oxidation of H_2S to elemental sulfur. These may be either liquid-phase processes using chemical oxidants such as vanadate ion or arsenate ion, or vapor-phase catalytic processes such as the Claus reaction. Some important H_2S oxidation reactions are summarized in Figure 1.9.2.

The chemistry of CO_2 in aqueous solutions is more complex than that of H_2S, and several scientific and industrially important aspects of it are not completely understood. Reaction of CO_2 (Lewis acid) with a chemical base B may be written in the generalized form

$$CO_{2(aq)} + B \rightleftharpoons CO_2 \cdot B$$

where $CO_2 \cdot B$ can be thought of as a "complex" formed between CO_2 and the base. The reactions of CO_2 with industrially important bases can be represented as follows:

water 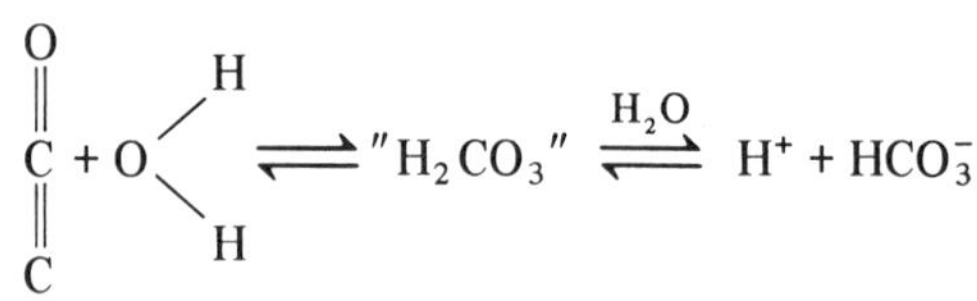 (1.9.3)

hydroxyl ion 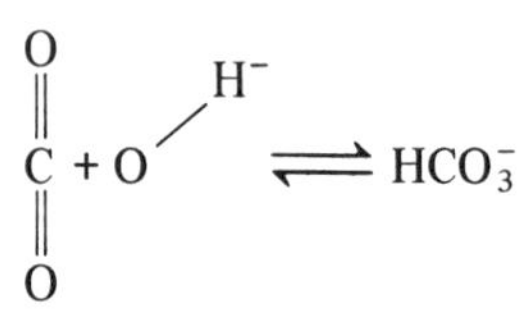 (1.9.4)

ammonia; primary or secondary amines 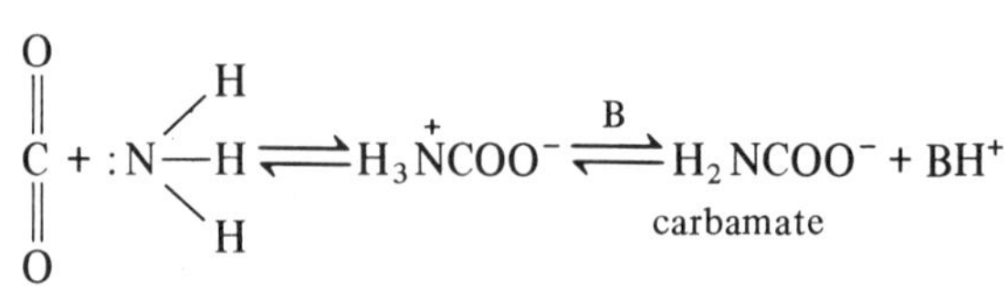 (1.9.5)

Although the existence of undissociated H_2CO_3 in carbonated water is well established, the existence of the zwitterion formed between CO_2 and ammonia or amines is not resolved at this time. Carbonate solutions are used for CO_2 removal; however, there is no direct reaction between CO_2 and carbonate ion, CO_3^{2-}, which only acts as a source of OH^-. However, the CO_3^{2-} ion can act directly as a base as in the second part of reaction 1.9.5.

LIQUID-PHASE PROCESSES (CHEMICAL OXIDANTS)

$$2\,H_2S_{(aq)} + 4\,VO_3^- \xrightarrow{CO_3^{--}/HCO_3^-} V_4O_9^{--} + 2OH^- + 2S + H_2O$$

(VANADATE)

$$3\,H_2S_{(aq)} + H_2AsO_3^- \xrightarrow{CO_3^{--}/HCO_3^-} H_2AsS_3^- + 3\,H_2O$$

(ARSENITE) (THIOARSENITE)

$$H_2AsS_3^- + 3\,H_2AsO_4^- \xrightarrow{SLOW} H_2AsO_3^- + 3\,H_2AsO_3S^-$$

(ARSENATE) (MONOTHIOARSENATE)

$$H_2AsO_3S^- \xrightarrow{H^+(CO_2)} H_2AsO_3^- + S$$

$$H_2AsO_3^- + 1\,1/2\,O_2 \longrightarrow H_2AsO_4^-$$

CATALYTIC, VAPOR-PHASE PROCESS (CLAUS REACTION)

$$H_2S + 3/2\,O_2 \longrightarrow SO_2 + H_2O$$

$$2H_2S + SO_2 \xrightarrow{CAT} 3S + H_2$$

FIGURE 1.9.2. Oxidation of hydrogen sulfide to sulfur.

CO_2 reaction with aqueous aminoalcohols is thought to be identical to the reaction of amines, there being virtually no reaction between CO_2 and the alcohol (—OH) groups. With primary and secondary amines, the step-wise reaction may be written in the form 1.9.5, with the only available base being the amine itself. The net result of these two steps is the overall reaction

$$CO_2 + 2\,:N\begin{smallmatrix} R \\ R' \\ H \end{smallmatrix} \rightleftharpoons RR'NCOO^- + RR'NH_2^+ \tag{1.9.6}$$

From the stoichiometry of reaction 1.9.6 it can be seen that the maximum pickup of CO_2 is limited to 0.5 moles of CO_2/mole of amine if carbamate ion formation is the only reaction. Although this may tend to be the case for amine forming carbamates of high stability, in general, CO_2 pickup in excess of $y = 0.5$ is possible owing to reversion of the carbamate

$$CO_2 + RR'NCOO^- + H_2O \rightleftharpoons RR'NH_2^+ + 2HCO_3^- \tag{1.9.7}$$

Even if there is a driving force for carbamate reversion, the rate of the reaction is, in general, extremely slow. Mass transfer of CO_2 into highly carbonated aqueous amine solutions ($y > 0.5$) proceeds extremely slowly.

With tertiary amines (which do not form carbamates), the amine acts as a base, producing hydroxyl ions with which CO_2 can react to form bicarbonate.

$$HOH + :N\begin{smallmatrix} R \\ R' \\ R'' \end{smallmatrix} \rightleftharpoons RR'R''NH^+ + OH^- \tag{1.9.8}$$

$$CO_2 + OH^- \rightleftharpoons HCO_3^- \tag{1.9.9}$$

Although carbamate formation with tertiary amines is not possible, the possibility of a zwitterion formation with tertiary aminoalcohols has been suggested as a possibility to explain their "catalytic" effect on CO_2 absorption rate:

$$CO_2 + R_3N \rightleftharpoons R_3\overset{+}{N}COO^- \overset{H_2O}{\rightleftharpoons} R_3NH^+ + HCO_3^- \tag{1.9.10}$$

Carbonate-bicarbonate buffers (solutions of potassium carbonate in water) are very important industrially for CO_2 absorption and desorption owing to the low steam requirement for regeneration. The overall stoichiometry is given by Eq. 1.3.1, but the rates of CO_2 absorption are determined by the kinetics of the reaction of CO_2 with molecular water and hydroxyl ions in solution.

$$CO_2 + HOH \rightleftharpoons HCO_3^- + H^+ \tag{1.9.11}$$

$$CO_2 + OH^- \rightleftharpoons HCO_3^- \tag{1.9.12}$$

The chemical rate law representing the sum of the two reactions can be written in the form

$$r = \frac{-d[CO_2]}{dt} = [CO_2]\,\{k_1 + k[OH^-]\} \tag{1.9.13}$$

when the reverse reaction is neglected. k_1 is the rate constant for the CO_2 hydration reaction (Eq. 1.9.11) and k, the second order rate constant for reaction of CO_2 with hydroxyl ion. The relative contributions of reactions 1.9.11 and 1.9.12 depend on the pH of the solution. For pH $<$ 7, reaction 1.9.11 is the principal mechanism. For pH $>$ 8, reaction 1.9.12 predominates. Both reactions are important in the intermediate pH range.

At all pH levels of practical interest, the CO_2 chemical reaction rate in solution is rather low owing mainly to the very small hydroxyl ion concentration (typically 10^{-3} molar or less). The low chemical reaction rate leads to low rates of CO_2 absorption into the solution from the gas phase. Methods for increasing the chemical reaction rate, such as homogeneous catalysis, are therefore of interest.

It is well known that certain Lewis bases ($\ddot{B}$) will catalyze the rate of CO_2 reaction to form bicarbonate ion. The mechanism of catalysis is a matter of conjecture, but it has been proposed that the catalytic chemistry involves formation of a $CO_2 \cdot B$ complex

$$CO_2 + B \rightleftharpoons CO_2 \cdot B \qquad (1.9.14)$$

which serves to "activate" the CO_2, thereby speeding its reaction with HOH which is present in large excess. Known homogeneous catalysts include the arsenite ion and the sulfite ion.

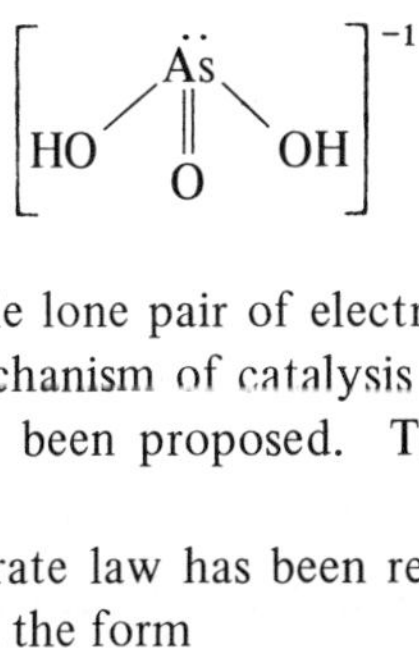

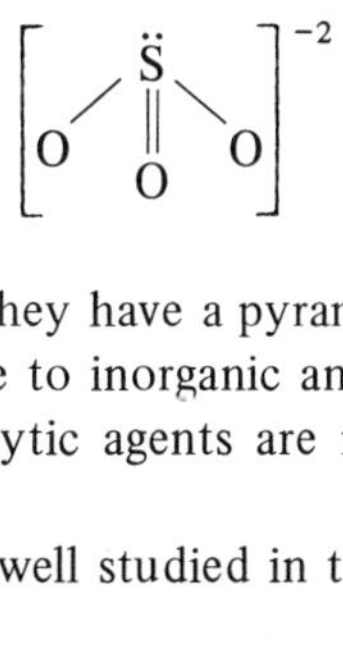

These have a single lone pair of electrons and they have a pyramidal structure like ammonia. A mechanism of catalysis applicable to inorganic and organic nitrogen base catalysts has been proposed. These catalytic agents are referred to as rate promoters.

The chemical rate law has been reasonably well studied in the case of arsenite promotion. It has the form

$$r = -\frac{d[CO_2]}{dt} = [CO_2]\,\{k_1 + k_2[OH^-] + k_3[H_2AsO_3]^-\} \qquad (1.9.15)$$

in which the contributions of the individual reactions are simply additive. Under practical conditions, the contribution of the $k_3[H_2AsO_3]^-$ term is large relative to the other two.

1.10. UNFINISHED PROBLEMS IN CHEMICAL GAS TREATING

Despite the great industrial importance of chemical solvents for gas treating, there are many areas where improved understanding is needed before a rational approach is available for the analysis and design of gas absorption processes. The final section

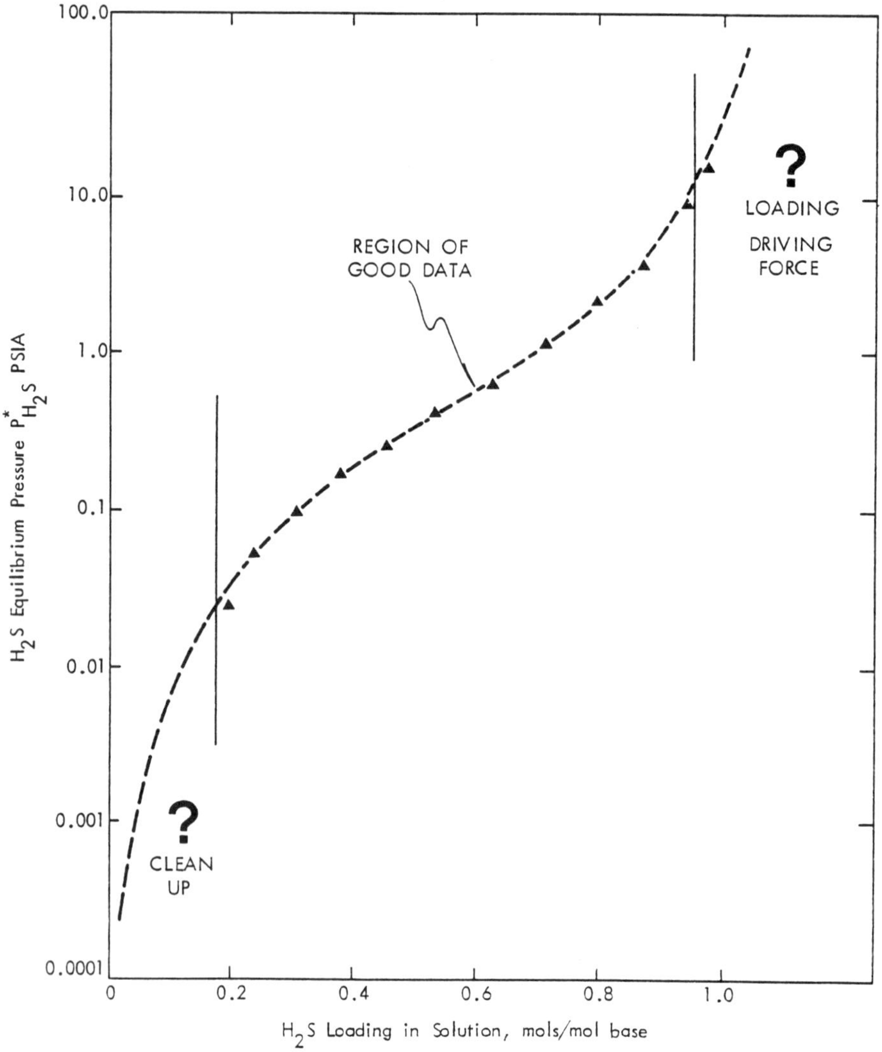

FIGURE 1.10.1. Modeling chemical solvent vapor–liquid equilibrium curves.

of this chapter is devoted to a brief review of just some of these "unfinished" problems which the remainder of the book will address, at least in part.

Prediction and Correlation of Vapor–Liquid Equilibria

Rational design requires that vapor–liquid equilibrium curves be available covering the full range of acid gas loadings and temperatures of industrial processes. Typically operating temperatures range from about 313K to over 393K, whereas loadings of acid gas in solution are in the range $0.01 < y < 1.0$ plus. Experimental techniques for measuring VLE data are such that accurate data may be available only in the middle range of loadings and pressures. Sound VLE models are needed that will correlate these data and allow accurate extrapolations to the extremes of low pressures and loadings needed to establish acid gas cleanup limits in treating processes. Similarly, accurate extrapolation of data to high pressures and loadings is needed to establish the thermodynamic limits to rich loading in the absorber. Figure 1.10.1 illustrates these points. An appropriate thermodynamic model must express the relationship between pressure and loading in terms of Henry's constants, ionization, chemical reaction equilibrium constants, and activity coefficients. A methodology for tackling problems of this nature is presented in Chapter 2.

Mass Transfer Under "Pinch" Conditions

Although most commercial towers operate under small driving force or "pinched" conditions, essentially all of the theory of mass transfer with simultaneous chemical reaction has been developed for the "irreversible" case where driving forces are

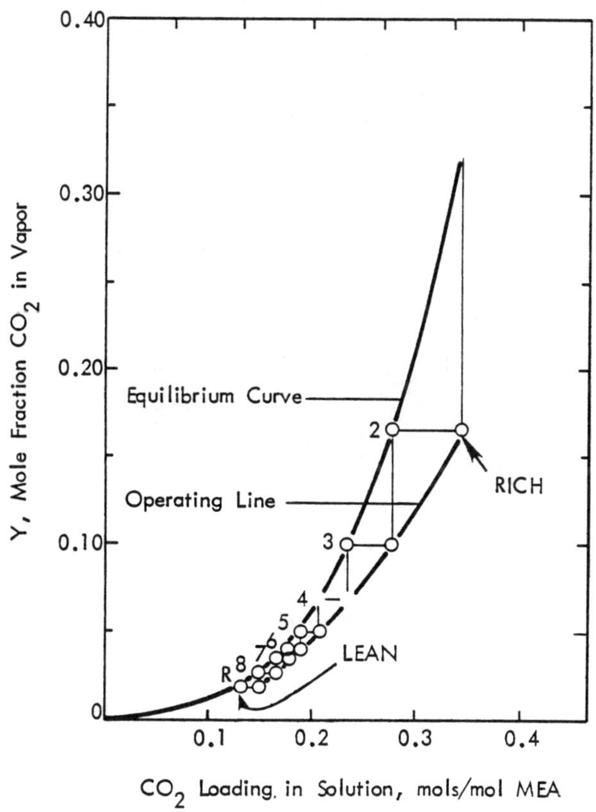

FIGURE 1.10.2. Mass transfer under pinch conditions (carbon dioxide desorption from aqueous monoethanolamine).

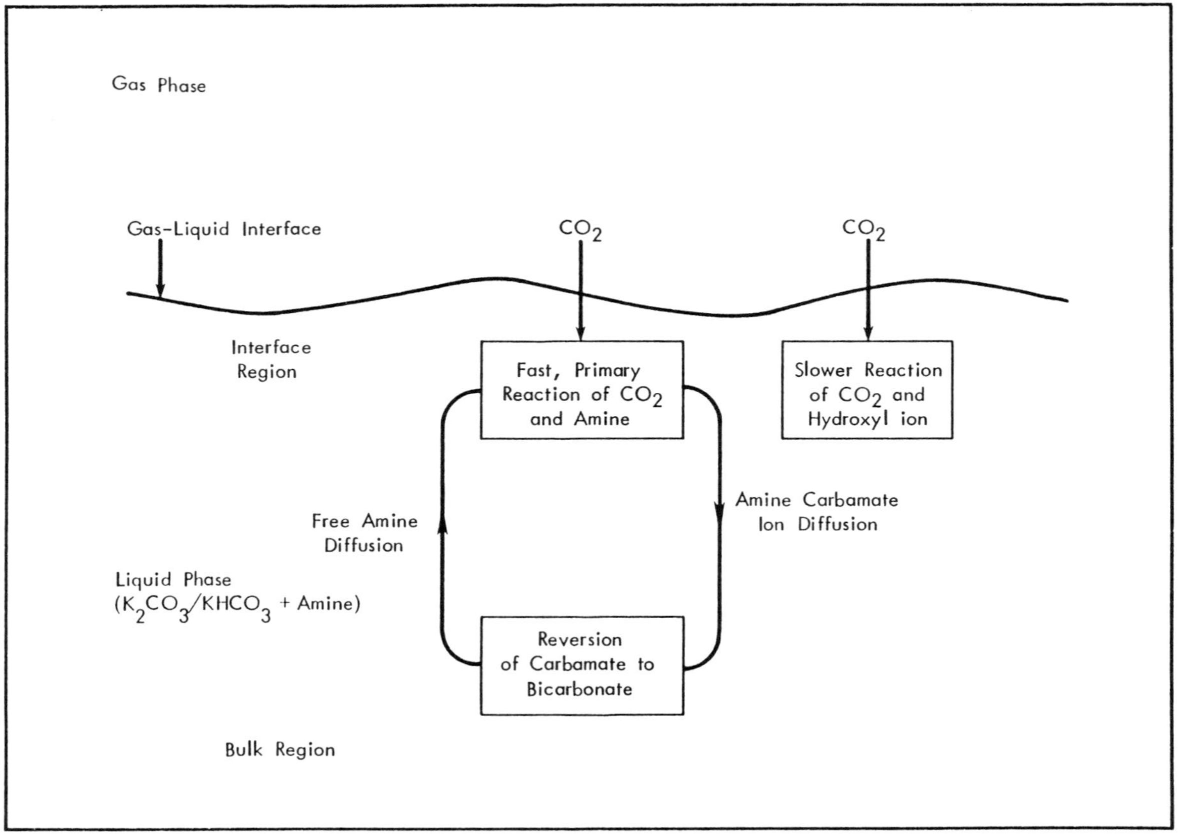

FIGURE 1.10.3. Mechanism proposed to explain enhancement of carbon dioxide absorption rate in amine-promoted carbonate solutions at low temperature.

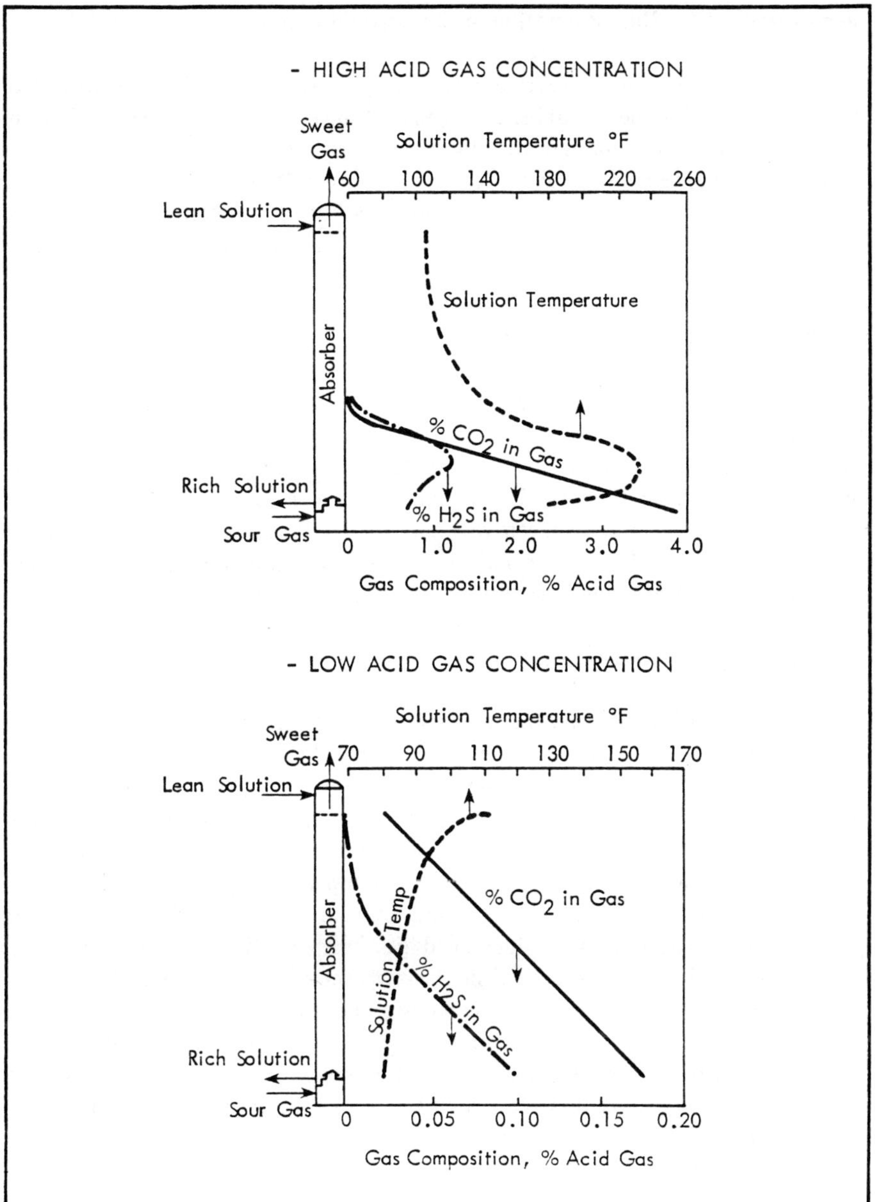

FIGURE 1.10.4. Countercurrent absorber temperature profiles.

necessarily very large. An example for CO_2 desorption (stripping) from aqueous MEA solution is shown in Figure 1.10.2. It is clear that a pinched condition exists over nearly the entire height of the stripping tower. Chapter 6 presents an analysis of mass transfer rate enhancement applicable to the very small driving force situation.

Mechanism of CO_2 Rate Promotion in Carbonate Solutions

Promoted hot carbonate processes are widely used for bulk CO_2 removal from gases, but little has been published on mass transfer in these systems. The only studies available are limited to low temperature absorptions where reaction-diffusion mechanisms of the type shown in Figure 1.10.3 have been proposed. In Chapter 10, a general theory of promoted hot carbonate is presented and compared with data obtained in laboratory units under conditions simulating commercial contacting.

Simultaneous Absorption of Two Gases

Probably the most commonly encountered industrial gas treating problem involves the simultaneous absorption of two or more gases. Despite this, few thermodynamic (VLE) data are available for systems containing two acid gases, and the interference between the two during simultaneous mass transfer has rarely been studied. Chapter 2 presents a general methodology for handling the VLE of mixed acid gas systems and Chapters 8 and 11 discuss theoretical and industrial aspects of simultaneous absorption and desorption.

Non-Isothermal Absorber Operation

In counterflowing absorbers and regenerators in which there is substantial heat release, unusual temperature profiles often taking the form of a temperature bulge in the lower part of the column are obtained. No full analysis of these thermal phenomena has been published and yet the effects of non-isothermal operation have important implications for both the thermodynamic and the mass transfer behavior of the system. Figure 1.10.4 shows temperature profiles measured during the simultaneous absorption of H_2S and CO_2 in monoethanolamine/diethylene glycol/water solutions. With high acid gas concentrations, a temperature rise exceeding 100°F is observed and, in the high temperature zone of the column, H_2S appears to be desorbed from the solution despite its tendency to be absorbed faster than CO_2. Chapter 15 presents an approach for handling gas treating problems involving non-isothermal absorption. Chapter 8 discusses mechanisms of the "reverse direction" mass transfer that may occur during simultaneous absorption of two acid gases.

PART TWO: THEORY

2 Thermodynamics of Gas Treating

A good many times I have been present at gatherings of people who by standards of the traditional culture are thought highly educated and who have with considerable gusto been expressing their incredulity at the illiteracy of scientists. Once or twice I have been provoked and have asked the company how many of them could describe the Second Law of Thermodynamics.

The response was cold; it was also negative. Yet I was asking something which is about the equivalent of: "Have you ever read a work of Shakespeare's?"

C. P. SNOW

The thermodynamics of gas treating is more subtle than that of most other chemical engineering processes because of the two-phase nature of the systems considered. Inasmuch as there are two phases to take into account, as well as chemical reactions taking place, one needs to analyze coupled problems of phase equilibrium and chemical equilibrium.

In this chapter we present a general treatment of the subject, in the sense that the methodology of thermodynamic analysis is illustrated. The methodology can then be applied to specific cases as the need arises. We do discuss a few relevant special cases and present thermodynamic data for these.

It should be pointed out that there are, in fact, two distinct thermodynamic problems to be considered. The first is simply related to the overall vapor-liquid equilibrium (VLE) behavior of the system, and may be stated as follows: determine the partial pressure(s) of the volatile components that equilibrate any given composition of the liquid phase. Knowledge of the VLE behavior is, in fact, crucial to the selection and design of a chemical gas treating process.

The second problem is related to the fact that during an actual process of chemical absorption or desorption (although the gas and liquid phases are, of course, not at equilibrium with each other), chemical equilibrium does, in fact, prevail in the bulk of the liquid phase under most industrial conditions. That this is so will be shown in the chapters dedicated to mass transfer; here we simply state the conclusion without proof. Since equilibrium prevails in the bulk of the liquid phase, the following thermodynamic problem arises: determine the concentration(s) of the dissolved volatile components in the bulk of the liquid corresponding to any given concentration distribution of the non-volatile components. These concentrations will, in general, *not* correspond to those that equilibrate the gas-phase partial pressures; the difference provides the driving force for the absorption or desorption process.

2.1. ELEMENTARY CHEMICAL THERMODYNAMICS

Before starting the analysis of gas treating systems, it is useful to recall some elementary results of chemical thermodynamics. We will use the symbols A and B, sometimes with suffixes, to identify chemical components, and the corresponding lower case symbols a and b to identify their concentrations in the liquid phase. We will use the symbol ν for stoichiometric coefficients of chemical reactions. Consider a chemical reaction such as

$$A + B_1 \rightleftharpoons B_2 \tag{2.1.1}$$

The condition for chemical equilibrium is expressed in terms of the *activities* of the components considered, say, for a reaction such as 2.1.1,

$$\frac{a_{B_2}}{a_{B_1} a_A} = \mathrm{K}_a \tag{2.1.2}$$

where K_a is an activity-based (hence the suffix a) thermodynamic constant, the value of which depends only on temperature, whereas a_A, a_{B_1}, and a_{B_2} are the activities of components A, B_1, and B_2, respectively. However, calculations based on activities are very cumbersome and it is preferable to express the result in Eq. 2.1.2 in terms of the concentrations a, b_1, b_2, rather than in terms of the activities.

The activity of a component A is related to its concentration by

$$a_A = \gamma_A a \tag{2.1.3}$$

where γ_A is the activity coefficient of component A. One may define for reaction 2.1.1 a concentration-based equilibrium constant K as follows:

$$\mathrm{K} = \frac{b_2}{b_1 a} \tag{2.1.4}$$

which is related to K_a by

$$\mathrm{K} = \mathrm{K}_a \frac{\gamma_{B_1} \gamma_A}{\gamma_{B_2}} = \frac{\mathrm{K}_a}{\mathrm{K}_\gamma} \tag{2.1.5}$$

where

$$K_\gamma = \frac{\gamma_{B_2}}{\gamma_A \gamma_{B_1}} \tag{2.1.6}$$

In a liquid phase, the activity coefficients are generally very weak functions of pressure; however, they depend strongly on the *composition* of the liquid phase. This point is worth a careful discussion.

In a *dilute* solution–i.e., a solution where the mole fraction of one particular component, called the solvent, is close to unity–the activity coefficients are not functions of composition; therefore, in dilute solutions, the constant K depends only on temperature as does K_a. In a concentrated solution, however, this is not true and K as defined in Eq. 2.1.4 is a strong function of chemical composition. Inasmuch as chemical composition is measured not by a single parameter but by a possibly large number of parameters, a thermodynamic analysis based on a full description of the dependency of the K values on composition is hopelessly complex for gas treating systems. However, it should be borne in mind that chemical solvents used in gas treating *are* concentrated solutions and that, therefore, regarding K as a constant having the value that holds in a dilute solution is an unacceptable approximation.

The difficulty can be circumvented by assuming that K depends on some single parameter that is a yardstick for the liquid phase composition; often the ionic strength I_i is chosen as such a yardstick. Therefore, thermodynamic analysis of gas treating systems is based on equilibrium conditions such as Eq. 2.1.4, with K regarded as a function of only temperature and another parameter such as ionic strength.

This is, of course, only an approximation, and a rigorous analysis should take into account fully the dependency of the individual activity coefficients on the composition of the liquid phase. However, there is generally not enough information available for carrying out such rigorous analyses; if activity coefficients are not evaluated with great accuracy, the more complete kind of analysis yields predictions that fit experimental data no better than the simple analysis based on the assumption that ionic strength is the only relevant parameter.

In the balance of this chapter, attention is focused on the problem of constructing an approximate thermodynamic model of a gas-treating system. Such models are very useful for the following reasons:

1. They allow easy interpolation and smoothing of vapor-liquid equilibrium data, and provide a basis for the calculation of both the limiting liquid circulation rate and the minimum steam requirements at the regenerator reboiler.
2. They allow extrapolation of experimental data to the regions of very lean and very rich solutions where experimental accuracy is hard to obtain.
3. They can be used to classify categories of chemical solvents in a logical way, thus making the preliminary choice of possible chemical solvents easier.

There is, of course, a separate thermodynamic problem to consider; namely, the determination of the values of the thermodynamic equilibrium constants (in the development of a model, often only the order of magnitude of several of the relevant constants is required). This problem can be dealt with in two ways. The first is simply to *measure* the equilibrium concentration distribution and calculate the thermodynamic values from such measurements. This experimental procedure will be discussed in Chapter 13. The other approach is to determine the values of the thermodynamic constants from theoretical chemistry (e.g., from calculations based on such concepts as bond strengths, intermolecular interaction forces, etc.). This approach, though very important from a qualitative viewpoint, does not, at the present-day state of the art, produce results of satisfactory quantitative accuracy. Of course, this statement, like all generalizations, should be regarded as admitting several counterexamples.

2.2. CHEMICAL THERMODYNAMICS OF SINGLE-PHASE SYSTEMS

In this section we introduce a few elementary concepts of chemical thermodynamics, restricting attention to the simple case where only one phase needs to be considered. In particular, we consider the case of a *liquid* phase, the density of which is independent of composition, so that number of moles and molar concentration are proportional to each other. The components present in the phase are indicated by the symbol B_j ($j = 1, 2, \ldots, N$), and their molar concentrations by the symbol b_j.

We begin by considering the very simple case where only one chemical reaction may take place. A chemical reaction is usually written in the following form:

$$\overline{\nu}_1 B_1 + \overline{\nu}_2 B_2 + \cdots \rightleftharpoons \overline{\nu}_3 B_3 + \overline{\nu}_4 B_4 + \cdots \qquad (2.2.\overline{1})$$

where all the stoichiometric coefficients $\overline{\nu}_j$ are positive. If all the $\overline{\nu}_j$s are multiplied by an arbitrary positive scalar, the same reaction is being described.

The components B_1, B_2 appearing on the left-hand side are usually referred to as reactants, the ones appearing on the right-hand side as products. However, from the equilibrium viewpoint, the distinction between reactants and products is irrelevant–the same chemical reaction is described if one switches the left and right-hand sides, as suggested by the double arrow sign. The only relevant point is that some of the components ($B_1, B_2, \ldots$) are on one side of the $\rightleftharpoons$ sign, and the other ones ($B_3, B_4, \ldots$) on the other side. This distinction can most easily be kept track of by choosing a new set of stoichiometric coefficients ν_j, which have the same absolute values as the $\overline{\nu}_j$s, but are taken as positive for the (arbitrarily chosen) reactants and negative for the products. With such a convention (which offers very strong advantages in both the equilibrium and the rate analysis and which will be used consistently in the following), Eq. 2.2.$\overline{1}$ can be written in the following equivalent form:

$$\sum_j \nu_j B_j = 0 \qquad (2.2.1)$$

The condition for chemical equilibrium can be written as

$$K = \sum_j b_j^{-\nu_j} \tag{2.2.2}$$

where the Π symbol has the usual meaning of the product, say,

$$\ln\left(\prod_j \psi_j\right) = \sum_j \ln \psi_j \tag{2.2.3}$$

Equation 2.2.2, however, is not sufficient to determine the equilibrium composition, since it is only one equation for the N (>1) unknowns b_j. Therefore, additional equations are required which are obtained as follows.

The physical significance of Eq. 2.2.1 is the following. As long as the composition of the considered phase changes only as a result of the chemical reactions, the variations of the number of moles of the individual components B_j (and hence also of their concentrations) are not independent of each other but are related by the following *stoichiometric condition:*

$$\frac{\Delta b_1}{\nu_1} = \frac{\Delta b_2}{\nu_2} = \frac{\Delta b_3}{\nu_3} = \cdots = -\Delta\xi \tag{2.2.4}$$

with $\Delta\xi$ a variable independent of the index j. The variable $\Delta\xi$ is usually referred to as the "variation of the extent of reaction."

Equation 2.2.4 relates to each other two compositions of the phase considered, one of which can be reached from the other only through the occurrence of the chemical reaction considered. It is now useful to introduce the concept of an "admissible composition" of the phase considered. This is a set of concentrations b_j^0 ($j = 1, 2, \ldots, N$) which is admissible in the sense that it is compatible with the bounds imposed on the problem considered by mass balances. The choice of a particular admissible composition is again somewhat arbitrary; it may be the actual composition of the phase considered at some particular instant in time, or any other one stoichiometrically compatible with an actual one. Once the admissible composition b_j^0 has been chosen, one may assign the value zero to the extent of reaction in the admissible composition, so that Eq. 2.2.4 becomes

$$b_j = b_j^0 - \nu_j\xi \tag{2.2.5}$$

which holds for all js; i.e., Eq. 2.2.5 is a set of N equations. With ξ an arbitrary scalar, which does not make negative any one of the b_js, Eq. 2.2.5 exhausts all possible admissible compositions.

Equations 2.2.2 and 2.2.5 represent a set of $N + 1$ equations for the $N + 1$ unknowns b_j ($j = 1, 2, \ldots, N$) and ξ. This allows us to calculate the equilibrium composition of the phase considered. Indeed, the problem is rather simple to solve; substitution of Eq. 2.2.5 into 2.2.2 yields

$$K = \prod_j (b_j^0 - \nu_j\xi)^{-\nu_j} \tag{2.2.6}$$

which is a polynomial equation for ξ. Let the degree of the polynomial, which depends on the set ν_j, be ω; Eq. 2.2.6 has, in general, ω roots. Of these, only one will be both real and such that, when substituted into 2.2.5, it produces all non-negative values of the b_js; that particular root represents the solution sought.

We now turn attention to the case where more than one chemical reaction may take place, say, a set of M independent chemical reactions

$$\sum_j \nu_{jk} B_j = 0 (k = 1, 2, \ldots, M) \tag{2.2.7}$$

The set of M reactions such as 2.2.7 is said to be *independent* if no one of them is a linear combination of the other ones, say, if

$$\text{rank } \|\nu_{jk}\| = M \tag{2.2.8}$$

(We recall that the rank of an $M \times N$ matrix with $M \leqslant N$ equals the number of rows of the largest square submatrix that has a non-zero determinant).

The set of stoichiometric coefficients ν_{jk} has two degrees of arbitrariness. The first is simply the generalization of the fact that the individual sets $\nu_{j1}, \nu_{j2}, \ldots, \nu_{jM}$ are defined to within an arbitrary multiplicative scalar; for example, the set $\beta_k \nu_{jk}$ with β_k being M arbitrary non-zero scalars identifies the same chemical reactions as ν_{jk}. The other degree of arbitrariness is related to the fact that any set of M linear combinations of the ν_{jk}s also describes the same chemical reactions provided they are independent. Mathematically, given an $M \times M$ matrix β_{pk}, the matrix ν'_{jk} given by

$$\|\nu'_{jk}\| = \|\nu_{jp}\| \cdot \|\beta_{pk}\| \tag{2.2.9}$$

is equivalent to ν_{jk} provided that

$$\text{rank } \|\nu'_{jk}\| = M \tag{2.2.10}$$

Once a particular choice for ν_{jk} has been made (and, as will be seen later, a great deal of ingenuity is required in order to choose the set that makes the calculation easiest), the equilibrium conditions are written as

$$\mathrm{K}_k = \prod_j b_j^{-\nu_{jk}} \tag{2.2.11}$$

Equation 2.2.11 is a set of M independent equations, which is not, however, sufficient to determine the equilibrium composition (i.e., the set of N ($>M$) concentrations b_j).

The generalization of Eq. 2.2.5 to the case of several chemical reactions is straightforward:

$$b_j = b_j^0 - \sum_k \nu_{jk} \xi_k \tag{2.2.12}$$

where ξ_k is the extent of the kth reaction. Substitution of Eq. 2.2.12 into 2.2.11 produces a set of M polynomial equations for the M unknowns ξ_k, which needs to be solved for the one set of ξ_k values that are all real and when substituted into Eq. 2.2.12 give all non-negative values of the b_js.

The procedure outlined above is, in principle, of general applicability. In practice, however, actual solution of the highly non-linear resulting problem may reserve subtle difficulties, particularly if M exceeds 2 or 3, and very often great ingenuity is required in order to reduce the problem to a tractable form.

The thermodynamics of gas treating systems induces the additional difficulty of the existence of two phases. This implies that at least one of the components may be absorbed or desorbed and, therefore, its total content in the liquid phase does not change only due to chemical reactions: Eq. 2.2.12 *does not* apply to components that can be transferred between two phases. These additional difficulties are discussed in the next section.

2.3. GAS TREATING SYSTEMS WITH ONLY ONE VOLATILE COMPONENT

In this section we focus attention on a system consisting of a chemical solvent used for eliminating *one* component from a gas stream. Let A be the component to be eliminated. All other components of the gas stream are regarded as non-condensable, and all other components in the liquid phase as non-volatile; therefore, only component A may be present in both phases. Although this is, of course, a sweeping simplification, consideration of such a simple system illustrates all the essential aspects of the thermodynamic problem.

We first introduce two important concepts: the *molarity* of the chemical solvent and the *degree of saturation*. By molarity m the total equivalent concentration of species that may react equimolarly with component A is intended. The molarity m is the actual concentration of reactive species that could be attained in the chemical solvent if, in the desorption unit, all chemically combined A could be stripped from the solvent.

The degree of saturation y is defined as follows: ym is the total concentration of chemically combined A in the liquid phase. The variable y takes values that may range between zero and unity if the molarity m has been defined appropriately.

Notice that ym is *not* the total content of A in the liquid, since A may also be present in the physically dissolved form, with concentration a. The total content of A is, therefore,

$$\alpha = a + ym = \bar{y}m \tag{2.3.1}$$

Equation 2.3.1 defines the *total* saturation $\bar{y}$; often the distinction between $\bar{y}$ and y is not important since, in general, $a \ll ym$. However, it is important to realize that most techniques of chemical analysis yield the value of $\bar{y}$, not of y.

Inasmuch as the solvent considered is a chemical one, component A may undergo, in the liquid phase, one or more chemical reactions with the liquid-phase components B_j. Suppose a set of such independent reactions has been written which also includes reactions that the B_j components may undergo among themselves. If the stoichiometric coefficient of A is non-zero in more than one reaction, one can always, by taking appropriate linear combinations of the original reactions, transform the set of reactions to an equivalent one where the stoichiometric coefficient of A is unity in one reaction and zero in all other ones, say, to the following general form:

$$A + \sum_j \nu_j B_j = 0 \qquad (2.3.2)$$

$$\sum_j \nu_{jk} B_j = 0 (k = 1, 2, \ldots, M) \qquad (2.3.3)$$

Notice that, since the stoichiometric coefficient of A in 2.3.2 is +1, we have (arbitrarily) chosen to regard reaction 2.3.2 to proceed in the forward direction during absorption and in the reverse direction during desorption.

The thermodynamic problem we wish to solve is as follows: given the values of m and y, calculate the composition of the liquid phase at equilibrium, as well as the equilibrium partial pressure of A.

The advantage of having written the reactions in the form of Equations 2.3.2 and 2.3.3 is that one may first solve the problem represented by the set of M equations 2.3.3, which only involves non-volatile components for which Eq. 2.2.2 holds true. The actual solution procedure is as follows.

First, both m and ym can always be expressed as a linear combination of the b_js, say,

$$m = \sum_j \lambda_j b_j \qquad (2.3.4)$$

$$ym = \sum_j \mu_j b_j \qquad (2.3.5)$$

The interpretation of the coefficients λ_j and μ_j is straightforward. The λ_j are (proportional to) the number of atoms in B_j of some element that is not present in component A. The μ_j are the number of moles of component A chemically combined in component B_j. (The CO_2-amine worked example in Section 2.4 will further clarify the issues involved.)

It is always possible to find a simple set of values b_j^0 such that Eqs. 2.3.4 and 2.3.5 are satisfied for any given couple of values of m and y. The simplest choice is obtained by assuming that all chemically combined A is present in one particular form B_j, (e.g., B_2); then

$$b_2^0 = \frac{ym}{\mu_2} \qquad (2.3.6)$$

whereas $b_j^0 = 0$ for all other B_js characterized by a non-zero μ_j. Such a set of b_j^0 values is an admissible composition, and the equilibrium composition must, therefore, be a set of b_j values related to the b_j^0s by Eq. 2.2.2, say,

$$b_j = b_j^0 - \sum_j \nu_{jk} \xi_k \qquad (2.3.7)$$

Substitution of Eq. 2.3.7 into the M equilibrium conditions for reactions 2.3.3 yields the following set of M polynomial equations:

$$\mathrm{K}_k = \prod_j (b_j^0 - \nu_{jk}\xi_k)^{-\nu_{jk}}, \qquad (k = 1, 2, \ldots, M) \qquad (2.3.8)$$

Equation 2.3.8 can be solved for the ξ_ks; these are then substituted into Eq. 2.3.7 to calculate the values of the b_js. This completes the solution of the subproblem of the set of reactions 2.3.3.

Once the b_js have been calculated, the value of a is calculated from the equilibrium condition for reaction 2.3.2:

$$\mathrm{K} = \left(\frac{\prod_j b_j^{-\nu_j}}{a} \right) \qquad (2.3.9)$$

(notice that we use the symbol K without suffixes for the equilibrium constant of reaction 2.3.2).

With this, the chemical composition of the liquid phase is calculated. The values of α and of $\bar{y}$ can be obtained directly from Eq. 2.3.1.

The equilibrium partial pressure of A, p^*, can finally be calculated from the requirement that equilibrium must prevail between the gaseous and physically dissolved forms of A, say, from the equilibrium condition for the following "reaction":

$$A(g) \rightleftharpoons A(\ell) \qquad (2.3.10)$$

As was discussed in Chapter 1, often such an equilibrium condition can be written in the form of Henry's Law; say,

$$p^* = \mathrm{H}a \qquad (2.3.11)$$

where H is Henry's constant. Evaluation of the value of H for chemical solvents presents important difficulties which will be discussed in a later section.

The procedure briefly outlined above allows, given y and m, the calculation of both $\bar{y}$ and p^*. The latter two quantities are directly accessible to measurement and, in fact, vapor–liquid equilibrium data are often available in the form of an empirically determined plot of p^* versus $\bar{y}$. In Section 2.4, two specific examples of the procedure are discussed in detail.

2.4. TWO SIMPLE SPECIFIC SYSTEMS

In this section we illustrate the application of the procedure outlined in Section 2.3 to two specific gas-treating systems, both involving the removal of CO_2 from a gas stream. The two chemical solvents considered are aqueous solutions of potassium carbonate and of any monoamine.

The chemistry involved in the case of potassium carbonate is extremely simple. There are only two non-volatile species to be considered, namely, the carbonate and the bicarbonate ions:

$$B_1 = CO_3^{2-} \tag{2.4.1}$$

$$B_2 = HCO_3^- \tag{2.4.2}$$

Of course, the liquid phase also contains other ionic species, namely, the potassium cation K^+, OH^-, and H^+, as well as water. However, K^+ does not participate in any of the chemical reactions involved and can, therefore, be regarded as an inert. Even at the highest possible concentration of either K_2CO_3 or $KHCO_3$, or both, the molar concentration of water is large enough so that at least as a first-order approximation, it may be regarded as essentially constant. The concentrations of OH^- and of H^+ are always very much smaller than either b_1 or b_2 or both, unless the desorption unit is so efficient as to result in the formation of some KOH, which is highly unlikely. Therefore, for the basic thermodynamic problem to be considered, one may limit attention to only B_1 and B_2.

The molarity and the saturation are easily identified as

$$m = b_1 + \tfrac{1}{2}b_2 \tag{2.4.3}$$

$$ym = \tfrac{1}{2}b_2 \tag{2.4.4}$$

Indeed, when $y = 0$, there is only carbonate in the solution and its concentration under these conditions is the molarity. When CO_2 is absorbed into the solution, it undergoes the following reaction:

$$CO_2 + H_2O + CO_3^{2-} \longrightarrow 2HCO_3^- \tag{2.4.5}$$

and, therefore, $\frac{1}{2}$ mole of CO_2 is chemically combined in each mole of HCO_3^-.

Since there are only two non-volatile species to consider, Eq. 2.4.3 and Eq. 2.4.4 are sufficient to determine their values. It follows that there are no reactions such as 2.3.3 to take into account, whereas reaction 2.4.5 plays the role of 2.3.2. In view of Eqs. 2.4.3 and 2.4.4, the concentrations b_1 and b_2 are

$$b_1 = m(1 - y) \tag{2.4.6}$$

$$b_2 = 2my \tag{2.4.7}$$

The matrix of the coefficients λ_j, μ_j, and ν_j (there are no ν_{jk}s to consider) is as follows:

	λ_j	μ_j	ν_j
B_1	1	0	1
B_2	$\frac{1}{2}$	$\frac{1}{2}$	-2

(2.4.8)

The equilibrium condition for reaction 2.4.5 is given by

$$\mathrm{K} = \frac{b_2^2}{b_1 a} \tag{2.4.9}$$

The question of the actual values of K will be discussed in a later section; for the purpose of the problem considered here, it is sufficient to state that, over the whole range of temperatures of industrial interest (30–120°C), K is larger than 10^3.

Combination of Eqs. 2.4.6, 2.4.7, and 2.4.9 yields

$$a = \frac{4m}{\mathrm{K}} \frac{y^2}{1-y} \tag{2.4.10}$$

which shows that a is about 1000 times smaller than m.

The values of α and $\bar{y}$ are obtained from Eq. 2.3.1;

$$\alpha = my\left(1 + \frac{4}{\mathrm{K}} \frac{y}{1-y}\right) \tag{2.4.11}$$

$$\bar{y} = y\left(1 + \frac{4}{\mathrm{K}} \frac{y}{1-y}\right) \tag{2.4.12}$$

Equation 2.4.12 shows that, for all practical purposes, $\bar{y}$ may be identified with y.

Finally, the equilibrium partial pressure p^* is obtained from Eqs. 2.3.11 and 2.4.10 as

$$p^* = \frac{4\mathrm{H}}{\mathrm{K}} m \frac{y^2}{1-y} \tag{2.4.13}$$

The available VLE data are plotted in Figure 2.4.1 as p^* versus $y^2/(1-y)$ on log-log paper. A straight line at 45° correlates the data with satisfactory accuracy, showing that the analysis given above correctly predicts the dependency of p^* on y.

A final point should be mentioned. The concentrations of OH^- and H^+ can be calculated from the equilibrium conditions for the following two reactions:

$$HCO_3^- + OH^- = CO_3^{2-} + H_2O \tag{2.4.14}$$

$$H_2O = H^+ + OH^- \tag{2.4.15}$$

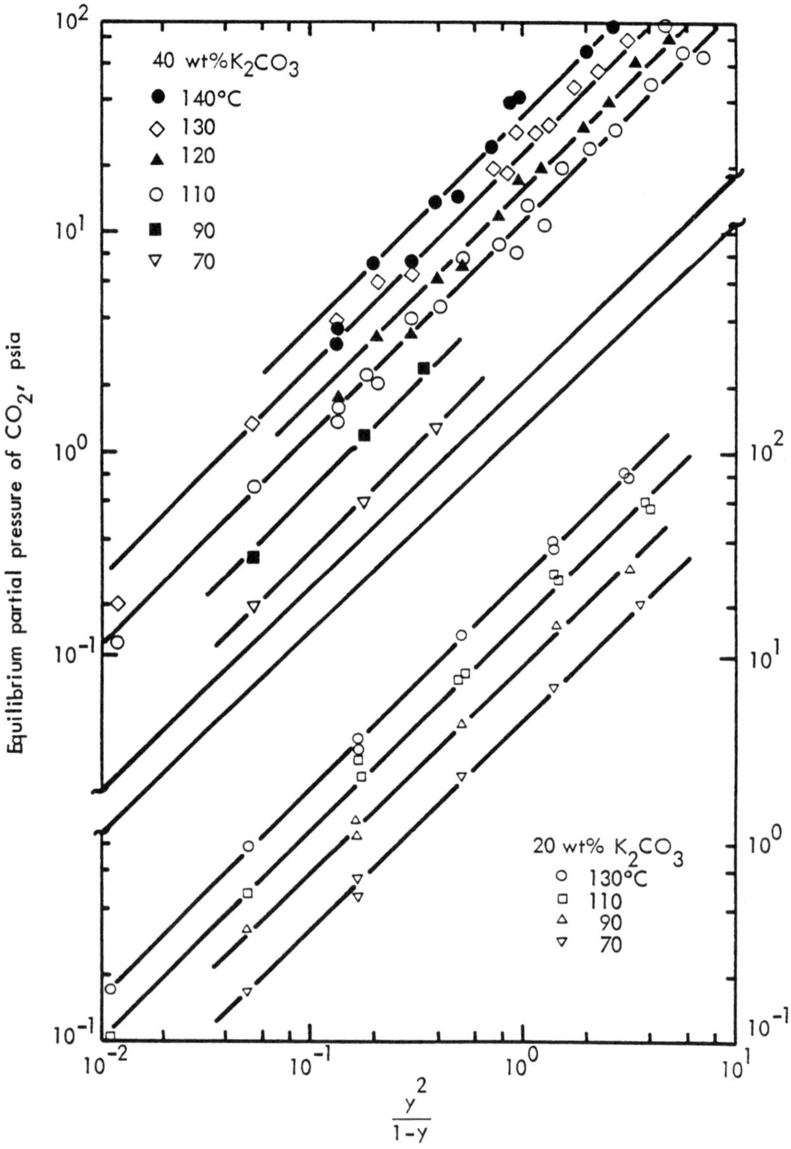

FIGURE 2.4.1. CO_2 equilibrium vapor pressures over potassium carbonate solutions.

as

$$[OH^-] = K_1 \frac{b_1}{b_2} \tag{2.4.16}$$

$$[H^+] = \frac{K_w}{K_1} \frac{b_2}{b_1} \tag{2.4.17}$$

where K_1 and K_w are the equilibrium constants for reactions 2.4.14 and 2.4.15. Since K_1 is of order 10^{-4} gmol/liter, and K_w of order 10^{-14} gmol/liter, both $[OH^-]$

and $[H^+]$ are negligibly small as compared to either b_1 or b_2 or both; therefore, the original restriction of the analysis to only the species B_1 or B_2 is justified.

We now turn attention to the case of an aqueous solution of a monoamine, which is considerably more complex. Let the monoamine be RNH; there are now five non-volatile species to consider, namely:

$$B_1 = \mathrm{RNH} \tag{2.4.18}$$

$$B_2 = \mathrm{RNH_2^+} \tag{2.4.19}$$

$$B_3 = \mathrm{RNCOO^-} \tag{2.4.20}$$

$$B_4 = \mathrm{HCO_3^-} \tag{2.4.21}$$

$$B_5 = \mathrm{CO_3^{2-}} \tag{2.4.22}$$

Again, the molarity and saturation are easily identified as

$$m = b_1 + b_2 + b_3 \tag{2.4.23}$$

$$ym = b_3 + b_4 + b_5 \tag{2.4.24}$$

Equation 2.4.24 simply reflects the fact that CO_2 may be chemically combined in the three different forms of the carbamic, carbonic, and bicarbonic ion. The equilibrium problem to be considered is essentially the splitting of the chemically combined CO_2 among the three forms.

In this case, one may write several different reactions involving CO_2; we arbitrarily choose to consider the formation of the carbamate, say,

$$\mathrm{CO_2 + 2RNH = RNH_2^+ + RNCOO^-} \tag{2.4.25}$$

Equation 2.4.25 plays the role of Eq. 2.3.2. The reactions to be considered in the role of Eq. 2.3.3 are

$$k = 1: \qquad \mathrm{RNH + HCO_3^- = RNH_2^+ + CO_3^{2-}} \tag{2.4.26}$$

$$k = 2: \qquad \mathrm{RNH + HCO_3^- = H_2O + RNCOO^-} \tag{2.4.27}$$

the equilibrium of which determines the splitting of CO_2 between the three possible forms B_3, B_4, and B_5. The matrix of the coefficients λ_j, μ_j, ν_j and ν_{jk} is as follows:

	λ_j	μ_j	ν_j	ν_{j1}	ν_{j2}	
B_1	1	0	2	1	1	
B_2	1	0	−1	−1	0	
B_3	1	1	−1	0	−1	(2.4.28)
B_4	0	1	0	1	1	
B_5	0	1	0	−1	0	

We now need to choose an admissible composition, and we do so by assuming that *all* chemically combined CO_2 is in the form of the bicarbonate. Therefore, the values of the b_j^0s are:

$$b_1^0 = m(1 - y) \tag{2.4.29}$$

$$b_2^0 = my \tag{2.4.30}$$

$$b_3^0 = 0 \tag{2.4.31}$$

$$b_4^0 = my \tag{2.4.32}$$

$$b_5^0 = 0 \tag{2.4.33}$$

The logic of the equilibrium calculation is as follows. We first assume that CO_2 is all in the bicarbonate form, and then allow reactions 2.4.26 and 2.4.27 to proceed to the right until chemical equilibrium is reached. This has the advantage that the equilibrium values of the extents of reaction ξ_1 and ξ_2 will be positive.

Equation 2.3.7 now takes the following forms:

$$b_1 = m(1 - y) - \xi_1 - \xi_2 \tag{2.4.34}$$

$$b_2 = my + \xi_1 \tag{2.4.35}$$

$$b_3 = \xi_2 \tag{2.4.36}$$

$$b_4 = my - \xi_1 - \xi_2 \tag{2.4.37}$$

$$b_5 = \xi_1 \tag{2.4.38}$$

The equilibrium values of ξ_1 and ξ_2 must satisfy the following constraints in order that all the b_js are non-negative:

$$\xi_1 \geqslant 0 \tag{2.4.39}$$

$$\xi_2 \geqslant 0 \tag{2.4.40}$$

$$\xi_1 + \xi_2 \leqslant my \tag{2.4.41}$$

$$\xi_1 + \xi_2 \leqslant m(1 - y) \tag{2.4.42}$$

When Eqs. 2.4.34-2.4.38 are substituted into the equilibrium conditions for reactions 2.4.26 and 2.4.27, one obtains

$$\mathrm{K}_1 = \frac{(my + \xi_1)\,\xi_1}{[m(1 - y) - \xi_1 - \xi_2]\,(my - \xi_1 - \xi_2)} \tag{2.4.43}$$

$$K_c = \frac{\xi_2}{[m(1-y) - \xi_1 - \xi_2]\,(my - \xi_1 - \xi_2)} \tag{2.4.44}$$

(In view of the importance of the value of the equilibrium constant of reaction 2.4.27, it is assigned the symbol K_c and will be referred to as the carbamate stability constant.)

The system 2.4.43 and 2.4.44 is a fourth-order polynomial system for ξ_1 and ξ_2. Only one of the four roots satisfies the constraints 2.4.39–2.4.42. When this solution is substituted into Eqs. 2.4.34–2.4.38, the values of the b_js are obtained. The value of a is then calculated from the equilibrium condition for reaction 2.4.25:

$$a = \frac{b_2 b_3}{b_1^2 K} \tag{2.4.45}$$

where K is the equilibrium constant for reaction 2.4.25. The values of p^* is again obtained from Eqs. 2.3.11 and 2.4.45.

The procedure as discussed above produces the complete solution to the problem. However, in order to illustrate the simplifications that are often possible when dealing with equilibrium problems of the type considered here, we consider the special case of an aqueous solution of monoethanolamine.

Taking the ratio of Eqs. 2.4.43 and 2.4.44, one obtains

$$\frac{K_1}{K_c} = \frac{(my + \xi_1)\,\xi_1}{\xi_2} \geqslant \frac{\xi_1^2}{\xi_2} \tag{2.4.46}$$

Consideration of Eqs. 2.4.41 and 2.4.42 shows that ξ_2 cannot exceed the value 0.5m. Therefore,

$$\xi_1 \leqslant \sqrt{\frac{0.5mK_1}{K_c}} \tag{2.4.47}$$

For monoethanolamine at room temperature, K_1/K_c is of order 2×10^{-3} gmol/liter, whereas values of m of practical interest are of order 1 gmol/liter; therefore,

$$\xi_1 \leqslant 0.03 \tag{2.4.48}$$

and an acceptable first-order approximation is $\xi_1 = 0$; i.e., *no CO_2 is chemically combined in the carbonate form.*

It is instructive to investigate how general this conclusion is (i.e., for which amines it holds true). Notice that reaction 2.4.26 describes the HCO_3^-/CO_3^{2-} equilibrium in an amine solution. That equilibrium is, of course, displaced to the CO_3^{2-} form with increasing alkalinity. It is also apparent that, when an acid gas such as CO_2 is absorbed into an amine solution, the alkalinity decreases and,

TABLE 2.4.1. Room Temperature *p*K for Different Amines[a]

Primary amines		
n-butylamine	10.6	(*n*BA)
ethylenediamine	9.9	(EDA)
monoethanolamine	9.6	(MEA)
monoisopropanolamine	9.4	(MIPA)
Secondary amines		
2-ethylaminoethanol	9.8	(EAE)
2-methylaminoethanol	9.7	(MAE)
diethanolamine	8.9	(DEA)
diisopropanolamine	8.8	(DIPA)
Tertiary amines		
n-methyl piperidine ethanol	9.5	(MPE)
methyldiethanolamine	8.5	(MDEA)
triethanolamine	7.8	(TEA)

[a] $pK = -\log_{10} K_p$, $K_p = \dfrac{[Am]\,[H^+]}{[AmH^+]}$.

therefore, that the HCO_3^-/CO_3^{2-} equilibrium will be most strongly displaced to the CO_3^{2-} form in very lean solutions (solutions lean in CO_2). If even a solution with $y = 0$ (i.e., containing only the amine) has a sufficiently weak alkalinity to displace the HCO_3^-/CO_3^{2-} equilibrium towards the bicarbonic form, practically no CO_3^{2-} ion will be formed. Indeed, K_1 is simply the ratio of the second dissociation constant of carbonic acid, $K_{c2} = [CO_3^{2-}]\,[H^+]/[HCO_3^-]$, to the protonation constant of the amine, $K_p = [RNH]\,[H^+]/[RNH_2^+]$.

At room temperature, the pK for the second dissociation of H_2CO_3 is 10.33. It follows that, if the amine has a room temperature $pK = -\log K_p$ less than about 10, no carbonate will be formed. A list of pK values for several amines is given in Table 2.4.1.

One may notice that all secondary and tertiary amines have pK values well below 10. Of the primary amines, the ones that, like monoethanolamine, have an alcoholic group, have pK values less than 10; the other ones are strong enough alkalis that formation of the carbonate should be taken into account, at least for lean solutions ($y \ll 1$).

Turning attention back to the thermodynamic model for monoethanolamine, neglecting the carbonate ion concentration reduces Eq. 2.4.44 to

$$K_c = \frac{\xi_2}{[m(1-y) - \xi_2]\,(my - \xi_2)} \qquad (2.4.49)$$

the solution of which (satisfying constraints 2.4.41 and 2.4.42) is

$$\xi_2 = \frac{m + 1/K_c - \sqrt{(m + 1/K_c)^2 - 4m^2 y(1-y)}}{2} \qquad (2.4.50)$$

For monoethanolamine at room temperature, K_c is of order 50 1/gmol and, therefore, $1/K_c$ is much less than m under all conditions of interest. Therefore, the right-hand side of Eq. 2.4.50 can be expanded with within $0\,(1/K_c m)$ to yield

$$\xi_2 = \frac{m(1-\delta)}{2} + \frac{1}{2K_c}\left(1 - \frac{1}{\delta}\right) \tag{2.4.51}$$

where

$$\delta = |1 - 2y| \tag{2.4.52}$$

The second term on the right-hand side of Eq. 2.4.51 is negligible as compared to the first term unless $\delta \to 0$; more exactly, if

$$\delta \gg \frac{1}{K_c m} \tag{2.4.53}$$

then

$$\xi_2 \simeq \frac{m(1-\delta)}{2} \tag{2.4.54}$$

Equation 2.4.54 can be written as follows:

$$y < 0.5, \qquad \xi_2 = my \tag{2.4.55}$$

$$y > 0.5, \qquad \xi_2 = m(1-y) \tag{2.4.56}$$

which, of course, hold only as long as 2.4.53 is satisfied, say, with the exception of a small region around $y = 0.5$ (since $1/K_c m$ is of order 10^{-2} for values of m of practical interest).

Again, it is instructive to investigate the generality of this conclusion. The conclusion is, of course, related to the fact that K_c is significantly larger than the molarity m, say, that the carbamate is a very stable form for monoethanolamine. Figure 2.4.2 is an Arrhenius plot of $1/K_c$ for monoethanolamine and for two secondary aminoalcohols. The values have been obtained by fitting a thermodynamic model to VLE data.

First of all, K_c decreases as temperature increases and, therefore, at high temperature the carbamate is not so stable, even in monoethanolamine solutions. The carbamate of secondary amines is less stable. Of course, tertiary amines cannot possibly form a carbamate; for such amines one could set $K_c = 0$. The result in Eqs. 2.4.55 and 2.4.56 *is not* of general validity; however, it is worthwhile to discuss it in some detail inasmuch as MEA is a commonly used chemical solvent for CO_2.

Equation 2.4.36 shows that ξ_2 is the carbamic ion concentration. Therefore,

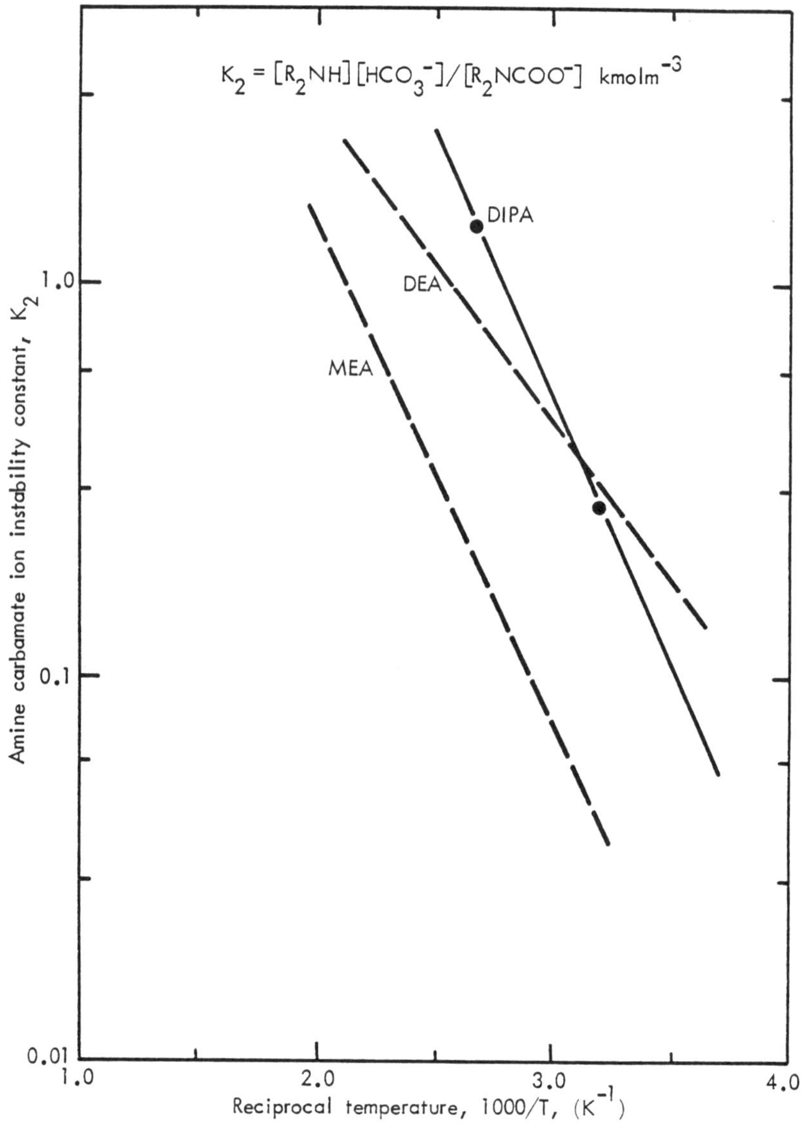

FIGURE 2.4.2. Carbamate instability constants for aminoalcohol solutions.

Eq. 2.4.55 implies that, as long as $y < 0.5$, essentially *all* the chemically combined CO_2 is in the form of the carbamate $RNH_2^+ \cdot RNCOO^-$. However, since in the carbamate two moles of amine are combined with one mole of CO_2, the largest saturation that can be reached with all CO_2 in the carbamate form is 0.5. When this value is reached, any additional CO_2 that is chemically combined must result in the reversion of the carbamate to the bicarbonate, which is reflected by Eq. 2.4.56.

Indeed, from Eq. 2.4.37 one obtains, at $y > 0.5$,

$$b_4 = m(2y - 1) \tag{2.4.57}$$

for the bicarbonate concentration. Of course, the transition between the two regions is not abrupt, as reflected by the requirement that condition 2.4.53 be satisfied; in a neighborhood of the stoichiometric transition point $y = 0.5$, the simple forms of Eq. 2.4.55 and 2.4.56 break down.

One may notice that the approximate solution discussed above is satisfactory for the concentrations b_j that are calculated as being non-zero; the other ones (such as, over the whole range of values of y, the carbonate concentration b_5) need to be calculated from the appropriate equilibrium conditions. The results are summarized in the following.

	$y < 0.5$	$y > 0.5$
b_1	$m(1 - 2y)$	$\frac{1-y}{2y-1}\frac{1}{K_2} \simeq 0$
b_2	my	my
b_3	my	$m(1 - y)$
b_4	$\frac{1}{K_2}\frac{y}{1-y} \simeq 0$	$m(2y - 1)$
b_5	$\frac{K_1}{K_2} \simeq 0$	$\frac{K_1}{K_2}\frac{1-y}{y} \simeq 0$

(2.4.58)

A comparison between the b_j values calculated from the complete numerical solution of the problem and the values reported in Eqs. 2.4.58 is presented in Figures 2.4.3 and 2.4.4. Values that are $\simeq 0$ in 2.4.58 are calculated with a procedure discussed in Section 2.5.

The CO_2 concentration is given by Eq. 2.4.45 and the corresponding equilibrium partial pressure is

$$p^* = \frac{H}{K}\frac{b_2 b_3}{b_1^2} \tag{2.4.59}$$

If the approximate equations in 2.4.58 are used, the equation for p^* is

$$y < 0.5, \qquad p^* = \frac{H}{K}\left(\frac{y}{1-2y}\right)^2 \tag{2.4.60}$$

$$y > 0.5, \qquad p^* = \frac{K_c^2 \cdot H}{K} m^2 \frac{y(2y-1)^2}{1-y} \tag{2.4.61}$$

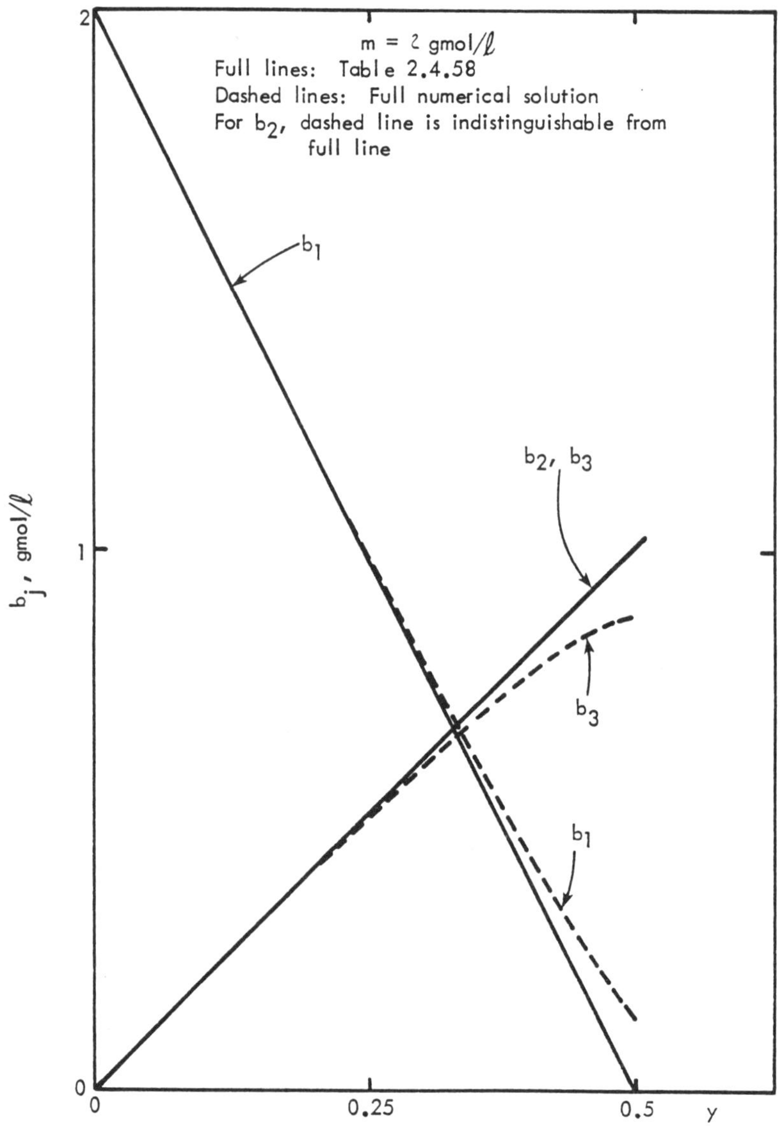

FIGURE 2.4.3. Liquid-phase composition versus CO_2 load for $0 < y < 0.5$.

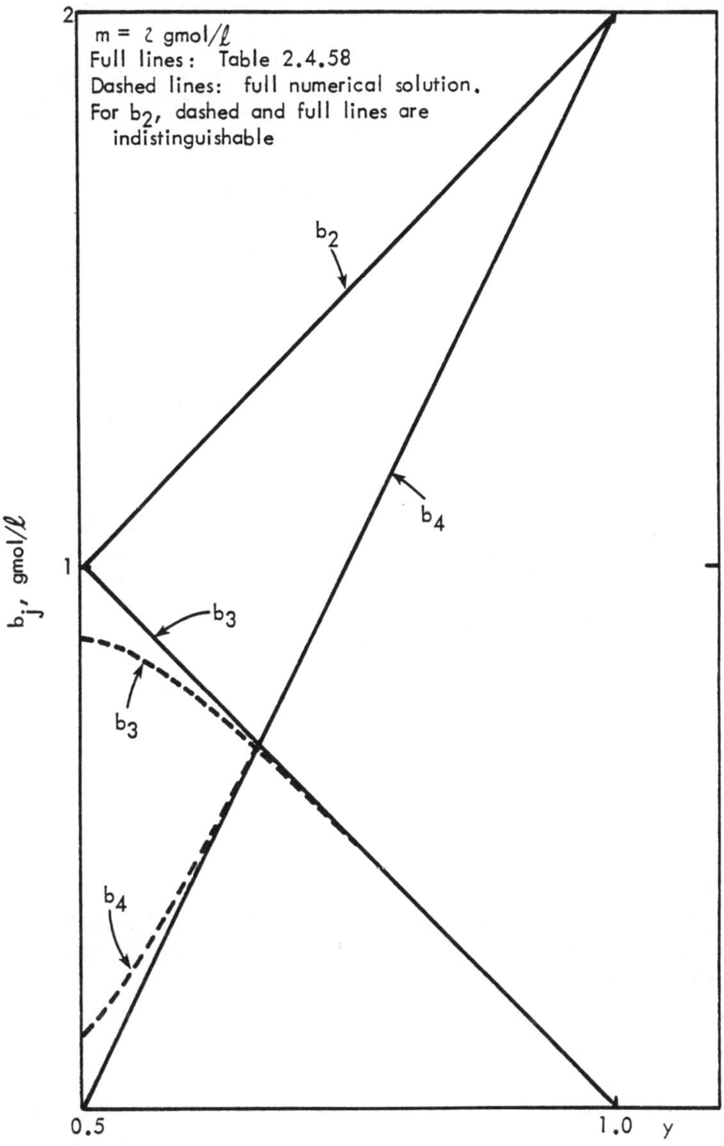

FIGURE 2.4.4. Liquid-phase composition versus CO_2 load for $0.5 < y < 1.0$.

The analysis given above for the two systems considered illustrates the use of the procedure for the thermodynamic modeling of gas treating systems. Certain general features of this procedure, which are of help in the modeling of any given system, are discussed in Section 2.5.

2.5. GENERAL FEATURES OF THERMODYNAMIC MODELING

The development of a thermodynamic model for a gas treating system is based on the procedure outlined in Section 2.4. However, that procedure allows a great flexibility and within such flexibility different choices of the equations to set up result in very different mathematical structures of the final model. It is, therefore, important to have some guidelines for selecting those equations that result in the simplest possible mathematical model.

First, one should consider the type of mathematical problem that needs to be solved. The equations that define the molarity m and the saturation y (Eqs. 2.3.4 and 2.3.5, or, in the two specific examples discussed in Section 2.4, Eqs. 2.4.3-2.4.4 and 2.4.23-2.4.24) are *linear.* In addition to these two equations, one has M equilibrium conditions for the M independent chemical reactions not involving the volatile species A, and $N - M - 2$ stoichiometric equations of the type of Eq. 2.3.7, where N is the number of non-volatile species B_j entering the problem:

Definitions of m and y	2	
Stoichiometric equations	$N - M - 2$	
Equilibrium conditions	M	
Total number of equations	N	(2.5.1)
Unknowns: $b_1, b_2, \ldots, b_n$	N	

The stoichiometric equations are again *linear*, but the equilibrium equations are not; the equilibrium condition for reaction k is a polynomial equation of degree ζ_k which is related to the set of stoichiometric coefficients ν_{jk}; more exactly, ζ_k is equal to the sum of either all the positive ν_{jk} or all the absolute values of the negative ν_{jk}, whichever is largest.

This consideration suggests a first general guideline. Since the set ν_{jk} can be substituted with any equivalent one ν'_{jk} such as given in Eq. 2.2.9, it is very often (but not necessarily always) possible to *select a specific set such that no ζ_k is larger than 2.* Equations 2.4.26 and 2.4.27 are an example of such a selection.

If, indeed, all the ζ_ks are equal to 2, the system of M equilibrium conditions is equivalent to a polynomial equation of degree 2^M, which has, in general, 2^M roots. Of these, only one is physically significant, and must be selected on the basis of constraints such as those in Eqs. 2.4.39-2.4.42.

It is apparent that the mathematical model's complexity increases very rapidly with increasing M and, therefore, guidelines are needed in order to set up a model with as low a value of M as possible.

The first guideline is the following. If possible, *include in the list of the b_js* *only those that appear in the definitions of m and y.* In the two examples discussed in Section 2.4, this guideline has been followed; the concentrations of other non-volatile components (specifically, the OH^- and H^+ ions) are then to be calculated independently once the main problem has been solved (see, e.g., Eqs. 2.4.16 and 2.4.17).

Even when this guideline is followed, the model may easily turn out to be remarkably complex. As an example, consider the comparatively simple case of CO_2 absorption into an aqueous solution of a symmetric diamine HNRNH (by symmetric we mean that the two NH groups are indistinguishable). The species that need to be considered because they enter the definitions of m and y are the following eight:

$$
\begin{array}{ll}
\text{HNRNH} & (\text{R}) \\
\text{HNRNH}_2^+ & (\text{R}^+) \\
{}^+\text{H}_2\text{NRNH}_2^+ & (\text{R}^{2+}) \\
\text{HNRNCOO}^- & (\text{R}^-) \\
{}^-\text{OOCNRNCOO}^- & (\text{R}^{2-}) \\
{}^+\text{H}_2\text{NRNCOO}^- & (\text{R}^{+-}) \\
\text{HCO}_3^- & \\
\text{CO}_3^{2-} &
\end{array}
\qquad (2.5.2)
$$

The definitions of m and y are:

$$m = \text{R} + \text{R}^+ + \text{R}^{2+} + \text{R}^- + \text{R}^{2-} + \text{R}^{+-} \qquad (2.5.3)$$

$$my = \text{R}^- + \text{R}^{2-} + \text{R}^{+-} + \text{HCO}_3^- + \text{CO}_3 \qquad (2.5.4)$$

Following by analogy the procedure discussed for the monoamine case, one would write

$$\text{CO}_2 + \text{R} = \text{R}^+ + \text{HCO}_3^- \qquad (2.5.5)$$

as the reaction involving CO_2. One would then regard $\text{R}^+ \cdot \text{HCO}_3^-$ as the only carbonated species in the admissible composition, and the following five independent chemical reactions would be required:

$$
\begin{array}{c}
\text{R} + \text{HCO}_3^- = \text{R}^+ + \text{CO}_3^{2-} \\
\text{R} + \text{HCO}_3^- = \text{R}^- \\
2\text{R}^- = \text{R} + \text{R}^{2-} \\
2\text{R}^+ = \text{R} + \text{R}^{2+} \\
\text{R}^+ + \text{R}^- = \text{R} + \text{R}^{+-}
\end{array}
\qquad (2.5.6)
$$

which indeed give all $\zeta_k = 2$. The problem is thus equivalent to a polynomial equation of degree $2^5 = 32$. Should the diamine not be symmetric, there would be three more chemical equilibria to be considered since there would be two different forms possible for the species R^+, R^-, and R^{+-}. This example shows how the complexity of the mathematical problem may easily become a major difficulty.

It follows that simplifications of the type discussed in Section 2.4 following Eq. 2.4.45 are often very useful. These simplifications consist essentially of the *uncoupling* of the system of equilibrium equations, as discussed below.

The conditions of equilibrium produce a set of coupled polynomial equations insofar as the definitions of m and y and the stoichiometric equations of the type 2.3.7 provide coupling between the different b_js. However, if one (or more) of the b_js are considerably smaller than the others (such as b_5 in the monoethanolamine example), the linear equations are *approximately* satisfied by setting those b_js as equal to zero. This reduces the value of M (in the monoethanolamine example, from 2 to 1); once the (simplified) main problem has been solved, the b_js having small values are then calculated from the (by now uncoupled) relevant equilibrium condition (see the entries for b_5 in 2.4.59). The uncoupling procedure can be iterated, as again exemplified in the monoethanolamine problem, where at $y < 0.5$, the value of b_4 becomes uncoupled and at $y > 0.5$, the value of b_1 becomes uncoupled.

It may be argued that, with modern computing facilities, such approximation schemes are not needed inasmuch as solution of a high-degree polynomial system is not a prohibitive task. However, there are two considerations that suggest that, in fact, approximation schemes are very important.

The first consideration is that, in the numerical solution of a problem where the orders of magnitude of different unknowns are very different, problems of convergence and/or of numerical stability may easily arise.

The second consideration is related to the need for an appropriate thermodynamic model in the *rate* analysis. The analysis of the rate behavior of gas treating systems is sufficiently complex in its own right that the availability of a possibly crude but simple thermodynamic model is crucial. In particular, it is very important to establish, at least over certain ranges of y values, which is the main reaction taking place during absorption or desorption: say, in the monoethanolamine example, the $CO_2 \rightleftharpoons$ carbamate reaction at $y < 0.5$ and the CO_2 + carbamate $\rightleftharpoons$ bicarbonate reaction at $y > 0.5$.

The concept of "main reaction" is very important, both for the thermodynamic analysis (as will be discussed in Section 2.7) and for the rate analysis to be discussed in Chapter 3. However, it should be kept in mind that the chemistry of a gas treating system may be complex enough to result in more than one main reaction. An example of such a system is discussed in the following.

Consider CO_2 removal with an amine solution characterized by a moderately stable carbamate, say, one where $K_c m$ is of order unity. The value of ξ_2 (i.e., the carbamate concentration) needs to be calculated from Eq. 2.4.50 and no significant simplifications are possible. Figure 2.5.1 is a plot of the solution of Eq. 2.4.50 for the case where $K_c m = 1$.

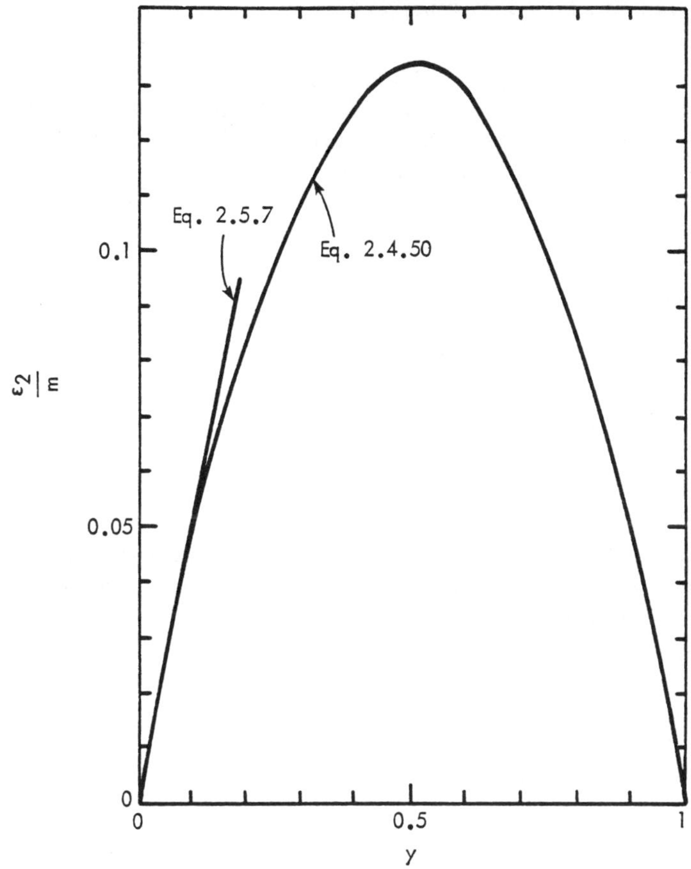

FIGURE 2.5.1. Carbamate concentration versus CO_2 load; $K_c m = 1$.

It is obvious from this figure that not all of the stoichiometrically possible combined CO_2 is in the carbamate form, and yet the carbamate concentration is not negligible. In this system, at $y < 0.5$, the formation of both the carbamate and the bicarbonate are main reactions and at $y > 0.5$, both the direct bicarbonate formation and the carbamate → bicarbonate shift are main reactions.

For systems such as the one described above, the relative importance of the two main reactions changes with changing y; it is often useful to develop asymptotic equations valid for very lean solutions (say, at $y \ll 1$). For the amine case, one gets, at $y \ll 1$, $K_c m = 1$,

$$\frac{\xi_2}{m} = \frac{y}{2} \tag{2.5.7}$$

which means that in lean solutions, the two main reactions take place in equal amounts.

2.6. TWO VOLATILE COMPONENTS

It is common in gas treating that two components (rather than only one) are simultaneously removed from a gas stream. The typical example is the one where both CO_2 and H_2S are removed; inasmuch as both are acid gases, most alkaline solutions are efficient chemical solvents for both.

The thermodynamic analysis of a system with two volatile components is not conceptually different from the one relative to only one volatile component. Let the two volatile components be A, A'. The chemical reactions to be considered can always be reduced to an equivalent set having the following structure:

$$A + \sum \nu_j B_j = 0 \tag{2.6.1}$$

$$A' + \sum \nu_j' B_j = 0 \tag{2.6.2}$$

$$\sum \nu_{jk} B_j = 0, \qquad (R = 1, 2, \ldots, M) \tag{2.6.3}$$

i.e., the same set as in Eqs. 2.3.2 and 2.3.3, with one more reaction (Eq. 2.6.2).

Furthermore, one can define a degree of saturation y', analogous to y, and a total content of A' in the liquid:

$$\alpha' = a' + y'm = \overline{y}'m \tag{2.6.4}$$

The molarity m and the fractional saturation y are still given by Eqs. 2.3.4 and 2.3.5, whereas y' is given by

$$y'm = \sum_j \mu_j' b_j \tag{2.6.5}$$

Again, one can easily define an admissible composition b_j^0 (i.e., a set of values b_j^0 compatible with given values of m, y, y'). The actual values b_j will again be given by Eq. 2.3.7 and can, therefore, be calculated by solving the system of Eqs. 2.3.8. Once the values of b_j are calculated, the value of a is again given by Eq. 2.3.9, and analogously the value of a' is given by

$$\mathrm{K}' = \frac{\prod_j (b_j)^{-\nu_j'}}{a'} \tag{2.6.7}$$

where K' is the equilibrium constant of reaction 2.6.2. Finally, p^* is given again by Eq. 2.3.11, and $p^{*\prime}$ by

$$p^{*\prime} = \mathrm{H}'a' \tag{2.6.8}$$

where H' is the Henry constant for component A'. At a given molarity m, both p^* and $p^{*\prime}$ will depend on *both* y and y'.

Two examples may be useful to illustrate the problem of thermodynamic modeling for systems with two volatile components. Both examples refer to CO_2 and H_2S as the volatile components; the chemical solvents considered are carbonate and monoethanolamine.

In a carbonate solution, the molarity m is the total K_2CO_3 concentration. The second dissociation constant of H_2S is so low that, at the pH values of interest, practically no S^{2-} is formed and all the chemically combined H_2S is in the form of the HS^- ion. Therefore, assigning the primed quantities to H_2S,

$$my' = b_3 \tag{2.6.9}$$

where $B_3 = HS^-$, whereas B_1 and B_2 are still given by Eqs. 2.4.1 and 2.4.2.

There is a subtlety involved with the definition of the fractional CO_2 saturation. The value $y = 0$ is assigned to a solution containing only K_2CO_3 (i.e., to a solution where $b_2 = b_3 = 0$). Furthermore, Eq. 2.4.4 should hold whenever there is no chemically combined H_2S in the liquid (i.e., when $b_3 = 0$). Now consider a solution with $b_3 = 0$ and $y \neq 0$; Eqs. 2.4.6 and 2.4.7 hold in such a solution. Suppose some H_2S is added to the solution; the following reaction will take place:

$$H_2S + CO_3^{2-} = HS^- + HCO_3^- \tag{2.6.10}$$

Reaction 2.6.10 will cause both b_1 and b_2 to change, *but not their sum.* Since, of course, addition of H_2S cannot change the value of y, it follows that Eqs. 2.4.6 and 2.4.7 do not hold, but their sum does. Hence, one has

$$m(1 + y) = b_1 + b_2 \tag{2.6.11}$$

The molarity is given by

$$m = b_1 + \tfrac{1}{2}b_2 + \tfrac{1}{2}b_3 \tag{2.6.12}$$

as can be obtained either by simple stoichiometric considerations, or by considering the requirement of electrical neutrality (with the H^+ and OH^- concentrations regarded as negligible).

The two main reactions to be considered are reaction 2.4.5 and reaction 2.6.10. Since Eqs. 2.6.9, 2.6.11, and 2.6.12 completely determine the values of the b_1, b_2, and b_3 once m, y, and y' have been assigned, there is no need to consider additional reactions of the type 2.6.3.

The concentrations b_1 and b_2 are given by

$$b_1 = m(1 - y - y') \tag{2.6.13}$$

$$b_2 = m(2y + y') \tag{2.6.14}$$

Equations 2.6.13 and 2.6.14, of course, degenerate into 2.4.6 and 2.4.7 when $y' = 0$.

It is important to realize that, although y' is necessarily non-negative no matter what the value of y may be, the converse is not true; y may well be negative, inasmuch as its minimum value is given by

$$y_{\min} = -\tfrac{1}{2}y' \tag{2.6.15}$$

A solution where $y = y_{\min}$ is a solution of K_2CO_3 and KHS; if even a very minute amount of $KHCO_3$ is added to such a solution, it will exhibit a small but finite vapor pressure of both CO_2 and H_2S (as will be seen in more detail below). This poses some limitations to the use of carbonate solutions for simultaneous removal of CO_2 and H_2S, as will be discussed in Chapter 11.

For completeness, we give below the matrix of the coefficients λ_j (which enter the definition of molarity, see Eq. 2.3.4), μ_j, μ_j', ν_j, and ν_j'.

	λ_j	μ_j	μ_j'	ν_j	ν_j'
B_1	1	0	0	1	1
B_2	$\frac{1}{2}$	$\frac{1}{2}$	0	-2	-1
B_3	$\frac{1}{2}$	0	1	0	-1

(2.6.16)

Notice that 2.6.16 degenerates properly into 2.4.8 when no H_2S is present.

The equilibrium of reaction 2.4.5 is given by Eq. 2.4.9, that of reaction 2.6.10 by

$$K' = \frac{b_3 b_2}{b_1 a'} \tag{2.6.17}$$

where K' is the equilibrium constant for reaction 2.6.10, i.e., the ratio of the first dissociation constant for H_2S to the second dissociation constant of H_2CO_3. Consequently, the equilibrium vapor pressures are given by

$$p^* = \frac{Hm}{K} \frac{(2y + y')^2}{1 - y' - y} \tag{2.6.18}$$

$$p^{*\prime} = \frac{H'm}{K'} \frac{y'(2y + y')}{1 - y' - y} \tag{2.6.19}$$

It is of interest to evaluate the ratio $p^*/p^{*\prime}$ of the two vapor pressures:

$$\frac{p^*}{p^{*\prime}} = \frac{H}{H'} \frac{K'}{K} \left(2\frac{y}{y'} + 1\right) \tag{2.6.20}$$

The quantities appearing on the right-hand side of Eq. 2.6.20, in addition to the composition term $2(y/y') + 1$, are the Henry's constant ratio H/H', which is a measure of the physical solubility ratio, and the thermodynamic constant K'/K, which is the ratio of the first dissociation constant of H_2S to the first dissociation constant of H_2CO_3; since this particular constant enters many problems, we assign it the symbol K_{cs}. Figure 2.6.1 is an Arrhenius plot for K_{cs}, which is seen to be a weak function of temperature. The ratio H/H' is to a very good approximation

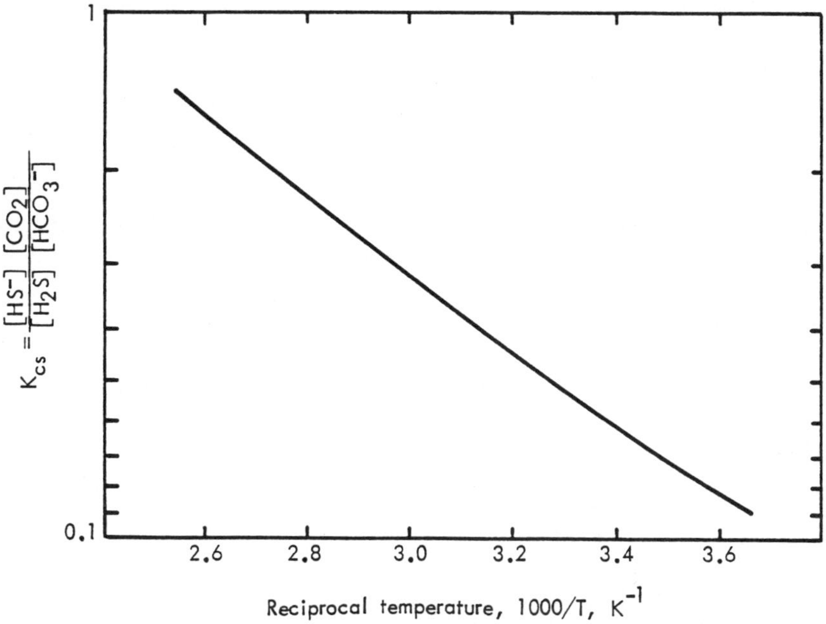

FIGURE 2.6.1. Arrhenius plot for K_{cs}.

independent of temperature and of liquid phase composition in aqueous solutions; its value is 3.05.

We now turn attention to monoethanolamine solutions. The thermodynamic modeling starts from the list of non-volatile components, which are the ones given in Eqs. 2.4.18–2.4.22, in addition to

$$B_6 = HS^- \tag{2.6.21}$$

Equations 2.4.23 (for the molarity) and 2.4.24 (for y) still hold. In addition to these, one has

$$b_6 = my' \tag{2.6.22}$$

The two basic reactions are reaction 2.4.25 and

$$H_2S + RNH = RNH_2^+ + HS^- \tag{2.6.23}$$

In addition to these, reactions 2.4.26 and 2.4.27 should be considered as reactions of the type 2.6.3. The following procedure, in general, would be parallel to the one discussed in Section 2.4; the admissible composition is given by

$$b_1^0 = m(1 - y - y') \tag{2.6.24}$$

$$b_2^0 = m(y + y') \tag{2.6.25}$$

$$b_3^0 = 0 \tag{2.6.26}$$

$$b_4^0 = my \tag{2.6.27}$$

$$b_5^0 = 0 \tag{2.6.28}$$

$$b_6^0 = my' \tag{2.6.29}$$

based on the assumption that *all* the chemically combined CO_2 is in the HCO_3^- form. Notice that Eqs. 2.6.24–2.6.29 degenerate into Eqs. 2.4.29–2.4.33 when $y' = 0$.

A model based on the approach described above has been developed in detail; it correlates available vapor–liquid equilibrium data with satisfactory accuracy. However, rather cumbersome numerical techniques are involved and (particularly for applications in rate analysis) simplifications of the type discussed in Section 2.4 are desirable. These simplifications are as follows.

The equilibrium of reaction 2.4.26 is displaced to the left strongly enough to make the CO_3^{2-} concentration negligible (at all values of y and y'). This reduces the problem to the solution of a quadratic equation of the type of Eq. 2.4.50.

$$\xi_2 = \frac{m(1-y') + (1/K_c) - \sqrt{[m(1-y') + (1/K_c)]^2 - 4m^2y(1-y-y')}}{2} \tag{2.6.30}$$

Furthermore, $K_c m \gg 1$, which is equivalent to the statement that reaction 2.4.27 is sufficiently displaced to the right to make the concentration of either RNH or HCO_3^- negligible. For the case at hand, this implies that, if $1 - 2y - y' > 0$, practically all the chemically combined CO_2 is in the carbamate form, whereas if $1 - 2y - y' < 0$, there is practically no free amine and the carbamate-bicarbonate ratio is determined by stoichiometry. The approximate equations for the b_js are:

	$2y + y' < 1$	$2y + y' > 1$
b_1	$m(1 - 2y - y')$	$\frac{1}{K_2}\frac{1-y-y'}{2y+y'-1}$
b_2	$m(y + y')$	$m(y + y')$
b_3	my	$m(1 - y - y')$
b_4	$\frac{1}{K_2}\frac{y}{1-2y-y'}$	$m(2y + y' - 1)$
b_5	$\frac{K_1}{K_2}\frac{y}{y+y'}$	$\frac{1}{K_2}\frac{1-y-y'}{y+y'}$
b_6	my'	my'

(2.6.31)

which, of course, degenerate into 2.4.59 when $y' = 0$.

The equilibrium vapor pressures are calculated from the equilibrium equations for reactions 2.4.25 (with thermodynamic constant K) and 2.6.23 (with thermodynamic constant K′). The results based on the values 2.6.31 are:

	$2y + y' < 1$	$2y + y' > 1$	(2.6.32)
p^*	$\frac{H}{K}\frac{y(y+y')}{(1-2y-y')^2}$	$\frac{H}{K}K_2^2 m^2 \frac{(y+y')(2y+y'-1)^2}{1-y-y'}$	
$p^{*\prime}$	$m\frac{H'}{K'}\frac{y'(y+y')}{1-2y-y'}$	$\frac{H'}{K'}K_2 m^2 \frac{y'(y+y')(2y+y'-1)}{(1-y-y')}$	
$\frac{p^*}{p^{*\prime}}$	$\frac{H}{H'}\frac{K'}{K_m}\frac{y}{(1-2y-y')y'}$	$\frac{H}{H'}\frac{K'}{K}\frac{2y+y'-1}{y'}$	

Figures 2.6.2 and 2.6.3 are plots of $p^{*\prime}$ and $p^*/p^{*\prime}$ of the form suggested by

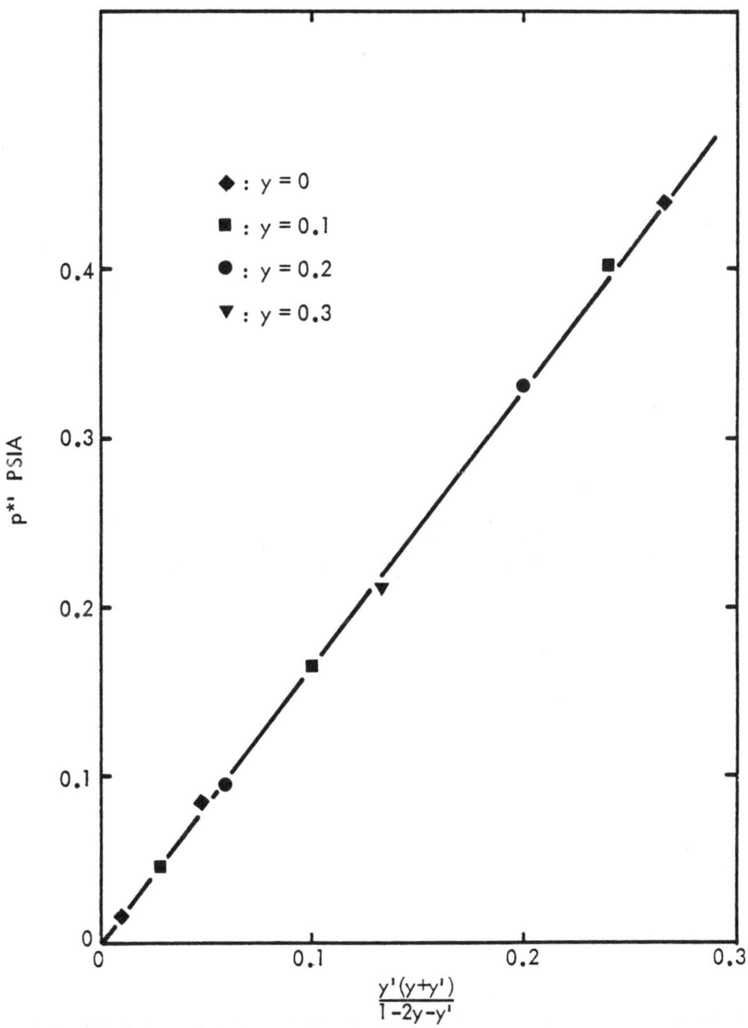

FIGURE 2.6.2. Comparison of theoretical model for p^* with experimental data.

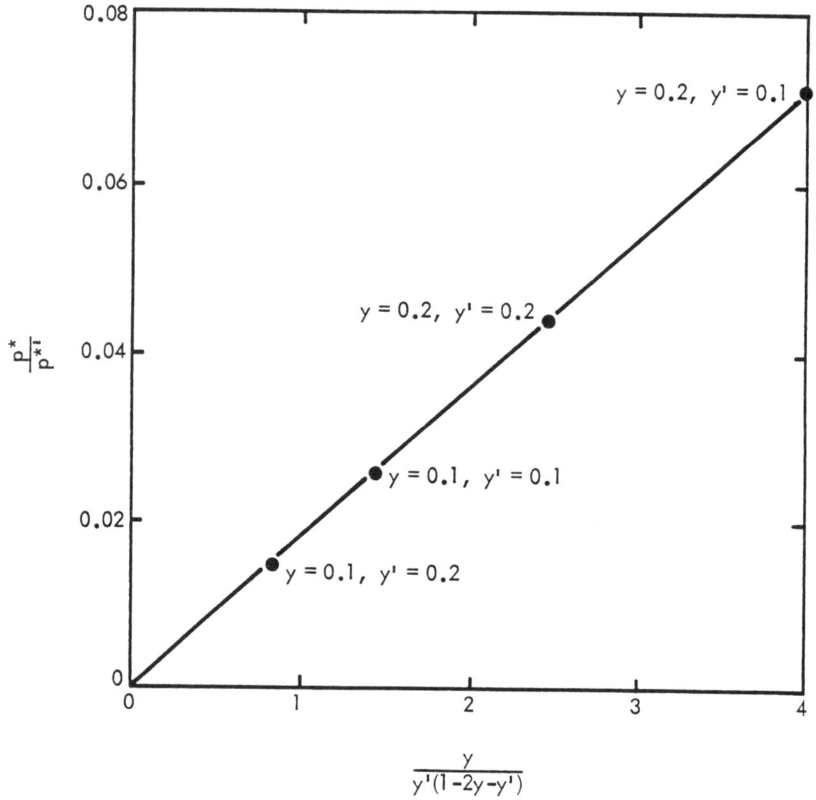

FIGURE 2.6.3. Comparison of theoretical model for p^*/p^* with experimental data.

Eq. 2.6.32. It can be seen that straight lines through the origin correlate the data extremely well, thus confirming the validity of the thermodynamic model leading to Eq. 2.6.32.

2.7. THERMODYNAMIC SELECTIVITY

In processes involving simultaneous removal of CO_2 and H_2S, selectivity for H_2S is always desirable and sometimes necessary. The specifications on H_2S content in the sweet gas are invariably stricter than those for CO_2 content; in some processes such as H_2S removal from low heating value gas, there are no specifications at all for CO_2—one in fact wishes to remove as little CO_2 as possible, since any CO_2 absorbed will increase the heating and pumping costs.

The actual selectivity achievable in a given process is related to both the thermodynamics and the kinetics of the system. In this section we limit attention to the thermodynamic selectivity; kinetic selectivity will be discussed in Chapters 8 and 11.

The thermodynamic selectivity S_T of a solvent is defined as follows:

$$S_T = \frac{p^*}{p^{*\prime}} \frac{\alpha'}{\alpha} \tag{2.7.1}$$

where primed quantities refer to H_2S. When $S_T > 1$, the ratio of the H_2S content to the CO_2 content is, at equilibrium, larger in the liquid phase than it is in the gas phase.

For physical solvents, the thermodynamic selectivity simply equals the ratio of Henry's Law constants:

$$S_T = \frac{\mathrm{H}}{\mathrm{H}'} \tag{2.7.2}$$

since $\alpha = a$, $\alpha' = a'$. S_T is 3.05 in water (independent of temperature). For non-aqueous solvents, higher selectivities are possible (see Figure 2.7.1). However, physical solvents, in addition to poor capacity, have practically no kinetic selectivity, so that the thermodynamic selectivity is the best one can obtain.

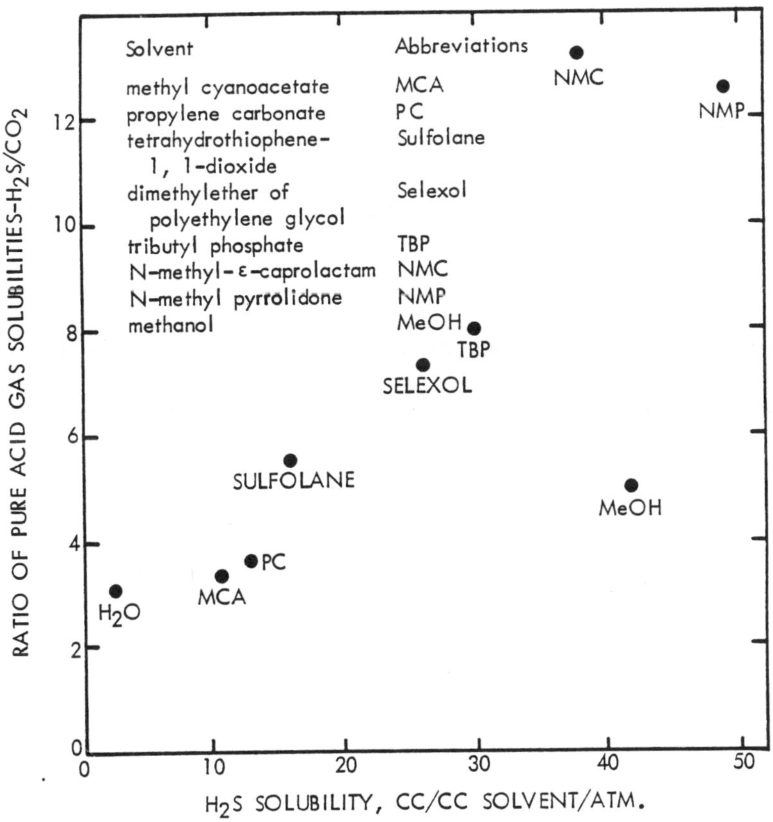

FIGURE 2.7.1. H_2S/CO_2 solubility ratio for organic physical solvents.

For chemical solvents, if a, a' are neglected in comparison to my, my' (which is always a reasonable approximation), Eq. 2.7.1 reduces to

$$S_T = \frac{p^*}{p^{*\prime}} \frac{y'}{y} = \frac{\mathrm{H}}{\mathrm{H}'} \frac{a}{a'} \frac{y'}{y} \tag{2.7.3}$$

All chemical solvents for H_2S and CO_2 are alkaline and, therefore, the following reaction is always possible:

$$H_2S + HCO_3^- = HS^- + CO_2 + H_2O \tag{2.7.4}$$

The equilibrium constant of reaction 2.7.4 is K_{cs}, which is plotted in Figure 2.6.1. Thus

$$K_{cs} = \frac{[HS^-]\,a}{[HCO_3^-]\,a'} \tag{2.7.5}$$

Substitution of Eq. 2.7.5 into 2.7.3 gives

$$S_T = \frac{\mathrm{H}}{\mathrm{H}'} K_{cs} \frac{[HCO_3^-]}{[HS^-]} \frac{y'}{y} \tag{2.7.6}$$

The chemically combined H_2S is almost invariably all in the form of the HS^- ion, since the second dissociation constant of H_2S is extremely small (see Figure 2.7.2) and, therefore, no S^{2-} is formed in any but extremely strong alkaline solutions. One may therefore write

$$my' = [HS^-] \tag{2.7.7}$$

In contrast with this, the chemically combined CO_2 may exist in forms, such as the CO_3^{2-} ion or the carbamate ion, other than HCO_3^-. Therefore, with the exception of K_2CO_3 solutions (and it is easy to convince oneself that carbonate solutions are the only possible exception), the following inequality holds true:

$$my \geqslant [HCO_3^-] \tag{2.7.8}$$

When Eqs. 2.7.7 and 2.7.8 are substituted into Eq. 2.7.5, one obtains

$$S_T \leqslant \frac{\mathrm{H}}{\mathrm{H}'} K_{cs} \tag{2.7.9}$$

Equation 2.7.9 is very important and should be discussed in detail. First of all, for chemical solvents that are aqueous solutions, inasmuch as $K_{cs} < 1$, the thermodynamic selectivity is less than 3.05.

The second point is that the thermodynamic selectivity increases as more and more of the chemically combined CO_2 is in the bicarbonate form. Since the chemical solvents to be considered are alkaline, one alternate possible form of chemically combined CO_2 is the carbonate ion CO_3^{2-}. Therefore, in order to obtain a high thermodynamic selectivity, the formation of the CO_3^{2-} ion should not be favored; say, the pK of the solution should be less than the pK for the second dissociation of carbonic acid (which is 10.33 at room temperature). *Very strong alkalis have poor thermodynamic selectivity.*

Of course, CO_3^{2-} is not the only form of chemically combined CO_2 other than the bicarbonate, unless the alkalinity is of inorganic nature; with amine solutions, the carbamate is another form of chemically combined CO_2. It follows that the

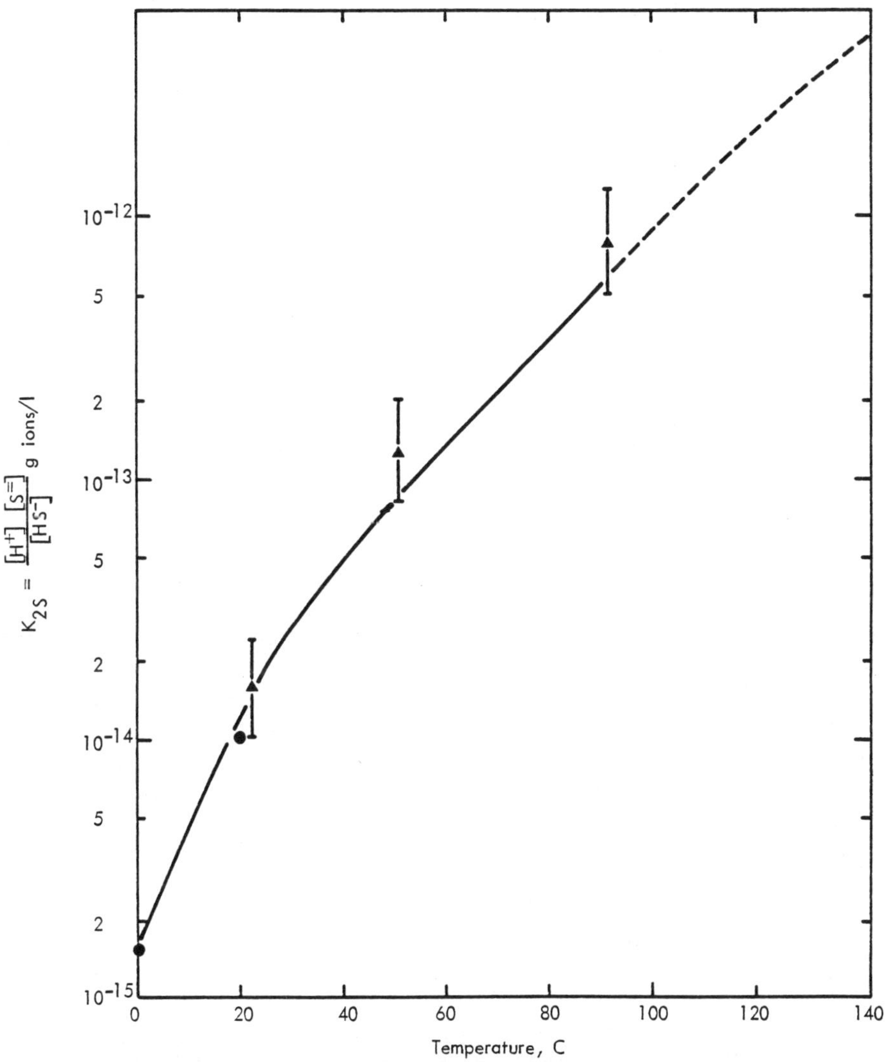

FIGURE 2.7.2. Second dissociation constant of H_2S.

thermodynamic selectivity will be higher the less stable the carbamate and, in fact, highest for tertiary amines.

In tertiary amine solution, and in solutions where the alkalinity is of inorganic nature (e.g., potassium phenolate), provided the pK is not too high, the chemically combined CO_2 will exist only in the HCO_3^- form. It follows that for such solutions the thermodynamic selectivity is the highest possible one; say,

$$S_T = \frac{\mathrm{H}}{\mathrm{H}'}\,\mathrm{K}_{cs} \tag{2.7.10}$$

In contrast with this, solutions of amines having stable carbamates such as monoethanolamine will exhibit very low thermodynamic selectivity (see Figure 2.6.3). In spite of this, monoethanolamine solutions can be used for simultaneous removal of CO_2 and H_2S because they exhibit a strong kinetic selectivity for H_2S.

In physical solvents, as well as in some chemical solvents, the thermodynamic selectivity is independent of liquid phase composition, (see Eqs. 2.7.2 and 2.7.10). However, this is not always the case; see, e.g., Eq. 2.6.32 which, at $y < 0.5$, yields

$$S_T = \frac{\mathrm{H}}{\mathrm{H}'}\,\frac{\mathrm{K}'}{\mathrm{K}}\,\frac{1}{1 - 2y - y'} \tag{2.7.11}$$

For such systems, it is useful to define a limiting thermodynamic selectivity, S_{T0}, corresponding to the value of S_T when both y and y' go to zero:

$$S_{T0} = \lim_{\substack{y \to 0 \\ y' \to 0}} S_T \tag{2.7.12}$$

The limiting thermodynamic selectivity is relevant insofar as it determines the behavior of the system at the lean end of the absorption unit, where the gas specifications must be met. For monoamine solutions, the value of S_{T0} is given by

$$S_{T0} = \frac{\mathrm{H}}{\mathrm{H}'}\,\mathrm{K}_{cs}\,\frac{1}{1 + \mathrm{K}_c m} \tag{2.7.13}$$

where K_c is the carbamate stability constant (i.e., the equilibrium constant of reaction 2.4.27). Equation 2.7.13 of course degenerates into Eq. 2.7.10 when $\mathrm{K}_c \to 0$ (tertiary amines), and it shows how the thermodynamic selectivity decreases with increasing carbamate stability. (For amines with pK in excess of 10, at very low y values some carbonate may form and the selectivity S_{T0} will be less than that given by Eq. 2.7.13.)

2.8. HENRY'S CONSTANT IN REACTIVE LIQUIDS

The experimental measurement of the value of Henry's Law constant H is comparatively straightforward in the case of physical solvents. All that is needed is an experimentally determined equilibrium curve, p^* versus α; in the simplest case, H is simply the slope of such a curve near the origin. Even in the case of a moderately strong electrolyte such as SO_2, the value of H can be extracted from the equilibrium curve, as was discussed in Section 1.8.

In the case of chemical solvents, the situation is considerably more complex, because the value of H *cannot* be extracted from VLE data. All equations discussed in Sections 2.3–2.6 show that p^* is always proportional to the ratio of H to some thermodynamic constant and, therefore, VLE data cannot yield separately the values of both H and the appropriate thermodynamic constant. However, it will be seen in Chapters 3–8 that the actual value of H needs to be known in order to carry out the rate analysis.

Let H^0 represent the value of H relative to the pure solvent of a chemically reactive solution (say water in the case of aqueous solutions). The value of H in the reactive solution will be, in general, of the same order of magnitude as, yet quite appreciably different from, H^0. This is due to the fact that reactive solutions are not dilute, and the value of $\gamma_{A\infty}$ in Eq. 1.7.7 depends on the solution composition. In general, $\mathrm{H} > H^0$ (i.e., the physical solubility is *lower* than in the pure solvent). This effect is sometimes referred to as the "salting out" effect.

There is some theoretical basis in support of the following equation for the ratio H/H^0:

$$\log \frac{\mathrm{H}}{H^0} = \mathrm{K}_i I_i \tag{2.8.1}$$

which is sometimes referred to as the Setchenov equation. I_i is the ionic strength and K_i is a constant that probably depends on the *nature* of the ions present in the solution, but not on their concentrations. Furthermore, there is some evidence to show that K_i has the same value for all moderately soluble gases. On this basis, one could estimate the value of H/H^0 by measuring the solubility of an *inert* gas (i.e., one that does not react in the liquid phase) in the reactive solution.

Other methods for measuring the value of H/H^0 are discussed in Chapter 14. The system for which the best data are available is the CO_2-K_2CO_3 one; for this system, Eq. 2.8.1 applies, with $\mathrm{K}_i = 0.062$ liter/gmol.

However, in a K_2CO_3 solution, the ionic strength is given by

$$I_i = m(3 - y) \tag{2.8.2}$$

and, therefore, cannot be changed very significantly at a fixed value of m. The data can, in fact, be correlated equally well by an equation of the following form:

$$\log \frac{\mathrm{H}}{H^0} = \mathrm{K}_m m \tag{2.8.3}$$

with $\mathrm{K}_m = 0.025$ liter/gmol.

It has been suggested that the value of K_i in Eq. 2.8.1 can be estimated for any system by means of an "ion contribution" procedure. The equation for K_i is

TABLE 2.8.1. Values of k_i

Ion	k_i
Na^+	0.094
K^+	0.071
NH_4^+	0.031
CO_3^{2-}	0.021
HCO_3^-	0.021
OH^-	0.061
SO_4^{2-}	0.021

simply

$$K_i = \sum k_i \tag{2.8.4}$$

where the k_i are the contributions of all ions in the solution. Table 2.8.1 reports recommended values of k_i.

2.9. THE DEPENDENCY OF p^* ON m AND y

Thermodynamic models of the type discussed in the preceding sections produce a relationship between the equilibrium partial pressure p^*, the molarity m, and the fractional saturation y which also contains one or more thermodynamic parameters, typically the ratio of the Henry's Law constant H and the thermodynamic constant of the main reaction K. In general, this relationship has the following form:

$$p^* = Am^q F(y) \tag{2.9.1}$$

with A proportional to H/K. In this section we discuss in some detail the significance of the exponent q, the form of the function $F(y)$, and the values of the constant A.

Thermodynamic models are particularly useful insofar as they produce the form of the function $F(y)$. In fact, for all three main uses discussed in Section 2.1 (interpolation of data, extrapolation of data, comparison of different solvents) the dependency of p^* on y at fixed m is the crucial bit of information required.

It is fortunate that, in spite of the gross simplifications on which the models are based (namely, lumping of all activity coefficients into the constant A), they are usually surprisingly effective in predicting the correct form of the function $F(y)$.

The function $F(y)$ is always a more than linearly increasing function, i.e., both dF/dy and d^2F/dy^2 are positive (a proof of this statement can be developed for the case where there is only one main reaction; the algebra involved is rather complex). This gives support to the discussion given in Section 1.5, and to the argument to be developed in Section 15.2. This should be contrasted with the case of physical solvents, where p^* is a linear function of load in the first-order approximation (i.e., when Henry's Law applies).

The ionic strength of the solution is, in general, a function of y; hence the assumption that the constant A depends on ionic strength implies that it should not be regarded as independent of y, as implicitly done in the development of the models. It turns out, however, that often the data can be fitted better by assuming that A does not depend on y, than by allowing it to depend on y through the ionic strength. This point should be checked for each specific system; there is some indication that for some systems a better fit is obtained if A is allowed to be a weak function of y (see, e.g., Eq. 10.2.8).

The value of the exponent q in Eq. 2.9.1 depends only on the stoichiometry of the main reaction involved; in fact, the following equation applies:

$$q = -\sum \nu_j \tag{2.9.2}$$

where ν_j are the stoichiometric coefficients of the main reaction (for those systems for which there is more than one main reaction to be considered, the thermodynamic model may produce an equation more complex than 2.9.1 which may not allow factoring out the dependencies on m and on y). Although the thermodynamic models, in general, correctly predict the *sign* of q, they do not correctly predict its value. For example, for the CO_2-K_2CO_3 system the model predicts $q = 1$, but actual data show that, in fact, $q = 0.4$. Sometimes (but not in the CO_2-K_2CO_3 case) a better prediction is obtained by taking into account the fact that as m is increased, the concentration of water is decreased; in other words, by not setting equal to zero the stoichiometric coefficient of water, even if it is a reactant in the main reaction.

The fact that $F(y)$ is a more than linearly increasing function of y is responsible for the fact that the cyclic capacity of a solvent is invariably an increasing function of the molarity, even when $q > 0$. In fact, given a value of p_B in Eq. 1.5.4, the value of y_{max} will decrease only slightly as m is increased and, correspondingly, the cyclic capacity as given by Eq. 1.5.2 will increase with increasing m, though less than linearly.

Finally, the actual values of the constant A cannot, of course, be predicted by a thermodynamic model of the type discussed in this chapter. Such values can, of course, be extracted from VLE data; if an independent estimate of the Henry's constant is available, a value of the thermodynamic constant can be calculated. (Values of K_c reported in Figure 2.4.1 have been obtained by such a procedure.) The *order of magnitude* of A can be estimated from values of H and K at infinite dilution; such values may be seriously in error, by as much as a factor of four.

3 Mass Transfer with Chemical Reaction: Basic Concepts and Mathematical Layout

Every aspect of nature can be approached by poetry or experiment, as well as by reason; and indeed this is the usual order in history.

C. A. TRUESDELL

This chapter presents a general analysis of the *rate* of mass transfer within an agitated liquid phase in contact with a gas phase. The analysis covers the transfer in the liquid phase of a dissolved component when diffusion and chemical reaction are occurring simultaneously.

It should be borne in mind that the gas-side resistance to mass transfer, though occasionally negligible, is never zero and, therefore, the overall transfer rate is, in general, influenced by it. In this chapter, we restrict attention to the liquid phase and regard the value of the *interface* partial pressure of the transferring component as a known quantity. The equations that are derived should be coupled with the equation for mass transfer in the gas phase (Eq. 1.6.3) in order to calculate the overall transfer rate.

The aim in this chapter is to introduce basic concepts and to establish the general mathematical background. Therefore, although the detailed analysis is postponed to Chapters 4-8, attention is restricted here to those conditions of simultaneous diffusion and chemical reaction that can be regarded as limiting or asymptotic cases. The term asymptotic is used to indicate certain limiting conditions where the mathematical analysis is considerably simplified. Significant simplifications may arise, for example, when the chemical reaction is extremely fast or extremely slow, as compared to diffusion phenomena.

3.1. BASIC CONCEPTS

Consider the case where a chemical solvent is used in gas treating, so that the liquid phase contains one or more components B_j $(j = 1, 2, \ldots, N)$ that may react chemically with the component A which is being either absorbed or desorbed. Usually the components B_j are non-volatile, so that all the chemical reactions take place in the liquid phase. Consequently, the phenomenon taking place in the gas phase is still simply a mass transfer one and Eq. 1.6.3 still holds. The assumption embodied in Eq. 1.6.5 is still valid.

The occurrence of the chemical reactions has two distinct effects on the overall behavior of the system. The first one is, for the case of absorption, as follows. When component A is absorbed into the liquid phase, it is consumed by the chemical reactions and, therefore, its concentration in the bulk of the liquid a_0 is kept low. This, in turn, implies that the driving force for additional absorption remains higher than it would be if no chemical reaction were taking place. Conversely, in desorption the chemical reaction continuously *produces* the component to be desorbed, thus providing a high concentration of it in the liquid and a large driving force for the desorption. Summarizing, in both absorption and desorption, the first effect of the chemical reaction is to maintain a high driving force for mass transfer in the liquid phase.

The second effect is more subtle. At a given level of driving force, the actual rate of mass transfer may be significantly larger when chemical reactions are taking place than it would be in the absence of chemical reactions. The rate enhancement may be very large, up to two orders of magnitude or even more. Indeed, the enhancement may be so large as to actually reduce the mass transfer resistance in the liquid phase to the point at which it is negligible as compared to the resistance in the gas phase, even for a system where condition 1.6.9 is satisfied.

The concept of *rate enhancement* introduced above is formalized as follows. In the absence of chemical reactions, the mass transfer rate in the liquid phase is given by Eq. 1.6.4, which is repeated below:

$$N = \mathrm{k}_L^0 (a_i - a_0) \tag{3.1.1}$$

The actual rate in the presence of chemical reactions may be *larger* than the value given by Eq. 3.1.1; a "chemical" mass transfer coefficient k_L, can be defined as follows:

$$N = \mathrm{k}_L (a_i - a_0) \tag{3.1.2}$$

The *rate enhancement factor I* is defined as the ratio of the actual rate and the rate that would be observed under the same driving force in the absence of chemical reactions:

$$I = \frac{N}{\mathrm{k}_L^0 (a_i - a_0)} = \frac{\mathrm{k}_L}{\mathrm{k}_L^0} \tag{3.1.3}$$

It is important to realize that the value of the mass transfer coefficient k_L^0 is determined by the fluid mechanics of the liquid phase in the neighborhood of the gas-liquid interface; the fact that its value cannot, in general, be predicted from first principles is due to the complexity of the fluid mechanics problem involved. Since predicting the value of k_L would involve an even more complex problem, there is, in practice, no hope of calculating k_L from first principles.

Fortunately, however, although both k_L^0 and k_L are strongly influenced by the details of the fluid mechanics involved, their ratio I turns out to be almost independent of it. Therefore, the theory of coupled mass transfer and chemical reaction can be developed on the basis of very crude models of the fluid mechanics involved. The kind of models that are commonly used are discussed in Section 3.2.

As will be discussed in detail in Chapters 4-8, the rate enhancement factor is, in general, a complicated function of the composition of the gas and liquid phases. However, simplified equations emerge from the consideration of limiting conditions. The basic concepts underlying these simplifications are discussed below.

The intrinsic rate of a chemical reaction is measurable by means of a yardstick called the *reaction time* t_r. A precise definition of t_r will be given in Section 3.3; at this stage, it is sufficient to say that t_r is a measure of the time *required* by the chemical reaction in order to change by a significant amount the concentration of the limiting reactant.

Diffusion phenomena can also be characterized by a time scale, the *diffusion time* t_D. The exact definition of t_D will be given in Section 3.2. The diffusion time is a measure of the time *available* for molecular diffusion phenomena to take place before mixing of the liquid phase makes the concentration uniform. Therefore, t_D decreases as the mixing or turbulence of the liquid phase is increased.

Having established the time scales t_D and t_r, their dimensionless ratio Φ is the measure of the *relative* rates of reaction and diffusion:

$$\Phi = \frac{t_D}{t_r} \tag{3.1.4}$$

When $\Phi \ll 1$, the reaction is too slow to have any significant influence on diffusion phenomena, and essentially no rate enhancement will take place. This is the situation referred to in the following as the *"slow reaction regime."* Of the two effects of chemical reactions discussed above, only the first takes place in the slow reaction regime.

Conversely, if $\Phi \gg 1$, the reaction is fast enough to result in a significant rate enhancement. However, there is an upper bound to the possible rate enhancement, which is discussed in qualitative terms in the following.

The assumption embodied in Eq. 1.6.5 (i.e., that fugacity is continuous across an interface) holds also for chemically reactive systems. Therefore, the value of a_i the interface concentration of *physically dissolved* transferring component is related to the interface partial pressure p_i by Eq. 1.6.5 (i.e., *physical equilibrium* prevails at the interface

$$a_i = \frac{p_i}{\mathrm{H}} \tag{3.1.5}$$

where H is Henry's Law constant).

However, the liquid at the interface is, in general, not in *chemical* equilibrium with the gas phase; the value of α, the total concentration of the transferring component (in both physically dissolved and chemically combined forms) at the interface is not the value α_i^* corresponding to equilibrium with p_i:

$$\alpha_i \neq \alpha_i^* \tag{3.1.6}$$

This is because chemical reactions proceed at a finite rate and chemical equilibrium is not established instantaneously.

However, in the limit of infinitely fast chemical reactions (mathematically, when $\Phi \rightarrow \infty$), chemical equilibrium is established instantaneously and, therefore, α_i will indeed be equal to α_i^*. This is the condition referred to in the following as the *"instantaneous reaction regime,"* and it represents the upper bound to the rate enhancement, which is attained when all resistance to mass transfer due to chemical kinetics has been eliminated.

In the instantaneous reaction regime, the transferring component can diffuse in both its physically dissolved and its chemically combined form, with no kinetic resistance to the transformation from one form to the other. Therefore, if diffusion in either form is equally easy (i.e., if the diffusivities of all species are equal), the rate of mass transfer will be governed by a driving force measured in terms of the *total* concentration α of the transferring component, rather than in terms of the concentration of its physically dissolved form

$$N = \mathrm{k}_L^0 (\alpha_i^* - \alpha_0) \tag{3.1.7}$$

The hypothesis that all diffusivities are equal results in the value of k_L^0 in Eq. 3.1.7 being the same as in Eq. 3.1.1. The quantity α_0 is the value of α in the bulk of the liquid. Equation 3.1.7 represents an upper bound for the rate of mass transfer. The corresponding value of the rate enhancement factor I_∞ (i.e., the value of I when $\Phi \rightarrow \infty$) is calculated from Eqs. 3.1.3 and 3.1.7:

$$I_\infty = \frac{\alpha_i^* - \alpha_0}{a_i - a_0} \tag{3.1.8}$$

Values of I_∞ calculated from Eq. 3.1.8 can be very large indeed and are typically of the order 10^2 to 10^4.

We have qualitatively established so far that, when $\Phi \ll 1$, the rate enhancement factor is unity, whereas when $\Phi \rightarrow \infty$, its value is very large and is given by Eq. 3.1.8. Clearly, an intermediate region exists where $\Phi \gg 1$ and, therefore, I is appreciably larger than unity, yet the reaction is not so fast as to be instantaneous and, therefore, I is appreciably less than I_∞. This intermediate region will be referred to in the following as the *"fast reaction regime."*

Although the issue of chemical kinetics can be avoided in the slow reaction regime (since the rate of reaction is so slow that its actual value needs not be known) and in the instantaneous reaction regime (since the rate of reaction is so fast that, again, its value needs not be known), it cannot be avoided in the case of the fast reaction regime. However, as will be discussed in Section 3.4, considerable simplifications arise also in this limiting case, and the following simple equation is obtained for the enhancement factor I:

$$I = \sqrt{\Phi} \tag{3.1.9}$$

3.2. HYDRODYNAMICS

Mass transfer in an agitated liquid phase is obviously influenced by the hydrodynamics of the liquid phase near the gas-liquid interface. Consider a gas-liquid interface such as sketched in Figure 3.2.1; let v be the velocity component in the plane of the interface and x be the distance from the interface. At the interface itself, the tangential stress must be continuous; in other words, the drag that the gas exerts on the liquid must be equal in value and opposite in sign to the drag that the liquid exerts on the gas. This leads to the following equation:

$$\mu_G \left.\frac{dv}{dx}\right|_G = \mu_L \left.\frac{dv}{dx}\right|_L \tag{3.2.1}$$

where μ is the viscosity and the subscripts G and L refer to the gas and the liquid phases, respectively. Since the viscosity of the gas μ_G is much less than that of the liquid (typically, $\mu_G/\mu_L = 10^{-3}$-10^{-2}), it follows that

$$\left.\frac{dv}{dx}\right|_L \ll \left.\frac{dv}{dx}\right|_G \tag{3.2.2}$$

and, as a first-order approximation, one may take

$$\left.\frac{dv}{dx}\right|_L = 0, \quad \left.\frac{dv}{dx}\right|_G = \gamma \tag{3.2.3}$$

at the interface itself. It follows that if v_i is the interface velocity, the velocity *near* the interface is

$$\text{Liquid:} \quad v = v_i + 0(x^2) \tag{3.2.4}$$

$$\text{Gas:} \quad v = v_i + \gamma x + 0(x^2) \tag{3.2.5}$$

where the symbol $0(x^2)$ means a term of order x^2.

These equations imply that, to within $0(x)$ (i.e., sufficiently near the interface) the velocity in the liquid phase is constant, whereas that in the gas phase is a linear function of the distance from the interface. Correspondingly, hydrodynamic

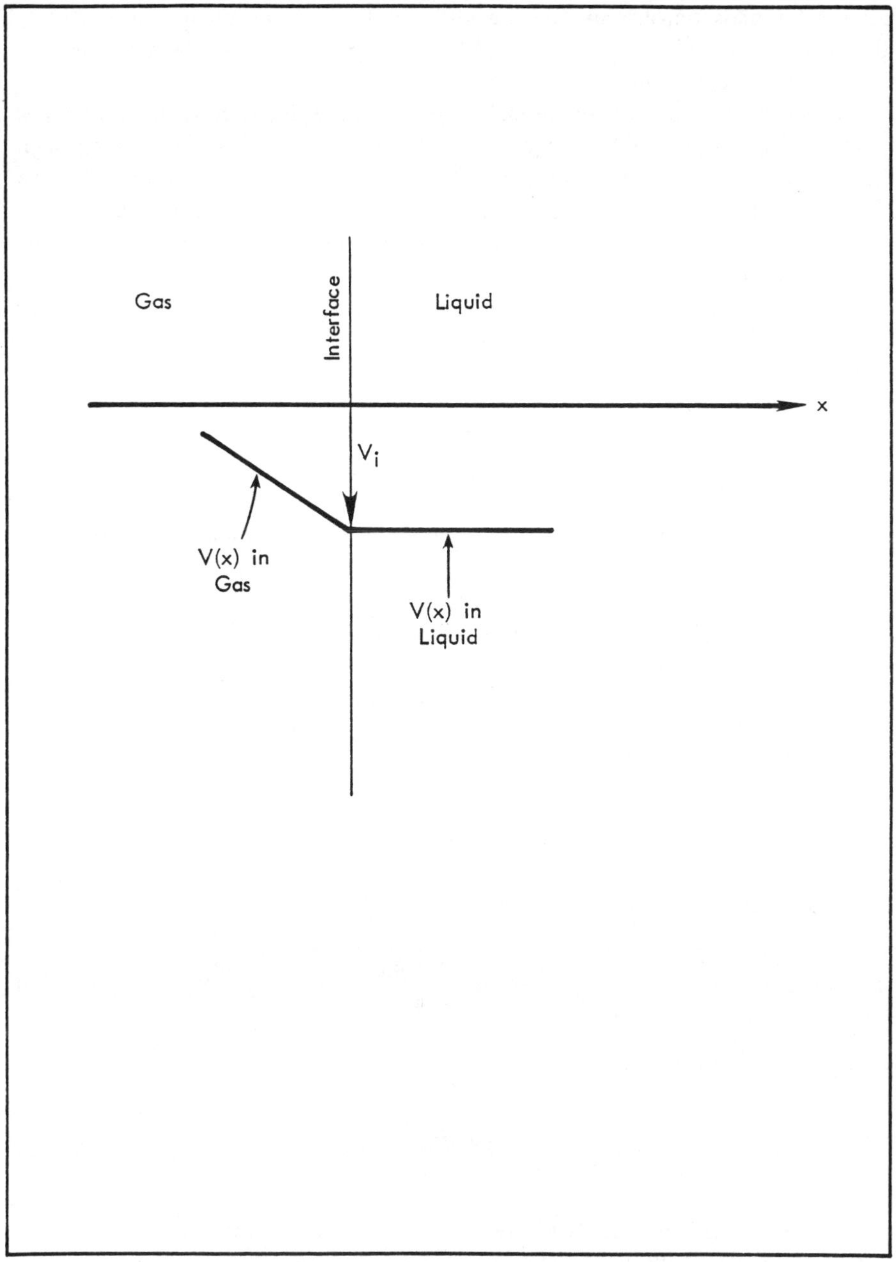

FIGURE 3.2.1. Velocity distribution near a gas–liquid interface.

models for mass transfer analysis are different for the two phases; in particular, those for the liquid phase are all based on Eq. 3.2.4 and they lead to the very important concept of the diffusion time t_D.

The simplest hydrodynamic model proposed in the literature is the *film theory* model. This assumes the existence, near the gas-liquid interface, of a stagnant film of thickness δ through which mass transfer can only take place by molecular diffusion. The rest of the liquid phase is assumed to be perfectly well mixed. Therefore, the concentration at a depth δ from the interface is equal to the bulk-liquid concentration for every species.

In the absence of any chemical reaction, the concentration profile of the volatile component in the film is linear, inasmuch as the differential equation for mass transfer reduces to

$$D \frac{d^2a}{dx^2} = 0 \tag{3.2.6}$$

where D is the diffusivity of the volatile component. The boundary conditions on Eq. 3.2.6 are

$$x = 0, \qquad a = a_i \tag{3.2.7}$$

$$x = \delta, \qquad a = a_0 \tag{3.2.8}$$

and the solution is, trivially,

$$a = a_i + (a_0 - a_i) \frac{x}{\delta} \tag{3.2.9}$$

(see Figure 3.2.2). Consequently, the mass transfer flux is

$$N = D \frac{a_i - a_0}{\delta} \tag{3.2.10}$$

or, comparing with Eq. 1.6.4,

$$k_L^0 = \frac{D}{\delta} \tag{3.2.11}$$

Equation 3.2.11 does not have predictive value, since the value of the film thickness δ (into which the whole ignorance about the true fluid mechanics has been lumped) is not known. However, if a problem of mass transfer with chemical reaction is analyzed on the basis of the film theory model, the value of I will usually turn out to depend on δ, and Eq. 3.2.11 can then be used to express I as a function of k_L^0. The value of the latter will, of course, need to be estimated on the basis of available correlations for classical mass transfer.

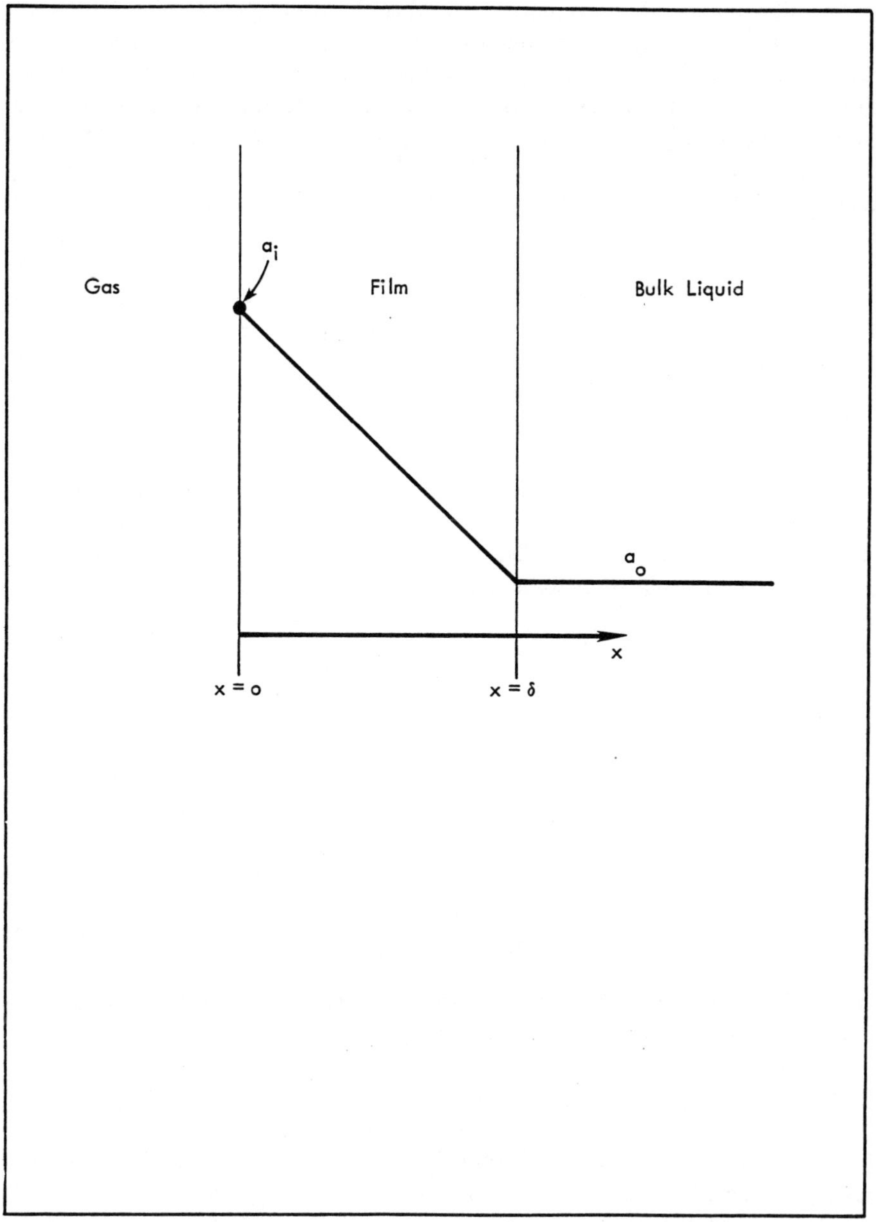

FIGURE 3.2.2. Concentration profiles in liquid. Film theory model, no chemical reaction.

The film theory has an important drawback. Although the value of δ is not known, one should regard it as uniquely determined by the hydrodynamics of the liquid phase. On this basis, Eq. 3.2.11 would predict k_L^0 to be proportional to the diffusivity D. Empirical mass transfer coefficient correlations available in the literature for a liquid in contact with a gas consistently indicate that, in fact, k_L^0 is proportional to the square root of D. Therefore, analyses based on the film theory model are not expected to predict correctly the influence of diffusivity values on the enhancement factor I.

A more complex model of the fluid mechanics involved is the *penetration theory* model. It is assumed that small liquid elements are continuously brought from the bulk of the liquid phase to the interface, where they stay for some time t^*; after that, the elements are brought back to the bulk of the liquid where they get mixed with it. During the time that an element stays at the interface, it essentially behaves as a rigid body (in the sense that there are no appreciable velocity gradients within it), because the gas phase exerts a negligible drag on the liquid phase. Therefore, absorption into the surface elements takes place by molecular diffusion alone; in this case, however, the diffusion phenomenon is unsteady, rather than steady as in the film theory model.

The differential equation for unsteady diffusion, in the absence of any chemical reaction, is

$$D \frac{\partial^2 a}{\partial x^2} = \frac{\partial a}{\partial t} \tag{3.2.12}$$

where t is time elapsed from the instant the element considered was brought to the interface. The boundary conditions to be imposed are as follows.

Since at $t = 0$ the element has been brought to the interface from the well mixed bulk where the concentration is a_0, one has

$$t = 0, \qquad a = a_0 \tag{3.2.13}$$

At the interface, the concentration has the value of a_i corresponding to equilibrium with the interface partial pressure of gas p_i:

$$x = 0, \qquad a = a_i \tag{3.2.14}$$

The concentration profile in the element at time t^*, when the element itself is taken off the interface, may look somewhat like that sketched in Figure 3.2.3. Let λ represent the "depth of penetration," as indicated in Figure 3.2.3. If λ is much less than the thickness of the surface element, the last boundary condition can be written as

$$x \to \infty, \qquad a \text{ is bounded} \tag{3.2.15}$$

The solution to the system of Eqs. 3.2.12-3.2.15 is straightforward since the

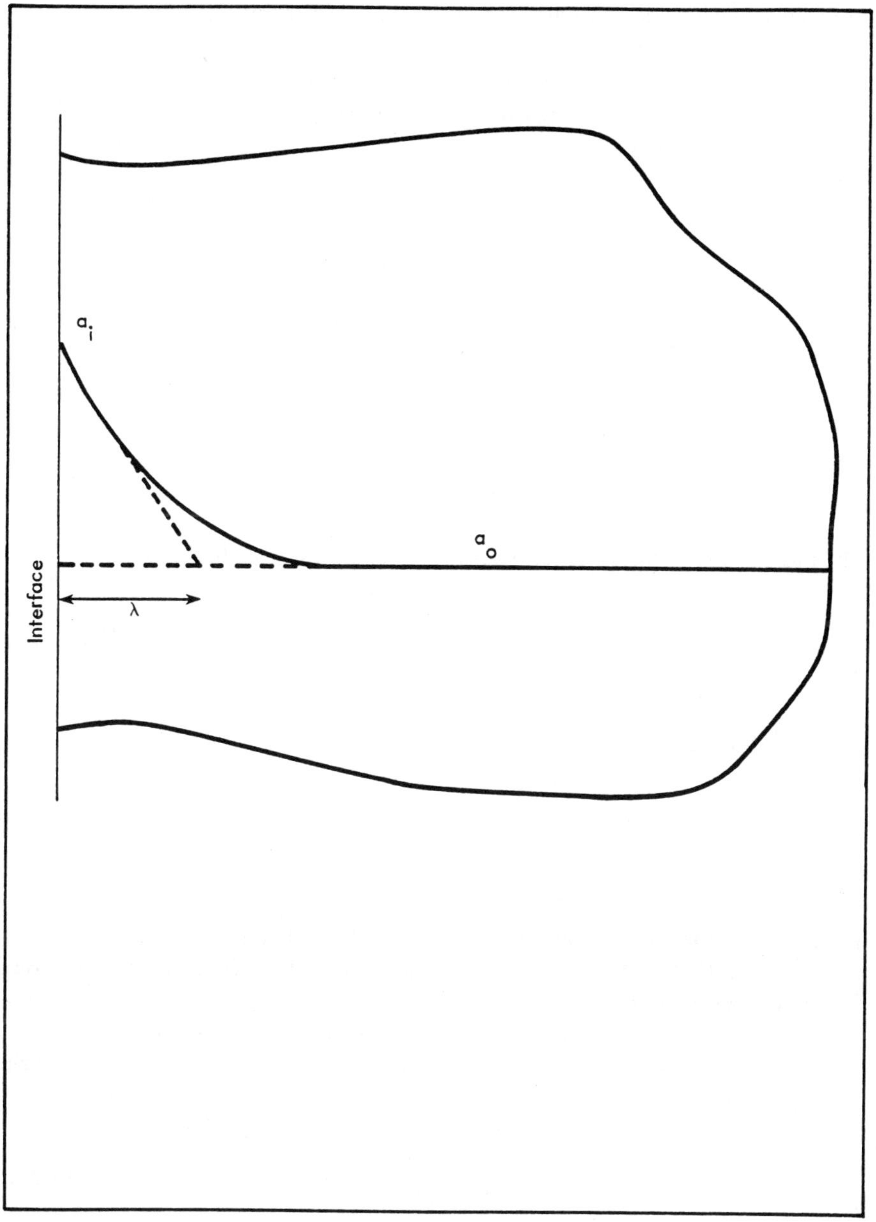

FIGURE 3.2.3. Concentration profile at time t^* in a surface element. Penetration theory model, no chemical reaction. λ is the depth of penetration.

problem is linear. The concentration distribution is given by

$$\frac{a - a_0}{a_i - a_0} = \operatorname{erfc}\left(\frac{x}{2\sqrt{Dt}}\right) \tag{3.2.16}$$

where

$$\operatorname{erfc} u = 1 - \frac{2}{\sqrt{\pi}} \int_0^u e^{-u^2}\, du \tag{3.2.17}$$

Notice that the instantaneous rate of mass transfer into the surface element is time dependent:

$$N_{\text{inst}}(t) = -D\left(\frac{\partial a}{\partial x}\right)_{x=0} = \sqrt{\frac{D}{\pi t}}\,(a_i - a_0) \tag{3.2.18}$$

The average rate of mass transfer over the time interval from 0 to t^* is

$$N = \frac{1}{t^*}\int_0^{t^*} N_{\text{inst}}(t)\, dt = 2\sqrt{\frac{D}{\pi t^*}}\,(a_i - a_0) \tag{3.2.19}$$

and, therefore, by comparison with Eq. 1.6.4,

$$k_L^0 = 2\sqrt{\frac{D}{\pi t^*}} \tag{3.2.20}$$

Notice that Eq. 3.2.20 predicts correctly that k_L^0 is proportional to $\sqrt{D}$, which indicates that the penetration theory is more realistic than the film theory.

Again, Eq. 3.2.20 has no predictive value, since t^* is (in general) not known. In a few cases, such as motion of gas bubbles through liquids, t^* can be calculated and, in those cases, Eq. 3.2.20 is well confirmed by experimental evidence. Indeed, one may observe that from classical dimensional analysis considerations, the value of k_L^0 must be given by an equation of the following form:

$$\frac{k_L^0}{v} = f\left(\frac{\mu}{\rho D}, \frac{Lv\rho}{\mu}\right) \tag{3.2.21}$$

where v is some characteristic velocity, L is a characteristic length, μ is the liquid viscosity, and ρ is the liquid density. The dimensionless groups appearing in Eq. 2.2.21 are the Stanton number k_L^0/v, the Schmidt number $\mu/\rho D$, and the Reynolds number $Lv\rho/\mu$. Since the characteristic time t^* in Eq. 3.2.20 (which again is the parameter into which the ignorance on true fluid mechanics has been lumped) can only be inversely proportional to the characteristic velocity v, Eq. 3.2.20 implies that Eq. 3.2.21 should have the following form:

$$\frac{k_L^0}{v} = (\text{constant}) \left(\frac{\mu}{\rho D}\right)^{-0.5} \left(\frac{Lv\rho}{\mu}\right)^{-0.5} \tag{3.2.22}$$

This prediction is well confirmed by empirical correlations for k_L^0.

Again, if a problem of mass transfer with chemical reaction is analyzed on the basis of the penetration theory model, Eq. 3.2.20 can be used to express the result of the analysis as a relationship between I and k_L^0.

There is one feature of the penetration theory model that is unsatisfactory, namely, the assumption that all surface elements stay on the surface for the same length of time t^*. Although this may be at least approximately true in some cases, it is easy to envisage cases where the hypothesis is grossly violated. This has led to the formulation of the "surface renewal" model.

In the surface renewal model, the interface is regarded as being formed by a variety of elements, each one of which has been brought to the surface some time t before the instant of observation. The distribution of surface elements is described by a distribution function $\psi(t)$, defined by the condition that $\psi(t)\,dt$ is the fraction of the interface area formed by elements with "age" comprised between t and $(t + dt)$. The distribution function must of course satisfy the condition

$$\int_0^\infty \psi(t)\,dt = 1 \tag{3.2.23}$$

Since in an element of "age" t the instantaneous rate of mass transfer is given by Eq. 3.2.18, the average rate of mass transfer over the whole interface is given by

$$N = \int_0^\infty N_{\text{inst}}(t)\,\psi(t)\,dt = \sqrt{\frac{D}{\pi}}\,(a_i - a_0) \int_0^\infty \frac{\psi(t)}{\sqrt{t}}\,dt \tag{3.2.24}$$

One may notice that Eq. 3.2.24 predicts k_L^0 to be proportional to $\sqrt{D}$ for any distribution function. The penetration theory model is a special case of the surface renewal model, corresponding to a distribution function given by

$$\begin{aligned} \psi(t) &= \frac{1}{t^*}, \qquad t < t^* \\ \psi(t) &= 0, \qquad t > t^* \end{aligned} \tag{3.2.25}$$

The distribution function in Eq. 3.2.25 emerges from the special assumption that all surface elements stay for the same time at the interface.

An almost diametrically opposite assumption can be made, namely, that the probability for a surface element to be brought off the interface is entirely independent of its age (in the penetration theory, this probability is zero at any age other than t^*). This leads to the following distribution function:

$$\psi(t) = se^{-st} \tag{3.2.26}$$

where s is the "rate of surface renewal." Substitution of Eq. 3.2.26 into 3.2.24 and comparison with Eq. 1.6.4 yields

$$k_L^0 = \sqrt{Ds} \tag{3.2.27}$$

Distribution functions for the penetration and surface renewal models are plotted in Figure 3.2.4.

One may notice that Eq. 3.2.27 is indistinguishable from Eq. 3.2.20 if the rate of surface renewal s is identified with $4/\pi t^*$. Inasmuch as both s and t^* are little more than fudge parameters, one may conclude that the surface renewal and the penetration theory models are, to all practical purposes, equivalent to each other.

Equations 3.2.11, 3.2.20, and 3.2.23, although based on very different physical models, have the common character of relating the value of k_L^0 to the diffusivity D and to a parameter (δ, t^*, s) which in some sense is a yardstick for the fluid mechanics involved. Indeed, it is useful to introduce the concept of "diffusion time" t_D *defined* from the following equation:

$$k_L^0 = \sqrt{\frac{D}{t_D}} \tag{3.2.28}$$

Comparison with Eqs. 3.2.11, 3.2.20, and 3.2.25 shows the relationship between the diffusion time t_D and the empirical parameters δ, t^*, and s as

$$t_D = \frac{1}{s} = \frac{\pi t^*}{4} = \frac{\delta^2}{D} \tag{3.2.29}$$

The concept of diffusion time, as will be seen in the following, is very useful in establishing conditions of asymptotic behavior of phenomena of coupled mass transfer and chemical reaction. These will be discussed in the next sections.

Values of t_D in industrial mass transfer units may be estimated either directly by actual inspection of the hydrodynamic conditions of the liquid phase or, more usually, from Eq. 3.2.28 and empirical correlations for the mass transfer coefficient k_L^0. When this is done, the values of t_D turn out to be in a comparatively narrow range:

$$4 \times 10^{-3} < t_D < 4 \times 10^{-2} \text{ sec} \tag{3.2.30}$$

for all units where the liquid phase is reasonably well mixed.

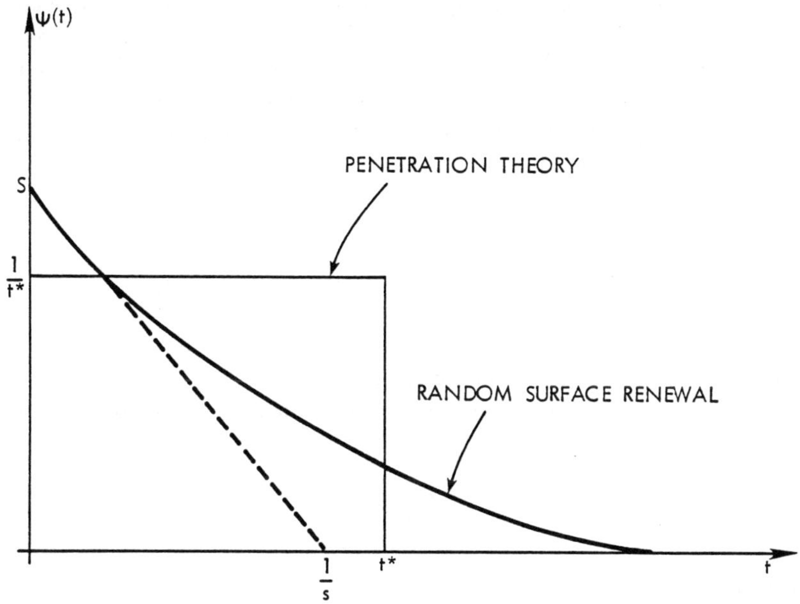

FIGURE 3.2.4. Distribution function of age of surface elements for penetration theory and random surface renewal models.

3.3. SLOW REACTION

We now turn attention to the analysis of coupled mass transfer and chemical reaction and, in particular, we try to establish conditions of asymptotic behavior. For the sake of simplicity, we begin by considering the case where only one chemical reaction may take place in the liquid; let r be the rate at which the reaction takes place (i.e., the number of moles of component A that are consumed per unit time and per unit volume). Notice that, by definition, r is positive an absorption and negative in desorption. In general, the rate r will be given by some kinetic equation of the following general form:

$$r = r(a, b_1, b_2, \ldots, b_N) \tag{3.3.1}$$

where b_j is the concentration of component B_j (recall that the B_js are non-volatile components present in the liquid phase).

Equation 3.3.1 implies that the rate of reaction is not constant throughout the liquid since its value depends on the local composition of the liquid phase. In the presence of the chemical reaction, the diffusion equation for component A becomes

$$D \frac{d^2 a}{dx^2} = r \tag{3.3.2}$$

for the film theory model, and

$$D \frac{\partial^2 a}{\partial x^2} = \frac{\partial a}{\partial t} + r \tag{3.3.3}$$

for the penetration theory and surface renewal models.

Let us first consider the film theory model. Whereas, in the absence of chemical reactions, Eq. 3.2.6 implies that the concentration profile has a zero curvature, Eq. 3.3.2 shows that the curvature will be non-zero when reactions take place. In particular, the curvature will be positive in absorption and negative in desorption; the concentration profiles are qualitatively sketched in Figure 3.3.1.

One may notice from Figure 3.3.1 that in both absorption and desorption, the effect of the chemical reaction is to make the concentration gradient at the interface *steeper* than it would be in the absence of chemical reactions. Hence one expects an enhancement factor larger than unity.

However, there are conditions where the enhancement effect is negligible, namely, when the curvature of the concentration profile is very small, say,

$$\text{Curvature} \ll \frac{\text{average gradient}}{\text{film thickness}} \tag{3.3.4}$$

Equation 3.3.2 shows that the curvature is r/D (i.e., it is not constant throughout

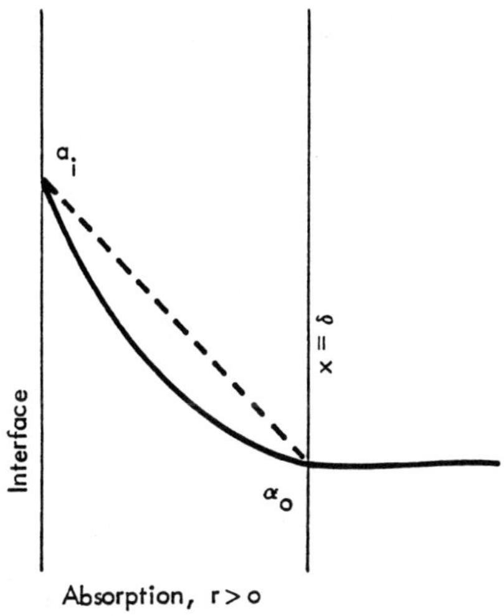

FIGURE 3.3.1. Concentration profiles for chemical absorption and desorption. Film theory model.

the film). However, an average value of the reaction rate can be defined as follows:

$$r_{avg} = \frac{1}{a_i - a_0} \int_{a_0}^{a_i} r\, da \tag{3.3.5}$$

which will be negative in desorption and positive in absorption. The average curvature is r_{avg}/D, whereas the average gradient is $(a_i - a_0)/\delta$; therefore, Eq. 3.3.4 reduces to

$$\frac{a_i - a_0}{r_{avg}} \gg \frac{\delta^2}{D} = t_D \tag{3.3.6}$$

When condition 3.3.6 is satisfied, no significant enhancement is expected; this condition will be referred to as the slow reaction regime, where the enhancement factor is expected to be unity:

$$I = 1 \tag{3.3.7}$$

The left-hand side of Eq. 3.3.6 can be interpreted as a "reaction time" t_r. Indeed, it represents the time required by the reaction to change the concentration over the interval $|a_i - a_0|$ should the reaction take place at the average rate over that interval. Therefore, we define a reaction time as follows:

$$t_r = \frac{a_i - a_0}{2r_{avg}} = \frac{(a_i - a_0)^2}{2\displaystyle\int_{a_0}^{a_i} r\, da} \tag{3.3.8}$$

The factor 2 is introduced so that, for a simple first-order reaction, t_r is simply the inverse of the kinetic constant. Condition 3.3.6 implies that the reaction requires much more time than is available during the diffusion phenomenon—hence the reaction does not appreciably influence the diffusion and no enhancement is observed.

The same conclusion is reached from the penetration theory or surface renewal model. In Eq. 3.3.3, the term $\partial a/\partial t$ can be estimated as

$$\frac{\partial a}{\partial t} = 0\left(\frac{|a_i - a_0|}{t_D}\right) \tag{3.3.9}$$

Therefore, if condition 3.3.6 is satisfied, the second term on the right-hand side of Eq. 3.3.3 can be neglected, and the equation reduces to Eq. 3.2.12. Therefore, Eq. 3.3.7 is obtained again.

In the slow reaction regime, since there is no enhancement effect, the only effect of the chemical reactions is their influence on the driving force. The phenomena of mass transfer and chemical reaction are in fact uncoupled. The regime will be discussed in detail in Chapter 4.

3.4. FAST REACTION

If condition 3.3.6 is *not* satisfied, the term r cannot be neglected in Eqs. 3.3.2 and 3.3.3. Yet the term r depends not only on a but also on the concentrations b_j of the non-volatile components: hence, in principle, Eqs. 3.3.2 and 3.3.3 are coupled with the diffusion equations of all other liquid phase components. The problem formulated in such a general form is very difficult to solve.

However, a major simplification arises from the following considerations. The concentrations b_j in the region near the interface (be it the film thickness of the film theory model, or the "penetration depth" of the surface renewal models) will in general be different from b_{j0}, and the difference is likely to be greatest at the interface itself. However, suppose that the following condition is satisfied for all the b_js:

$$|b_{ji} - b_{j0}| \ll b_{j0} \tag{3.4.1}$$

Condition 3.4.1 implies that the concentrations b_j can, at all points in the liquid phase, be approximated with their bulk-liquid values. Therefore, the term r in Eqs. 3.3.2 and 3.3.3 could be approximated with:

$$r = r(a, b_{10}, b_{20}, \ldots, b_{N0}) = r_0(a) \tag{3.4.2}$$

so that the diffusion equation for a would be uncoupled from those for the b_js.

The question that arises is whether condition 3.4.1 is likely to be fulfilled. One may notice that the analogous condition for a is certainly *not* fulfilled, since if it were the liquid phase composition would be uniform everywhere and no mass transfer would take place. However, the chemical capacity of the liquid by far exceeds the physical solubility of component A and, therefore, the bulk-liquid concentrations b_{j0} are likely to be much larger than either a_i or a_0; therefore, condition 3.4.1 is not an unrealistic one. Of course, the requirements to be satisfied in order that condition 3.4.1 hold remain to be determined. This point will be discussed in detail in Chapter 6.

Inasmuch as we are now considering the case where condition 3.3.6 is not satisfied, we may go one step further and assume that

$$t_r \ll t_D \tag{3.4.3}$$

The asymptotic behavior observable when both 3.4.1 and 3.4.3 hold will be referred to as the "fast reaction regime."

When 3.4.3 holds, the term $\partial a/\partial t$ in Eq. 3.3.3 can be neglected. Therefore, both the film theory equation (3.3.2) and the surface renewal equation (3.3.3) reduce to

$$D \frac{d^2 a}{dx^2} = r_0(a) \tag{3.4.4}$$

The boundary condition at $x = 0$ is Eq. 3.2.14. The second boundary condition is best written in the form

$$a = a_0, \qquad \frac{da}{dx} = 0 \tag{3.4.5}$$

which applies somewhere in the liquid, at a distance from the interface well in excess of the depth of penetration.

For the film theory model, the second boundary condition is Eq. 3.2.8. However, if Eq. 3.4.3 is satisfied, the curvature of the concentration profile in the film is very large; consequently, the concentration gradient at $x = \delta$ is certainly very small. It follows that Eq. 3.4.5 is, in fact, satisfied at $x = \delta$ and can therefore be used as a boundary condition also for the film theory model.

It is important to realize that Eq. 3.4.4, together with the boundary conditions 3.2.14 and 3.4.5, does not contain any parameter related to the fluid mechanics involved: *in the fast reaction regime, the rate of mass transfer is independent of the fluid mechanics.* This conclusion is confirmed by the fact that different models of the fluid mechanics, such as the film theory and surface renewal models, produce the same differential equation and boundary conditions.

The solution of Eq. 3.4.4 is straightforward. A simple change of variables is called for, namely,

$$y = \frac{da}{dx} \tag{3.4.6}$$

so that Eq. 3.4.4 reduces to

$$Dy \frac{dy}{da} = r_0(a) \tag{3.4.7}$$

which integrates directly to

$$\left(\frac{da}{dx}\right)^2 = \frac{2}{D} \int_{a_0}^{a} r_0(a)\, da \tag{3.4.8}$$

when boundary condition 3.4.5 is taken into account. The right-hand side of Eq. 3.4.8 is always positive; in absorption, $r_0(a) > 0$ and $a_i > a_0$; in desorption, $r_0(a) < 0$

and $a_i < a_0$. From Eq. 3.4.8, the concentration gradient at the interface is

$$\left.\frac{da}{dx}\right|_{x=0} = \pm\sqrt{\frac{2}{D}\int_{a_0}^{a_i} r_0(a)\,da} \tag{3.4.9}$$

where the plus sign refers to desorption and the minus sign to absorption. The chemical mass transfer coefficient k_L is therefore given by

$$k_L = \sqrt{\frac{2D\int_{a_0}^{a_i} r_0(a)\,da}{|a_i - a_0|}} \tag{3.4.10}$$

Comparison with Eq. 3.3.8 shows that Eq. 3.4.10 can be cast in the following instructive form:

$$k_L = \sqrt{\frac{D}{t_r}} \tag{3.4.11}$$

Correspondingly, the enhancement factor I is given by:

$$I = \sqrt{\frac{t_D}{t_r}} = \sqrt{\Phi} \tag{3.4.12}$$

One may recall that the values of t_D of practical interest are in a rather narrow range (see Eq. 3.2.30). In contrast with this, reaction times are very different for different reactions; t_r may well be, for some fast reactions, of order 10^{-1} sec or even less. Correspondingly, Eq. 3.4.12 predicts enhancement factors of order 10 or more.

3.5. INSTANTANEOUS REACTION

Equation 3.4.11 indicates that, in the fast reaction regime, the mass transfer rate increases with increasing rate of the chemical reactions, though less than linearly. However, this is true only as long as the condition in Eq. 3.4.1 is fulfilled. As the rate of reaction becomes very large, the concentrations of the non-volatile components near the interface become appreciably different from their bulk-liquid values, and the analysis in the preceding section ceases to be valid.

It is comparatively easy, however, to envisage the asymptotic behavior of the process considered in the limit of extremely large rates of chemical reactions. As the kinetic constants keep increasing, a situation is reached where chemical equilibrium will prevail *everywhere* in the liquid phase, rather than only in the bulk of the liquid. Further increase of the kinetic constants will then not have any effect, and the rate of mass transfer is, therefore, expected to become independent of the chemical kinetics (i.e., I is expected to reach some asymptotic upper limit I_∞ which is independent of the reaction time t_r).

The calculation of I_∞ offers, in the general case, some rather subtle mathematical difficulties, as can be understood by considering the essentially singular character of the limit $t_r \to 0$, and the non-linearity induced by the requirement of chemical equilibrium. Some of these difficulties will be discussed in Chapter 6; in this section we focus attention on a rather simple special case, the analysis of which, however, reveals the essential features of the instantaneous reaction regime.

Consider a system where the main reaction can be written as

$$A + \sum_j \nu_j B_j = 0 \tag{3.5.1}$$

The equilibrium condition for reaction 3.5.1 is

$$\mathrm{K} = \frac{\prod_j (b_j^{-\nu_j})}{a} \tag{3.5.2}$$

In the bulk of the liquid, the concentrations b_{j0} are determined by the local values of the molarity m_0 and the saturation y_0 (see Chapter 2); Eq. 3.5.2 then determines the bulk-liquid value of a_0. In the instantaneous reaction regime, Eq. 3.5.2 will hold at all points in the liquid phase and not only in the bulk.

We first consider the film theory model, for which the diffusion equation for the non-volatile species B_j is

$$D_j \frac{d^2 b_j}{dy^2} = \nu_j r \tag{3.5.3}$$

where r is the reaction rate. Notice that, although chemical equilibrium prevails everywhere in the liquid, r is not zero; the fallacy of such a conclusion has been discussed.

At every point in the liquid, the molarity m is given by Eq. 2.3.4:

$$m = \sum_j \lambda_j b_j \tag{3.5.4}$$

The coefficients λ_j and ν_j are not independent of each other since the progressing of reaction 3.5.1 *by itself* (i.e., in the absence of diffusion phenomena) does not change the molarity; hence

$$\sum_j \lambda_j \nu_j = 0 \tag{3.5.5}$$

If Eq. 3.5.4 is differentiated twice with respect to x and Eqs. 3.5.3 are substituted into the result, one obtains

$$\frac{d^2m}{dx^2} = r \sum_j \frac{\lambda_j \nu_j}{D_j} \tag{3.5.6}$$

Notice that, in general, the right-hand side of Eq. 3.5.6 is *not* zero; this implies that the molarity is not constant in the liquid phase. This induces a very significant difficulty, since the solution of the equilibrium equations requires m to be known.

A considerable simplification, however, arises if *all the* D_j *are equal.* Inasmuch as molecular diffusivities in ordinary liquids are not very different from each other, in this section we restrict attention to the case where

$$D_1 = D_2 = \cdots = D' \tag{3.5.7}$$

so that Eqs. 3.5.5 and 3.5.6 imply that

$$\frac{d^2m}{dx^2} = 0 \tag{3.5.8}$$

Since the boundary conditions on 3.5.8 are

$$x = \delta, \qquad m = m_0 \tag{3.5.9}$$

$$x = 0, \qquad \frac{dm}{dx} = 0 \tag{3.5.10}$$

the solution is

$$m = m_0 \tag{3.5.11}$$

i.e., *if the diffusivities of all the non-volatile species are equal, the molarity is a constant throughout the liquid phase.* Notice that the same conclusion is reached if one considers the surface renewal differential equations.

We now recall the definition of the total content of A, α, given in Eq. 2.3.1. Also considering Eq. 2.3.5, one has

$$\alpha = a + \sum_j \mu_j b_j \tag{3.5.12}$$

Again, the coefficients ν_j and μ_j are not independent of each other since the progressing of reaction 3.5.1 does not by itself change the value of α. Hence

$$1 + \sum_j \nu_j \mu_j = 0 \tag{3.5.13}$$

The film theory diffusion equation for a is

$$D \frac{d^2 a}{dx^2} = r \tag{3.5.14}$$

Differentiating Eq. 3.5.12 twice with respect to x, and substituting Eqs. 3.5.3 and 3.5.14, one obtains

$$\frac{d^2 \alpha}{dx^2} = r \left[\frac{1}{D} + \frac{1}{D'} \sum_j \nu_j \mu_j \right] \tag{3.5.15}$$

We now introduce an additional simplification, namely, the assumption that

$$D = D' \tag{3.5.16}$$

If Eq. 3.5.16 holds, Eqs. 3.5.13 and 3.5.15 imply that

$$\frac{d^2 \alpha}{dx^2} = 0 \tag{3.5.17}$$

Notice that, as said before, should the chemical reaction take place in the *absence* of diffusion phenomena, it would not change the value of α, though, of course, the values of a and b_j would change.

When the chemical reaction takes place in the presence of diffusion phenomena, however, the value of α could change. In fact, suppose, for example, that $D' > D$; before the reaction takes place, component A must diffuse in the physically dissolved form (i.e., with diffusivity D); after some reaction has taken place, some of component A will be in a chemically combined form and can, therefore, diffuse more easily. If, however, Eq. 3.5.16 holds true, diffusion of A in any form is equally easy and thus the diffusion equation for α becomes independent of the kinetic term r.

Inasmuch as the diffusion equation for α has the same form as that for a in the absence of chemical reaction, the rate of mass transfer in the presence of the chemical reaction will be equal to the rate without chemical reaction that would be observed under a driving force $|\alpha_i - \alpha_0|$; i.e.,

$$I = \frac{\alpha_i - \alpha_0}{a_i - a_0} \tag{3.5.18}$$

Subject to the condition that Eq. 3.5.16 holds true, the result in Eq. 3.5.18 is of general validity; however, it cannot, in general, be used for predictive purposes, since the value of α_i is, in general, not known. However, in the instantaneous reaction limit, α_i can be calculated from the requirement that equilibrium should prevail at the interface; i.e.,

$$I_\infty = \frac{\alpha_i^* - \alpha_0}{a_i - a_0} \tag{3.5.19}$$

where α_i^* is the value of α corresponding to chemical equilibrium with the interface partial pressure p_i. Equation 3.5.19 is very important and will be discussed in some detail.

The difference $\alpha_i^* - \alpha_0$ represents the total amount of component A that would need to be absorbed (or desorbed of $\alpha_i^* < \alpha_0$) in order to bring the liquid to *chemical* equilibrium with the gas. In this sense, $\alpha_i^* - \alpha_0$ can be regarded as the "chemical driving force," as compared to the *physical* driving force $a_i - a_0$. Turning back to the definition of I in Eq. 3.1.3, one sees that the rate of mass transfer in the instantaneous reaction regime is given by

$$N = I_\infty k_L^0 (a_i - a_0) = k_L^0 (\alpha_i^* - \alpha_0) \tag{3.5.20}$$

i.e., the rate of mass transfer is the rate of *physical* mass transfer that would be observed under a driving force equal to the *chemical* driving force.

In a sense, this result could have been foreseen without any algebra. When the chemical reactions are infinitely fast, there is no kinetic resistance to the chemical transformation from one species to another, so that mass transfer of A may take place in any of the forms in which A may be chemically combined. If, furthermore, the diffusivities of all species are equal, mass transfer for any component is equally easy; therefore, the total mass transfer rate is enhanced by a factor equal to the ratio of chemical to physical driving forces.

It remains to be discussed how the value of α_i^* may be calculated. In order to do so, we begin by introducing the following definition of ξ_j:

$$b_j = b_{j0} - \nu_j \xi_j \tag{3.5.21}$$

which upon substitution into Eqs. 3.5.3, and taking into account Eq. 3.5.7, yields

$$-D' \frac{d^2 \xi_j}{dx^2} = r \tag{3.5.22}$$

The boundary conditions on Eqs. 3.5.22 are as follows. At $x = \delta$, quite obviously,

$$x = \delta, \qquad \xi_j = 0 \tag{3.5.23}$$

The boundary condition at $x = 0$ is more subtle. Although the flux of B_j from the interface to the gas phase is obviously zero, this does not imply that the flux towards the interface from the liquid is zero: in fact, since the kinetic constants tend to infinity, the rate of reaction at the interface, in spite of the fact that the latter has zero volume, may have some finite value r_0. Therefore, one has

$$x = 0, \qquad D' \frac{db_j}{dx} = \nu_j r_0 \tag{3.5.24}$$

which, in view of Eq. 3.5.21, reduces to

$$x = 0, \qquad -D' \frac{d\xi_j}{dx} = r_0 \tag{3.5.25}$$

Since both the differential equation for ξ_j (Eq. 3.5.22) and the two boundary conditions (Eqs. 3.5.23 and 3.5.25) are the same for all values of the index j, the solution *must* necessarily be the same, say,

$$\xi_1 = \xi_2 = \cdots = \xi \tag{3.5.26}$$

where ξ can be interpreted as an *extent of reaction with respect to the bulk-liquid composition.* It is important to realize that the result in Eq. 3.5.26 is based *only* on the assumption that Eq. 3.5.7 holds true; i.e., *if the diffusivities of the non-volatile species are all equal, their concentrations are everywhere related to each other by Eqs. 3.5.21 and 3.5.26.*

This result is particularly significant in the case of the instantaneous reaction regime. In fact, the equilibrium condition (Eq. 3.5.2) can be written at the interface as

$$\mathrm{K}a_i = \prod_j (b_{j0} - \nu_j \xi_i)^{-\nu_j} \tag{3.5.27}$$

Equation 3.5.27 can be solved in general for ξ_i, the extent of reaction at the interface, as a function of a_i. Once ξ_i has been calculated, the value of I_∞ is calculated as follows.

From Eqs. 3.5.12, 3.5.21, and 3.5.26, one obtains

$$\alpha_i - \alpha_0 = a_i - a_0 - \xi_i \sum_j \nu_j \mu_j \tag{3.5.28}$$

and considering Eq. 3.5.13,

$$\alpha_i - \alpha_0 = a_i - a_0 + \xi_i \tag{3.5.29}$$

Therefore, in view of Eq. 3.5.18,

$$I = 1 + \frac{\xi_i}{a_i - a_0} \tag{3.5.30}$$

With ξ_i^* the solution of Eq. 3.5.27, we finally obtain

$$I_\infty = 1 + \frac{\xi_i^*}{a_i - a_0} \tag{3.5.31}$$

It is instructive to consider the limiting behavior when $K \to \infty$ (i.e., when the *absorption* reaction is irreversible). We choose to assign the index 1 to the particular component of B_j that, among those for which $\nu_j > 0$, has the lowest value of b_{j0}/ν_j. This is the *stoichiometrically limiting reactant in absorption* (*SLRA*). When $K \to \infty$, the solution of Eq. 3.5.27 with $a_i \neq 0$ is

$$\xi_i^* = \frac{b_{10}}{\nu_1} \tag{3.5.32}$$

which implies that $b_1 = 0$ at the interface. The physical interpretation of this conclusion is very simple; since equilibrium prevails everywhere and the reaction is irreversible, A and B_1 cannot coexist anywhere. At the interface $a \neq 0$ and, therefore, $b_1 = 0$; conversely, in the bulk of the liquid $b_1 \neq 0$, and therefore, $a_0 = 0$. Therefore,

$$I_\infty = 1 + \frac{b_{10}}{\nu_1 a_i} \tag{3.5.33}$$

A simple physical interpretation of Eq. 3.5.33 can be obtained by considering the film theory model. The concentration profiles are given in Figure 3.5.1. Since the reaction is irreversible, either b_1 or a must be zero. The film is divided into two parts: one near the interface where $a > 0$ and $b_1 = 0$, and one near the bulk where $b_1 > 0$ and $a = 0$. In both regions no reaction takes place since one of the reactants is absent; therefore, the concentration profiles in both regions are linear. The reaction takes place only at the reaction plane that divides the two regions, located at $x = \lambda$, and it takes place at a rate controlled by the rate at which the two reactants reach the reaction plane by diffusion. The diffusive fluxes of A and B towards the reaction plane must satisfy the obvious stoichiometric condition

$$\nu_1 \frac{a_i}{\lambda} = \frac{b_{10}}{\delta - \lambda} \tag{3.5.34}$$

from which one gets

$$\frac{\delta}{\lambda} = 1 + \frac{b_{10}}{\nu_1 a_i} \tag{3.5.35}$$

The interface concentration gradient for a would be, in the absence of the

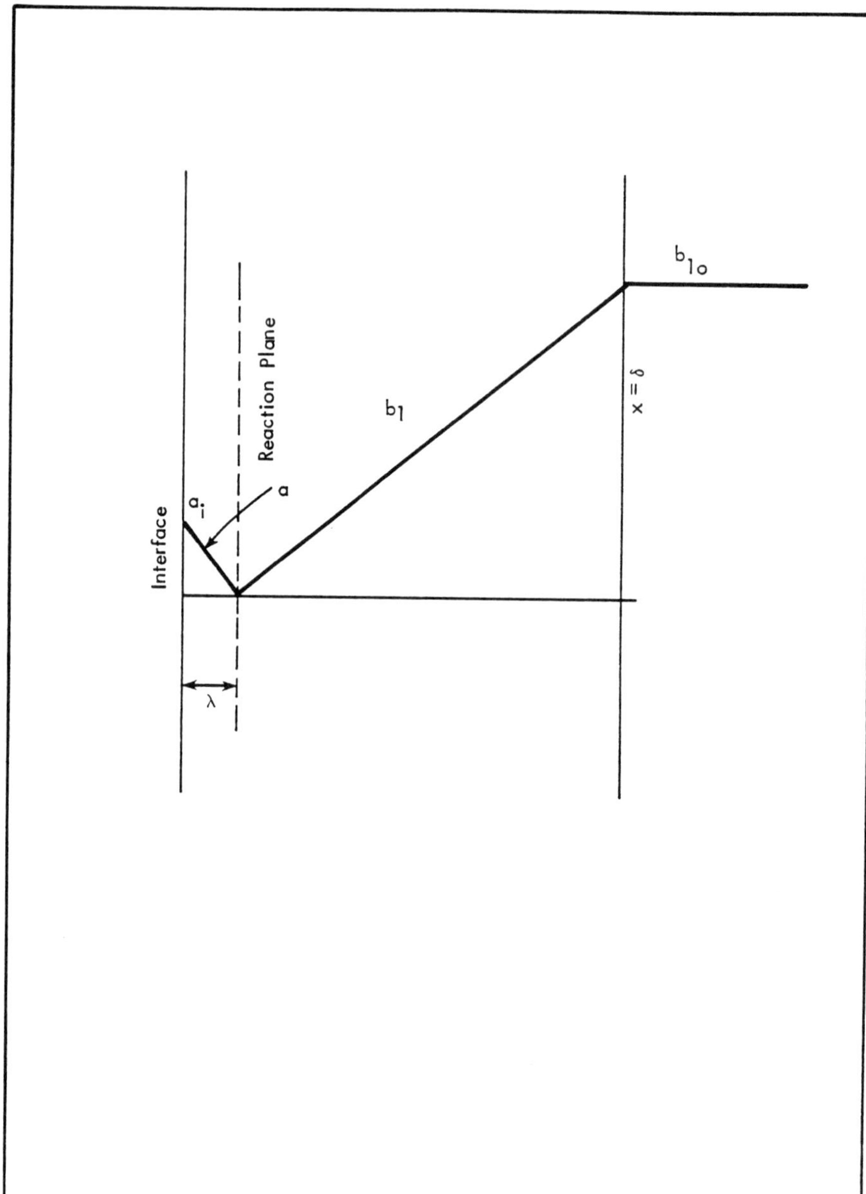

FIGURE 3.5.1. Instantaneous irreversible reaction in chemical absorption. Concentration profiles, film theory model.

chemical reaction, a_i/δ; the actual concentration gradient is a_i/λ; therefore,

$$I = \frac{\delta}{\lambda} = 1 + \frac{b_{10}}{\nu_1 a_i} \tag{3.5.36}$$

which is, of course, the same as Eq. 3.5.33.

Notice that, in general, b_{10}/ν_1 is very much larger than a_i; in fact, b_{10}/ν_1 is of the order of the chemical capacity of the liquid, whereas a_i is only the physical capacity. Therefore, Eq. 3.5.36 predicts values of I that can be significantly larger than unity, say, easily of the order of 10^2 to 10^3.

Turning back attention to Figure 3.5.1, one may notice that, since in actual cases $\lambda \ll \delta$, the chemical reaction essentially takes place at the interface itself. The rate-determining phenomenon, therefore, becomes the diffusion of B_1 from the bulk to the interface, rather than the diffusion of A from the interface to the bulk. Inasmuch as the driving force for diffusion of B_1 (which is b_{10}/ν_1) is so much larger than the driving force for diffusion of A, a very significant enhancement will be observed.

No analogous simple irreversible limit arises in the case of desorption. The irreversible limit for desorption could be regarded as the one where $K \to 0$; however, Eq. 3.5.2 shows that, in this limit, the bulk-liquid concentration a_0 would tend to infinity. The bulk-liquid concentration a_0 cannot exceed the total pressure divided by Henry's constant, for if it does, the excess volatile component would simply boil off—that is essentially what happens in a flash desorption.

However, an analog of an irreversible limit is obtained for desorption in the limit $a_i \to 0$. Let the index 2 be assigned to that B_j, among those that have a negative ν_j, characterized by the lowest value of $b_{j0}/|\nu_j|$. This is the *stoichiometrically limiting reactant in desorption, SLRD.* When $a_i \to 0$, the solution of Eq. 3.5.27 is

$$\xi_i^* = \frac{b_{20}}{\nu_2} \tag{3.5.37}$$

Notice that in this case ξ_i^* is *negative*; this is not surprising, since the definition of ξ is such that it increases towards the interface in absorption and vice versa. Since $a_i = 0$, Eq. 3.5.31 reduces to

$$I_\infty = 1 - \frac{b_{20}}{\nu_2 a_0} \tag{3.5.38}$$

which produces $I_\infty > 1$ since $\nu_2 < 0$. The enhancement is, again, very large; the rate-determining step is diffusion of the SLRD towards the interface. However, no simple concentration profile of the type of Figure 3.5.1 can be sketched for this limiting desorption behavior.

3.6. SUMMARY

Three asymptotic behaviors have been discussed: slow, fast, and instantaneous reaction. These are best summarized by considering the parameter Φ defined by

$$\Phi = \frac{t_D}{t_r} \tag{3.6.1}$$

The parameter Φ increases with increasing rate of the chemical reactions.

As long as Φ is significantly smaller than unity, no enhancement takes place (i.e., $I = 1$). When Φ exceeds unity by a significant amount, the enhancement factor is given by Eq. 3.4.12 (i.e., $I = \sqrt{\Phi}$). However, an upper bound to the value of I is reached in the instantaneous reaction regime; i.e., when $\Phi \to \infty$, I approaches some asymptotic value I_∞ to be calculated from the solution of the appropriate instantaneous reaction problem. The value of I_∞ depends only on the liquid-phase composition and the value of a_i, but is independent of the fluid mechanics of the liquid phase and of the parameter Φ.

Given a value of a_i and a composition of the liquid phase (e.g., given the conditions at a particular point of an industrial unit), the value of I_∞ is fixed and I may be regarded as a unique function of Φ. The asymptotic behavior of this function is reported in Figure 3.6.1. Of course, the *actual* behavior of the function is more complex since the two cusps at $\Phi = 1$ and $\Phi = I_\infty^2$ are in fact smoothed out; these transition regions will be discussed in Chapter 7. However, particularly if $I_\infty \gg 1$, the three asymptotic equations $I = 1$, $I = \sqrt{\Phi}$, and $I = I_\infty$ will be quite acceptable approximations over rather wide ranges of values of Φ; the details of these asymptotic regions are discussed in Chapters 4, 5, and 6.

The value of I_∞, as can be seen by considering Eq. 3.5.19, is of the order of magnitude of the ratio of the chemical capacity to the physical capacity of the liquid phase. Therefore, in most industrial cases, I_∞ will be in the range 10^2 to 10^3. With diffusion times typically of order 10^{-2} sec, the three asymptotic behaviors are to be expected in the following range of values of reaction times (based on $I_\infty = 100$):

Slow reaction	$t_r > 10^{-2}$ sec
Fast reaction	10^{-2} sec $> t_r > 10^{-6}$ sec
Instantaneous reaction	$t_r < 10^{-6}$ sec

As will be discussed in more detail in Chapter 5, few if any commercial gas treating processes are performed under slow reaction conditions. A typical example of a process taking place under fast reaction conditions is the hot-carbonate process for CO_2 absorption; values of the reaction time for that process are in the range 10^{-2}–10^{-5} sec over the temperature range 45–100°C.

Reaction times low enough for a process to be in the instantaneous reaction regime are typical of proton transfer reactions such as H_2S generally undergoes when absorbed in an alkaline solution.

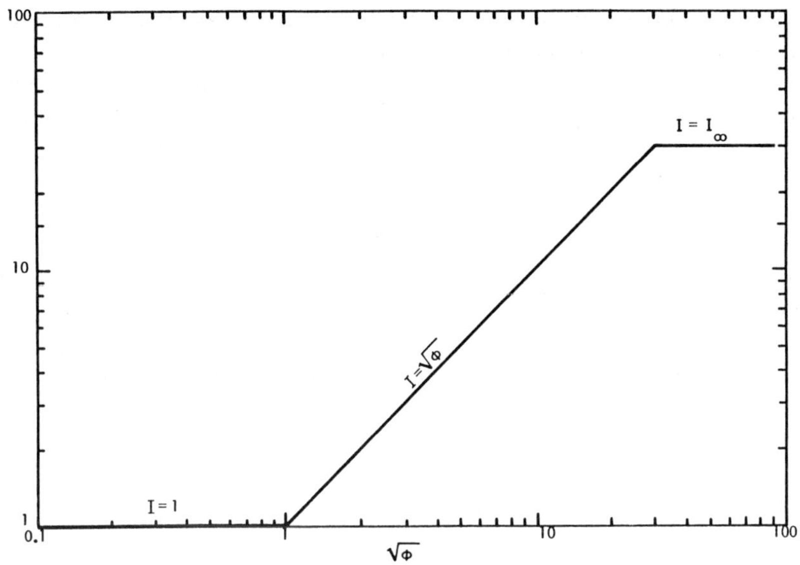

FIGURE 3.6.1. Asymptotic behavior for I (Φ).

Borderline cases are encountered in the absorption of CO_2 into strongly alkaline and lean amine-water solutions, where reaction times of the order of 10^{-5}–10^{-7} sec are common.

3.7. THE EFFECTIVENESS CONCEPT

Rates of absorption and desorption from chemical solvents are typically described in terms of the enhancement factor I, as was discussed in Section 3.6. The enhancement factor I is defined by Eq. 3.1.3; i.e., its definition is based on the *physical* driving force $a_i - a_0$ (as contrasted to the *chemical* driving force $\alpha_i^* - \alpha_0$).

However, a different viewpoint may be taken, which is sometimes a more useful one. This alternate viewpoint leads to the concept of the effectiveness of rate enhancement η.

Consider a gas phase and a liquid phase that are not at equilibrium with each other, and let the interface partial pressure of component A be such that its interface concentration in the liquid phase is a_i, whereas the concentration in the bulk is a_0. Since the two phases are not at equilibrium, $a_i \neq a_0$. For the purpose of the present discussion, we consider the case of absorption ($a_i > a_0$); the argument applies to desorption as well.

Consider the case where the liquid phase is a physical solvent. In order to reach equilibrium with the interface partial pressure, an amount $a_i - a_0$ (per unit volume) of component A would need to be absorbed into the liquid. Therefore, the quantity $a_i - a_0$, in addition to being the driving force for mass transfer, also measures the *capacity* of the liquid—i.e., the amount that can be absorbed into the liquid per unit volume before equilibrium is reached.

It follows that, in the case of a physical solvent, the following identity holds at all times:

$$\frac{\text{specific rate}}{\text{capacity}} = \frac{N}{a_i - a_0} = k_L^0 \tag{3.7.1}$$

Now consider the case of a *chemical* solvent, with the same values of a_i and a_0 as in physical absorption. In this case, the capacity of the liquid is much larger than $a_i - a_0$, and is, in fact, given by $\alpha_i^* - \alpha_0$. Addition of a reactive component to the liquid phase results in a large increase in capacity, which is the main reason why chemical solvents are used in gas treating processes; the increase in capacity is given by

$$\frac{(\text{capacity})_{\text{chemical}}}{(\text{capacity})_{\text{physical}}} = \frac{\alpha_i^* - \alpha_0}{a_i - a_0} \tag{3.7.2}$$

Although the increase in capacity is given by Eq. 3.7.2, the increase in rate due to addition of a reactive component is measured by the enhancement factor I. It follows that, if I is less than the right-hand side of Eq. 3.7.2, it will take a chemical solvent *longer* to reach equilibrium than it takes a physical solvent to do so. Notice that I is always less than $(\alpha_i^* - \alpha_0)/(a_i - a_0)$ unless instantaneous reaction conditions are reached [in Chapter 6 it will be seen that even in the instantaneous reaction regime, I may, in fact, be somewhat less than $(\alpha_i^* - \alpha_0)/(a_i - a_0)$].

The concept of effectiveness of rate enhancement is suggested by the idea that, in order to be as fast as a physical solvent in reaching equilibrium, a chemical solvent should exhibit a rate enhancement as large as the capacity enhancement. Therefore, the effectiveness η may be defined as follows:

$$\eta = \frac{\text{rate enhancement}}{\text{capacity enhancement}} = \frac{I(a_i - a_0)}{\alpha_i^* - \alpha_0} \tag{3.7.3}$$

Chemical solvents characterized by η values very significantly less than unity are, from the rate viewpoint, not effective; apparent efficiencies of trays, e.g., will be very low. Since for many gas treating processes instantaneous reaction conditions cannot be reached, such low rate effectiveness is not uncommon for chemical solvents.

4 Slow Reaction Regime

Hence Achilles can never reach the turtle.

ZENO

In Chapter 3 it was established that, when the chemical reactions taking place in the liquid phase are slow enough, no rate enhancement will be observed; this situation has been termed the slow reaction regime.

In an industrial gas treating process where a chemical solvent is used, the capacity of the solvent is considerably larger than the physical solubility of the component to be removed from the gas phase. Therefore, should the absorption process take place in the slow reaction regime, the effectiveness of rate enhancement, introduced in Section 3.7, would be unacceptably low. From this viewpoint, one could consider it useless to discuss this regime of mass transfer at all. There are, however, several reasons why the analysis to be presented in this chapter is of interest for gas treating processes.

1. Even if one wishes to avoid slow reaction conditions, understanding of this asymptotic region is needed for the very purpose of avoiding it.
2. In some cases of gas treating where selective removal of one component is sought, it is useful that the other component undergoes inefficient removal such as takes place in the slow reaction regime.
3. As will be seen in Section 4.6, for processes where gas treating is accomplished by slurried solids, the slow reaction regime is sometimes unavoidable.
4. Sometimes slow reaction conditions need to be established in laboratory experiments, such as those aimed at measuring the value of Henry's Law constant in reactive solvents.
5. Industrial gas-liquid processes which take place in the slow reaction regime do exist; these are, however, not gas treating processes, but, as in the case of aerobic fermentation, processes the aim of which is the production of the reaction product.

4.1. QUASI-STEADY-STATE APPROXIMATION

In this chapter, attention is restricted to the case where the condition for "slow reaction" is satisfied; i.e.,

$$t_r \gg t_D \tag{4.1.1}$$

When 4.1.1 is satisfied, the enhancement factor is equal to unity and the rate of mass transfer is given by

$$N = \mathrm{k}_L^0 (a_i - a_0) \tag{4.1.2}$$

The effectiveness of rate enhancement is very low; from Eq. 3.8.3 one obtains

$$\eta = \frac{a_i - a_0}{\alpha_i^* - \alpha_0} \ll 1 \tag{4.1.3}$$

In the slow reaction regime, the effect of chemical reactions is simply to keep the value of a_0 low (in absorption) or high (in desorption) in the bulk of the liquid. The rate of mass transfer can be calculated from Eq. 4.1.2, provided the value of a_0 is known. The latter, in turn, is determined by the rate at which chemical reactions take place in the liquid phase and, therefore, kinetics need to be taken into account. In order to discuss how this can be accomplished, attention will be focused on a specific problem, the analysis of which permits the introduction of the significant concepts. These concepts can then be used also in the analysis of different processes.

The specific process considered is the following one. A volume V of a chemically reactive liquid is kept in a stirred vessel in the presence of a gas at some fixed pressure p_0. The gas is made up of pure component A; for the sake of simplicity, we assume that all the constituents of the liquid phase, including the solvent, are nonvolatile. The system is allowed to reach equilibrium. At time $t = 0$, the pressure of the gas is suddenly brought to some different value p_1 and held at that value thereafter. We wish to analyze the rate at which mass transfer will take place, up to the point where a new equilibrium is reached.

Since at time zero the system is in equilibrium, the bulk-liquid concentration of A at time zero is the one corresponding to equilibrium with the pressure p_0, say,

$$a_0(0) = \frac{p_0}{H} \tag{4.1.4}$$

If A is the area of the gas-liquid interface, V is the liquid volume, and k is the kinetic constant (for the sake of simplicity, we consider the simple case of a first-order reaction), an instantaneous mass balance yields

$$\mathrm{k}_L^0 (a_i - a_0) A = \mathrm{k}V(a_0 - a^*) + V \frac{da_0}{dt} \tag{4.1.5}$$

where a^* is the concentration of a that would be in chemical equilibrium with the actual concentrations b_{j0} in the bulk of the liquid. In Eq. 4.1.5, both a_0 and a^* are functions of time; in particular, at time 0.

$$a^*(0) = \frac{p_0}{H} \tag{4.1.6}$$

since both physical and chemical equilibrium prevailed before the sudden change in pressure. In contrast with this, a_i is a constant given by

$$a_i = \frac{p_1}{H} \tag{4.1.7}$$

If we recall the definition of the total content of component A in the liquid α, an additional mass balance can be written in the following form:

$$\mathrm{k}_L^0 A(a_i - a_0) = V\frac{d\alpha_0}{dt} \tag{4.1.8}$$

which simply states that whatever A is (say) absorbed must accumulate in either the physically dissolved or the chemically combined form, or both.

First consider the behavior of such a system should no chemical reaction take place; mathematically, this corresponds to setting k = 0 in Eq. 4.1.5. The solution of Eq. 4.1.4 is then

$$a_0(t) = a_0(0) + [a_i - a_0(0)]\,[1 - e^{-\mathrm{k}_L^0 A t/V}] \tag{4.1.9}$$

Equation 4.1.9 is plotted in Figure 4.1.1. One may notice that physical saturation (i.e., the concentration in the bulk approaching the equilibrium value a_i) is reached over a time scale t_{ps} given by

$$t_{ps} = \frac{V}{\mathrm{k}_L^0 A} \tag{4.1.10}$$

which is calculated by dividing the total concentration change $a_i - a_0(0)$ by the inital rate of change of concentration.

Turning attention to the case where chemical reactions do take place, the time scale for *chemical* saturation can be estimated from Eq. 4.1.8. The value of α_0 needs to change from some initial value $\alpha_0(0)$ corresponding to chemical equilibrium under a partial pressure p_0, to some final value $\alpha_0(\infty)$ corresponding to chemical equilibrium under a partial pressure p_1. This change will, in general, be much *larger* than $a_i - a_0$, since the (much larger) concentrations b_j of non-volatile reactants need to change significantly. Yet Eq. 4.1.8 shows that the initial rate of change of α_0 is equal to the initial rate of change of a_0 when no chemical reaction

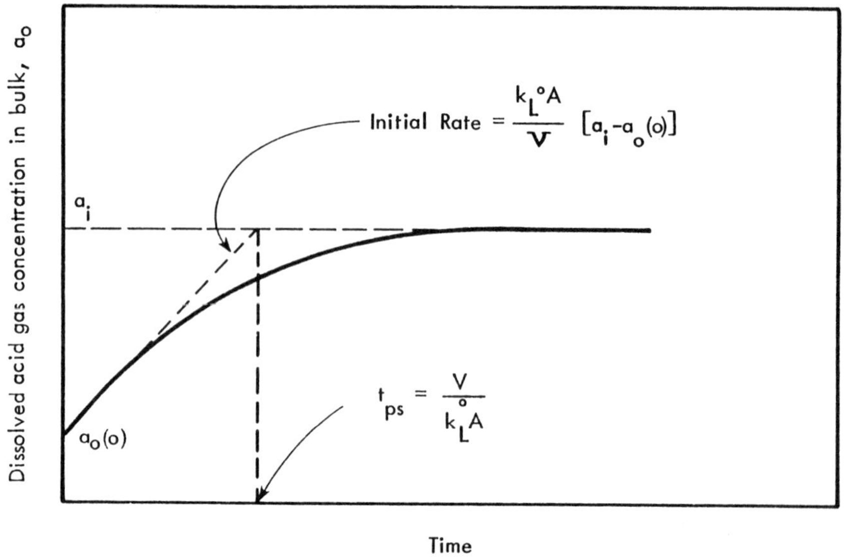

FIGURE 4.1.1. Physical saturation.

takes place. It follows that the time scale for chemical saturation t_{cs} by far exceeds the time scale for physical saturation; indeed, it will be shown in the next sections that

$$t_{cs} \geqslant t_{ps} \frac{\alpha_0(\infty) - \alpha_0(0)}{a_i - a_0(0)} \tag{4.1.11}$$

and that, indeed, t_{cs} may be much larger than the right-hand side of Eq. 4.1.11.

This consideration justifies the "quasi-steady-state" approximation. Turning attention back to Eq. 4.1.5, the term $V(da_0/dt)$ is seen to become negligible after an initial induction time of order t_{ps}, so that Eq. 4.1.5 can be written as

$$t \gg t_{ps}, \quad Ak_L^0(a_i - a_0) = kV(a_0 - a^*) \tag{4.1.12}$$

Physically, this means that, after the induction time has elapsed, the rates of mass transfer from the interface and of chemical reaction in the bulk essentially balance each other, since further accumulation or depletion in the bulk (da_0/dt) takes place over a much larger time scale given by t_{cs}.

Equation 4.1.12 is algebraic and can be solved directly for the bulk concentration a_0. With the quasi-steady-state approximation, a_0 is calculated from Eq. 4.1.12 and then substituted into Eq. 4.1.8 in order to calculate the evolution in time of the chemical composition of the liquid phase. An example of this procedure is given at the end of this chapter.

4.2. THE KINETIC SUBREGIME

Consideration of the quasi-state Eq. 4.1.12, shows that the total driving force for the process $|a_i - a^*|$ is divided in two parts: $|a_i - a_0|$, is the driving force for mass transfer between interface and bulk, and $|a_0 - a^*|$, is the driving force for the chemical reaction in the bulk. The ratio of these two driving forces (which always have the same sign) is

$$\frac{a_0 - a^*}{a_i - a_0} = \frac{A\mathrm{k}_L^0}{V\mathrm{k}} = \frac{t_r}{t_{ps}} \tag{4.2.1}$$

where t_{ps} is given by Eq. 4.1.10 and t_r is the reaction time $t_r = 1/\mathrm{k}$.

Suppose the following condition is fulfilled:

$$t_r \gg t_{ps} \tag{4.2.2}$$

i.e., the reaction is so slow that the reaction time by far exceeds the time scale for physical saturation. If that is the case, Eq. 4.2.1 shows that the driving force for the chemical reaction $|a_0 - a^*|$ will be almost equal to the total driving force $|a_i - a^*|$; physically, if condition 4.2.2 is satisfied, the chemical reaction becomes the rate-determining step, and mass transfer takes place under a negligibly small driving force. The liquid phase, after the induction time, is physically saturated with component $A(a_0 \simeq a_i)$. This will be referred to as the kinetic subregime.

In the kinetic subregime, the total rate of mass transfer is given by

$$NA = \mathrm{k}V(a_i - a^*) \tag{4.2.3}$$

Several important observations need to be made concerning Eq. 4.2.3.

First of all, one may notice that the total rate of mass transfer is independent of k_L^0 (i.e., is independent of the intensity of mixing). This is, of course, due to the fact that since the mass transfer step is not rate-determining, any change of the efficiency of mass transfer is irrelevant as long as Eq. 4.2.2 is satisfied.

The second point is that the kinetic constant k can be calculated from Eq. 4.2.3 and can, therefore, be deduced directly from the measured rate of mass transfer. Only in the kinetic subregime is the (measurable) rate of mass transfer directly proportional to the kinetic constant k (or, more generally, to the rate of the chemical reaction). This, in turn, implies that a direct measure of the rate of a gas–liquid reaction can only be obtained if the reaction is slow enough for condition 4.2.2 to be satisfied.

The third observation is related to the *condition* for the validity of Eq. 4.2.3 (i.e., with Eq. 4.2.2). It is useful to estimate the range of values in which pragmatically relevant values of t_{ps} may be found.

Values of the mass transfer coefficient k_L^0 are typically in the range $0.01 \div 0.05$ cm/sec. The ratio of the liquid volume per unit interface area V/A is not likely to be much less than 0.1 cm, though it can, of course, be considerably more than that.

Therefore, one may write

$$t_{ps} > 2 \text{ sec} \tag{4.2.4}$$

This result leads to two conclusions. First, since t_{ps} is always much larger than the diffusion time t_D, condition 4.2.2 is much more stringent than condition 4.1.1. Second, only for moderately slow chemical reactions can kinetic data be gathered directly from measurements of total rates of mass transfer.

For the kinetic subregime, the time scale for chemical saturation can be calculated as follows. From Eqs. 4.1.8, 4.1.12, and 4.2.1 one obtains

$$\frac{d\alpha_0}{dt} = \mathrm{k}[a_i - a_0(0)] \tag{4.2.5}$$

and, therefore,

$$t_{cs} = t_r \frac{\alpha_0(\infty) - \alpha_0(0)}{a_i - a_0(0)} \tag{4.2.6}$$

In view of Eq. 4.2.2, one sees that the time for chemical saturation is, in fact, much longer than the right-hand side of Eq. 4.1.11.

It is important to realize that, if condition 4.2.2 is satisfied, $a_0 = a_i$ is a very good approximation whatever the form of the kinetic equation may be; i.e., the general conclusions in this section are not limited to the simple first-order kinetics. In the general case where the rate of reaction is given by Eq. 3.4.1, one has instead of Eq. 4.2.3,

$$NA = Vr(a_i, b_{10}, \ldots, b_{N0}) \tag{4.2.7}$$

A final point concerning the kinetic subregime is concerned with the question of how to establish experimentally whether a process is actually taking place in this subregime. This can be ascertained by noting that the (measurable) total mass transfer rate NA is, in the kinetic subregime:

1. Independent of mixing rate of the liquid phase, i.e., independent of k_L^0.
2. Independent of interface area.
3. Proportional to the liquid volume.

It will be seen in the following that the kinetic subregime is the *only* case where 2 and 3 hold true. This allows easy experimental verification of the hypothesis that a given process takes place in the kinetic subregime.

A final point should be discussed, namely, the length of the induction time required in order to reach the quasi-steady-state conditions assumed in Eq. 4.1.11. This is easily obtained by considering that, in Eq. 4.1.5, the term $\mathrm{k}V(a_0 - a^*)$ is zero at time zero. Hence one obtains in general,

$$t_{\text{ind}} = \frac{(a_0)\text{qss} - a^*}{(da_0/dt)_{t=0}} = t_{ps}\frac{(a_0)\text{qss} - a^*}{a_i - a^*} \tag{4.2.8}$$

where (a_0)qss is the value of a_0 calculated from the quasi-steady-state hypothesis. Since in the kinetic subregime $(a_0)\text{qss} \cong a_i$, one obtains

$$t_{\text{ind}} = t_{ps} \tag{4.2.9}$$

i.e., the quasi-steady-state conditions are reached after a time of the order of the time required for physical saturation of the liquid. For a continuous absorption or desorption tower, the induction period would be of the order of the height corresponding to one equilibrium stage. This point will be discussed in more detail in Section 4.5.

4.3. THE DIFFUSIONAL SUBREGIME

Consider the case where both the following inequalities hold true:

$$t_D \ll t_r \ll t_{ps} \tag{4.3.1}$$

In order for that to be possible, one of course needs that

$$t_{ps} \ggg t_D \tag{4.3.2}$$

However, Eq. 4.3.2 is, in fact, true in practically all cases of pragmatic interest (see Eqs. 3.2.30 and 4.2.4). Of course, Eq. 4.3.2 *allows* but does not *imply* Eq. 4.3.1.

If Eq. 4.3.1 is satisfied, the process takes place in the slow reaction regime and, moreover, from Eq. 4.2.1, the value of a_0 is almost equal to a^*. These conditions will be referred to as the diffusional subregime. Physically, the diffusional subregime is that condition where the chemical reaction is slow enough not to result in any enhancement ($t_r \gg t_D$, hence $I \cong 1$), and yet fast enough to maintain chemical equilibrium in the bulk of the liquid ($t_r \ll t_{ps}$, hence $a_0 \cong a^*$). The rate of mass transfer is given by

$$N = \text{k}_L^0(a_i - a^*) \tag{4.3.3}$$

One may notice that, contrary to the kinetic subregime, the total rate of mass transfer NA is:

Dependent on the intensity of mixing of the liquid phase, i.e., proportional to k_L^0.
Proportional to interface area.
Independent of liquid volume.

Futhermore, the rate NA is proportional to $a_i - a^*$; this would be true in the kinetic subregime only in the case of simple first-order kinetics. Finally, it should be pointed out that although in the kinetic subregime the rate NA will, in general be a strong function of temperature, it is only a weak function of temperature in the diffusional subregime.

In the diffusional subregime, the rate of mass transfer is independent of the kinetics of the chemical reaction (see Eq. 4.3.3). However, the fact that chemical reactions do occur has a very important effect, namely, that of keeping the bulk liquid concentration a_0 equal to its equilibrium value, so that mass transfer keeps taking place at an essentially constant rate.

The induction time in the diffusional regime is very short, since the quasi-steady-state value of a_0 almost coincides with its initial value. Indeed from Eqs. 4.2.8 and 4.2.1 one obtains

$$t_{\text{ind}} = t_r \tag{4.3.4}$$

and, in view of Eq. 4.3.1, $t_{\text{ind}} \ll t_{ps}$. In the diffusional subregime, the induction phenomenon can be neglected altogether and the quasi-steady-state approximation can be used from time zero onwards.

4.4. TRANSITION REGION

In this section we consider a generalization of the quasi-steady-state equation by allowing the rate of reaction r to be given by a power-law kinetic equation, say,

$$r = k_n(a - a^*) \, |a - a^*|^{n-1} \tag{4.4.1}$$

(The power-law is written in the form of Eq. 4.4.1 to allow for both absorption, $r > 0$ and desorption, $r < 0$). The kinetic equation may, of course, not be of the simple power-law type, but it can usually be approximated by it rather satisfactorily.

The average rate of reaction is given by

$$r_{\text{avg}} = \frac{1}{a_i - a^*} \int_{a^*}^{a_i} r \, da = \frac{\text{k}_n(a_i - a^*) \, |a_i - a^*|^{n-1}}{n+1} \tag{4.4.2}$$

and, therefore, consistent with the definition in Eq. 3.3.8, the reaction time is given by

$$t_r = \frac{|a_i - a^*|^{1-n}}{k_n} \left(\frac{n+1}{2} \right) \tag{4.4.3}$$

The quasi-steady-state equation takes the form

$$\text{k}_L^0 A(a_i - a_0) = \text{k}_n V(a_0 - a^*) \, |a_0 - a^*|^{n-1} \tag{4.4.4}$$

It is now useful to introduce the following dimensionless quantities. The first is the fraction ξ of the overall driving force $|a_i - a^*|$ which is required by the chemical reaction in the bulk, say,

$$\xi = \frac{a_0 - a^*}{a_i - a^*} \tag{4.4.5}$$

The second quantity is the ratio Ω between the time for physical saturation and the reaction time:

$$\Omega = \frac{t_{ps}}{t_r} = \frac{k_n V}{k_L^0 A \,|a_i - a^*|^{1-n}} \left(\frac{2}{n+1}\right) \tag{4.4.6}$$

Equation 4.4.4 reduces to

$$\frac{n+1}{2}\,\Omega = \frac{1-\xi}{\xi^n} \tag{4.4.7}$$

which, in turn, reduces to Eq. 4.2.1 in the special case where $n = 1$.

Equation 4.4.7 is easily solved for the two limiting asymptotic cases, namely $\Omega \gg 1$ (diffusional subregime) and $\Omega \ll 1$ (kinetic subregime). In fact,

$$\Omega \gg 1: \qquad \xi = \left[\frac{2}{(n+1)\Omega}\right]^{1/n} \ll 1 \tag{4.4.8}$$

$$\Omega \ll 1: \qquad \xi = 1 - \frac{n+1}{2}\,\Omega \simeq 1 \tag{4.4.9}$$

Figure 4.4.1 is a plot of Eq. 4.4.7 for several values of n. From that figure, one may notice that the transition region (i.e., the range of values of Ω over which the asymptotic equations are poor approximations) is larger the larger the value of n and, furthermore, that the error induced by using the asymptotic equations is larger the larger is n. For $n = 1$, the kinetic subregime equation is acceptable (less than 10% error) when $\Omega < 0.1$, and the diffusional subregime equation when $\Omega > 10$.

It is also useful to introduce the following dimensionless quantity:

$$\Psi = \frac{t_{ps}}{t_D} \tag{4.4.10}$$

which, as discussed in Section 4.2, is always much larger than unity (in fact, at least of order 100). It is important to realize that Ψ depends only on the hydrodynamics

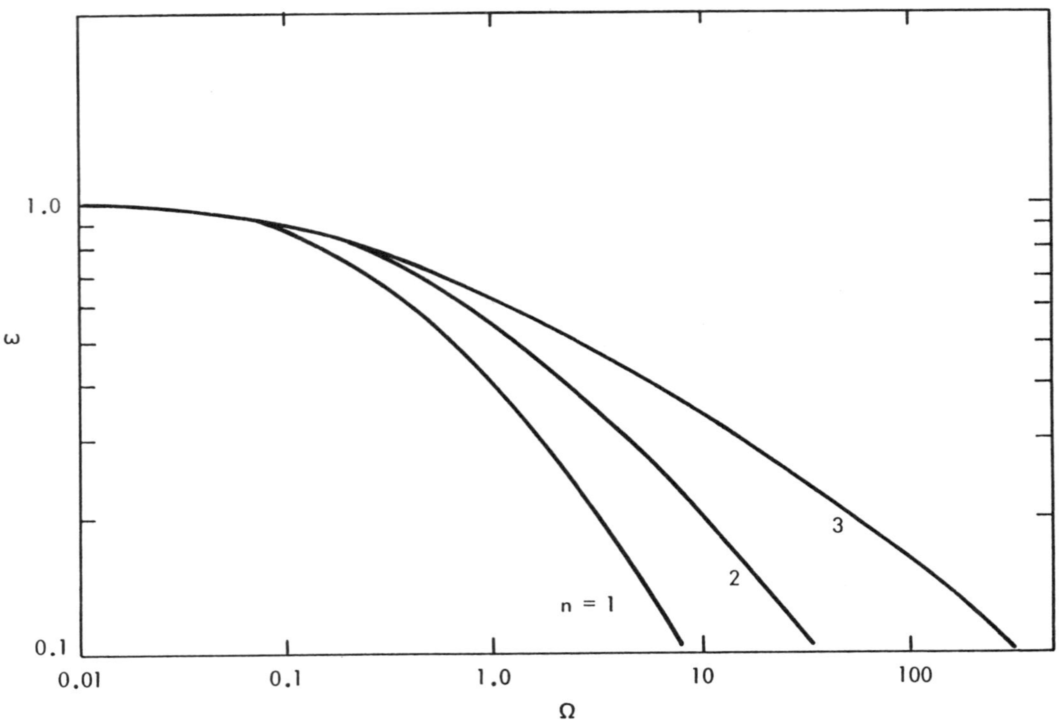

FIGURE 4.4.1. Plot of fraction of overall driving force required by chemical reaction in bulk against ratio of physical saturation time to reaction time.

of the liquid phase and not on the kinetics of the reaction. The quantity $\Phi = t_D/t_r$ introduced in Chapter 3 is related to Ψ and Ω by

$$\Phi = \frac{\Omega}{\Psi} \tag{4.4.11}$$

4.5. PACKED TOWERS

The process considered in Section 4.1 is essentially a batch process; however, continuous processes are common in industrial gas treating operations. The concepts developed so far are easily applied to continuous operations by simply measuring time t along the liquid path from the feed point. If ϵ represents the fractional liquid holdup in a packed tower, Z axial distance measured from the liquid feed section, and V_L the superficial liquid velocity, one has

$$t = \frac{\epsilon Z}{V_L} \tag{4.5.1}$$

The value of t_{ps} can be estimated as follows. The height of a transfer unit for liquid-side mass transfer control H_L is given by

$$H_L = \frac{V_L}{\mathrm{k}_L^0 \tilde{a}} \tag{4.5.2}$$

where $\tilde{a}$ is the interface area per unit volume. The liquid volume per unit interface area V/A is clearly $\epsilon/\tilde{a}$. Therefore,

$$t_{ps} = \frac{\epsilon}{\mathrm{k}_L^0 \tilde{a}} = \frac{\epsilon H_L}{V_L} \tag{4.5.3}$$

From Eq. 4.5.1, this corresponds to

$$Z_{ps} = \frac{V_L}{\epsilon} t_{ps} = H_L \tag{4.5.4}$$

which shows that the axial length required for physical saturation is of the order of the height of a transfer unit (or one equilibrium stage for a plate column). The induction height Z_{ind} is of order H_L in the kinetic subregime, and much less than that in the diffusional subregime. This shows that the quasi-steady-state approximation is valid over most of the tower.

However, in an industrial tower, the aim is to approach *chemical* saturation of the liquid phase so as to keep the rate of circulation of the liquid as low as possible. Therefore, the total axial length of the tower Z would need to exceed Z_{ps} by a factor of the order of t_{cs}/t_{ps}, say,

$$Z_1 \cong Z_{ps} \frac{t_{cs}}{t_{ps}} = H_L \frac{t_{cs}}{t_{ps}} \tag{4.5.5}$$

Equation 4.5.5 shows that the ratio t_{cs}/t_{ps} is a measure of the number of transfer units (or of equilibrium stages) required in order to perform a gas treating operation in the slow reaction regime. Unfortunately, the ratio t_{cs}/t_{ps} turns out to be an exceedingly large number and inordinately large heights of packing would be required. The physical reason for this is that, in the slow reaction regime, the rate of mass transfer is governed by the physical solubility a_i and, in fact, is the same as would be obtained without any chemical reaction. However, since chemical conversion of the liquid is required, one needs a very large total mass transfer in the unit.

The situation just discussed almost rules out the feasibility of a chemical gas treating process in the slow reaction regime, say, under conditions where no rate enhancement is established.

It should be noticed that the calculation of t_{cs} for a packed tower is more involved than the one worked out in the example at the end of this chapter, since simultaneous variations of the gas-phase composition along the tower axis need to be taken into account. A complete solution is available in the literature for the case of a first-order reaction, with t_r constant along the tower axis. This solution is not very important industrially, since in any practical case t_r would change along the tower axis; however, the solution represents a minimum value of the number of transfer units required and even this minimum is very large.

Of course, industrial processes taking place in the slow reaction regime do exist; many aerobic fermentation processes fall into this category. However, these are not gas treating processes. In gas treating, one may select the appropriate chemical solvent; one of the specifications required is that the rates of reaction should be large enough to result in significant mass transfer rate enhancements.

4.6. SLURRY REACTORS

A slurry reactor is one where a gas-phase component (e.g., A) is absorbed in the liquid phase where it reacts with the suspended solid. In gas treating, slurry reactors are commonly used for flue gas desulfurization, with the solid phase being a cheap alkali, the sulfite and sulfate of which can be easily disposed.

In the analysis of slurry reactors, two basic subcases should be considered. Let B_1 be the solid-phase reactant; this component may be either moderately soluble in the liquid phase, or almost insoluble.

In the first case, the concentration b_{10} of component B_1 in the liquid phase is non-negligible and the reaction between A and B_1 will take place in the liquid itself. Therefore, the absorption of A may proceed in any of the regimes discussed in Chapter 3 according to the rate of the liquid-phase reaction. The suspended solids' only role is to keep the concentration of dissolved B_1 in the liquid phase from being depleted by the reaction itself.

It is, however, important to realize a fundamental difference between a homogeneous liquid chemical solvent and a slurry. In a homogeneous liquid, the total capacity is related to the total concentration of B_1 in the lean solvent fed to the absorber, thus a high capacity can only be achieved with large values of b_{10}. In contrast with this, the value of b_{10} in a slurry reactor is limited (it cannot exceed the solubility of B), and the high capacity is obtained because the suspended solid will keep dissolving into the liquid as long as B_1 is consumed by the reaction.

As discussed in Section 3.5, the largest possible enhancement factor I_∞ increases with increasing b_{10} and, therefore, in the case of a slurry reactor, the value of I_∞ may be only marginally larger than unity (see Eq. 3.5.33; if the solubility of B_1 is less than the physical solubility of A, I_∞ is less than 2). In this case, even if the reaction is fast, the rate enhancement is only minor and to all practical purposes absorption of A may be regarded as taking place in the slow reaction regime.

Now turn attention to the case where B_1 is almost insoluble so that the reaction needs to take place essentially at the solid-liquid interface between physically dissolved A and solid B_1. The question arises whether the reaction at the liquid-solid interface may or may not influence the mass transfer rate of A from the gas-liquid interface to the bulk of the liquid.

In order to discuss this problem, it is useful to refer back to Figure 3.3.1. Any chemical reaction taking place in the liquid phase can influence the mass transfer rate only if it takes place appreciably in the gas-liquid interface region, say, at $x < \delta$. Values of δ of practical interest can be calculated from Eqs. 3.2.29 and 3.2.30; taking for D a typical value of 10^{-5} cm^2/sec, one obtains

$$2 \times 10^{-4} < \delta < 4 \times 10^{-4} \text{ cm} \tag{4.6.1}$$

A reaction taking place on the liquid-solid interface could, therefore, result in appreciable rate enhancement only if a significant number of solid particles are present at distances from the gas-liquid interface less than 10^{-4} cm. This, in turn, would require particle diameters no more than 10^{-5} cm, an unrealistically low value. It therefore appears that, whenever the reaction takes place at the liquid-solid interface, no significant rate enhancement will be observed for the gas-liquid mass transfer process: the latter will essentially proceed in the slow reaction regime.

In conclusion, one may say that mass transfer from the gas-liquid interface to the bulk of the liquid in slurry reactors may be regarded as taking place in the slow reaction regime unless the solubility of the solid is appreciably larger than the physical solubility of the gas. Of course, the value of a_0 in a slurry reactor is determined by the kinetics of both solid dissolution and chemical reaction; these will be discussed in some detail in Chapter 12.

Example 4.1

One liter of an aqueous solution of 1.5 moles/liter of K_2CO_3, 0.5 moles/liter of $KHCO_3$ is in equilibrium with a gas phase consisting of CO_2 and water vapor at a temperature of 20°C. At time zero, the gas pressure is brought to 1 atm by addition of CO_2, and kept at that level thereafter. The liquid is continuously and vigorously stirred; the value of k_L^0 is 0.03 cm/sec, and the gas–liquid interface area is 10^3 cm^2. We wish to calculate the time of chemical saturation.

The value of $a_0(0)$ is calculated from the equilibrium of the reaction

$$CO_2 + CO_3^{2-} + H_2O \rightleftharpoons 2HCO_3^- \tag{A4.1}$$

the equilibrium constant of which is, at 20°C, 1.0×10^4. Hence,

$$a_0(0) = \frac{1.67 \times 10^{-5} \text{ gmol}}{\text{liter}} \tag{A4.2}$$

Neglecting the water vapor pressure, the value of a_i is given by

$$a_i = \frac{p_1}{H} = \frac{1}{40.5} = \frac{2.55 \times 10^{-2} \text{ gmol}}{\text{liter}} \tag{A4.3}$$

The rate of reaction for this system can be expressed as

$$r = \mathrm{k}\frac{b_1}{b_2}a \tag{A4.4}$$

where $b_1 = [CO_3^{2-}]$ and $b_2 = [HCO_3^-]$. The constant k is 4.0 sec^{-1} at 20°C. Therefore, the reaction time t_r at the start is

$$t_r = \frac{0.5}{1.5 \times 4.0} = 0.133 \text{ sec} \tag{A4.5}$$

and will be larger than that at any later time. The diffusion time t_D is calculated from Eq. 4.3.4 and the known value of $D(1.8 \times 10^{-5}$ cm^2/sec) as

$$t_D = \frac{1.8 \times 10^{-5}}{9 \times 10^{-4}} = 0.02 \text{ sec} \tag{A4.6}$$

Hence condition 4.1 is fulfilled; the process proceeds in the slow reaction regime. The value of t_{ps} is calculated from Eq. 4.1.10 as

$$t_{ps} = \frac{10^3}{0.3 \times 10^3} = 33 \text{ sec} \tag{A4.7}$$

Hence, at least initially, the process takes place in the diffusional subregime. The induction time equals the reaction time, say,

$$t_{\text{ind}} = 0.133 \text{ sec} \tag{A4.8}$$

and, therefore, the induction phenomenon can be neglected altogether.

At time zero, the value of a^* equals a_0:

$$a^* = \frac{1.67 \times 10^{-5}\ \text{gmol}}{\text{liter}} \tag{A4.9}$$

and can be neglected with respect to a_i. Thus Eq. 4.1.11 takes the following form at low times:

$$t_{cs} = t_{ps} \frac{\alpha_0(\infty) - \alpha_0(0)}{a_i} \tag{A4.10}$$

The quantity α is, for the case at hand,

$$\alpha = a + b_1 + b_2 \tag{A4.11}$$

and, therefore,

$$\alpha_0(0) = 1.5 + 0.5 + 1.67 \times 10^{-5} = \frac{2.00\ \text{gmol}}{\text{liter}} \tag{A4.12}$$

At the final equilibrium, the concentrations will be $a = a_i$, $b_1 = 0.044$ gmol/liter, $b_2 = 3.412$ gmol/liter, and, therefore, $\alpha_0(\infty) = 3.48$ gmol/liter. Therefore,

$$t_{cs} = 33 \frac{3.48 - 2.00}{2.55 \times 10^{-2}} = 1914\ \text{sec} \tag{A4.13}$$

The actual time required for chemical saturation will, in fact, be larger since when $b_2 = 3.334$ gmol/liter, and $b_1 = 0.083$ gmol/liter, $t_r = t_{ps}$. When that level of saturation has been reached, final approach to equilibrium will take place in the much slower kinetic subregime.

Notice that for the conditions considered in this example, the relevant time scales t_D, $t_r(0)$, t_{ps}, and t_{cs} are spaced very widely apart. Unfortunately, situations may often arise where at least two time scales are of comparable magnitude and the problem becomes more complicated.

5 Fast Reaction Regime

Chemistry is unique among engineering sciences, because chemical laws are generally non-linear even in their simplest possible formulation.

G. ASTARITA

Although the effectiveness of rate enhancement in the fast reaction regime is appreciably less than unity, in many gas treating processes, mass transfer takes place in this regime. In particular, CO_2 mass transfer very often takes place in the fast reaction regime because the reactions that CO_2 undergoes in an alkaline aqueous solution are not extremely rapid; this is related to the fact that CO_2 is a Lewis acid which needs to hydrate before reacting with an alkali.

It is a peculiarity of the fast reaction regime of mass transfer that the rate of mass transfer is independent of the rate of mixing of the liquid phase. This peculiarity is the cause for some problems that are not yet completely solved and which will be discussed in Section 5.5.

5.1. THE UPPER BOUND

We recall from Chapter 3 that chemical absorption and desorption proceed in the fast reaction regime if the conditions expressed by Eqs. 3.4.1 and 3.4.3 are both satisfied. These conditions are reported again here:

$$\Phi = \frac{t_D}{t_r} \gg 1 \tag{5.1.1}$$

$$|b_{ji} - b_{j0}| \ll b_{j0} \tag{5.1.2}$$

We also recall the definition of reaction time t_r given in Eq. 3.4.8:

$$t_r = \frac{(a_i - a_0)^2}{2\int_{a_0}^{a_i} r\,da} \tag{5.1.3}$$

where $r = r(a, b_1, \ldots, b_N)$ is the rate of the main reaction taking place in the liquid, taken as positive in absorption ($a_i > a_0$) and negative in desorption ($a_i < a_0$).

Condition 5.1.1 represents, in order of increasing rate of reaction, a lower bound. In order to ascertain whether condition 5.1.1 is fulfilled, the value of t_r needs to be calculated at the bulk-liquid composition, i.e.,

$$r = r(a, b_{10}, \ldots, b_{N0}) \tag{5.1.4}$$

and, therefore, the fulfillment of condition 5.1.1 is easily ascertained if enough physico-chemical information is available.

The fulfillment of condition 5.1.2 (which represents, in order of increasing rate of reaction, an upper bound) is less easily ascertained. The argument that needs to be developed is given below. Since only an order-of-magnitude estimate is required, the argument is developed on the basis of the simple film theory model and all the diffusivities are assumed to be equal.

On the basis of these assumptions, the diffusion equations for the volatile component A and for the non-volatile components B_j take the following form:

$$D \frac{d^2 a}{dx^2} = r \tag{5.1.5}$$

$$D \frac{d^2 bj}{dx^2} = \nu_j r \tag{5.1.6}$$

where ν_j are the stoichiometric coefficients of the main reaction taking place during absorption or desorption. (This main reaction should *not* be confused with the (arbitrarily chosen) reaction involving A in the thermodynamic model. As an example, if we refer to the CO_2–mono-ethanolamine system discussed in Section 2.4 at $y < 0.5$ the main reaction is the amine → carbamate reaction, not reaction 2.4.25.)

Eliminating r between Eqs. 5.1.5 and 5.1.6 and integrating once, one obtains

$$\frac{db_j}{dx} = \nu_j \left[\frac{da}{dx} - \left(\frac{da}{dx} \right)_{x=0} \right] \tag{5.1.7}$$

where the non-volatility condition $(db_j/dx)_{x=0} = 0$ has been used. Notice that

$$-D \left[\frac{da}{dx} \right]_{x=0} = N = Ik_L^0 (a_i - a_0) \tag{5.1.8}$$

Substitution of Eq. 5.1.8 into 5.1.7 yields

$$\frac{db_j}{dx} = \nu_j \left[\frac{da}{dx} + \frac{Ik_L^0}{D} (a_i - a_0) \right] \tag{5.1.9}$$

Integration of Eq. 5.1.9 from $x = 0$ to $x = \delta$ yields

$$b_{j0} - b_{ji} = \nu_j \left[(a_0 - a_i) + \frac{I k_L^0 \delta}{D} (a_i - a_0) \right] \tag{5.1.10}$$

Since by definition $k_L^0 \delta / D = 1$, one has finally

$$b_{j0} - b_{ji} = \nu_j (a_i - a_0)(I - 1) \tag{5.1.11}$$

Equation 5.1.11 should be discussed in some detail. It shows that, near the interface, reactants are always depleted and products are accumulated. In absorption, $a_i - a_0 > 0$ and reactants ($\nu_j > 0$) are depleted; and vice versa. (There is a subtlety involved for those components for which $\nu_j = 0$; this will be discussed later. Here it is enough to say that Eq. 5.1.11 should be used only when $\nu_j \neq 0$.) Also, one may notice that, since $|\nu_j|$ is of order unity $|b_{j0} - b_{ji}|$ will, in general, be *larger* than $|a_i - a_0|$ even in the fast reaction regime; this, however, does not imply that condition 5.1.2 is violated.

Indeed, substitution of Eq. 5.1.11 into 5.1.2 yields the following form for the upper bound condition:

$$(I - 1)|a_i - a_0| \ll \frac{b_{j0}}{|\nu_j|} \tag{5.1.12}$$

which needs to be satisfied for *all* the b_js characterized by $\nu_j \neq 0$. In practice, condition 5.1.12 needs to be satisfied for that component B_j that is characterized by the *lowest* value of $b_{j0}/|\nu_j|$; this component may be a reactant or a product.

If the "irreversible" limits of the instantaneous reactions equations as given by Eqs. 3.5.36 and 3.5.38 are considered, one realizes that condition 5.1.12 is very often (but not necessarily always) equivalent to the following:

$$I \ll I_\infty \tag{5.1.13}$$

or, finally, considering Eqs. 3.4.12 and 3.6.1,

$$\sqrt{\Phi} = \sqrt{\frac{t_D}{t_r}} \ll I_\infty \tag{5.1.14}$$

In the form of Eq. 5.1.14, the upper bound for the fast reaction regime is as easy to calculate as the lower bound.

5.2. PROPERTIES OF FAST REACTION REGIME

We recall that, in the fast reaction regime (i.e., when both conditions 5.1.1 and 5.1.2 are satisfied) the enhancement factor I is given by Eq. 3.4.12, reported below:

$$I = \sqrt{\Phi} = \sqrt{\frac{t_D}{t_r}} \tag{5.2.1}$$

The reaction time t_r is given by Eq. 5.1.3, with r calculated from Eq. 5.1.4.

First of all, we prove the following result: *the bulk-liquid concentration a_0 can be assumed to coincide with the equilibrium value a_0^*.* Indeed, the quasi-steady-state condition takes the following form:

$$Ik_L^0(a_i - a_0)A = V\mathrm{k}(a_0 - a_0^*) = \frac{V(a_0 - a_0^*)}{t_r} \tag{5.2.2}$$

Equation 5.2.2 is written for the case of first-order kinetics in the bulk; however, the result to be obtained is easily generalized for arbitrary kinetics.

Substitution of Eqs. 5.2.1, 4.4.10, and 4.1.10 into Eq. 5.2.2 yields

$$\frac{a_0 - a_0^*}{a_i - a_0} = \frac{1}{\Psi\sqrt{\Phi}} \tag{5.2.3}$$

Since $\Psi \gg 1$ always and $\Phi \gg 1$ in the fast reaction regime, $a_0 - a_0^* \ll a_i - a_0$.

Physically, the reason for this result is easy to understand. In the diffusional subregime, chemical equilibrium prevails in the bulk of the liquid. As the reaction rate increases (say, t_r decreases), the rate of absorption increases proportionally to only the square root of the reaction rate (see Eq. 5.2.1) whereas the kinetics in the bulk, of course, increase linearly. Therefore, if a reaction is fast enough to cause enhancement of the mass transfer rates, it is *a fortiori* fast enough to maintain chemical equilibrium in the bulk of the liquid.

Having established that, in the fast reaction regime, chemical equilibrium prevails in the bulk of the liquid and the concentrations b_j are constant throughout the liquid phase (i.e., condition 5.1.2 is fulfilled), it is convenient to define a *displacement from equilibrium* concentration $\hat{a}$ as

$$\hat{a} = a - a_0 \tag{5.2.4}$$

which will be non-negative everywhere in absorption, and non-positive in desorption. The reaction rate as given by Eq. 5.1.4 has values that depend unequivocally on $\hat{a}$; this implies that a new kinetic function $\hat{r}(\cdot)$ can be defined as follows:

$$\hat{r}(\hat{a}) = r(a, b_{10}, \ldots, b_{N0}) \tag{5.2.5}$$

The sign of the value of the function $\hat{r}(\cdot)$ is, of course, equal to the sign of the argument.

With these definitions, the value of t_r in Eq. 6.1.3 can be calculated as

$$t_r = \frac{\hat{a}_i^2}{2\int_0^{a_i} \hat{r}(\hat{a})\,d\hat{a}} \tag{5.2.6}$$

Equation 5.2.6 shows that, at a given value of $|\hat{a}_i|$, *the value of t_r is independent* *of the sign of* $\hat{a}_i$ *only if* $\hat{r}(\cdot)$ *is an odd function, say, if*

$$\hat{r}(\hat{a}) = -\hat{r}(-\hat{a}) \tag{5.2.7}$$

This result is very important since $|\hat{a}_i|$ represents the driving force in both absorption and desorption; the difference between the two cases is described by the *sign* of $\hat{a}_i$, which is positive in absorption and negative in desorption. Equation 5.2.1, of course, holds in both absorption and desorption; however, at a given value of the driving force $|\hat{a}_i|$, it will produce the same value of I only if condition 5.2.7 is satisfied. *At a given value of the driving force, the enhancement factor is equal in absorption and desorption if and only if the reaction rate is an odd function of the displacement from equilibrium.*

It is interesting to ask oneself the conditions under which Eq. 5.2.7 is satisfied. The simplest case is, of course, the case where the rate of reaction is first-order with respect to the concentration of the volatile component A; say,

$$\hat{r} = \mathrm{k}_0 \hat{a} \tag{5.2.8}$$

where k_0 is a pseudo-first-order kinetic constant the value of which depends on the bulk-liquid chemical composition (say, on the values of the b_{j0}s). The first-order case is, in fact, a very realistic one; since the concentration of A is often the lowest of the concentrations of interest, it is very common that the slow step of the kinetic scheme involves a reaction where one molecule of A reacts with some non-volatile component.

However, the first-order kinetics is, in practice, the only realistic case where condition 5.2.7 is satisfied. Consider, for example, a case where the forward reaction is second order, such as

$$2A \rightleftharpoons A_2 \text{ (slow)} \tag{5.2.9}$$

$$A_2 + B \rightleftharpoons C \text{ (fast)} \tag{5.2.10}$$

A standard kinetic calculation based on the hypothesis that reaction 5.2.10 is always at equilibrium yields

$$r = \mathrm{k}_2 (a^2 - a_0^2) \tag{5.2.11}$$

where k_2 is the forward kinetic constant for reaction 5.2.9. Equation 5.2.10 shows that the reaction rate is not an odd function of the displacement from equilibrium $a - a_0$. Of course, it is mathematically easy to construct a non-linear function $\hat{r}(\cdot)$ that satisfies condition 5.2.7, yet such a function is hardly likely to represent a realistic chemical mechanism. In conclusion, one may say that from a pragmatic viewpoint, condition 5.2.7 is satisfied only in the (likely) case where $\hat{r}$ is given by Eq. 5.2.8. In this case, the value of I at a given driving force is the same in absorption and desorption.

When Eq. 5.2.8 is satisfied, the reaction time t_r equals the inverse of the pseudo-first-order kinetic constant k_0; Eq. 5.2.1 therefore reduces to

$$I = \sqrt{k_0 t_D} \tag{5.2.12}$$

or, equivalently,

$$k_L = \sqrt{D k_0} \tag{5.2.13}$$

5.3. MACROSCOPIC BEHAVIOR

In the fast reaction regime, the mass transfer coefficient k_L is given by Eq. 5.2.13. This implies that the rate of mass transfer is independent of t_D, say, of the intensity of mixing of the liquid phase. Furthermore, the rate of mass transfer is strongly dependent on temperature, with an apparent activation energy which is approximately one-half the activation energy of the chemical reaction involved. Finally, the mass transfer rates are independent of the liquid volume. The dependency on the driving force $|\hat{a}_i|$ depends on the form of the kinetic equation; in the common case where Eq. 5.2.8 applies, the rates of mass transfer are proportional to the driving force, say, from Eq. 5.2.13,

$$N = \sqrt{D k_0}\, \hat{a}_i \tag{5.3.1}$$

The fact that the rates of mass transfer are independent of the value of t_D (as long as condition 5.1.1 is satisfied) forms the basis for a method of measurement of interfacial areas in gas-liquid contacting equipment. In fact, suppose that mass transfer rates for a system where fast reaction regime conditions prevail have been measured in a laboratory unit of known interfacial area. From such data, values of k_L can be calculated. If now overall mass transfer rates are measured for the same system in a unit of unknown interfacial area, the latter can be calculated, since the value of k_L will be the same as in the laboratory unit. This technique can be used only if fast reaction regime conditions are known to prevail in both units; in fact, only in this case will the value of k_L be the same in the two units.

When a process takes place in the fast reaction regime, significant enhancement of the mass transfer rates occurs. However, consideration of condition 5.1.12 shows that the enhancement factor in the fast reaction regime is always significantly less than the ratio of the chemical and physical capacities of the liquid. In fact, the smallest value of $b_{j0}/|\nu_j|$ is, at most, equal to the chemical capacity, whereas $|a_i - a_0|$ is, of course, the physical capacity. This implies that the time required to reach chemical saturation (or the height of packing required), though significantly less than it would be should the process take place in the slow reaction regime, is still significantly more than t_{ps}. Industrial processes where fast reaction regime conditions prevail are common; however, such processes still offer room for additional improvement through larger enhancement of the mass transfer rates. These can be obtained by rate promotion additives; some examples will be discussed in later chapters.

5.4. KINETIC AND THERMODYNAMIC INFORMATION

Suppose a process is known to take place in the fast reaction regime; furthermore, suppose that the vapor–liquid equilibrium behavior of the system considered is known. This means that, given the composition of the bulk of the liquid (say, the values of m and y), the driving force $\hat{a}_i$ can be expressed as

$$\hat{a}_i = \frac{p_i - p^*(y_0)}{\mathrm{H}} \tag{5.4.1}$$

where p_i is the interface partial pressure of A. Consideration of Eqs. 5.3.2 and 5.4.1 shows that, in order to be able to evaluate the mass transfer rate, the ratio $\sqrt{\mathrm{k}_0}/\mathrm{H}$ needs to be known.

Several comments are in order with regard to this conclusion. First of all, the equilibrium constant of the main reaction does not influence the rate of mass transfer; this implies that the value of Henry's constant needs to be known, and not simply its ratio to the equilibrium constant. This justifies the interest in the ability to determine experimentally the value of H, with techniques that will be discussed in Chapter 14.

Another important consideration is that the equivalent kinetic constant k_0 will, in general, depend on the composition of the bulk liquid, say on the value of y_0. If a thermodynamic model is available, the concentrations b_{j0} in the bulk of the liquid can be calculated from the value of y_0, since, as discussed in Section 5.2, chemical equilibrium prevails in the bulk of the liquid. Once this is done, the value k_0 can be related to some true kinetic constant provided the kinetic mechanism is understood. The following example clarifies the issue.

In the absorption of CO_2 into potassium carbonate solutions, a simple thermodynamic model is available; see Section 2.4. The kinetic mechanism is well understood; the rate-determining step is the direct reaction of CO_2 with the hydroxyl ion and, therefore, the rate is given by

$$r = \mathrm{k}_2 a[\mathrm{OH}^-] \tag{5.4.2}$$

where k_2 is a true kinetic constant. The value of $[\mathrm{OH}^-]$ is calculated from the thermodynamic model (see Eqs. 2.4.6, 2.4.7, and 2.4.16):

$$[\mathrm{OH}^-] = \mathrm{K}_1 \frac{1 - y}{2y} \tag{5.4.3}$$

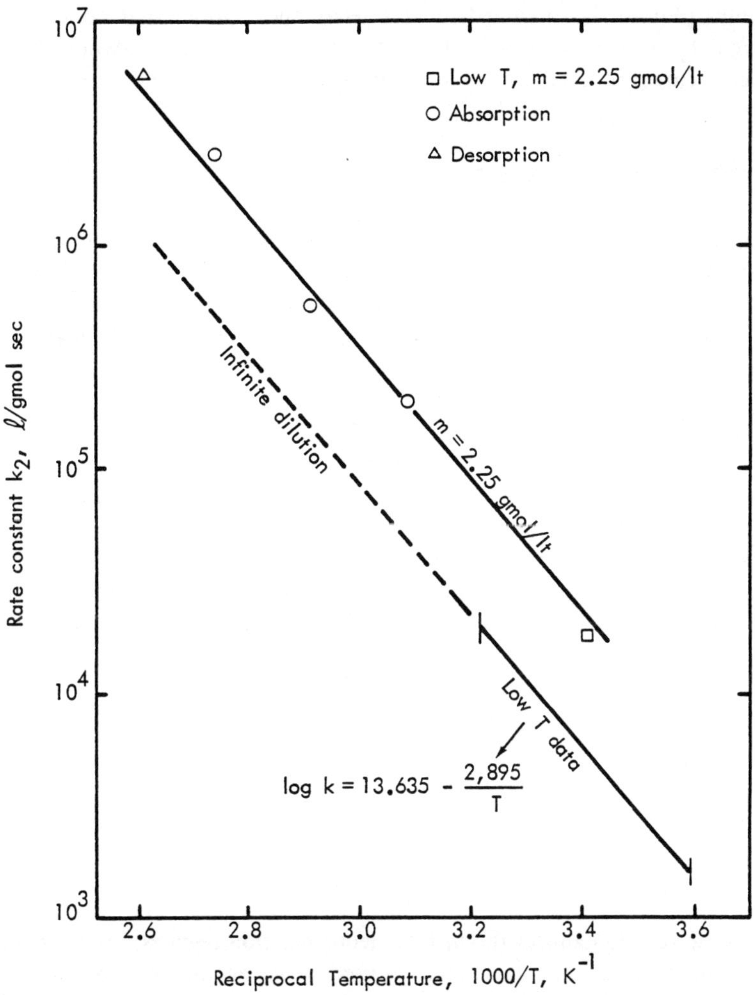

FIGURE 5.4.1. Kinetic constant for CO_2 reaction with hydroxyl ion.

and, therefore,

$$k_0 = k_2 K_1 \frac{1 - y}{2y} \tag{5.4.4}$$

Consideration of Eqs. 5.3.2 and 5.4.4 shows that, if the process takes place in the fast reaction regime, the mass transfer rate is proportional to $\sqrt{(1 - y)/y}$; this has been well confirmed experimentally at temperatures exceeding 323K. Since for this system the value of H is independently known, values of the kinetic constant k_2 can be extracted from the measured mass transfer rates. Figure 5.4.1 is an

Arrhenius plot for k_2 obtained by this technique. Notice that values as large as 10^7 l/gmol, sec can be measured; no direct kinetic technique would be reliable for measurement of kinetic constants as large as this.

5.5. SOME OPEN PROBLEMS

As was discussed in Section 5.3, the fact that the mass transfer rate in the fast reaction regime is independent of the rate of mixing of the liquid phase is the basis of a method of measurement of interfacial areas. This is helpful, inasmuch as interfacial areas are not easy to measure, but it also creates some problems which are discussed below.

First of all, the fact that interfacial areas can be measured implies that they also *need* to be measured. This can be realized by considering the following argument which is developed for the case of a packed tower but can obviously be generalized. Suppose that the value of $k_L^0\tilde{a}$ has been measured for a given packing, with $\tilde{a}$ the interfacial area per unit volume; it is well known that standard mass transfer measurements yield the value of the product $k_L^0\tilde{a}$ but not k_L^0 and $\tilde{a}$ separately. As long as only physical mass transfer takes place, this is sufficient information, since the rate of mass transfer per unit volume of packing N_v can be expressed as

$$N_v = (k_L^0\tilde{a})\,(a_i - a_0) \tag{5.5.1}$$

and only N_v needs to be known for design.

Now suppose that mass transfer in a chemical solvent is considered. The value of N_v will be given by

$$N_v = I k_L^0\tilde{a}(a_i - a_0) \tag{5.5.2}$$

In the slow reaction and in the instantaneous reaction regimes, I is independent of Φ and can, therefore, be calculated without knowing the value of k_L^0; all that is needed is an estimate of k_L^0 so that the order of magnitude of Φ can be evaluated. However, in the fast reaction regime, I depends on Φ and, therefore, the value of k_L^0 needs to be known in order to use Eq. 5.5.2 for design. The only way of calculating k_L^0 from known values of $k_L^0\tilde{a}$ is to measure $\tilde{a}$ independently.

The second problem that arises is the following one. In a packed tower, some quasi-stagnant pools of liquid may exist. These will contribute very little to mass transfer in physical solvents since the liquid in the pools will be almost at equilibrium with the gas phase. In the case of chemical solvents, such pools will be equally ineffective in the instantaneous reaction regime for the same reason. However, in the fast reaction regime, the time required for equilibration is much longer due to the low effectiveness of rate enhancement and quasi-stagnant pools may contribute to the mass transfer. This point raises the disturbing possibility that the *effective* interfacial area may be different in different mass transfer regimes. It also implies that interfacial areas measured with the chemical method discussed in Section 5.3

may be larger than the effective interfacial areas except for mass transfer in the fast reaction regime.

Finally, it should be noted that packings are often rated in terms of their performance in the absorption of CO_2 into aqueous hydroxide solutions. Such absorption is very likely to occur in the fast reaction regime and, therefore, packings are really being rated on the basis of their effective interfacial areas in the fast reaction regime. There is clearly ground for doubting that this is the best yardstick for comparison of performance of different packings.

6 Instantaneous Reaction Regime

. . . fortunately, the methods are brilliant in their simplicity and can be taught clearly without much need for advanced mathematics.

M. FEINBERG

Mass transfer in the instantaneous reaction regime is of particular interest in the removal of H_2S from gas streams with chemical solvents because H_2S undergoes virtually instantaneous proton transfer reactions in alkaline liquids. The effectiveness of rate enhancement in the instantaneous reaction regime is unity if all diffusivities are equal, as was discussed in Section 3.7; however, the effect of unequal diffusivities may be significant, as will be discussed in the next sections.

6.1. THE IRREVERSIBLE LIMIT IN ABSORPTION

In Chapter 3 the instantaneous reaction regime equations have been established for the special case where the diffusivities of all species are equal. We now wish to analyze the case where the diffusivities are *not* equal, and we begin by considering the simplest possible case, i.e., absorption accompanied by an instantaneous irreversible reaction. In Chapter 3, it was shown that the penetration theory model predicts the dependency of mass transfer rates on diffusivity better than the film theory model; it is, therefore, appropriate to approach the problem considered here on the basis of the penetration theory model.

Let the main reaction be

$$A + \nu_1 B_1 \longrightarrow \text{Products} \tag{6.1.1}$$

where the right-hand side need not be specified since the reaction is irreversible. As was discussed in Section 3.5, the concentrations a and b_1 cannot be both non-zero at any point within the liquid. Therefore, the liquid itself may be divided into two regions (see Figure 6.1.1). The first region, located near the interface ($0 \leqslant x \leqslant \lambda$),

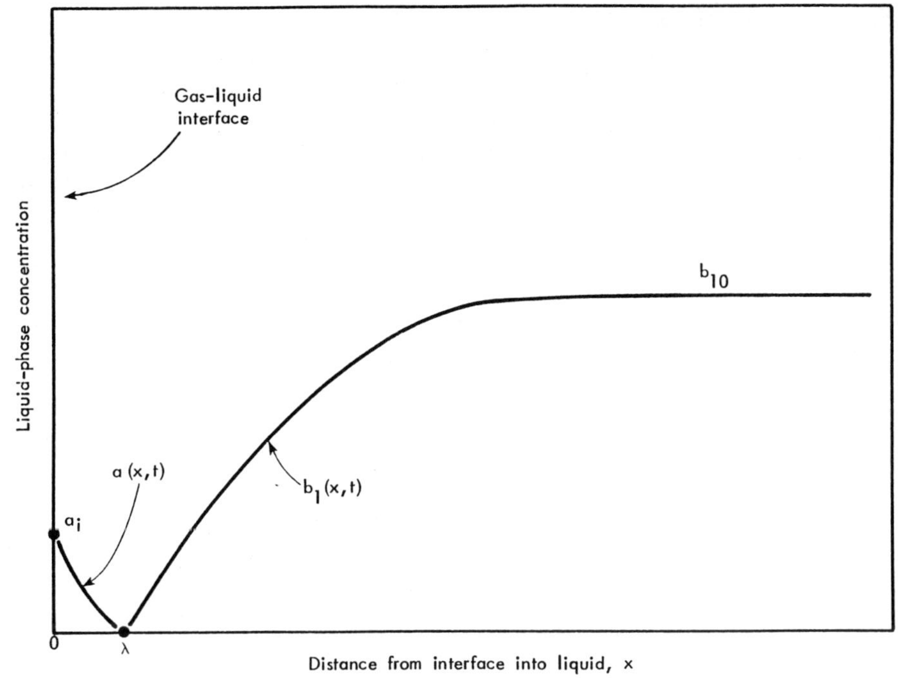

FIGURE 6.1.1. Irreversible limit for absorption in instantaneous reaction regime.

is the region where $a \neq 0$, $b_1 = 0$; the second region ($x \geqslant \lambda$) is the one where $a = 0$, $b_1 \neq 0$. Since the penetration theory model is considered, the depth at which the reaction plane is located is a function of time:

$$\lambda = \lambda(t) \tag{6.1.2}$$

The differential equations and boundary conditions to be considered are as follows. For the region $0 \leqslant x \leqslant \lambda(t)$,

$$D \frac{\partial^2 a}{\partial x^2} = \frac{\partial a}{\partial t} \tag{6.1.3}$$

$$x = 0, \qquad a = a_i \tag{6.1.4}$$

$$x = \lambda(t), \qquad a = 0 \tag{6.1.5}$$

which are sufficient since

$$t = 0, \qquad \lambda(t) = 0 \tag{6.1.6}$$

For the region $\lambda(t) \leqslant x$,

$$D_1 \frac{\partial^2 b_1}{\partial x^2} = \frac{\partial b_1}{\partial t} \tag{6.1.7}$$

$$t = 0, \qquad b_1 = b_{10} \tag{6.1.8}$$

$$x = \lambda(t), \qquad b_1 = 0 \tag{6.1.9}$$

and, of course, b_1 is bounded for all values of x. The two differential equations are coupled through a stoichiometric condition to be satisfied at the reaction plane:

$$x = \lambda(t), \qquad -\nu_1 D \frac{\partial a}{\partial x} = D_1 \frac{\partial b_1}{\partial x} \tag{6.1.10}$$

Finally, the motion of the reaction plane is determined by a mass balance on it (see Figure 6.1.2).

$$\frac{d\lambda}{dt} = -\left(\frac{\partial a/\partial t}{\partial a/\partial x}\right)_{x=\lambda(t)} \tag{6.1.11}$$

The system of Eqs. 6.1.3-6.1.11 constitutes a classical Stefan problem and can be solved by standard techniques. The solution is as follows:

$$\lambda = 2\sqrt{Kt} \tag{6.1.12}$$

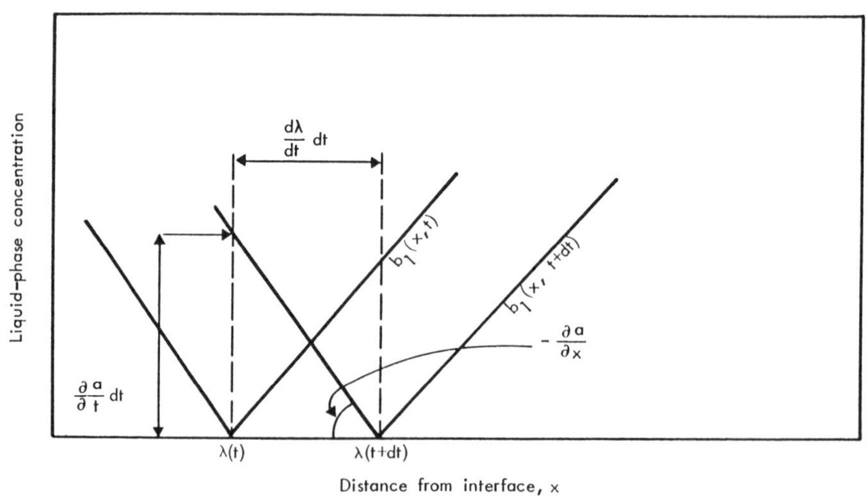

FIGURE 6.1.2. Motion of reaction plane for absorption with instantaneous reaction.

$$a = a_i \left[1 - \frac{\operatorname{erf}(x/2\sqrt{Dt})}{\operatorname{erf}\sqrt{K/D}} \right] \tag{6.1.13}$$

$$b_1 = b_{10} \left[1 - \frac{\operatorname{erfc}(x/2\sqrt{D_1 t})}{\operatorname{erfc}\sqrt{K/D_1}} \right] \tag{6.1.14}$$

$$I_\infty = 1/\operatorname{erf}\sqrt{K/D} \tag{6.1.15}$$

with the constant K being given by

$$\sqrt{\frac{D_1}{D}} \frac{b_{10}}{\nu_1 a_i} \exp\left(-\frac{K}{D_1} + \frac{K}{D} \right) \operatorname{erf} \sqrt{\frac{K}{D}} + \operatorname{erf} \sqrt{\frac{K}{D_1}} = 1 \tag{6.1.16}$$

These equations can be considerably simplified if attention is restricted to the (likely) case where

$$\frac{b_{10}}{\nu_1 a_i} \gg 1 \tag{6.1.17}$$

When Eq. 6.1.17 is satisfied, Eq. 6.1.16 implies that $K \ll D$; therefore, the transcendental functions appearing in it can be series-expanded to obtain the following result:

$$I_\infty = \sqrt{\frac{D}{D_1}} + \sqrt{\frac{D_1}{D}} \frac{b_{10}}{\nu_1 a_i} \tag{6.1.18}$$

Notice that the second term on the right-hand side of Eq. 6.1.18 is the leading one whenever condition 6.1.17 is satisfied. Also, notice that if $D = D_1$, Eq. 6.1.18 takes the form of Eq. 3.5.33 (when $D = D_1$, Eq. 6.1.18 is exact; i.e., it holds also if 6.1.17 does not hold).

The solution of the same problem for the film theory model is obtained trivially; with reference to Figure 3.5.1, when $D \neq D_1$, one obtains

$$I_\infty = 1 + \frac{D_1}{D} \frac{b_{10}}{\nu_1 a_i} \tag{6.1.19}$$

A very important feature of the equations discussed above is that, even in the very general form of Equation 6.1.15, no matter what the diffusion time is, the *ratio* of chemical to physical absorption rates is the same. This is due to the fact that the enhancement of absorption rate due to the reaction is simply to substitute, as a rate-controlling process, the diffusion of the second reactant from the bulk of the liquid towards the interface, to the diffusion of the absorbing component from the interface towards the bulk of the liquid; the former process takes place under a higher overall driving force.

Let us write both Equations 6.1.18 and 6.1.19 in the form

$$I = \left(\frac{D_1}{D_2}\right)^{1-m} + \left(\frac{D_2}{D_1}\right)^{m} \frac{b_{10}}{\nu_1 A_i} \tag{6.1.20}$$

the exponent m being 1 according to the film theory (which derives a linear dependency of mass transfer rates on molecular diffusivity) and 0.5 according to penetration theory, whatever the age distribution function (penetration theory predicts that mass transfer rates are proportional to the square root of diffusivity). If one considers the problem of mass transfer from a solid surface to a laminar boundary layer, accompanied by an instantaneous reaction, one obtains an equation reminiscent of Equation 6.1.20; the exponent m is, in this case $\frac{2}{3}$, as could be expected by considering that mass transfer rates in boundary layers are proportional to the $\frac{2}{3}$ power of molecular diffusivity.

6.2. FILM THEORY EQUATIONS

The procedure discussed in Section 3.5 for the case where all the liquid-phase diffusivities are equal can easily be generalized to the surface renewal diffusion equation. In particular, if ξ_j is defined as in Eq. 3.6.21, the differential equation for ξ_j becomes

$$-D' \frac{\partial^2 \xi_j}{\partial x^2} = -\frac{\partial \xi_j}{\partial t} + r \tag{6.2.1}$$

subject to the following boundary conditions:

$$t = 0, \qquad \xi_j = 0 \tag{6.2.2}$$

$$x = 0, \qquad -D' \frac{\partial \xi_j}{\partial x} = r_0 \tag{6.2.3}$$

$$x \to \infty, \qquad \xi_j \text{ is bounded} \tag{6.2.4}$$

This implies that ξ_j has the same value for all js (i.e., Eq. 3.5.26 applies).

However, when the diffusivities *are not* all equal, only the film theory equations can be solved in general. The procedure to do so is illustrated in the following.

It is evident that, in the instantaneous reaction regime, the concentration distributions are independent of the kinetics of the reaction, i.e., of the specific form of the function $r(\cdot)$. The distributions could be calculated in two distinct ways. The first one is to set up a specific form for the kinetic function $r(\cdot)$, and seek the asymptotic solution when the kinetic constants appearing in $r(\cdot)$ tend to infinity. The asymptote should, in fact, be independent of the form of $r(\cdot)$ and, therefore, this first procedure could be followed by setting up an arbitrary kinetic function $r(\cdot)$. A second procedure, which is the one developed in the following, is to eliminate the $r(\cdot)$ function from the problem and only impose the requirement that chemical equilibrium prevails everywhere in the liquid.

Eliminating r between Eqs. 5.1.5 and 5.1.6 gives

$$\nu_j D \frac{d^2 a}{dx^2} = D_j \frac{d^2 b_j}{dx^2} \tag{6.2.5}$$

In Chapter 5, Eq. 5.1.7 was obtained by setting $db_j/dx = 0$ at the interface. However, in the instantaneous reaction regime, the reaction rate at the interface r_0 need not be zero. The boundary condition at the interface is that the flux of the nonvolatile components is related stoichiometrically to the rate of reaction at the interface itself r_0:

$$x = 0, \qquad D_j \frac{db_j}{cx} = \nu_j r_0 \tag{6.2.6}$$

The flux N of component A from the bulk of the gas to the interface (which is, of course, negative in desorption) is the sum of the amount of A consumed by the reaction at the interface and the amount that diffuses from the interface towards the bulk of the liquid phase:

$$N = I_\infty \mathrm{k}_L^0 (a_i - a_0) = -D \left.\frac{da}{dx}\right|_{x=0} + r_0 \tag{6.2.7}$$

The reaction rate r_0 can be eliminated between Eqs. 6.2.6 and 6.2.7 to yield

$$x = 0, \qquad D_j \frac{db_j}{dx} = \nu_j \left[I_\infty \mathrm{k}_L^0 (a_i - a_0) + D \frac{da}{dx} \right] \tag{6.2.8}$$

Integration of Eq. 6.2.5 between $x = 0$ and a generic position $x \leqslant \delta$ yields

$$D_j \left(\frac{db_j}{dx} - \left.\frac{db_j}{dx}\right|_{x=0} \right) = \nu_j D \left(\frac{da}{dx} - \left.\frac{da}{dx}\right|_{x=0} \right) \tag{6.2.9}$$

which upon substitution of Eq. 6.2.8 reduces to

$$D_j \frac{db_j}{dx} = \nu_j D \frac{da}{dx} + \nu_j I_\infty \mathrm{k}_L^0 (a_i - a_0) \tag{6.2.10}$$

Finally, integration of Eq. 6.2.10 between $x = 0$ and $x = \delta$ yields

$$b_{ji} = b_{j0} - \nu_j \frac{D}{D_j} \bar{\xi} \tag{6.2.11}$$

where

$$\bar{\xi} = (I_\infty - 1)(a_i - a_0) \tag{6.2.12}$$

Equation 6.2.11 bears a superficial resemblance to the equation obtained in the case where all diffusivities are equal, say, Eq. 3.5.21 under the condition that Eq. 3.5.26 applies:

$$b_{ji} = b_{j0} - \nu_j \xi \tag{6.2.13}$$

However, although ξ in Eq. 6.2.13 can be interpreted as an extent of reaction, $\bar{\xi}$ in Eq. 6.2.11 cannot. The composition b_{ji} given by Eq. 6.2.11 could not be reached by a chemical reaction starting from the bulk-liquid composition; in particular, the interface value of the molarity will, in general, be different from the bulk-liquid value.

The last conclusion implies that the liquid-phase composition at the interface cannot be extracted from vapor-liquid equilibrium data, inasmuch as the latter are available as p^* versus y curves at a given molarity and cannot be used unless the molarity is known.

However, since chemical equilibrium must prevail everywhere in the liquid phase when the reaction is instantaneous, the condition of chemical equilibrium at the interface becomes, in view of Eq. 6.2.11,

$$\mathrm{K} = \frac{\prod_j \left(b_{j0} - \nu_j \dfrac{D}{D_j} \bar{\xi}\right)^{-\nu_j}}{a_i} \tag{6.2.14}$$

Equation 6.2.14 can, in general, be solved for $\bar{\xi}$, and the value of I_∞ can then be calculated from Eq. 6.2.12.

It is convenient to slightly transform the equations obtained above. Since chemical equilibrium prevails also in the bulk of the liquid, one may write

$$\mathrm{K} = \frac{\prod_j (b_{j0})^{-\nu_j}}{a_0} \tag{6.2.15}$$

Taking the ratio of Eqs. 6.2.14 and 6.2.15 yields

$$\psi = \frac{a_i}{a_0} = \prod_j \left(1 - \bar{\xi} \frac{\nu_j D}{D_j b_{j0}}\right)^{-\nu_j} \tag{6.2.16}$$

The advantage of Eq. 6.2.16 lies in the fact that $\bar{\xi}$ (and hence also I_∞) can be calculated if the bulk-liquid composition (b_{j0}) and the ratio $\psi = a_i/a_0$ are known. The quantity ψ is equal to the ratio of the interface partial pressure of A and the vapor pressure corresponding to equilibrium with the bulk-liquid composition and can, therefore, be calculated if vapor-liquid equilibrium curves are available, without need to know the value of Henry's Law constant.

The asymptotic behavior of the system 6.2.12–6.2.16 is easily established. The following cases are of interest:

1. $\psi \gg 1$ Large driving force in absorption.
2. $\psi \ll 1$ Large driving force in desorption.
3. $\psi = 1 + \delta$, with $|\delta| \ll 1$ Small driving force.

Of the components with $\nu_j > 0$, let $j = 1$ be assigned to the component having the lowest value of $D_j b_{jo}/\nu_j$; this is the diffusionally limiting reactant in absorption, DLRA. Conversely, of the components with $\nu_j < 0$, let $j = 2$ be assigned to the one that has the lowest value of $D_j b_{jo}/|\nu_j|$; this is the diffusionally limiting reactant in desorption, DLRD. Notice that the DLRA and the DLRD need not coincide with the SLRA and the SLRD defined in Section 3.5, though in most cases they will, in fact, coincide.

The solution to Eqs. 6.2.12 and 6.2.16 for cases 1 and 2 is:

$$\psi \gg 1, \qquad I_\infty = 1 + \frac{b_{10}}{\nu_1 a_i} \frac{D_1}{D} \tag{6.2.17}$$

$$\psi \ll 1, \qquad I_\infty = 1 - \frac{b_{20}}{\nu_2 a_0} \frac{D_2}{D} \tag{6.2.18}$$

Notice that Eq. 6.2.17 coincides with Eq. 6.1.19 which was obtained by considering a simple sketch of the concentration profiles. There is no shortcut procedure for obtaining Eq. 6.2.18.

We now turn attention to the low driving force asymptote. Although very little, if any, experimental evidence is reported in the literature on this case, it is, in fact, an important one in industrial practice since a large fraction of the total height of absorption and desorption towers is used for mass transfer under conditions of small driving forces (say, near pinches). The solution to Eq. 6.2.16 for case 3 is

$$\bar{\xi} = \frac{\delta}{\sum_j (\nu_j^2/b_{jo})(D/D_j)} \tag{6.2.19}$$

and, therefore, considering that $\delta = (a_i - a_0)/a_0$;

$$I_\infty = 1 + \frac{1}{Da_0 \sum (\nu_j^2/b_{jo}D_j)} \tag{6.2.20}$$

6.3. PENETRATION THEORY EQUATIONS

The diffusion equations for the penetration theory model, after elimination of the reaction rate r, take the form

$$D_j \frac{\partial^2 b_j}{\partial x^2} - \frac{\partial b_j}{\partial t} = \nu_j \left[D \frac{\partial^2 a}{\partial x^2} - \frac{\partial a}{\partial t} \right] \tag{6.3.1}$$

Let there be N components B_j. Consequently, there are N equations of the form 6.3.1, and $N + 1$ concentrations (b_j and a) to be determined. The last equation emerges from the requirement that chemical equilibrium prevails everywhere in the liquid phase, say,

$$a = \frac{\prod_j (b_j)^{-\nu_j}}{\mathrm{K}} \tag{6.3.2}$$

The set of Eqs. 6.3.1, with 6.3.2 substituted on the right-hand side, is a set of N non-linear differential equations for the functions $b_j\,(x, t)$. The boundary conditions are as follows.

At the interface, the value a_i is assigned, although conditions of the form of 6.2.6 (with $\partial b_j/\partial x$ instead of db_j/dx) apply. Hence

$$x = 0, \qquad \frac{D_1}{\nu_1} \frac{\partial b_1}{\partial x} = \frac{D_2}{\nu_2} \frac{\partial b_2}{\partial x} = \cdots \tag{6.3.3}$$

$$\prod_j (b_j)^{-\nu_j} = \mathrm{K} a_i \tag{6.3.4}$$

The other boundary conditions are

$$\left.\begin{matrix} t = 0 \\ x \to \infty \end{matrix}\right\} \quad b_j = b_{jo} \tag{6.3.5}$$

It is convenient to introduce the new variable

$$\eta = \frac{x}{2\sqrt{Dt}} \tag{6.3.6}$$

which reduces Eqs. 6.3.1 to ordinary differential equations

$$\frac{D_j}{D} \frac{d^2 b_j}{d\eta^2} + 2\eta \frac{db_j}{d\eta} = \nu_j \left[\frac{d^2 a}{d\eta^2} + 2\eta \frac{da}{d\eta} \right] \tag{6.3.7}$$

subject to boundary conditions

$$\eta = 0, \qquad \frac{D_1}{\nu_1}\frac{db_1}{d\eta} = \frac{D_2}{\nu_2}\frac{db_2}{d\eta} = \cdots \tag{6.3.8}$$

$$\prod_j (b_j)^{-\nu_j} = a_i \mathrm{K} \tag{6.3.9}$$

and to Eq. 6.3.5 at $\eta \to \infty$. The numerical solution of this problem is reasonably easy for any specific case. A general analytic solution, however, cannot be obtained.

However, the low driving force asymptote can be calculated provided the ratios a_0/b_{j0} are small numbers, as is likely to be the case. The algebra is somewhat cumbersome and we simply give the result without proof:

$$I_\infty = \frac{1 + \sum (\nu_j^2 D a_0 / D_j b_{j0})}{\sum (\nu_j^2 \sqrt{D/D_j}\, a_0 / b_{j0})} \tag{6.3.10}$$

Notice that the Σ term appearing in the numerator is much smaller than unity. Therefore, the leading term in Eq. 6.3.10 coincides with the leading term in Eq. 6.2.20 if $\sqrt{D/D_j}$ is substituted for D/D_j. The same is true if Eqs. 6.1.18 and 6.1.19 are compared with each other. This conclusion suggests that, as an engineering rule of thumb, the film theory equations with $\sqrt{D/D_j}$ substituted for D/D_j can be used in general.

6.4. MACROSCOPIC BEHAVIOR

In the instantaneous reaction regime, the mass transfer flux N is a highly non-linear function of the driving force $a_i - a_0$. In fact, N is given by Eq. 6.2.7. When ψ is close to unity, Eq. 6.3.10 (or 6.2.20) shows that I_∞ is *independent* of $(a_i - a_0)$ and, *therefore,* N is proportional to $(a_i - a_0)$. This situation will be referred to in the following as the small driving force, SDF, subregime; the value of I_∞ for the SDF subregime will be indicated as I_∞^S. Conversely, when ψ is either $\gg 1$ or $\ll 1$, (the large driving force subregime, LDF), Eqs. 6.2.17 and 6.2.18 show that the leading term in the expression for I_∞ is inversely proportional to the driving force $a_i - a_0$; consequently, the mass transfer flux N is almost independent of the driving force $a_i - a_0$. A qualitative plot of N/k_L^0 versus $a_i - a_0$ is given in Figure 6.4.1; the positive quadrant (which extends to $a_i - a_0 \to \infty$) corresponds to absorption, whereas the negative quadrant (which extends to $a_i - a_0 = -a_0$) corresponds to desorption. Values indicated on the figure correspond to the simple case where all diffusivities are equal.

By definition, I_∞ is given by

$$I_\infty = \frac{N/\mathrm{k}_L^0}{a_i - a_0} \tag{6.4.1}$$

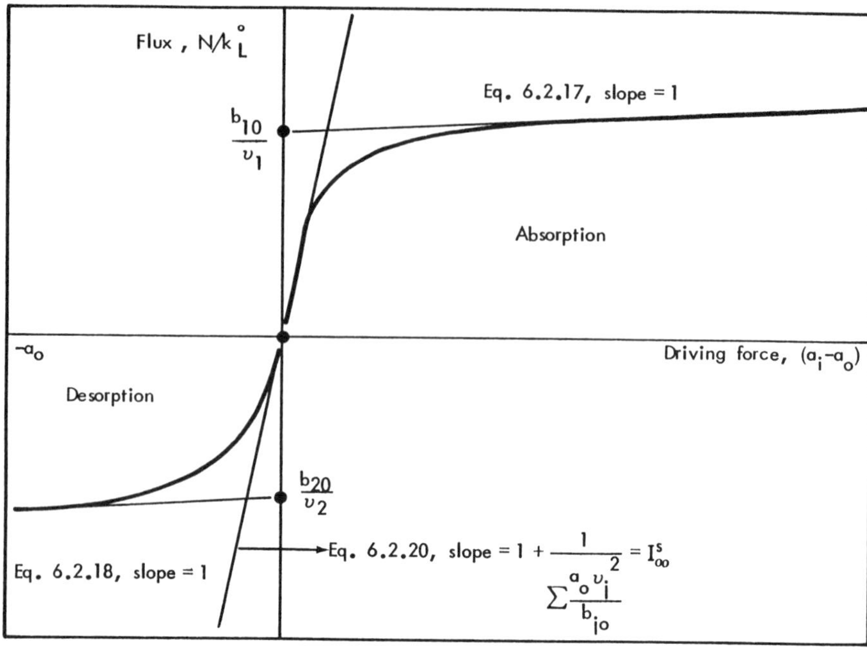

FIGURE 6.4.1. Dependence of mass transfer flux on driving force in the instantaneous reaction regime.

i.e., I_∞ is the slope of the straight line through the origin and any given point of the curve in Figure 6.4.1. Quite obviously, this slope is largest at $\psi = 0$; i.e., a plot of I_∞ versus $a_i - a_0$ has the qualitative shape reported in Figure 6.4.2.

The macroscopic behavior observed in the LDF and in the SDF subregimes is very different. In the SDF case, the value of $I_\infty^s - 1$ is inversely proportional to the bulk-liquid concentration a_0. The value of a_0, at a given molarity m and saturation y_0, is inversely proportional to the thermodynamic equilibrium constant of the main reaction K (see Eq. 6.2.15). Hence $I_\infty^s - 1$ is directly proportional to K and is, therefore, a strong function of temperature. Furthermore, a_0 is, in general, a very strong (increasing) function of the saturation y (see, e.g., Eq. 2.4.10) and, therefore, I_∞^s is a strongly decreasing function of y.

In contrast with this, in the LDF subregime, the value of I_∞ is independent of K (see Eqs. 6.2.17 and 6.2.18). Consequently, I_∞ is independent of temperature. The value of b_{10} is a (comparatively weak) decreasing function of y and so is I_∞ as given by Eq. 6.2.17. Conversely, b_{20} is an increasing function of y and so is I_∞ as given by Eq. 6.2.18.

With reference to industrial gas treating processes, it is instructive to consider the top part of the absorption tower and the bottom part of the desorption tower. In both, the *total* driving force $|p - p^*(y_0)|$ will be comparatively small: in the desorption tower, because $p^*(y_0)$ is small (lean solution); in the absorption tower, because p is small (lean gas). Therefore, these are the regions where the SDF sub-

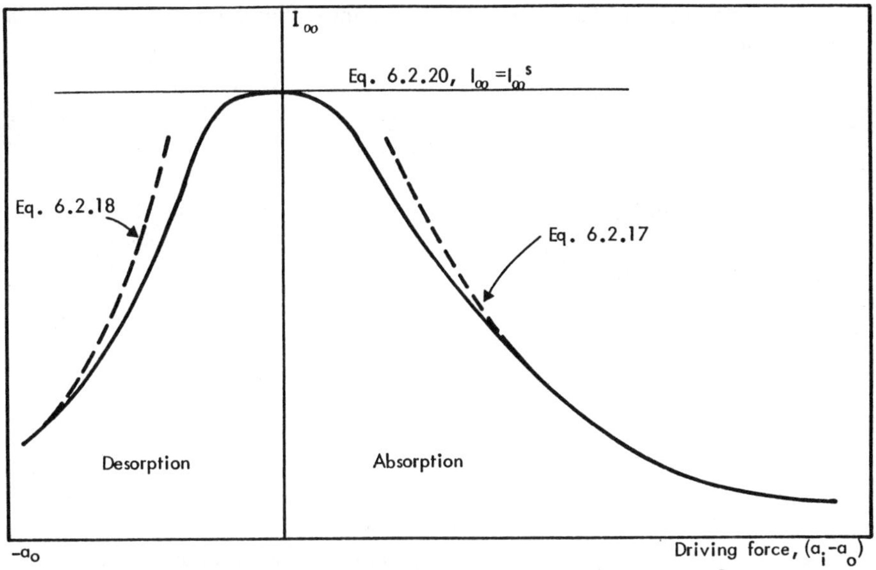

FIGURE 6.4.2. Mass transfer enhancement factor in the instantaneous regime as a function of driving force.

regime is more likely to occur. Furthermore, in these regions, y_0 is very small and, therefore, I_∞^s is very large indeed. A specific example is illuminating.

Consider the case of CO_2 absorption in carbonate solutions; we use the same notation as in Section 2.4. Again, assuming that all diffusivities are equal, Eq. 6.2.20 becomes for this case

$$I_\infty^s = 1 + \frac{1}{a_0[(1/b_{10}) + (4/b_{20})]} \tag{6.4.2}$$

Substitution of Eqs. 2.4.6, 2.4.7, and 2.4.10 and rearrangement yields

$$I_\infty^s = 1 + \frac{K(1 - y_0)^2}{4y_0(2 - y_0)} \tag{6.4.3}$$

At the lean-solution end of both the absorption and the desorption tower, the bulk-liquid saturation y_0 is likely to be of the order of 0.1 or even less; hence I_∞^s is of the same order of magnitude as the thermodynamic constant K (i.e., of the order of 10^4). This conclusion can, in fact, be generalized.

If Eq. 6.2.15 is substituted into Eq. 6.2.20 (with again all diffusivities taken as equal), one obtains

$$I_\infty^s = 1 + \frac{K}{\prod_j (b_{j0})^{-\nu_j} \cdot \sum_j (\nu_j^2/b_{j0})} \tag{6.4.4}$$

It is convenient to introduce the following definition of the dimensionless bulk-liquid concentrations $\hat{b}_{jo}$:

$$\hat{b}_{jo} = \frac{b_{jo}}{m_0} \tag{6.4.5}$$

where m_0 is the bulk-liquid molarity. The quantities $\hat{b}_{jo}$ are all of order unity, since so are the coefficients λ_j appearing in Eq. 2.3.4. When Eq. 6.4.5 is substituted into Eq. 6.4.4, one obtains

$$I_\infty^s = 1 + \frac{K(m_0)^{1+\Sigma\nu_j}}{\prod_j (\hat{b}_{jo})^{-\nu_j} \sum_j (\nu_j^2/\hat{b}_{jo})} \tag{6.4.6}$$

The leading term on the right-hand side of Eq. 6.4.6 is, of course, the second one. The denominator is always of the order of unity, whereas the numerator is a very large number in any case of practical interest.

It is also interesting to notice that the molarity m_0 appears in Eq. 6.4.6 raised to the power $1 + \Sigma\nu_j$. If the main reaction is equimolar (i.e., the number of moles of products equals the number of moles of reactants), the exponent $1 + \Sigma\nu_j$ equals 0 (the CO_2/carbonate case is in this category). It follows that, if the main reaction is equimolar, I_∞^s is independent of the molarity of the solution. In contrast with this, I_∞ as given by Eqs. 6.2.17 and 6.2.18 (i.e., in the LDF subregime) is invariably proportional to the bulk-liquid molarity.

The fact that I_∞^s is always a very large number has important implications. The first stems from the fact that, in order that the instantaneous reaction regime equations apply, the following condition must of course be met:

$$\sqrt{\Phi} = \sqrt{\frac{t_D}{t_r}} \gg I_\infty \tag{6.4.7}$$

When the condition $\psi = 1 + \delta$, $|\delta| \ll 1$ holds (i.e., when $I_\infty = I_\infty^s$), condition 6.4.7 is very difficult to satisfy. With t_D values of the order of 10^{-2} sec (see Eq. 3.2.25) and I_∞^s values of the order of 10^4, the reaction time t_r needs to be of the order of 10^{-11} sec for Eq. 6.4.7 to apply. Such extremely short reaction times are possible only for proton-transfer reactions such as are encountered for H_2S absorption (or desorption) in alkaline solutions. If the reactions involved are not extremely fast, condition 6.4.7 will be violated; at low values of ψ (say, at the lean end of absorption and desorption units) fast reaction conditions are most likely to occur.

Even if the reactions are proton-transfer ones so that Eq. 6.4.7 is satisfied, the very large values of I_∞^s may imply the following consequences. Combination of Eqs. 3.1.2, 3.1.4, and 6.2.7 yields

$$N = \frac{(p/H) - a_0}{(1/Hk_G) + (1/I_\infty k_L^0)} \tag{6.4.8}$$

Even if condition 3.1.8 is very abundantly satisfied (i.e., even if, in the absence of chemical reactions, mass transfer were entirely controlled by the liquid-side resistance), with very large values of I_∞ it may well be that

$$\frac{I_\infty \mathrm{k}_L^0}{\mathrm{k}_G \mathrm{H}} \gg 1 \tag{6.4.9}$$

so that, in fact, gas-side mass transfer control is likely to occur. This point should be discussed in some detail.

The parameter ψ, defined by Eq. 6.2.16, refers only to the liquid-side driving force. A parameter $\overline{\psi}$, referring to the *overall* driving force, may be defined as

$$\overline{\psi} = \frac{p}{p^*(y_0)} = \frac{p/\mathrm{H}}{a_0} \tag{6.4.10}$$

At the lean end of both absorption and desorption towers, the value of $\overline{\psi}$ is likely to be of order unity, say, to lay in the range $\frac{1}{2} < \overline{\psi} < 2$.

Now suppose the following (likely) condition is satisfied:

$$\frac{I_\infty^s \mathrm{k}_L^0}{\mathrm{k}_G \mathrm{H}} \gg 1 \tag{6.4.11}$$

Again combination of Eqs. 3.1.2, 3.1.4, and 6.2.7 yields

$$\psi - 1 = \frac{\overline{\psi} - 1}{1 + I_\infty \mathrm{k}_L^0/\mathrm{Hk}_G} \tag{6.4.12}$$

where I_∞ is a function of ψ. However, if Eq. 6.4.11 is satisfied, and if $\overline{\psi}$ is not very much different from unity, Eq. 6.4.12 will be satisfied with a value of ψ very close to unity, and hence $I_\infty = I_\infty^s$. The conclusion is therefore reached that, whenever Eq. 6.4.11 is satisfied in the lean-solution end of both absorption and desorption towers, mass transfer will be controlled by the gas-side resistance.

The conclusion reached above can be approached from a different viewpoint. Consider the overall driving force $p/\mathrm{H} - a_0$. If the gas-phase resistance to mass transfer is negligible, the flux N is related to $(p/\mathrm{H} - a_0)$ by a curve having the shape given in Figure 6.4.1 (see also Figure 6.4.3, curve marked LSC, liquid side control). If the liquid phase resistance to mass transfer is negligible, the flux N is given by

$$N = \mathrm{Hk}_G \left(\frac{p}{\mathrm{H}} - a_0\right) \tag{6.4.13}$$

Equation 6.4.13 is plotted in Figure 6.4.3 as the straight line through the origin marked GSC (gas side control). In an actual case where resistances on both sides are taken into account, the plot of N/k_L^0 versus $p/\mathrm{H} - a_0$ will be below both the

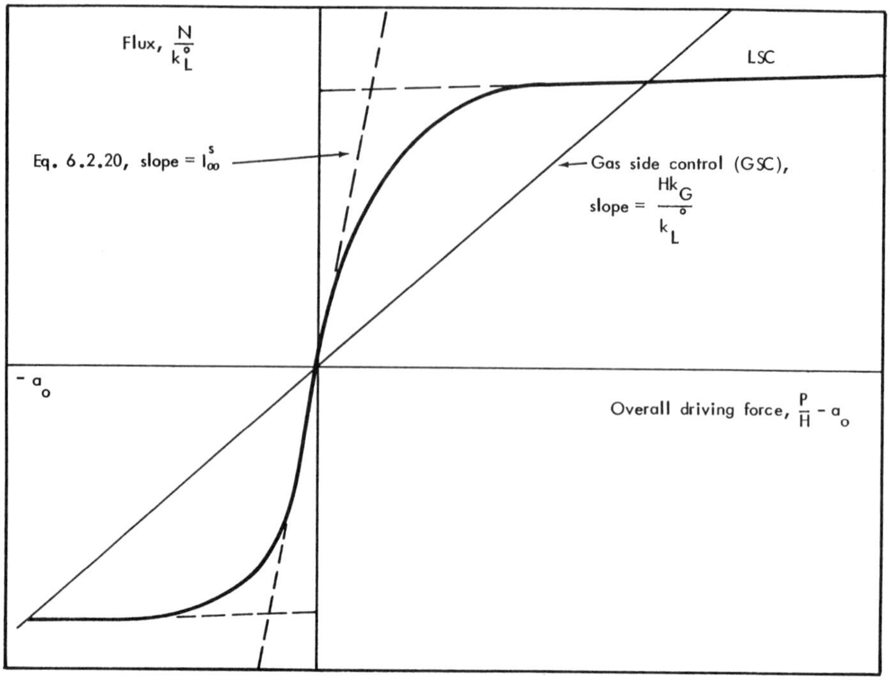

FIGURE 6.4.3. Mass transfer flux dependence on overall driving force for a partly gas-phase-controlled process.

GSC and LSC lines and asymptotically approaching both lines when the other gives a much higher value. If Eq. 6.4.11 is satisfied, the system will become gas-side controlled before the LSC curve has any significant deviation from the LDF asymptotes. If such is the case, a considerable simplification arises, which is discussed below for the case of absorption, though the desorption case can be analyzed along the same lines.

To simplify the analysis, we restrict attention to the case where all diffusivities are equal. The liquid-side transport equation becomes

$$N = \mathrm{k}_L^0 a_i \left(1 + \frac{b_{10}}{\nu_1 a_i}\right) = \mathrm{k}_L^0 \left(a_i + \frac{b_{10}}{\nu_1}\right) \tag{6.4.14}$$

The gas-side transport equation is

$$N = \mathrm{k}_G (p - \mathrm{H} a_i) \tag{6.4.15}$$

If N is eliminated between Eqs. 6.4.14 and 6.4.15, the value of a_i is calculated as

$$a_i = \frac{\mathrm{k}_G p - \mathrm{k}_L^0 b_{10}/\nu_1}{\mathrm{k}_L^0 + \mathrm{H}\mathrm{k}_G} \tag{6.4.16}$$

However, the value of a_i cannot be negative. Therefore, if $k_G p$ is larger than $k_L^0 b_{10}/\nu_1$, the flux is given by Eq. 6.4.15 with a_i calculated from Eq. 6.4.16; but if $k_G p$ is *smaller* than $k_L^0 b_{10}/\nu_1$, the flux is given simply by

$$N = k_G p \tag{6.4.17}$$

The latter situation has been referred to in the literature as the "surface reaction" regime. However, this name is somewhat misleading because it is based on a conception of the governing equation that disregards the fact that they only are an approximation and, in fact, one that is justified only if Eq. 6.4.11 is satisfied.

7 Transition Regions

Right and wrong are never separated by a sharp cut.

A. MANZONI

In the preceding three chapters, the asymptotic behavior of the $I(\Phi)$ curve has been discussed in detail. The three asymptotic regions are sketched in Figure 3.6.1. However, as was indicated in Section 3.6, the cusps appearing at the intersections of two asymptotes (at $\Phi = 1$ and at $\Phi = I_\infty^2$) are in fact smoothed out, and transition regions exist in the neighborhood of these points. The transition region near $\Phi = I_\infty^2$ is particularly important inasmuch as many gas treating processes take place under conditions that fall in this region.

7.1. THE SLOW-FAST TRANSITION

In this section we discuss the transition region around $\Phi = 1$. We assume that $I_\infty \gg 1$ and that, therefore, the transition curves around $\Phi = 1$ and around $\Phi = I_\infty^2$ do not significantly overlap. This assumption (which will be relaxed in a later section) allows us to deal with the slow-fast (SF) transition on the basis of the following simplifications.

In Section 3.3, as well as in Chapter 6, it was shown that in the fast reaction regime the concentrations of the non-volatile components are practically constant throughout the liquid phase; say, condition 3.4.1 is satisfied. This is, of course, also true at the lower Φ values corresponding to the SF transition which can, therefore, be handled by writing the kinetic equation in the form of Eq. 5.2.5:

$$\hat{r}(\hat{a}) = r(a, b_{10}, \ldots, b_{N0}) \tag{7.1.1}$$

with $\hat{a} = a - a_0$, and $\hat{r}$ having always the same sign as $\hat{a}$.

Notice that chemical equilibrium prevails in the bulk of the liquid in the SF transition unless it extends to such low values of Φ as to overlap with the kinetic-diffusional transition discussed in Chapter 3. As will be seen below, this is not the case and, therefore, $\hat{r}(0) = 0$.

We begin by considering the film theory model. The differential equation is still Eq. 3.4.4, but since the curvature of the concentration profile in the film need not be large (the boundary conditions are Eqs. 3.2.7 and 3.2.8 say), the mathematical problem can be cast in the following form:

$$D \frac{d^2 \hat{a}}{dx^2} = \hat{r}(\hat{a}) \tag{7.1.2}$$

$$x = 0, \qquad \hat{a} = \hat{a}_i \tag{7.1.3}$$

$$x = \delta, \qquad \hat{a} = 0 \tag{7.1.4}$$

A first step of integration yields

$$\frac{d\hat{a}}{dx} = \pm \sqrt{\frac{2}{D} \int_0^{\hat{a}} \hat{r}(\hat{a})\, d\hat{a} + \mathrm{M}} \tag{7.1.5}$$

where the – sign refers to absorption, and the + sign to desorption. The constant M equals the square of the concentration gradient at $x = \delta$, and is to be calculated from the following equation, which is obtained by integrating Eq. 7.1.5 between $x = 0$ and $x = \delta$.

$$\left| \int_0^{a_i} \frac{d\hat{a}}{\sqrt{(2/D) \int_0^{\hat{a}} \hat{r}(\hat{a})\, d\hat{a} + \mathrm{M}}} \right| = \delta \tag{7.1.6}$$

Once M has been calculated, the mass flux N is obtained from Eq. 7.1.5 as

$$N = -D \left. \frac{d\hat{a}}{dx} \right|_{x=0} = \pm \sqrt{2D \int_0^{\hat{a}_i} \hat{r}(\hat{a})\, d\hat{a} + \mathrm{M}D^2} \tag{7.1.7}$$

Taking into account Eq. 5.2.6, one obtains

$$N = \pm \sqrt{\frac{D}{t_r} \hat{a}_i^2 + \mathrm{M}D^2} \tag{7.1.8}$$

or, for the chemical mass transfer coefficient,

$$\mathrm{k}_L = \frac{N}{\hat{a}_i} = \sqrt{\frac{D}{t_r} + \frac{\mathrm{M}D^2}{\hat{a}_i^2}} \tag{7.1.9}$$

A substantial simplification arises in the case where the function $\hat{r}(\cdot)$ is linear, say,

$$\hat{r}(\hat{a}) = \mathrm{k}\hat{a} \tag{7.1.10}$$

Notice that the validity of Eq. 7.1.10 *does not* imply that the function $r(a, b_1, \ldots, b_N)$ is linear. Indeed, unless the slow step (in absorption) is a dimerization of component A before it undergoes any other reaction, Eq. 7.1.10 will apply. When Eq. 7.1.10 is substituted into Eq. 7.1.2, straightforward algebra yields the result

$$I = \frac{\sqrt{\Phi}}{\mathrm{tgh}\,\sqrt{\Phi}} \tag{7.1.11}$$

Eq. 7.1.11 obviously degenerates into Eq. 3.4.12 when $\Phi \gg 1$, and into Eq. 3.3.7 when $\Phi \ll 1$. A plot of Eq. 7.1.11 is given in Figure 7.1.1, lower curve.

Several observations are in order. First, I, as given by Eq. 7.1.11, is within 10% of the value of unity corresponding to slow reaction conditions at a Φ value of about 0.3. Since the value of Ψ, as discussed in Chapter 5, is at the very least 100, the value of Ω corresponding to $\Phi = 0.3$ is at least 30 (see Eq. 4.4.11). At such a large value of Ω, the conditions for the diffusional subregime to apply (i.e., for chemical equilibrium to prevail in the bulk of the liquid) are abundantly satisfied; the SF transition does not overlap with the kinetic-diffusional transition.

The second observation is as follows. Eqs. 7.1.6 and 7.1.9 are easily solved for a power-law kinetic equation such as in Eq. 4.4.1; it turns out that Eq. 7.1.11 is, to within a few percent, a satisfactory approximation also for this more general case. This suggests that the SF transition curve is little influenced by the specific functional form of $\hat{r}(\cdot)$.

The plot in Figure 7.1.1 shows that, in fact, the transition from the fast reaction

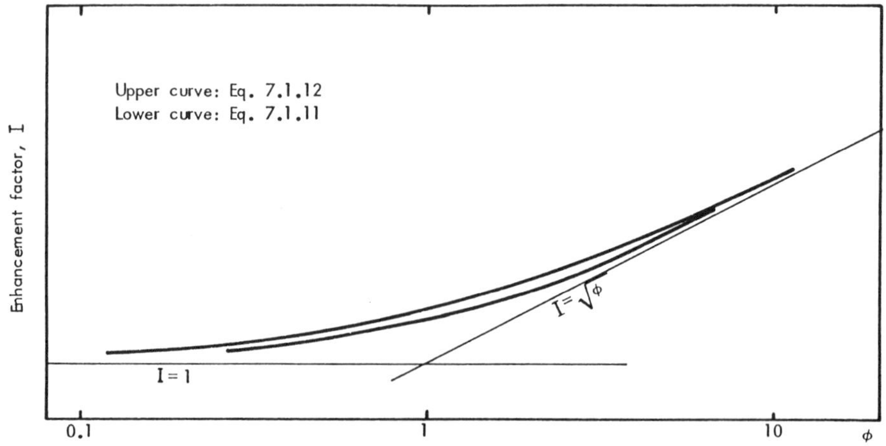

FIGURE 7.1.1. The slow–fast transition behavior.

to the slow reaction asymptotes is rather sharp, and that (based on the film theory model) even the very crude approximation of using Eq. 3.4.12 whenever $\Phi > 1$, and Eq. 3.3.7 whenever $\Phi < 1$, does not result in an error larger than 30%. This suggests the futility of an excess of mathematical sophistication when dealing with the SF transition.

It remains to be established whether use of the film theory model is a satisfactory approximation for the analysis of the SF transition. In order to discuss this point, attention may be limited to the case where Eq. 7.1.10 applies, for which analytic solutions to the surface renewal model equations are available. The algebra involved is considerable and we state here the result without proof:

$$I = \sqrt{1 + \Phi} \tag{7.1.12}$$

Equation 7.1.12 is plotted in Figure 7.1.1, upper curve. It is seen that the largest deviation from Eq. 7.1.11 is only 10%, which suggests that the SF transition curve is very little influenced by the details of the hydrodynamic conditions of the liquid phase. The same conclusion is reached from an analysis based on a very different hydrodynamic model.

The fact that the film theory model gives a satisfactory approximation to the $I(\Phi)$ curve in SF transition is related to the fact that the equations governing the $I(\Phi)$ curve are, in fact, independent of the value of the diffusivity. It will be seen that, as a general rule of thumb, when the diffusivities can be eliminated from the governing equations, the film theory model invariably produces a satisfactory approximation.

7.2. KINETICS

Before dealing with the fast-instantaneous transition, it is useful to analyze certain relevant features of the kinetic function $r(\cdot)$. The value of the $r(\cdot)$ function is the *net* rate of reaction and can, therefore, always be regarded as the difference between a forward rate $\vec{r}$ and a reverse rate $\overleftarrow{r}$, say,

$$r(\cdot) = \vec{r}(\cdot) - \overleftarrow{r}(\cdot) \tag{7.2.1}$$

where the arguments of $r(\cdot)$, $\vec{r}(\cdot)$, and $\overleftarrow{r}(\cdot)$ are the same (i.e., the set of concentrations values a, b_j).

A considerable simplification arises under the hypothesis that the dependency on a can be factored out of the forward rate function, say, if

$$\vec{r}(\cdot) = f(a)\, g(b_1, b_2, \ldots, b_N) \tag{7.2.2}$$

The hypothesis that Eq. 7.2.2 holds is a very mild one and most kinetic mechanisms for homogeneous reactions (such as are of interest in gas treating) will, in fact, lead to an expression for the forward rate of the type of Eq. 7.2.2.

Let asterisks indicate equilibrium concentrations. At equilibrium, the forward and reverse rates are equal; hence if Eq. 7.2.2 holds, one has

$$\vec{r}^* = \overleftarrow{r}^* = f(a^*)\, g(b_1^*, \ldots, b_N^*) \tag{7.2.3}$$

Since a^* is in equilibrium with the b_js, one has

$$a^* = \frac{\prod_j (b_j^*)^{-\nu_j}}{K} \tag{7.2.4}$$

We now make a second (equally mild) assumption, i.e., that the functional form for $\overleftarrow{r}(\cdot)$ is the same near equilibrium and far from equilibrium and thus we write

$$\overleftarrow{r} = f(\bar{a})\, g(b_1, \ldots, b_N) \tag{7.2.5}$$

where

$$\bar{a} = \frac{\prod_j (b_j)^{-\nu_j}}{K} \tag{7.2.6}$$

The quantity $\bar{a}$ is the concentration of A that would equilibrate the local concentrations b_j. Substitution of Eqs. 7.2.2 and 7.2.5 into 7.2.1 yields

$$r = [f(a) - f(\bar{a})]\, g(b_1, \ldots, b_N) \tag{7.2.7}$$

Several observations are in order. First, $f(a)$ is certainly a monotonical increasing function and, therefore, the term in square brackets on the right-hand side of 7.2.7 has the same sign as the driving force for the chemical reaction $a - \bar{a}$ (since the decomposition of r in Eq. 7.2.2 is not unique, we may always choose $g(\cdot)$ so that its values are positive). As obvious, this implies that $r > 0$ in absorption and $r < 0$ in desorption.

In the slow reaction regime, in the SF transition, and in the fast reaction regime, the b_js have everywhere almost the same value as in the bulk of the liquid phase and hence

$$g(\cdot) = g(b_{10}, \ldots, b_{N0}) = g_0 \tag{7.2.8}$$

where the (constant) g_0 may be regarded as an equivalent kinetic constant. However, when condition 3.4.1 is *not* satisfied (i.e., in the fast–instantaneous reaction regime), $g(\cdot)$ is not constant throughout the liquid phase. In this regard, it is important to realize that, among the arguments of the function $g(\cdot)$, concentrations b_j may appear of components that are characterized by a zero value of ν_j (i.e., of components that are not either reactants or products of the main reaction). This point deserves a detailed discussion.

Suppose that the slow step of the main reaction is

$$A + B_K \longrightarrow \text{products} \tag{7.2.9}$$

where B_K is a non-volatile component with $\nu_k = 0$. Correspondingly,

$$\vec{r} = k' a b_K \tag{7.2.10}$$

where k′ is the kinetic constant for reaction 7.2.9. The fact that $\nu_K = 0$ implies that once the slow step 7.2.9 has taken place, component B_K is replaced by a secondary fast reaction such as

$$\sum \nu_j' B_j - B_K = 0 \tag{7.2.11}$$

where now the ν_j's are all non-zero. Since reaction 7.2.11 is fast (as compared to 7.2.9), it will everywhere be at equilibrium, say,

$$b_K = K' \prod_j (b_j)^{-\nu_j'} \tag{7.2.12}$$

where K′ is the thermodynamic constant for reaction 7.2.11. (It is important to realize that the ν_j's are the stoichiometric coefficients of reaction 7.2.11, not to be confused with the ν_js). Substitution of Eq. 7.2.12 into 7.2.10 yeilds

$$\vec{r} = a k' K' \prod_j b_j^{-\nu_j'} \tag{7.2.13}$$

which has the form of Eq. 7.2.2, with

$$f(a) = a \tag{7.2.14}$$

$$g(\cdot) = k' K' \prod_j b_j^{-\nu_j'} \tag{7.2.15}$$

If this procedure is followed, the arguments of the $g(\cdot)$ function can always be made to all have non-zero ν_js. In the following, we will assume that the function $g(\cdot)$ has been so chosen.

In the example given above, the function $f(a)$ is *linear*, and the function $g(\cdot)$ has the form of Eq. 7.2.15; i.e., it is a *group*. Notice that the mathematical form of a group is special in that it contains only *one* dimensional parameter–in the case of Eq. 7.2.15, the parameter is k′K′. It is interesting to discuss the likelihood of $\vec{r}(\cdot)$ having this special form.

The forward rate $\vec{r}$ is determined by the rate of whatever is the slowest reaction of the sequence of which the overall reaction is formed. The slowest reaction is at most bimolecular, since the direct rate of trimolecular reactions is negligible (the probability for three molecules to come simultaneously in contact is practically zero). The possible cases are reviewed in the following.

1. The slow step is the combination of A with a component B_j characterized by a non-zero ν_j. In this case, both 7.2.14 and 7.2.15 apply, with $g(\cdot)$ having the particularly simple form of being proportional to a single concentration b_j.

2. The example discussed before, leading to Eq. 7.2.14 and 7.2.15, constitutes a case.

3. The slow step does not involve component A. In this case, A undergoes some fast reaction that is followed by the slow step. Therefore, the concentration of at least one of the reactants in the slow step is determined by the equilibrium of the fast step involving A. There are two possible subcases, namely:

   3a. In the fast step involving A, the stoichiometric coefficient of A is unity. In this case, 7.2.14 and 7.2.15 apply.

   3b. In the fast step involving A, the stoichiometric coefficient of A is not unity. In this case, $f(a)$ is not linear.

4. The slow step is the dimerization of A. In this case $f(a)$ is not linear.

As can be seen, Eq. 7.2.14 is likely to apply in a variety of cases. When $f(a)$ is linear, the reaction time t_r is independent of the driving force $a_i - a_0$.

A complication that may arise occasionally is that there are two possible parallel mechanisms for the slow step, and that the rates of these two mechanisms are comparable (if one of them is much slower, its contribution to the overall rate is negligible and it can, therefore, be neglected altogether). In this special case, if $f(a)$ is linear for *both* mechanisms, it is also linear for the overall rate; i.e., Eq. 7.2.14 applies. However, Eq. 7.2.15 *does not* apply since $g(\cdot)$ will take the form of the sum of two different groups and will contain two independent dimensional parameters. If $f(a)$ has a different mathematical form for the two mechanisms, it is impossible to express $\vec{r}(\cdot)$ in the form of Eq. 7.2.2. We do not know of any problem in gas treating where this complication arises, and in the following we consider only the case where 7.2.2 does apply.

7.3. THE FAST-INSTANTANEOUS TRANSITION

In the Fast-Instantaneous (FI) transition, the kinetics of the reaction need, of course, to be taken explicitly into account; therefore, a solution to the appropriate differential equations needs to be calculated for every specific kinetic function $r(\cdot)$. Furthermore, the b_js are *not* constant throughout the liquid phase and, therefore, the diffusion equations to be solved are those for all the b_js, rather than only the one for a. The mathematical problem is rather complex, and even for the simplest possible kinetic function, numerical methods are required.

However, an approximate procedure can be developed in general for the film theory model. The basic idea for the procedure was presented in 1948 for a very simple kinetic scheme; i.e., Case 1 discussed in Section 7.2 was analyzed. The procedure has been recently generalized to an arbitrary kinetic function. The basic approximation involved in this procedure is as follows.

Inasmuch as the differential equations need to be solved only in order to calculate the interface concentration gradient $(da/dx)_{x=0}$, the solution needs to be accurate only near the interface. More exactly, since an interface gradient needs to be calculated, the solution may be sought only to within $0(x)$, with terms of $0(x^2)$ being neglected.

As long as the rate of reaction is finite, the condition of non-volatility of the B_js implies that

$$\left.\frac{db_j}{dx}\right|_{x=0} = 0 \tag{7.3.1}$$

It follows that, near the interface,

$$b_j = b_{ji} + 0(x^2) \tag{7.3.2}$$

i.e., to within $0(x)$ the concentrations of the B_js may be taken as equal to their interface values. The differential equation for a becomes

$$D\frac{d^2a}{dx^2} = r(a, b_{1i}, b_{2i}, \ldots, b_{Ni}) \tag{7.3.3}$$

and is, therefore, uncoupled from the other ones, provided the values of the b_{ji}s can be calculated independently.

The procedure leading from Eq. 6.2.5 to Eq. 6.2.11 can be duplicated as shown below. Elimination of the reaction rate between the diffusion equations for A and B_j gives

$$\nu_j D\frac{d^2a}{dx^2} = D_j\frac{d^2b_j}{dx^2} \tag{7.3.4}$$

A first integration step, together with Eq. 7.3.1, yields

$$D_j\frac{db_j}{dx} = \nu_j\left[D\frac{da}{dx} + I\mathrm{k}_L^0(a_i - a_0)\right] \tag{7.3.5}$$

Notice that Eq. 7.3.5 has the same structure as Eq. 6.2.10, in spite of the fact that the latter was obtained on the basis of boundary condition 6.2.6 instead of 7.3.1. This fact removes the singularity connected with the boundary condition 6.2.6, i.e., with the fact that the rate of reaction on the interface r_0 is zero always except in the instantaneous reaction limit.

Integration of Eq. 7.3.5 between $x = 0$ and $x = \delta$ yields

$$b_{ji} = b_{jo} - \nu_j \frac{D}{D_j}(a_i - a_o)(I - 1) \tag{7.3.6}$$

Notice that Eq. 7.3.6 applies up to, *and including*, the instantaneous reaction limit (see Eqs. 6.2.11 and 6.2.12) which can be combined to yield

$$b_{ji}^* = b_{jo} - \nu_j \frac{D}{D_j}(a_i - a_o)(I_\infty - 1) \tag{7.3.7}$$

where b_{ji}^* is the value of b_{ji} relative to the instantaneous reaction limit. Combination of Eqs. 7.3.6 and 7.3.7 yields

$$b_{ji} = b_{jo} + \frac{I - 1}{I_\infty - 1}(b_{ji}^* - b_{jo}) \tag{7.3.8}$$

Equation 7.3.8 is very important and should be discussed in some detail. The most important point is that it does not contain the diffusivities D and D_j; i.e., once the values of I_∞ and b_{ji}^* are calculated with the procedures discussed in Chapter 7, b_{ji} as calculated from Eq. 7.3.8 is independent of the value of the diffusivities (or, in other words, the latter only influence the value of b_{ji}^* and I_∞). This conclusion is very important in view of the reliability of the film theory model whenever there is no problem with estimating the influence of the values of the diffusivities.

The second important point is that once the b_{ji} values given by Eq. 7.3.8 are substituted into Eq. 7.3.3, the latter can be integrated and the value of I can be calculated. The resulting equation for I will be highly non-linear and implicit, but algebraic.

We now restrict attention to the special but likely case where Eq. 7.2.14 applies. Equation 7.3.3 then takes the form

$$D \frac{d^2 a}{dx^2} = \mathrm{g_i}(a - \bar{a}_\mathrm{i}) \tag{7.3.9}$$

where the constants $\mathrm{g_i}$ and $\bar{a}_\mathrm{i}$ are given by

$$\mathrm{g_i} = g(b_{1i}, b_{2i}, \ldots, b_{\mathrm{Ni}}) \tag{7.3.10}$$

$$\bar{a}_\mathrm{i} = \frac{\prod_j (b_{ji})^{-\nu_j}}{\mathrm{K}} \tag{7.3.11}$$

The boundary conditions are

$$x = 0, \qquad a = a_{\mathrm{i}} \tag{7.3.12}$$

$$x = \delta, \qquad a = a_0 \tag{7.3.13}$$

The solution is straightforward and leads to

$$I = \frac{\Psi - \bar{a}_{\mathrm{i}}/a_0}{\Psi - 1} \frac{\gamma}{\operatorname{tgh} \gamma} - \frac{1 - \bar{a}_{\mathrm{i}}/a_0}{\Psi - 1} \frac{\gamma}{\sinh \gamma} \tag{7.3.14}$$

where

$$\gamma = \frac{\sqrt{D g_{\mathrm{i}}}}{k_L^0} = \sqrt{g_{\mathrm{i}} t_D} = \frac{\sqrt{\Phi g_{\mathrm{i}}}}{g_0} \tag{7.3.15}$$

$$\Psi = \frac{a_{\mathrm{i}}}{a_0} \tag{7.3.16}$$

Again, notice that the values of the diffusivities do not appear on the right-hand side of Eq. 7.3.14 except insofar as they determine the value of g_{i} through Eq. 7.3.8.

Equation 7.3.14 describes *both* the SF transition and the FI transition. Indeed, as long as condition 3.4.1 is satisfied, $g_{\mathrm{i}} \cong g_0$, $\bar{a}_{\mathrm{i}} = a_0$, and $\gamma = \sqrt{\Phi}$, so that Eq. 7.3.14 reduces to Eq. 7.1.11. Also, notice that when $\Phi \to \infty$, also $\gamma \to \infty$, and the second term on the right-hand side of Eq. 7.3.14 becomes negligible; the equation so reduced describes the FI transition.

Let β_j, β_j^* be defined as

$$\beta_j = \frac{b_{ji}}{b_{jo}} \tag{7.3.17}$$

$$\beta_j^* = \frac{b_{ji}^*}{b_{jo}} \tag{7.3.18}$$

The quantities β_j and β_j^* are related to each other through the following equation, which is obtained directly from Eq. 7.3.8:

$$\beta_j = 1 + \frac{I - 1}{I_\infty - 1} (\beta_j^* - 1) \tag{7.3.19}$$

The quantities a_i, b_{ji}^* are related to each other by the same equilibrium condition as the quantities a_0, b_{jo}, say, by Eq. 6.3.2. Hence the quantity $\bar{a}_{\mathrm{i}}/a_0$ can be expressed as

$$\frac{\bar{a}_{\mathrm{i}}}{a_0} = \prod_j (\beta_j)^{-\nu_j} \tag{7.3.20}$$

A considerable simplification of Eq. 7.3.14 arises when $g(\cdot)$ can be expressed in the form of Eq. 7.3.15. Indeed, if such is the case, one has

$$\sqrt{\frac{g_i}{g_0}} = \sqrt{\prod_j (\beta_j)^{-\nu'_j}} \tag{7.3.21}$$

and, in view of Eqs. 7.3.20 and 7.3.21, the right-hand side of Eq. 7.3.14 is seen to depend only on ψ, Φ, and β_j, but not separately on b_{ji} and b_{j0}.

The procedure discussed above is a generalization of the one originally presented. The simple kinetic scheme considered previously corresponds to $\nu'_1 = 1$, $\nu'_j = 0$ for $j \neq 1$ (i.e., a single bimolecular irreversible reaction). For the same kinetic scheme, the full partial differential equations emerging from the penetration theory model have been solved numerically by several authors. Comparison with the results of the procedure discussed above shows that, if agreement is forced for the asymptotic value I_∞ (i.e., if I_∞ is taken as given by Eq. 6.1.18 and, of course, b^*_{1i} is taken as zero), the largest discrepancy is only 8%. A similar comparison has been carried out recently for the more complex kinetic mechanism arising for the CO_2-carbonate system. Calculations carried out for both absorption and desorption, and under conditions where reversibility needs to be taken into account, again have shown that the procedure discussed in this section agrees to within 10% with the numerical results obtained from the full penetration theory equations.

A plot of the FI transition curves is reported in Figure 7.3.1. Two observations

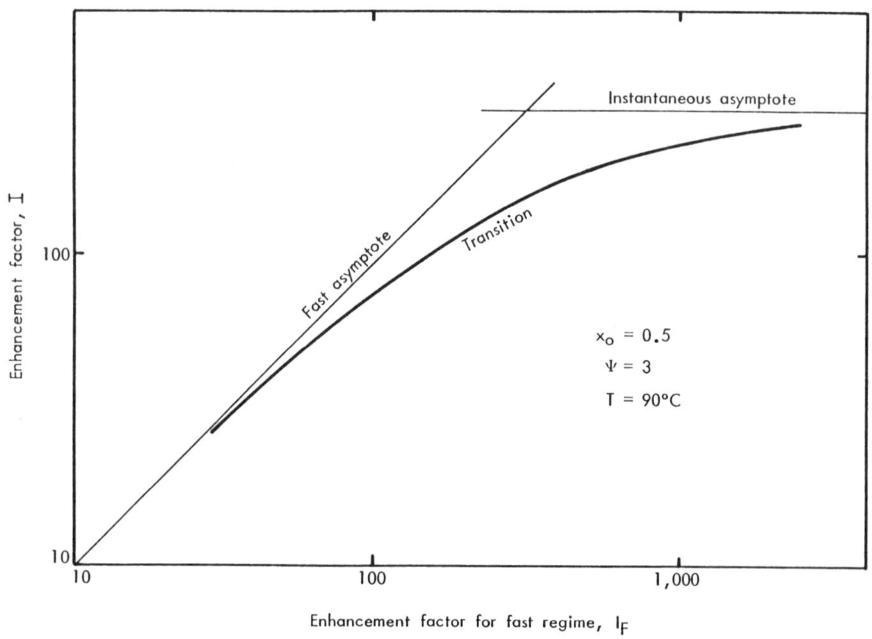

FIGURE 7.3.1. Fast–instantaneous transition curve.

are in order. The first is that the FI transition curve is smoother than the SF transition; i.e., conditions intermediate between the fast reaction and the instantaneous reaction asymptotes are likely to be encountered in practice. The second observation is that the FI transition curve becomes smoother with increasing complexity of kinetic behavior; the calculated curves are appreciably distant from the asymptotic lines over several orders of magnitude of the independent variable $\sqrt{\Phi}$.

7.4. AN APPROXIMATE EXPLICIT EQUATION

The procedure outlined in the preceding section produces, for any given kinetic scheme, an algebraic but implicit equation for the enhancement factor I. For purposes of repeated calculations such as are involved in the design of a continuous unit for absorption or desorption, an explicit equation is, of course, highly preferable. Such an equation should give I as an explicit function of I_∞ and Φ.

The following equation has been proposed:

$$I = \left[\left(1 + \frac{n^{n/(n-1)} (I_\infty - 1)^{1/(n-1)} I_\infty}{(1+\Phi)^{n/(2n-2)}}\right)^{1/n} - 1\right] \times \frac{(1+\Phi)^{n/(2n-2)}}{n^{1/(n-1)} (I_\infty - 1)^{1/(n-1)}} \qquad (7.4.1)$$

as appropriate wherever I_∞ is significantly larger than unity (which is the industrially relevant case). In Eq. 7.4.1, n is a constant the value of which depends on the specific kinetic scheme.

Equation 7.4.1 should be discussed in some detail. First, it has the correct asymptotic behavior. Indeed, if

$$\sqrt{\Phi + 1} \ll I_\infty \qquad (7.4.2)$$

it reduces to Eq. 7.1.12; i.e., it correctly describes the slow and fast reaction regimes as well as the slow-fast transition. Conversely, if

$$\sqrt{\Phi + 1} \gg I_\infty \qquad (7.4.3)$$

it reduces to $I = I_\infty$, as required.

The smoothness of the fast-instantaneous transition curve increases with increasing n. Indeed, if $n = 1$, the transition is abrupt, say, the cusp as in Figure 3.6.1 is described mathematically. When $n = 2$, it reduces essentially to the equation proposed earlier as a very good approximation for the simple bimolecular reaction scheme. For $n = 3$, Eq. 7.4.1 is a very good approximation for the case of the CO_2-carbonate system. It is, therefore, plausible that Eq. 7.4.1 may be used in general, though the value of n appropriate for the problem at hand must be established for each specific case. This is easily accomplished by following the procedure outlined in Section 7.3 for a few selected values of the parameters. A simple trial-and-error procedure allows the value of n that gives the best fit to be established. Once this has been done, Eq. 7.4.1 can be used with confidence for the direct explicit calculation of I for any set of values of the relevant parameters.

8 Simultaneous Mass Transfer

One woman in the house is peace; two women in the house is war.
CONFUCIUS

Gas treating processes where more than one volatile component is simultaneously removed from a gas stream are common. Mass transfer in such processes is a very complex phenomenon, and often the mass transfer of the different volatile components are coupled to each other because of common reactants participating in the different reactions taking place in the liquid phase. The most important case in industrial practice is that of simultaneous removal of CO_2 and H_2S; the complex coupling of the two mass transfer phenomena generally results in an often desirable selectivity for H_2S.

Much of the material in this chapter is either original, or has only recently appeared in the published literature. The reader should be aware that the material has not withstood the test of longtime public exposure, with the possibility of criticism and improvement that such exposure offers.

8.1. BASIC CONCEPTS AND DEFINITIONS

Let A, A' represent the two volatile components, and let the two main reactions that they undergo be written as

$$A + \sum_{1}^{M} \nu_j B_j = 0 \tag{8.1.1}$$

$$A' + \sum_{1}^{M} \nu'_j B_j = 0 \tag{8.1.2}$$

with K, K$'$ the corresponding equilibrium constants. We restrict attention to the (likely) case where there is only one non-volatile component endowed with a positive stoichiometric coefficient ν_j, *and* that the same component is also the only one endowed with a positive stoichiometric coefficient ν_j'. We agree on assigning the index 1 to this particular component, say,

$$\nu_1 > 0, \qquad \nu_1' > 0 \tag{8.1.3}$$

$$j > 1, \qquad \nu_j \leqslant 0, \qquad \nu_j' \leqslant 0 \tag{8.1.4}$$

In Section 8.6 we will discuss two special cases where Eqs. 8.1.3 and 8.1.4 are not satisfied.

If Eq. 8.1.2 is multiplied by ν_1/ν_1' (which is a positive quantity in view of Eq. 8.1.3) and subtracted from Eq. 8.1.1, one obtains

$$A - \frac{\nu_1}{\nu_1'} A' + \sum_2^M \left(\nu_j - \frac{\nu_1}{\nu_1'} \nu_j'\right) B_j = 0 \tag{8.1.5}$$

Reaction 8.1.5 will be called the "shift reaction;" notice that the stoichiometric coefficient of component 1 in the shift reaction is zero and the stoichiometric coefficient of A' in the shift reaction is negative. The equilibrium constant for the shift reaction K_s is related to K and K′ by

$$K_s = \frac{K}{K'^{\nu_1/\nu_1'}} \tag{8.1.6}$$

An example may be illuminating. Consider the system CO_2 (A)–H_2S (A')–aqueous monoethanolamine. As long as there is a non-negligible concentration of free amine in the bulk of the liquid, the two main reactions are

$$CO_2 + 2RNH = RNCOO^- + RNH_2^+ \tag{8.1.1′}$$

$$H_2S + RNH = RNH_2^+ + HS^- \tag{8.1.2′}$$

If RNH is identified with B_1, Eqs. 8.1.3 and 8.1.4 are satisfied. The shift reaction is

$$CO_2 + RNH_2^+ + 2HS^- = RNCOO^- + 2H_2S \tag{8.1.5′}$$

and, indeed, the free amine RNH has a zero stoichiometric coefficient in this shift reaction, whereas H_2S has a negative stoichiometric coefficient.

We now turn attention to the question of driving forces. The *physical* driving forces for A and A' are $a_i - a_0$ and $a_i' - a_0'$; the *chemical* driving forces are $\alpha_i - \alpha_0$ and $\alpha_i' - \alpha_0'$. When only one volatile component needs to be considered, the physical driving force and the chemical driving force always have the same sign; i.e., they are either both positive (absorption) or both negative (desorption). The situation is not so simple in the case where two volatile components are to be considered. Suppose, e.g., that both physical driving forces are positive; i.e., the interface partial pressures of both A and A' in the gas are *larger* than their equilibrium vapor pressures corresponding to the bulk-liquid composition. The *chemical* driving force for one of the components may well be negative, as can easily be verified by considering the equilibrium equations discussed in Chapter 3 which do not imply that an increase of both p^* and $p^{*\prime}$ necessarily corresponds to an increase of both y and y'.

This point has important consequences. In fact, as has been shown in the preceding chapters, the mass transfer rate may be proportional to either the physical or the chemical driving force, depending on the particular regime of mass transfer. In the case of only one volatile component, this is relevant in determining the rate of mass transfer but not its sign; when more than one volatile component is involved, even the sign of mass transfer may be determined by the regime that applies.

In view of the important additional complications that arise in the case of simultaneous chemical mass transfer, in this chapter we restrict attention to the case where all the liquid-phase diffusivities are equal. Consequently, we always refer to the film theory model, and we always regard the molarity m as constant throughout the liquid phase.

Once the bulk-liquid composition and the interface concentrations a_i, a_i' are established, it is possible to establish which regime of mass transfer would apply for each of the two volatile components *should the other one not be present.* This is accomplished with the methods discussed in the preceding chapters. However, the regime that actually holds in simultaneous mass transfer may not be the one so established due to the interference between the two processes (the chemical interference is, of course, related to the possibility of the shift reaction taking place). For purposes of notation, we refer to the no-interference regimes; so, e.g., an FI regime is used to indicate that component A would undergo a fast reaction phenomenon in the absence of A', and A' would undergo an instantaneous reaction phenomenon in the absence of A; which phenomenon they, in fact, undergo when simultaneous mass transfer takes place needs, of course, to be established.

The variety of conceivable reaction schemes in simultaneous chemical mass transfer is so large that it is almost impossible to examine all possible cases. Therefore, in this chapter we basically restrict attention to those cases that are of interest for the specific problems to be discussed in Chapter 12, with no attempt at exhausting all possible ones. However, it is hoped that the kind of difficulties that may be encountered with all problems of industrial interest can be handled with the techniques discussed in the following.

8.2. SIMULTANEOUS BUT UNCOUPLED MASS TRANSFER

Coupling between the chemical mass transfer phenomena of two volatile components may take place in principle both because diffusive phenomena are coupled and because of chemical coupling. Coupling of diffusive phenomena is generally of minor importance since the diffusivities are often rather weak functions of the detailed composition of the liquid phase, particularly when aqueous solutions are considered in which the major component is water. We therefore neglect all effects of diffusive coupling and concentrate attention on the much more important issue of chemical coupling. In this section we establish conditions under which the chemical coupling also can be neglected, so that simultaneous mass transfer may be dealt with as a simple superposition of two (or more) single component mass transfer phenomena.

Let t_r, t'_r be the reaction times for reactions 8.1.1 and 8.1.2. These are understood as calculated with respect to the bulk-liquid concentrations b_{j0} of the nonvolatile components B_j. It is important to establish the value of the reaction time for the shift reaction. Two cases are possible.

In the first case, the kinetic mechanism of the shift reaction implies a sequence of steps such that both reactions 8.1.1 and 8.1.2 need to take place. When this is the case, the reaction time for the shift reaction obviously coincides with the lowest one between t_r and t'_r; in other words, the shift reaction cannot be faster than the slower of the two main reactions.

In the second case, the shift reaction may take place through a direct kinetic mechanism that does not involve the actual occurrence of the two main reactions. In this "fast-shift" case, one would need to consider the possibility of the shift-reaction influencing kinetically the mass transfer rates of A and A', even if the main reactions are slow. We do not consider this case in the following.

As long as both main reactions are at most fast enough that mass transfer of A and A' would by itself take place in the fast reaction regime (i.e., in the SS, SF and FF regimes), there is no coupling between the two mass transfer phenomena. In fact, the diffusion equations are

$$D\frac{d^2a}{dx^2} = r(a, b_1, \ldots, b_N) \tag{8.2.1}$$

$$D\frac{d^2a'}{dx^2} = r'(a, b_1, \ldots, b_N) \tag{8.2.2}$$

and, therefore, as long as the approximation $b_j \simeq b_{j0}$ is justified, the diffusion equations are uncoupled, and their solution is the same as would be obtained in the absence of the other volatile components.

However, this is not true as soon as one approaches instantaneous reaction conditions for at least one of the volatile components. Notice that depletion of the liquid-phase reactant in absorption (say, component B_1 if Eqs. 8.1.3 and 8.1.4 apply) is likely to occur more easily in simultaneous absorption since that component is consumed by both chemical reactions. Hence deviations from fast reaction behavior are likely to occur at lower values of Φ (or Φ') in simultaneous absorption than in single-gas absorption.

8.3. IRREVERSIBLE INSTANTANEOUS ABSORPTION

In this section we consider the simplest possible case of an *I–I* phenomenon, i.e., simultaneous absorption of two volatile components that react instantaneously and irreversibly with the same non-volatile liquid phase component. The adverb "irreversibly" should, of course, be understood in the sense that the physical driving force for both A and A' is very large compared to their bulk-liquid concentrations, say,

$$\psi = \frac{a_i}{a_0} \gg 1, \qquad \Psi' = \frac{a_i'}{a_0'} \gg 1 \tag{8.3.1}$$

Equations 8.3.1 imply that since both reactions 8.1.1 and 8.1.2 are instantaneous and, therefore, at equilibrium everywhere, the film thickness δ is divided in two regions: the first, $0 \leqslant x \leqslant \lambda$, is a region where b_1 is negligibly small (as compared to its bulk-liquid value b_{10}), whereas a and a' are non-negligible (as compared to their interface values a_i and a_i'). Conversely, in the region $\lambda \leqslant x \leqslant \delta$, both a and a' are negligibly small, whereas b_1 is finite (see Figure 8.3.1).

In the region $\lambda \leqslant x \leqslant \delta$, neither reaction 8.1.1 nor reaction 8.1.2 can take place appreciably since $a \simeq 0$, $a' \simeq 0$. In the same region, and for the same reason, the shift-reaction also cannot take place appreciably. It follows that the only phe-

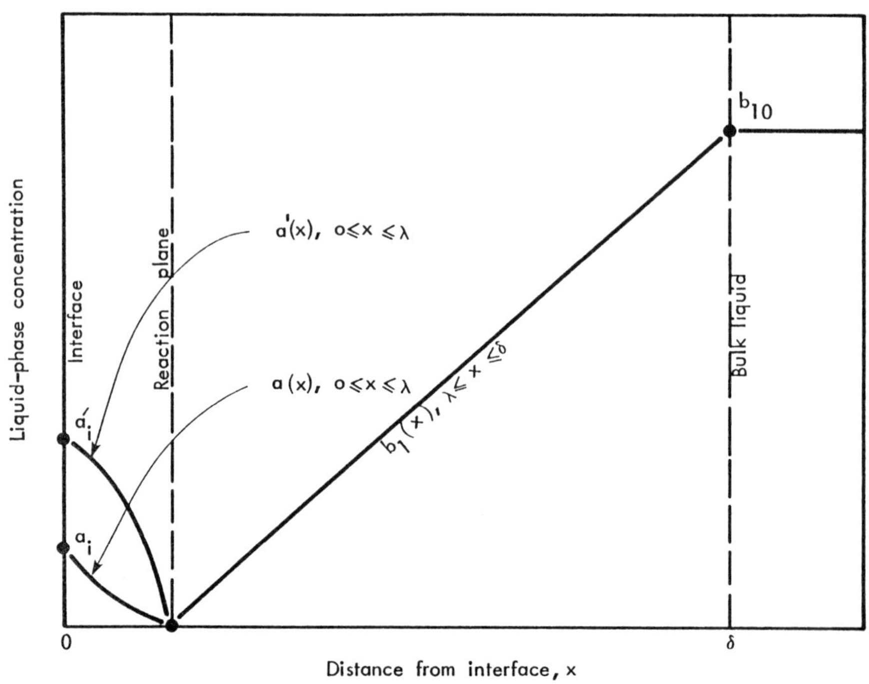

FIGURE 8.3.1. Irreversible instantaneous regime absorption of two reactive gases.

nomenon of interest in the region $\lambda \leqslant x \leqslant \delta$ is diffusion of B_1 from the bulk of the liquid to the reaction plane located at $x = \lambda$, which takes place at a rate given by

$$N_1 = D \frac{db_1}{dx} = \frac{Db_{10}}{\delta - \lambda} \tag{8.3.2}$$

The stoichiometric condition at the reaction plane $x = \lambda$ can be written as

$$\frac{N_1}{D} = \frac{b_{10}}{\delta - \lambda} = -\nu_1 \left.\frac{da}{dx}\right|_{x=\lambda} - \nu_1' \left.\frac{da'}{dx}\right|_{x=\lambda} \tag{8.3.3}$$

which expresses the condition that whatever B_1 reaches the reaction plane must be consumed there by either reaction 8.1.1 or reaction 8.1.2.

In the region $0 \leqslant x \leqslant \lambda$, neither reaction 8.1.1 nor reaction 8.1.2 can take place appreciably, because $b_1 \cong 0$. However, the shift reaction may, and does, take place in this region; it is not hindered by the absence of any significant amount of B_1 since the stoichiometric coefficient of B_1 in the shift reaction is zero. It follows that the concentration profiles of A and A' need not be linear (and will, in general, not be linear) in the region $0 \leqslant x \leqslant \lambda$ between the interface and the reaction plane. This, in turn, implies that the ratio of the mass transfer rates N/N' in general *will not* coincide with the ratio of physical driving forces a_i/a_i' and, in fact, one of the two mass transfer rates may well be *negative*; i.e., one of the two volatile components may be desorbed (the one that exhibits a negative curvature, say, with reference to Figure 8.3.1, component A'). The component that may desorb is the one that is *formed* by the shift reaction (i.e., the one chemically displaced from the liquid by the other).

The solution of the appropriate equations is rather difficult, even for the simple irreversible case in the film theory model considered here, if an arbitrary set of stoichiometric coefficients for the shift reaction is considered. We illustrate in the following a special case that has a very simple stoichiometry, say, the case where reactions 8.1.1 and 8.1.2 take the form

$$A + B_1 = B_2 \tag{8.3.4}$$

$$A' + B_1 = B_3 \tag{8.3.5}$$

so that the shift reaction is

$$A + B_3 = A' + B_2 \tag{8.3.6}$$

The analysis is as follows:

1. In the region $0 \leqslant x \leqslant \lambda$,

$$\frac{d^2}{dx^2}(a + b_2) = \frac{d^2}{dx^2}(a' + b_3) = 0 \tag{8.3.7}$$

and, therefore,

$$N = \frac{D}{\lambda}(a_i + b_{2i} - b_{2\lambda}) \tag{8.3.8}$$

$$N' = \frac{D}{\lambda}(a_i' + b_{3i} - b_{3\lambda}) \tag{8.3.9}$$

where $b_{2\lambda}, b_{3\lambda}$ are the values of b_2, b_3 at $x = \lambda$.

2. In the region $\lambda \leqslant x \leqslant \delta$, A is carried only in the chemically combined form B_2, and A' in the form B_3; hence

$$N = \frac{D}{\delta - \lambda}(b_{2\lambda} - b_{20}) \tag{8.3.10}$$

$$N' = \frac{D}{\delta - \lambda}(b_{3\lambda} - b_{30}) \tag{8.3.11}$$

If $b_{2\lambda}$ and $b_{3\lambda}$ are eliminated from Eqs. 8.3.8-8.3.11, one obtains

$$N = \mathrm{k}_L^0 (a_i + b_{2i} - b_{20}) \tag{8.3.12}$$

$$N' = \mathrm{k}_L^0 (a_i' + b_{3i} - b_{30}) \tag{8.3.13}$$

which are, of course, nothing but special cases of Eq. 3.5.20. These two equations show that the mass transfer rates are proportional to the *chemical* driving force, one of which may well be negative even if both physical driving forces are positive.

3. In the region $0 \leqslant x \leqslant \lambda$, $b_2 + b_3$ is constant and, therefore,

$$b_{2i} + b_{3i} = b_{2\lambda} + b_{3\lambda} \tag{8.3.14}$$

In the region $\lambda \leqslant x \leqslant \delta$, only diffusion takes place, and one mole of B_1 must diffuse to the reaction plane for every mole of $B_2 + B_3$ that diffuses towards the bulk of the liquid; hence

$$b_{2\lambda} + b_{3\lambda} = b_{10} + b_{20} + b_{30} \tag{8.3.15}$$

4. Chemical equilibrium must prevail at the interface, say,

$$\frac{a_i' b_{2i}}{a_i b_{3i}} = \mathrm{K}_s \tag{8.3.16}$$

Equations 8.3.14-8.3.16 determine the values of b_{2i} and b_{3i}; when these are substituted into Eqs. 8.3.12 and 8.3.13 one obtains, after algebraic rearrangement,

$$I_\infty = 1 + \frac{K_s m}{a_i' + K_s a_i} - \frac{b_{20}}{a_i} \tag{8.3.17}$$

$$I_\infty' = 1 + \frac{m}{a_i' + K_s a_i} - \frac{b_{30}}{a_i'} \tag{8.3.18}$$

where $m = b_{10} + b_{20} + b_{30}$ is the molarity of the solution.

It is important to realize that either I_∞ or I_∞' (but not both) may be negative. Suppose, e.g., that K_s is a relatively large number and that b_{30} is not very small; I_∞' will be negative (i.e., component A' will desorb to the gas phase in spite of the fact that $a_i' \gg a_0'$). The concentration profiles that would be established in the film in such a case are sketched in Figure 8.3.2.

Equation 8.3.18 shows that I_∞' will be negative if

$$b_{30} > a_i' + \frac{m}{1 + K_s(a_i/a_0)} \tag{8.3.19}$$

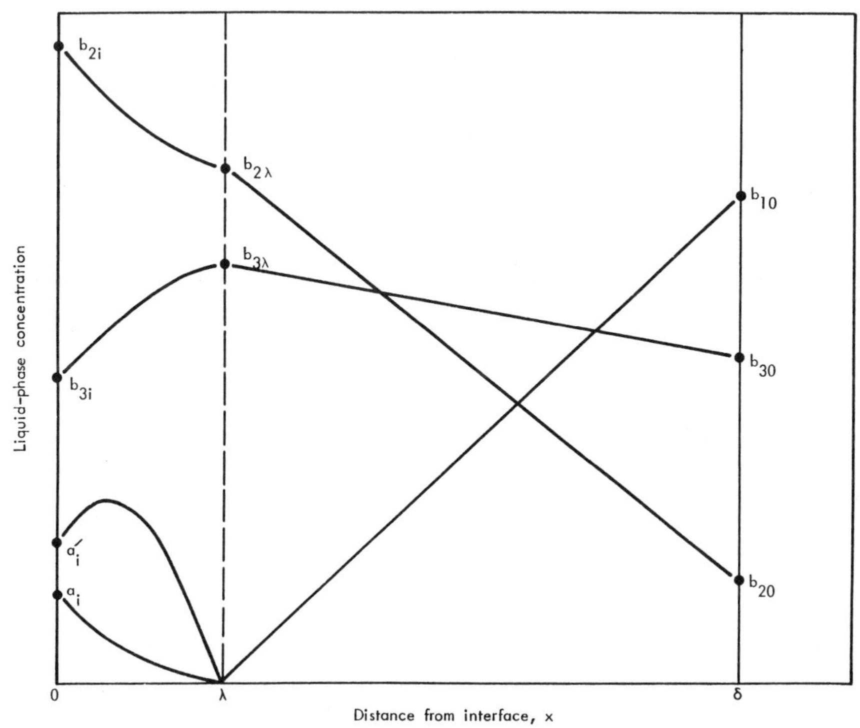

FIGURE 8.3.2. Simultaneous absorption of two gases in the instantaneous regime where the shift reaction is considered.

If the thermodynamic analysis of the system described by Eqs. 8.3.4-8.3.6 is carried out for the conditions assumed in this section (i.e., both K and $K' \to \infty$, but K_s is finite), Eq. 8.3.19 is identified as the condition for the *chemical* driving force $\alpha_i' - \alpha_0'$ to be negative. In the instantaneous reaction regime, mass transfer rates are proportional to the chemical driving force.

The simple example discussed in this section shows that, in the case of simultaneous absorption in the I-I regime, there is no special advantage in considering simple concentration sketches such as in Figures 8.3.1 and 8.3.2 since the occurrence of the shift reaction in the region $0 \leqslant x \leqslant \lambda$ makes the concentration profiles non-linear. It is, therefore, convenient to attack the problem directly from a more general viewpoint, by duplicating the approach discussed in Section 6.2. This approach is developed in Section 8.4.

8.4. GENERAL ANALYSIS OF THE INSTANTANEOUS-INSTANTANEOUS REGIME

We now turn attention again to the general system described by Eqs. 8.1.1 and 8.1.2. In the I-I regime, chemical equilibrium prevails everywhere in the liquid phase; say,

$$K = \prod_j \frac{b_j^{-\nu_j}}{a} \tag{8.4.1}$$

$$K' = \prod_j \frac{b_j^{-\nu_j'}}{a'} \tag{8.4.2}$$

The diffusion equations are, for the film theory model,

$$D \frac{d^2 a}{dx^2} = r \tag{8.4.3}$$

$$D \frac{d^2 a'}{dx^2} = r' \tag{8.4.4}$$

$$D \frac{d^2 b_j}{dx^2} = \nu_j r + \nu_j' r' \tag{8.4.5}$$

where r, r' are the local rates of reactions 8.1.1 and 8.1.2. If these are eliminated from Eqs. 8.4.3-8.4.5, one obtains

$$\frac{d^2 b_j}{dx^2} = \nu_j \frac{d^2 a}{dx^2} + \nu_j' \frac{d^2 a'}{dx^2} \tag{8.4.6}$$

At the interface, the reaction rates per unit area r_0, r'_0 may be non-zero; therefore,

$$x = 0, \qquad D\frac{db_j}{dx} = \nu_j r_0 + \nu'_j r'_0 \tag{8.4.7}$$

Let N, N' be the fluxes of A, A' from the bulk of the gas phase towards the interface; both may be either positive or negative. The fluxes N, N' are the sum of the amount consumed by the reactions at the interface plus the amount that diffuses from the interface towards the bulk of the liquid; say,

$$x = 0, \qquad N = r_0 - D\frac{da}{dx} = I_\infty k_L^0 (a_i - a_0) \tag{8.4.8}$$

$$x = 0, \qquad N' = r'_0 - D\frac{da'}{dx} = I'_\infty k_L^0 (a'_i - a'_0) \tag{8.4.9}$$

Integration of Eq. 8.4.6 between $x = 0$ and $x \leqslant \delta$ yields, taking into account Eq. 8.4.7,

$$\frac{db_j}{dx} = \nu_j \left[\frac{da}{dx} + \frac{I_\infty (a_i - a_0)}{\delta}\right] + \nu'_j \left[\frac{da'}{dx} + \frac{I'_\infty (a'_i - a'_0)}{\delta}\right] \tag{8.4.10}$$

Integration of Eq. 8.4.10 between $x = 0$ and $x = \delta$ yields

$$b_{ji} = b_{jo} - \nu_j \xi - \nu'_j \xi' \tag{8.4.11}$$

where

$$\xi = (I_\infty - 1)(a_i - a_0) \tag{8.4.12}$$

$$\xi' = (I'_\infty - 1)(a'_i - a'_0) \tag{8.4.13}$$

Notice that ξ and ξ' are the extents of reaction at the interface with respect to the bulk-liquid composition. This is, in a sense, an obvious result since when all diffusivities are equal, the interface composition must be stoichiometrically accessible from the bulk-liquid one.

Equations 8.4.1 and 8.4.2 apply both at the interface and in the bulk of the liquid phase. If b_{ji}, as given by Eq. 8.4.11, and then b_{jo} are substituted into Eqs. 8.4.1 and 8.4.2 and the appropriate ratios are taken, one obtains

$$\psi = \frac{a_i}{a_0} = \prod_j \left(1 - \frac{\nu_j \xi}{b_{jo}} - \frac{\nu'_j \xi'}{b_{jo}}\right)^{-\nu_j} \tag{8.4.14}$$

$$\psi' = \frac{a'_i}{a'_0} = \prod_j \left(1 = \frac{\nu_j \xi}{b_{jo}} - \frac{\nu'_j \xi'}{b_{jo}}\right)^{-\nu'_j} \tag{8.4.15}$$

Given the bulk-liquid composition b_{j0}, a_0, a'_0 and the interface concentrations a_i, a'_i, Eqs. 8.4.14 and 8.4.15 can be solved, in general, for the root $\{\xi, \xi'\}$ which makes all b_{ji}s as given by Eq. 8.4.11 non-negative. From these values, I_∞ and I'_∞ can be calculated through Eqs. 8.4.12 and 8.4.13.

The procedure outlined above can be applied, in general, for both absorption and desorption and is not limited to sets of stoichiometric coefficients that satisfy Eqs. 8.1.3 and 8.1.4. However, the procedure is limited to systems where only reactions 8.1.1 and 8.1.2 (as well as the shift reaction which is a linear combination of the latter) need be considered.

The important point to be considered is the following one. Equations 8.4.14 and 8.4.15 are obviously the analogues of Eq. 6.2.16 (in this chapter we make the assumption $D_j = D$), which holds when only one volatile component needs to be considered. However, for Eq. 6.2.16 (with $D = D_j$), the solution ξ is positive whenever $\psi > 1$ (absorption) and negative whenever $\psi < 1$ (desorption). For Eqs. 8.4.14 and 8.4.15, if $\psi > 1$ and $\psi' > 1$, the solution $\{\xi, \xi'\}$ need not be one where *both* ξ and ξ' are positive and if $\psi < 1$, $\psi' < 1$ the solution need not be one where both ξ and ξ' are negative. Important implications of this result will be discussed in Chapter 11.

8.5. THE FAST-INSTANTANEOUS REGIME

In this section we analyze the case where one of the volatile components (say, A) would by itself undergo a fast reaction phenomenon, whereas the other (say, A') would by itself undergo an instantaneous reaction process. The fast-instantaneous (FI) regime is, as will be seen in Chapter 12, particularly important in the case of simultaneous removal of CO_2 and H_2S from a gas stream, with CO_2 the component which by itself would undergo a fast reaction phenomenon.

We consider the case described by reactions 8.1.1 and 8.1.2. The second one is everywhere at equilibrium and since all diffusivities are equal, the interface composition must be stoichiometrically accessible from the bulk-liquid one, say,

$$b_{ji} = b_{j0} - \nu'_j \xi' \tag{8.5.1}$$

where ξ' is the interface extent of reaction 8.1.2 with respect to the bulk-liquid composition. Inasmuch as reaction 8.1.1 is not fast enough to be able to change significantly the concentrations of the non-volatile components, $\xi = 0$; i.e., Eq. 8.5.1 applies also for those components for which $\nu'_j = 0$. (In other words, the concentrations of those non-volatile components that do not participate in reaction 8.1.2 is constant throughout the liquid phase.)

Substituting Eq. 8.5.1 into the equilibrium condition 8.4.1, the value of ξ' can be calculated. The enhancement factor I'_∞ is then calculated from Eq. 8.3.13. Notice that, in following this procedure, the value of I'_∞ obtained is independent of the fact that component A is present; *the rate of mass transfer of component A' in the FI regime is not influenced by the presence of component A*. The converse, however, is certainly not true.

In order to analyze the rate of mass transfer of component A, the kinetics of reaction 8.1.1 need not be considered. The net rate r of consumption of component A may often (though not necessarily always) be expressed in the form

$$r = (a - \bar{a})\, g(b_1, \ldots, b_N) \tag{8.5.2}$$

where

$$\bar{a} = \frac{\prod_j b_j^{-\nu_j}}{K} \tag{8.5.3}$$

(see the discussion in Sections 7.2 and 7.3).

Should the component A undergo mass transfer in the absence of component A', the b_j values in Eqs. 8.5.2 and 8.5.3 could be taken as equal to the bulk-liquid values b_{j0}. However, when component A' is present, the concentrations b_j will, in general, *not* be equal to b_{j0} (see Eq. 8.5.1); indeed, the most relevant values would be the interface values b_{ji}, which are the most different from the bulk-liquid values b_{j0}. Therefore, the presence of component A' strongly influences the rate of reaction 8.1.1 and thus the rate of mass transfer of A. This influence is qualitatively discussed in the following.

Consider the case where A' is either absorbed or desorbed under a small driving force, say,

$$|\psi' - 1| \ll 1 \tag{8.5.4}$$

This condition applies whenever mass transfer of A' takes place near a pinch *and* whenever it is controlled by the gas-phase resistance. In this case, the solution of Eqs. 8.4.1–8.5.1 has the form of Eq. 6.2.19; say,

$$\xi' = \frac{\psi - 1}{a_0 \sum (\nu_j'^2 / b_{j0})} \tag{8.5.5}$$

and, therefore,

$$|\xi'| \ll b_{j0} \tag{8.5.6}$$

(i.e., the liquid-phase composition is approximately uniform). It follows that the rate of reaction 8.1.1 is approximately the same as it would be in the absence of component A'; *if A' undergoes a pinched instantaneous reaction phenomenon, it does not influence the rate of mass transfer of A.*

Next consider the case where A' is absorbed under a large driving force, say,

$$\psi' \gg 1 \tag{8.5.7}$$

In this case, the interface concentration of the stoichiometrically limiting reactant in absorption (say, component B_1 for the case where Eqs. 8.1.3 and 8.1.4 apply) will be almost zero. Now, independently of the specific form of the function $g(\cdot)$ appearing in Eq. 8.5.2, the fact that $b_{1i} \to 0$ implies that the value of $\bar{a}$ tends to infinity near the interface since $\nu_1 > 0$. This, in turn, implies that reaction 8.1.1 tends to take place in the direction of *formation* of component A which may, in fact, be *desorbed* even if its driving force is positive, say, even if $\psi > 1$. This conclusion will be true provided there is at least one component in the liquid phase that has both ν_j and ν_j' positive (i.e., a component that may react with both A and A'). In the case of CO_2 and H_2S, any alkaline component will play such a role.

The result above also can be seen from a different viewpoint. Consider the shift reaction taking place in the direction $A' \to A$. This can be regarded as consisting of two steps, namely, reaction 8.1.2 followed by reaction 8.1.1 taking place in the direction of the formation of A. The first step is instantaneous and takes place near the interface under a large driving force; hence a large driving force for the second step is established near the interface independently of what it may be in the bulk of the liquid. Notice that the chemical formation of A near the interface is a kinetic phenomenon, not a thermodynamic one; this is in contrast with the case discussed in Sections 8.3 and 8.4, where the direction in which the shift reaction takes place is determined solely by thermodynamics.

The converse situation arises if A' is *desorbed* under a large driving force, say, if

$$\psi' \ll 1 \tag{8.5.8}$$

In fact, in this case the stoichiometrically limiting reactant for desorption will have a vanishingly small concentration at the interface. However, this by itself would not necessarily influence greatly the value of $\bar{a}$, since components with a negative ν_j' may well have a zero ν_j (see, e.g., reactions 8.3.4 and 8.3.5; even if $b_{3i} \to 0$, the value of $\bar{a}$ near the interface would not be influenced). The final result will depend on the specific set of stoichiometric coefficients and on the form of the function $g(\cdot)$; in general it will be, if anything, in the direction of decreasing the value of $\bar{a}$ near the interface and, therefore, cause A to be chemically consumed near the interface. The shift reaction will have a tendency to take place in the $A \to A'$ direction near the interface.

Indeed, the whole discussion in this section may be summarized in the following conclusions. First, the rate of mass transfer of A' is governed by its own chemical driving force and is not influenced by the presence of A. Second, the shift reaction is kinetically driven in the direction of the chemical driving force of A'; hence absorption of A' will drive the shift reaction in the direction favoring desorption of A and vice versa.

8.6. SPECIAL CASES

In this section we briefly analyze two special cases. The first is the case of simultaneous absorption of CO_2 and H_2S into strong alkaline solutions, such as, say, aqueous KOH. For this system, reactions 8.1.1 and 8.1.2 would be written as

$$CO_2 + 2OH^- = CO_3^{2-} + H_2O \tag{8.6.1}$$

$$H_2S + OH^- = HS^- + H_2O \tag{8.6.2}$$

since from an equilibrium viewpoint practically no HCO_3^- is formed as long as there is an appreciable concentration of OH^-. If only reactions 8.6.1 and 8.6.2 are considered, the shift reaction would be written as

$$CO_2 + 2HS^- + H_2O = CO_3^{2-} + 2H_2S \tag{8.6.3}$$

and the system would appear to be of the same type as the one discussed in Section 8.3.

However, if both CO_2 and H_2S are, say, absorbed under a large driving force, the concentration of OH^- near the interface tends to zero and the possibility of formation of HCO_3^- near the interface cannot be neglected. In fact, the following reaction is likely to occur near the interface:

$$H_2S + CO_3^{2-} = HS^- + HCO_3^- \tag{8.6.4}$$

Reaction 8.6.4 is a proton-transfer reaction and is, therefore, so fast as to be everywhere at equilibrium. Furthermore, its equilibrium constant is very large and, therefore, it is essentially displaced to the right–say, the concentration of H_2S is negligibly small wherever that of CO_3^{2-} is appreciable and vice versa.

If reaction 8.6.1 is regarded as instantaneous (as appropriate under realistic conditions), as well as reaction 8.6.2 (as always appropriate), the system considered here can be analyzed easily for the simple "irreversible" case ($\psi \gg 1$, $\psi' \gg 1$). The concentration profiles in the liquid film are as sketched in Figure 8.6.1; reaction 8.6.1 takes place at $x = \lambda_2$, together with the following proton-transfer reaction:

$$HCO_3^- + OH^- = CO_3^{2-} + H_2O \tag{8.6.5}$$

Reaction 8.6.4 takes place at $x = \lambda_1$. In the region $\lambda_1 < x < \lambda_2$, CO_3^{2-} and HCO_3^- diffuse back and forth, forced by reactions 8.6.4 and 8.6.5 which take place at the two boundaries of this region.

The system considered above has some pragmatic interest in its own right, and it suggests the following general considerations.

Given a system where simultaneous chemical mass transfer may take place, the main reactions (say, reactions 8.1.1 and 8.1.2 or, in the case considered above, reactions 8.6.1 and 8.6.2) are established by analysis of an appropriate thermodynamic model, as was discussed in Chapter 3. In such a model, certain other reactions may have been excluded from consideration since they occur only negligibly *at the bulk-liquid composition* (say, reaction 8.6.4 in the example discussed above). However, during mass transfer accompanied by instantaneous reactions,

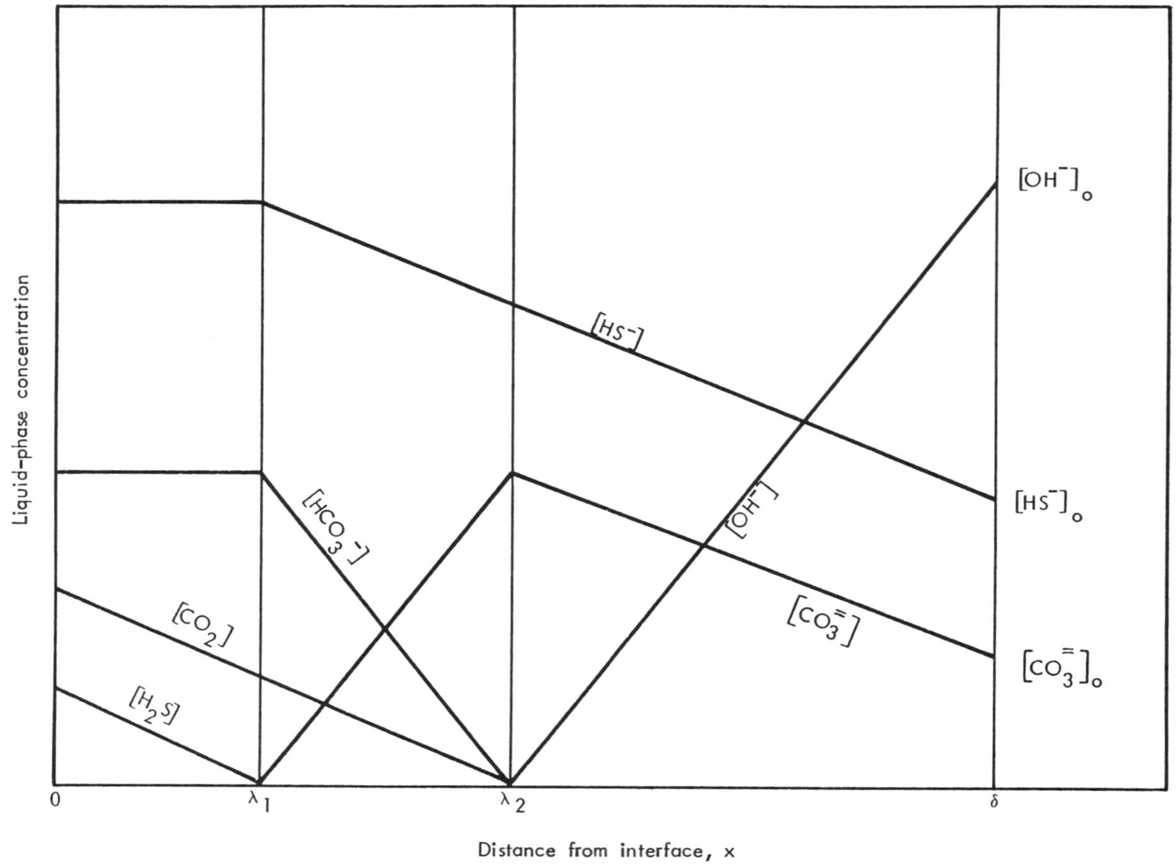

FIGURE 8.6.1. Simultaneous absorption of CO_2 and H_2S into hydroxide solution.

the interface composition may be very different from the bulk-liquid one and, therefore, reactions that are unlikely to occur in the bulk may be important near the interface. *The thermodynamic model should be reconsidered from the viewpoint of possible interface compositions.*

The second special case to be considered is that two volatile components A and A' may react *with each other*. (This case could be considered as one where in the shift reaction the stoichiometric coefficients of A and A' have the same sign; however, the name of shift reaction would obviously be inappropriate.) This case is of interest, e.g., the simultaneous absorption of CO_2 and NH_3.

The following analysis has been published in the literature. Let the chemical reaction be

$$A + \nu A' \longrightarrow \text{products} \tag{8.6.6}$$

with a rate given by

$$r = \mathrm{k}aa' \tag{8.6.7}$$

(i.e., the reaction may be regarded as irreversible). The reaction times at the interface for the two components are

$$A: \; t_r = \frac{1}{\mathrm{k}a_i'} \tag{8.6.8}$$

$$A': \; t_r' = \frac{1}{\mathrm{k}a_i} \tag{8.6.9}$$

If both t_r and t_r' are larger than t_D, both gases undergo a slow reaction phenomenon and the problem is trivial. If $\mathrm{k} \rightarrow \infty$, the reaction takes place instantaneously at the interface and the phenomenon in the liquid phase reduces to diffusion of the products from the interface towards the bulk. The interesting case is where at least one of the t_r's is less than t_D, but finite.

Consider the case where

$$a_i \ll a_i' \tag{8.6.10}$$

The concentration profiles will have the qualitative form in Figure 8.6.2. Therefore, A' will undergo essentially a slow reaction phenomenon (even if $t_r' \ll t_D$), whereas A will undergo a fast reaction phenomenon with

$$I = \frac{\sqrt{D\mathrm{k}a_i'}}{\mathrm{k}_L^0}; \qquad (I' = 1) \tag{8.6.11}$$

Conversely, consider the case where

$$\nu a_i = a_i' \tag{8.6.12}$$

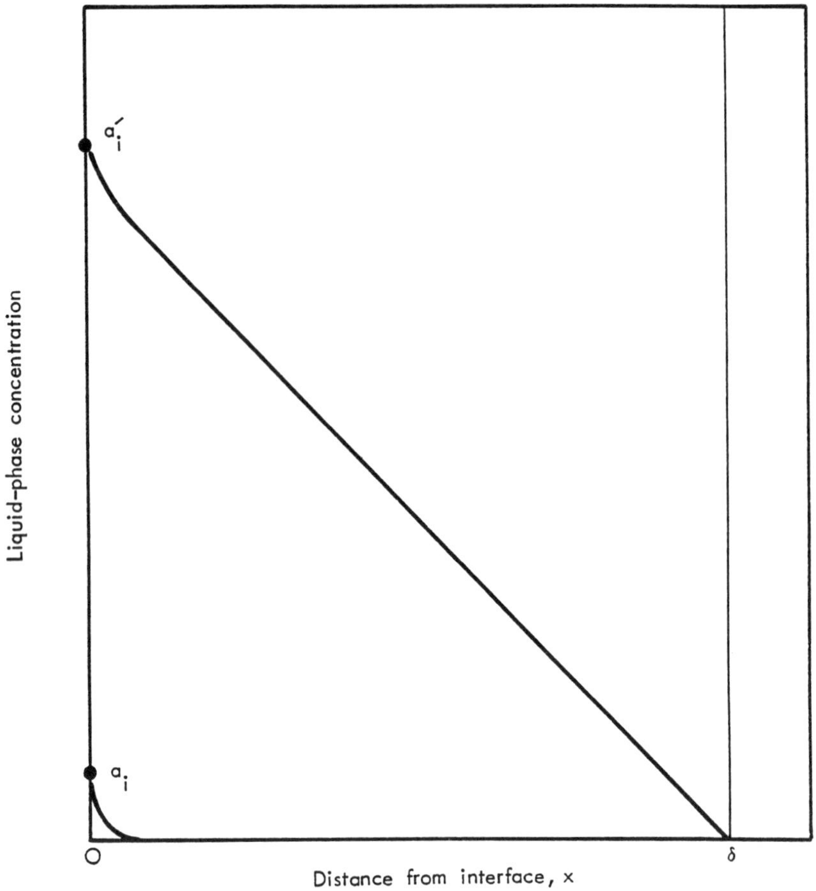

FIGURE 8.6.2. Concentration profile for the case where the two volatile components react with each other in the liquid.

If the reaction rate is eliminated between the diffusion equations for A and A', one obtains

$$\frac{d^2}{dx^2}(\nu a - a') = 0 \tag{8.6.13}$$

which has the trivial solution $\nu a = a'$ when Eq. 8.6.12 holds. Substituting this back into the diffusion equation for A (or A') and integrating, one obtains

$$I = I' = \frac{\sqrt{\frac{2}{3} a_i' \mathrm{k} D}}{\mathrm{k}_L^0} \tag{8.6.14}$$

Since the right-hand side of Eqs. 8.6.11 and 8.6.14 only differ by a factor 0.82, any interpolation formula will be appropriate. Situations such as the one discussed above are unlikely to occur in gas treating processes.

PART THREE: APPLICATIONS

9 Removal of Carbon Dioxide

Dry carbonic acid, mixed with dry ammonia, gives a solid white compound formed by union of one molecule of carbonic acid with two of ammonia. If heated, this solid gives off gases which, from their properties, the chemists have identified as a mixture of carbonic acid and ammonia.

P. DUHEM, 1895

Very large quantities of carbon dioxide (CO_2) must be removed from gases at high pressure in the manufacture of hydrogen, ammonia, and synthetic natural gas (SNG), and in the treating of natural gas. The partial pressure of CO_2 in the feed gas to treating is usually in the range 30–400 psia. CO_2 in the treated gas must be reduced to 1-2% in the case of pipeline natural gas, below 0.1% in the case of H_2 and NH_3 synthesis gas, and below 150 ppm in the case of natural gas that is a feedstock to an LNG plant. In the case of NH_3 synthesis gas, the treated gas specification is less than 10 ppm total carbon oxides (CO_2 + CO) to avoid poisoning the synthesis catalyst. This very low level is usually achieved by scrubbing the CO_2 down to about 0.1% with a liquid reagent. Final purification to remove traces of CO_2 and CO is done by methanation, a gas-phase catalytic reaction between H_2 and carbon oxides.

CO_2 removal requirements depend strongly on the plant feedstock and the process used to generate the gas. For NH_3 synthesis gas, the CO_2 removal requirement varies from 1.22 tons CO_2/ton of NH_3 for steam reforming of natural gas to 2.49 tons CO_2/ton NH_3 for gas made by the partial oxidation of heavy fuel oil. In large plants, the amounts of CO_2 to be removed are enormous, ranging from 2,000-10,000 tons/day for large H_2 and NH_3 plants to over 20,000 tons/day in a large coal gasification complex.

9.1. PROCESS SELECTION

Owing to the large volumes of gas and the large amounts of CO_2 to be scrubbed, processes that are adaptable to very large sizes, thus allowing CO_2 separation in single train plants, tend to be preferred. For any piece of equipment, there is usually an economic size limit beyond which one large unit will cost more than two small ones. Also, price and delivery are affected adversely when practical break points in manufacturing and transport capacity are passed.

An important issue in gas treating processes for CO_2 removal is *process integration.* CO_2 disposal, final purification to specification levels, and waste-heat utilization are the most significant "boundary conditions" imposed on the selection of the CO_2 removal process.

Separated CO_2 is often simply vented to the atmosphere. Alternatively, CO_2 may be recovered for use in the manufacture of urea or for sale to the carbonated beverage industry after additional purification. Recently, a new use for CO_2 in enhanced oil recovery by the miscible flood technique has arisen. In some cases, it is attractive to operate the CO_2 regenerator at high pressure (e.g., 2-3 atm) to minimize compression for supplying CO_2 to a urea plant or an enhanced oil recovery field. This causes the regenerator to run hot, which may introduce problems.

Final purification of the treated gas by methanation is usually attractive provided the gas to the methanator contains less than about 0.1% CO_2.

The most important element in process integration is heat recovery. In hydrogen and ammonia plants, it is generally desirable to use the sensible and latent heat in the gas leaving the low-temperature shift converter to reboil the CO_2-removal system. The excellent fit between the heat available in the shift effluent and the reboiler heat requirements is one of the main reasons why promoted hot carbonate and high strength MEA processes are used in this service.

Heat recovery can be obtained in several ways. The flowsheet in Figure 9.1.1*a* can be used. The gases are cooled against a flow of condensate in a direct contact tower filled with packing. The heat thus transferred to the condensate is then flashed off as steam into the base of the regenerator before the condensate, so cooled, is pumped back to the direct contact tower. This system can be used only on solutions employing relatively non-volatile absorbents that remain in the solution and are not carried over in the stripping steam from the top of the regenerator. Any steam fed to a regenerator as live steam cannot be returned to the solution; therefore, any volatile component associated with it in the overheads from the regenerator is irrecoverably lost.

Limitation of the inlet temperature of the heating medium entering the reboiler is required with MEA and promoted carbonate processes in order to avoid corrosion of the tubes. It may be sufficient in these circumstances to take condensate from the acid gas system into a reboiler accepting the high-temperature gases and so raise steam from condensate rather than solution. A small flow of condensate in excess of the steam to be raised would be passed to the reboiler to ensure that it runs full and does not concentrate any salt that might be near saturation in the liquor. The amount of heat removed from the gases in this way is limited, however, to the amount of actual stripping steam leaving the top of the regenerator, as the solution would otherwise be diluted.

The arrangement shown in Figure 9.1.1*b* is an alternative that eliminates the need for a boiler with its expensive fittings. Condensate and steam circulate in a

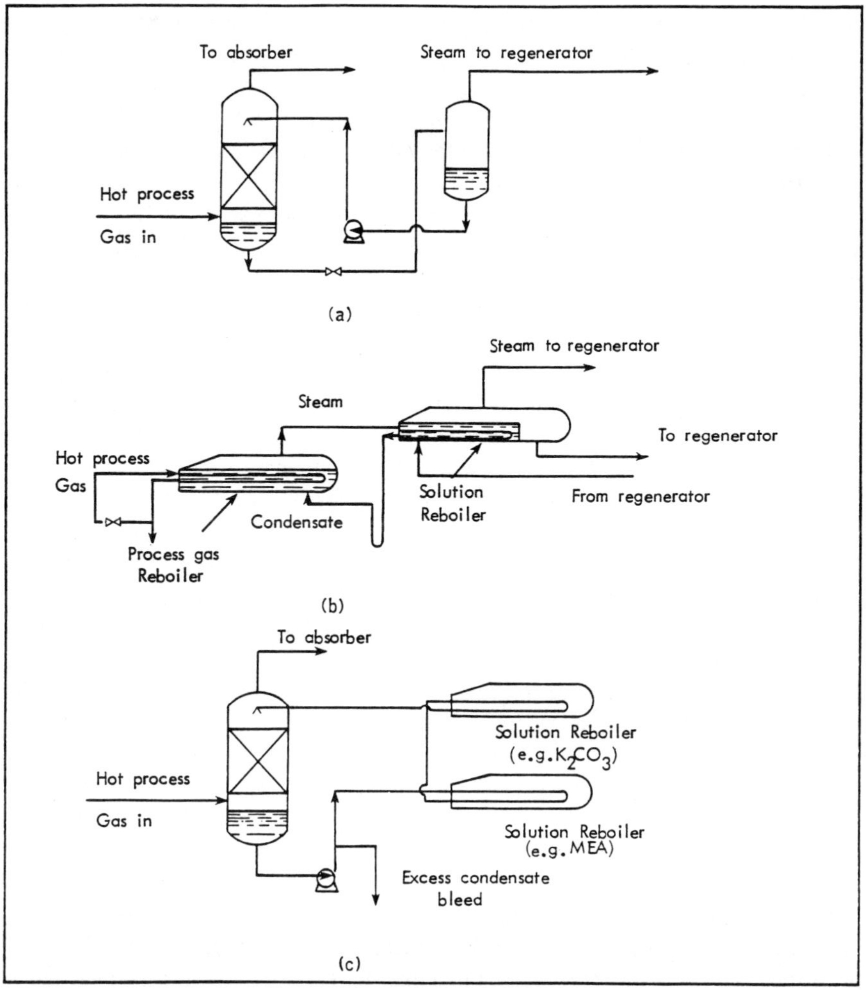

FIGURE 9.1.1. Heat recovery arrangements.

closed system between the process gas reboiler and the solution reboiler. The process gas reboiler is situated below the solution reboiler so that condensate return is by gravity. Control is accomplished simply by bypassing process gas around the process gas reboiler.

In another variation, shown in Figure 9.1.1*c*, the gases are cooled in a packed tower with condensate. The condensate, which is thus heated to a level suitable for entry into reboilers, is then pumped from the bottom of the packed tower through reboilers, delivering the necessary heat for regeneration to the solution, and from these back to the top of the tower.

Physical solvents are used for CO_2 removal. Physical solvent processes do not utilize process waste heat for solvent regeneration; however, this does not imply that there is no heat recovery process integration factor, since waste heat is available for other duty. Recently, several new ammonia flowsheets have been designed around physical solvents, the waste heat being used to operate the dilute ammonia still.

Two important CO_2 removal processes using physical solvents are the propylene carbonate process and the chilled methanol process.

A flowsheet of the type generally used for propylene carbonate CO_2 removal is shown in Figure 9.1.2. The gas is usually first compressed to 800–2500 psig before passing to the absorber. This increases the partial pressure of CO_2 in the feed gas and hence the capacity of the solvent for CO_2. The absorber may also be cooled, below ambient, to further increase absorption capacity for CO_2.

The synthesis gas, after shift conversion and heat recovery, is compressed in a centrifugal compressor and passed to the base of the absorber. The propylene carbonate flowing down through the absorber removes essentially all the carbon dioxide. The remaining traces are removed by methanation in the usual manner. Purified gas is returned to the centrifugal compressor to be raised to the desired pressure for use.

The small quantities of dissolved H_2, N_2, and any light hydrocarbons are released by flashing the rich solvent at high pressure. Flash gas may be recycled back to the centrifugal compressor suction to prevent loss. The solvent is then flashed at a lower pressure to release a major portion of the absorbed carbon dioxide as a high-purity stream. Flash regeneration may be the only regeneration needed if the CO_2 cleanup requirement in the treated gas is not severe. For high CO_2 cleanup, a thorough regeneration of the solvent may be achieved with a mild stripping operation using any available gas (except gas containing an appreciable concentration of carbon dioxide). Lean solvent from the base of the stripper is pumped back to the absorber. The flashing carbon dioxide cools the solvent by autorefrigeration, thereby reducing or, in most cases, eliminating the need for other refrigeration.

A chilled methanol unit for treating sulfur-free gas from a steam reforming plant is shown in Figure 9.1.3. The absorber operates at temperatures below 273K. Gas from the shift converter section of the ammonia plant enters the system at the gas cooler E-1 where it is cooled by indirect heat exchange with the purified gas stream and by part of the evolved carbon dioxide stream. Part of the methanol is added before the exchange to prevent freezing out of water vapor. The cooled gas enters scrubbing tower T-1 where the carbon dioxide is absorbed by recycled stripped methanol entering at the top and partially stripped methanol entering at the mid-section. The heat of carbon dioxide absorption is removed by the refrigerated cooling coil E-5 in the lower part of the tower.

The bulk of the methanol is cooled in heat exchanger E-3 and then flashed by partial release of pressure in vessel D-1 to release hydrogen absorbed from the feed gas. The flashed hydrogen is compressed and recycled to the inlet feed gas through exchanger E-2 and compressor C-1. A sidestream of loaded methanol from the bottom of the tower is sent to a separation unit (not shown) in which water condensed from the gas is removed.

The methanol is stripped of carbon dioxide in stripper tower T-2 at low pressure.

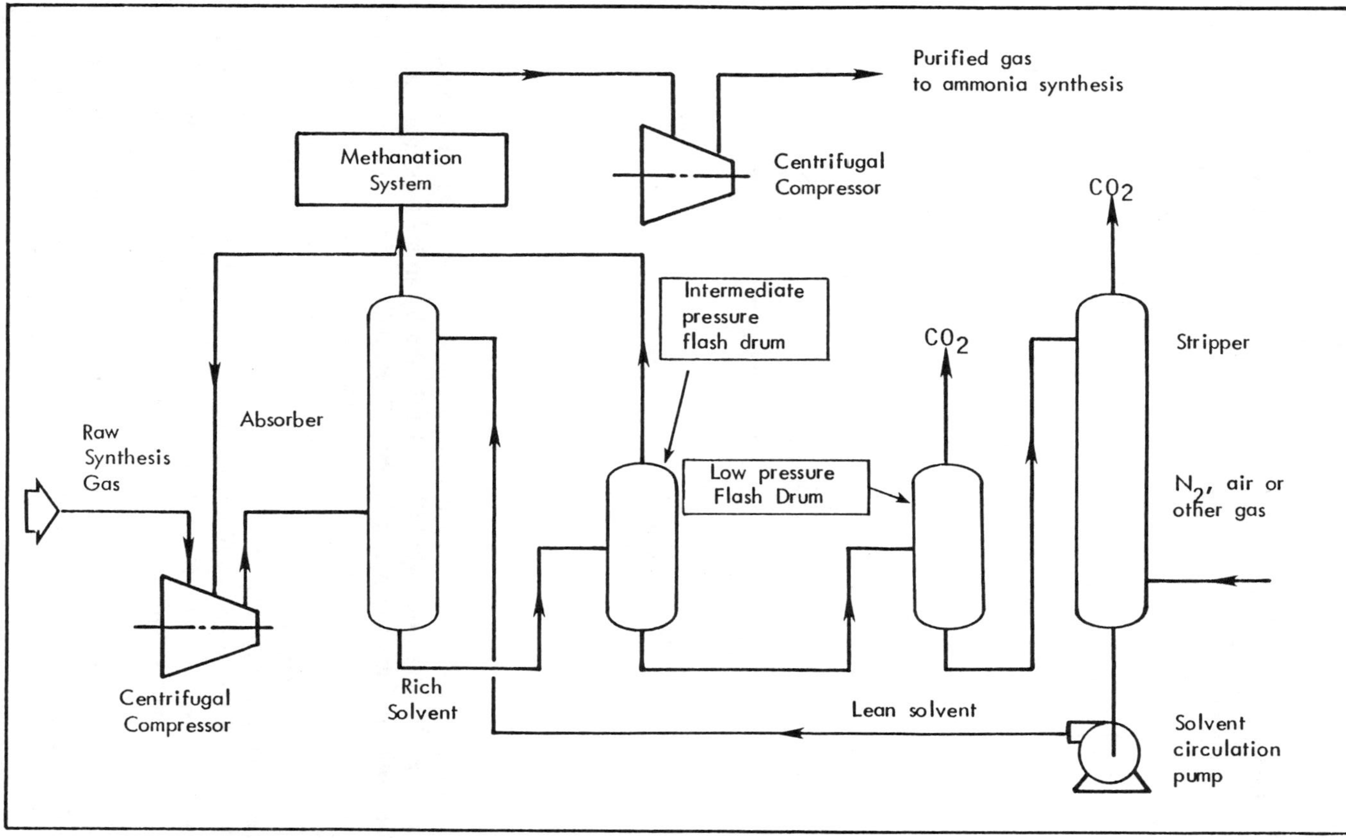

FIGURE 9.1.2. Propylene carbonate process for removing carbon dioxide from ammonia synthesis gas.

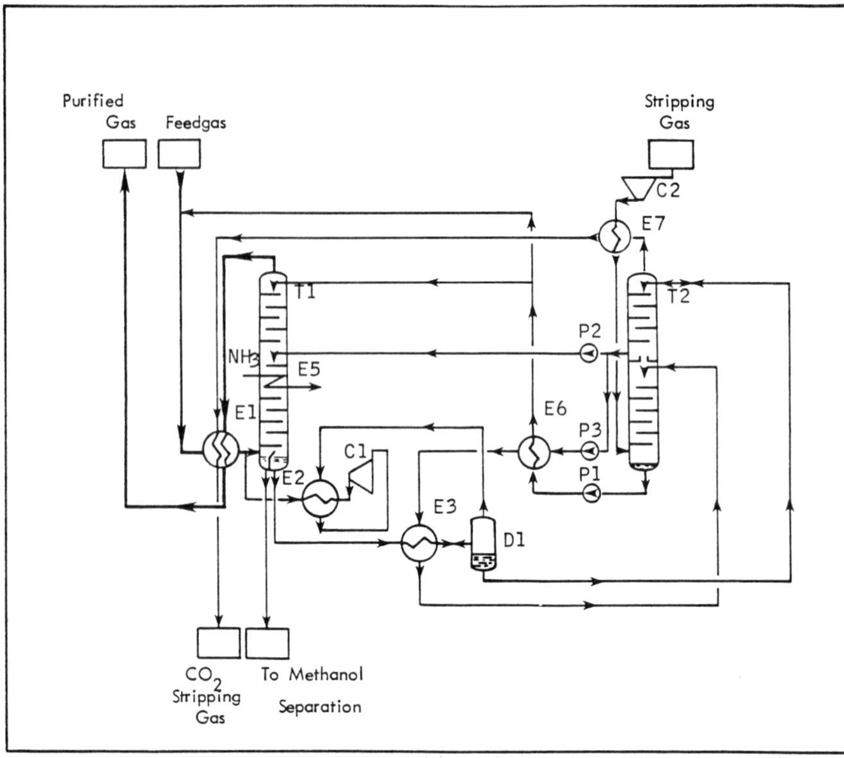

FIGURE 9.1.3. Chilled methanol unit for treating sulfur-free reformed gas for CO_2 removal in ammonia synthesis.

The stripping is accomplished by the pressure release, by heat picked up by solvent cycled through heat exchangers E-6 and E-3, and by an inert stripping gas introduced through blower C-2 and exchanger E-7. Partially stripped solvent is collected at the midsection of the stripper and pumped to the midsection of the absorption tower, and fully stripped liquid from the stripper bottom passes to the top of the absorber after cooling in E-6. The inert stripping gas can be air when the feed gas is free of sulfur compounds.

The capacity of physical solvents for CO_2 is, in general, unacceptably low unless the partial pressure of CO_2 in the feed gas is very large (see the discussion in Section 1.8). Ammonia synthesis is carried out at high pressure and, therefore, physical solvents are attractive because compression of the H_2/N_2 synthesis gas is required at some point (only compression of the CO_2 itself can be debited to the CO_2 removal process). In many other situations, where the partial pressure of CO_2 in the feed gas is not very large and the treated gas is not required to be at high pressure, chemical solvents are preferred because of their larger capacity. CO_2 removal by chemical solvents is discussed in detail in the next sections.

Figure 9.1.4 shows the approximate ranges of application of different types

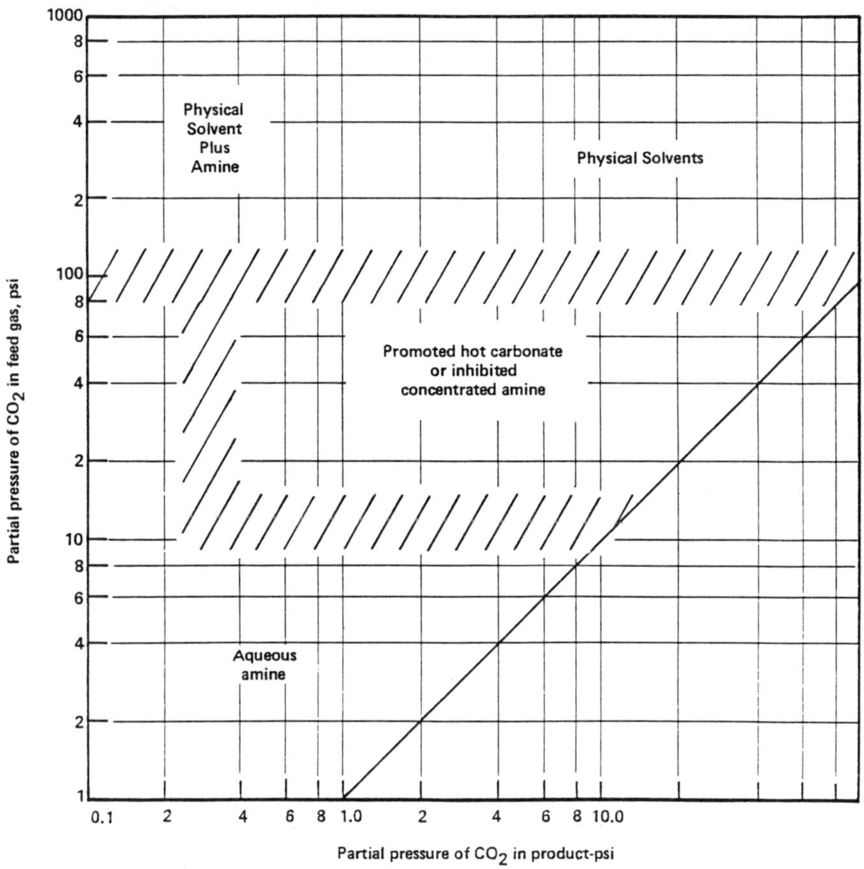

FIGURE 9.1.4. Range of gas treating process application for CO_2 removal with no H_2S or other acidic impurities present.

of gas treating processes for CO_2 removal with no H_2S or other acidic impurities present in the gas. Amine-containing chemical solvents are generally preferred when the partial pressure of CO_2 in the feed gas is relatively low and/or when CO_2 must be reduced to a very low level in the treated gas. Physical solvents are used at high CO_2 pressures in the feed and when deep CO_2 removal is not required. Promoted potassium carbonate processes are used at intermediate CO_2 partial pressures in the feed gas.

9.2. GENERAL THERMODYNAMICS AND CHEMISTRY

CO_2 is an acid gas, and any alkaline solution is eligible as a chemical solvent; in practice, chemical solvents in common use are predominantly aqueous solutions. When CO_2 is dissolved in an aqueous solution, it can hydrate to form the carbonic

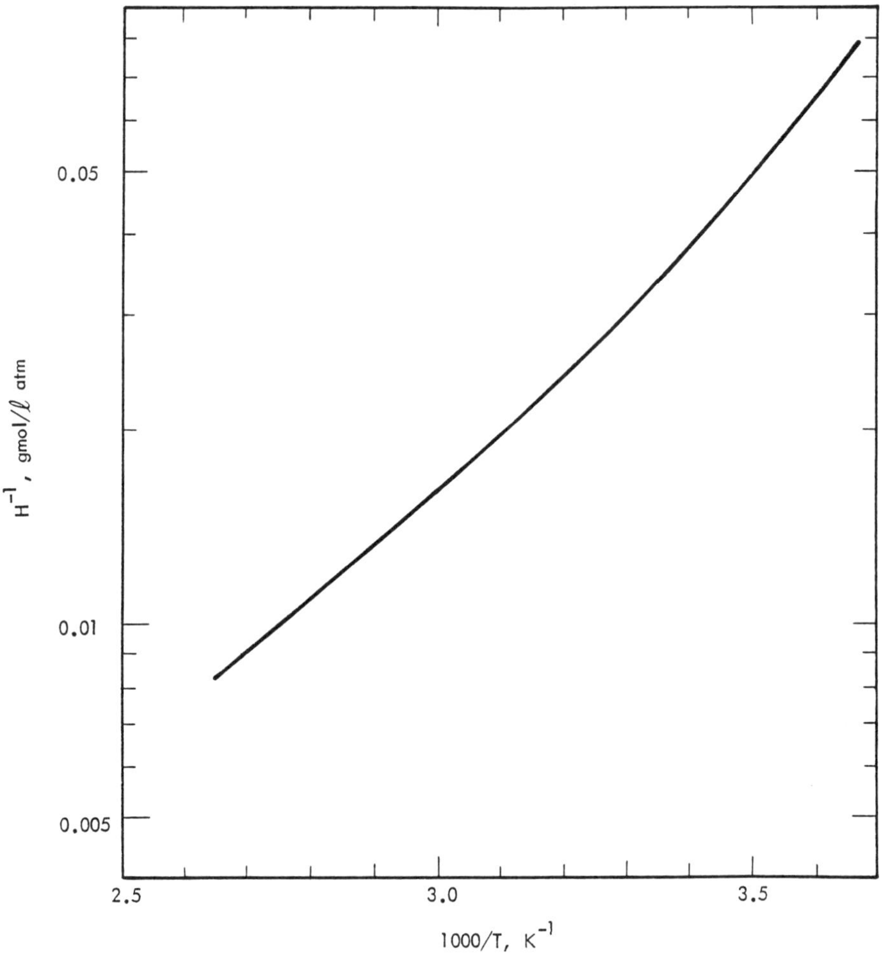

FIGURE 9.2.1. Henry's constant for CO_2 in water.

acid H_2CO_3; however, as was discussed in Section 1.8, there is no need to keep track of the distinction between carbonic acid and physically dissolved carbon dioxide.

Figure 9.2.1 is an Arrhenius plot for the physical solubility of CO_2 in water, as measured by the inverse of Henry's Law constant. It should be borne in mind that the physical solubility of CO_2 into aqueous solutions is not equal to that in water (see the discussion in Section 2.8). Methods for determining the physical solubility of CO_2 in reactive aqueous solutions are discussed in Section 13.3.

Figure 9.2.2 is an Arrhenius plot for the first and second dissociation constants for H_2CO_3. The *p*K for the second dissociation is high enough so that only in strongly alkaline solutions (i.e., solutions with non-neligible concentrations of free hydroxide) will CO_2 tend to transform into the carbonate ion. Such strongly alka-

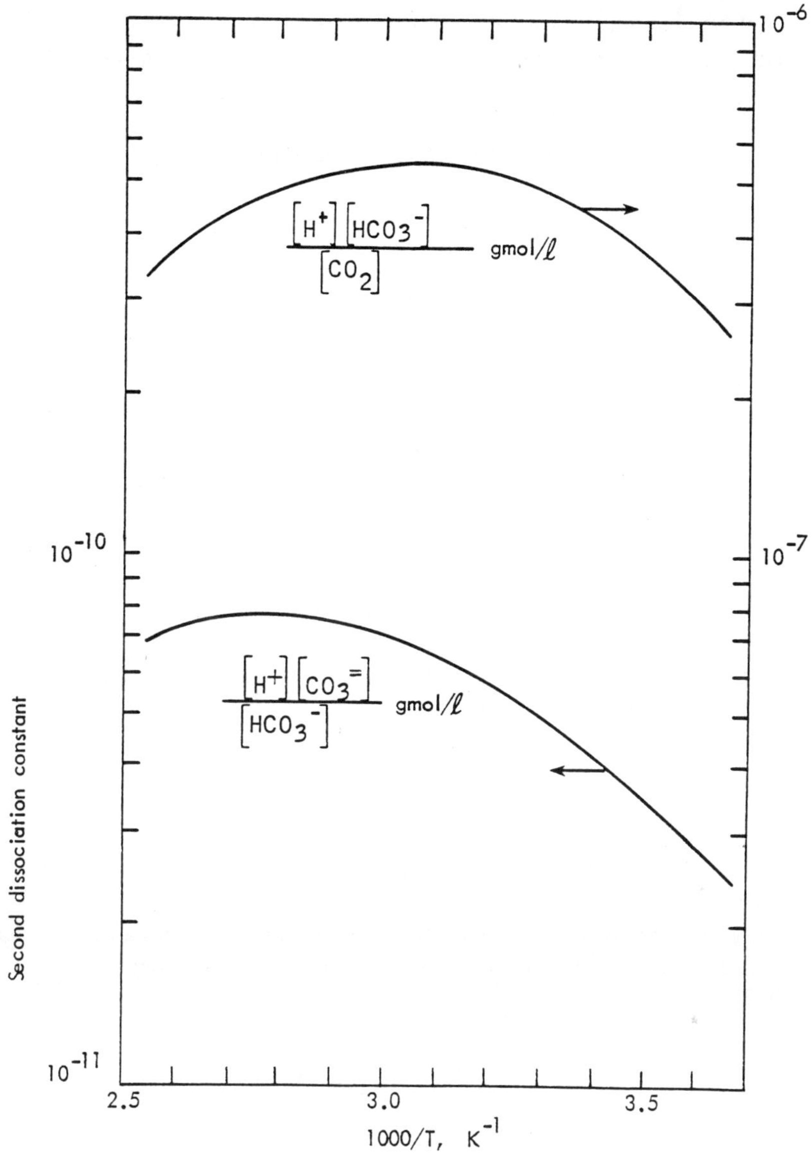

FIGURE 9.2.2. First and second dissociation constants of carbonic acid.

line solutions cannot be regenerated, as will be discussed in Section 9.3, and are, therefore, not used widely as chemical solvents.

Since all chemical solvents for CO_2 are alkaline, a very important reaction is the formation of the bicarbonate ion, i.e.,

$$CO_2 + OH^- \rightleftharpoons HCO_3^- \tag{9.2.1}$$

Reaction 9.2.1 may take place via two different mechanisms. The first, and most important, is the direct one where reaction 9.2.1 is the actual kinetic step, with rate given by

$$r = \mathrm{k}[OH^-]\,([CO_2] - \overline{[CO_2]}) \tag{9.2.2}$$

where

$$\overline{[CO_2]} = \frac{[HCO_3^-]}{[OH^-]\,K_2} \tag{9.2.3}$$

with K_2 the thermodynamic constant for reaction 9.2.1.

The kinetic constant k in Eq. 9.2.2 has been measured by a variety of techniques in dilute aqueous solutions at low temperatures; the data are well represented by the following equation:

$$\log \mathrm{k} = 13.635 - \frac{2895}{T} \tag{9.2.4}$$

which correlates the infinite dilution data very well in the range 0–40°C. At room temperature, the value of k is a strong increasing function of ionic strength; the following equation is recommended:

$$\log \frac{\mathrm{k}}{\mathrm{k}_{\mathrm{inf.\,dil}}} = 0.08\, I_i \tag{9.2.5}$$

The value of k has been measured in the range 45–110°C for an ionic strength of 7.75 gion/liter. The activation energy is the same as at infinite dilution. In summary, the following equation for k is reasonably well established and can be used with confidence at all ionic strengths of interest, as well as in the temperature range 0-110°C.

$$\log \mathrm{k} = 13.635 - \frac{2895}{T} + 0.08\, I_i \tag{9.2.6}$$

The second mechanism for reaction 9.2.1 is believed to be the direct reaction of CO_2 with water, followed by fast dissociation of the carbonic acid. The forward rate is given by

$$r = \hat{\mathrm{k}}[CO_2] \tag{9.2.7}$$

However, the value of $\hat{\mathrm{k}}$ is rather low, so that only at pH values below 8 has this second mechanism a rate comparable to that of the direct mechanism.

The formation of the bicarbonate ion can be catalyzed by a large variety of dissolved species. The most efficient catalyst is the carbonic anhydrase enzyme which is present in the human lung and serves to speed up the transfer of CO_2 from the blood to the lung cavities. Catalysis of the bicarbonate formation reaction is of great importance in many gas treating processes and will be discussed in more detail in the next sections.

9.3. HYDROXIDE SOLUTIONS

The main reaction of CO_2 in hydroxide solutions is

$$CO_2 + 2OH^- = CO_3^{2-} + H_2O \tag{9.3.1}$$

Formation of the bicarbonate ion is negligible as long as there is an appreciable concentration of free hydroxide. Only two non-volatile components need be considered: $B_1 = OH^-$ and $B_2 = CO_3^{2-}$, with the molarity and the fractional saturation given by

$$m = b_1 + 2b_2 \tag{9.3.2}$$

$$my = b_2 \tag{9.3.3}$$

Notice that the fractional saturation cannot exceed the value 0.5.

The equilibrium partial pressure of CO_2 is easily calculated from the equilibrium of reaction 9.3.1; one obtains, if K is the equilibrium constant for reaction 9.3.1,

$$p = \frac{\mathrm{H}}{\mathrm{K}m} \frac{y}{(1 - 2y)^2} \tag{9.3.4}$$

Even at a temperature as high as 100°C, and at the comparatively low molarity of 1 gmol/liter, the quantity $\mathrm{H}/\mathrm{K}m$ is only 6.7×10^{-8} atm. This has two important consequences. It is practically impossible to regenerate a carbonated hydroxide solution by stripping the CO_2; hence hydroxide solutions can only be used in "once through" operations. The second point is that hydroxide solutions can be used to lower the partial pressure of CO_2 in a gas stream to exceedingly low values; thus they can be used for removing the last traces of CO_2 from a gas from which the bulk of the CO_2 content has been eliminated by some other process.

Reaction 9.3.1 takes place via reaction 9.2.1 followed by

$$HCO_3^- + OH^- = CO_3^{2-} + H_2O \tag{9.3.5}$$

Reaction 9.3.5 only involves a proton transfer and is, therefore, instantaneous. Therefore, the rate of reaction 9.3.1 is given by Eq. 9.2.2; the reverse rate is negligible since the concentration of HCO_3^- is negligible. The reaction time is given by

$$t_r = \frac{1}{kb_{10}} = \frac{1}{km_0(1 - 2y_0)} \tag{9.3.6}$$

At low temperatures, t_r is of the order of 10^{-4} sec and, therefore, the value of Φ is typically of the order of 100 (here, and everywhere in the following, we regard 10^{-2} sec as a typical value for the diffusion time t_D). With increasing temperature, the value of Φ increases significantly and may be as large as 10^4.

As said before, hydroxide solutions are of industrial interest only for gases containing very low concentrations of CO_2, say, partial pressures of at most 10^{-3} atm. These correspond to a_i values of the order of 10^{-5} gmol/liter or less. Therefore, the values of I_∞, to be calculated from

$$I_\infty = \sqrt{\frac{D_{CO_2}}{D_{OH^-}}} + \sqrt{\frac{D_{OH^-}}{D_{CO_2}}}\,\frac{m_0(1 - 2y_0)}{a_i} \tag{9.3.7}$$

are typically of the order of 10^5. For conditions of industrial interest, I_∞ is always exceedingly larger than $\sqrt{\Phi}$ and, therefore, the process proceeds in the fast reaction regime; i.e.,

$$I = \sqrt{\Phi} = \sqrt{km_0(1 - 2y_0)\, t_D} \tag{9.3.8}$$

Values of I calculated from Eq. 9.3.8 are of the order of 10–100 and, therefore, liquid-side control of mass transfer is likely.

CO_2 absorption into hydroxide solutions is often used to test the relative merits of different types of packings. As long as the conditions discussed above are realized in such tests, what is really being measured is the interfacial area on the packing, but not the mixing conditions of the liquid phase, since in fast reaction regime the rate of mass transfer is independent of the hydrodynamics of the liquid phase. However, the interfacial area is possibly the most important parameter for characterizing the quality of packings.

Laboratory experiments can, of course, be performed under conditions where I_∞ is not exceedingly large as compared to $\sqrt{\Phi}$ and, therefore, instantaneous reaction conditions may be achieved. Careful experiments carried out under conditions of instantaneous reaction have confirmed the result in Eq. 9.3.7. Experiments in the region of transition from fast to instantaneous reaction have provided experimental verification of the theoretical transition curve for I calculated following the procedures discussed in Section 8.3. Several other laboratory experiments on this system have been reported in the literature; they all confirm the theoretical analysis.

9.4. AMINE SOLUTIONS

The chemistry of CO_2 reactions in amine solutions is remarkably complex and cannot be said to be entirely understood. A very important parameter is the stability of the carbamate, as measured by the value of the thermodynamic constant K_c. Values of K_c for three different aminoalcohols are plotted in Figure 2.4.2.

We first establish the main reactions that take place when CO_2 is absorbed in an amine solution. If the formation of CO_3^{2-} is neglected (even in solutions of the most alkaline amines, CO_3^{2-} will tend to form only at very low y values), Eq. 2.4.50 applies for all monoamine solutions. This equation is best rewritten in the following form:

$$\hat{b}_3 = \frac{P - \sqrt{P^2 - 4y(1-y)}}{2} \tag{9.4.1}$$

where $\hat{b}_3$ is the dimensionless carbamate concentration (i.e., the fraction of the molarity present in the carbamate ion form), and P is defined by

$$P = 1 + \frac{1}{K_c m} \tag{9.4.2}$$

The parameter P may range from 1 (extremely stable carbamate; the MEA example discussed in Section 2.4 was approximated by this limiting behavior) to ∞ (extremely unstable carbamate; tertiary amines are the limiting example). Figure 2.5.1 is a plot of the solution to Eq. 9.4.1 for $P = 2$.

First of all, it may be noticed that the solution to Eq. 9.4.1 is symmetrical with respect to $y = 0.5$, no matter what the value of P may be. This implies that, for all monoamines, at $y < 0.5$ the carbamate concentration tends to increase with increasing y, whereas at $y > 0.5$ it decreases with increasing y. When $P \to 1$, carbamate formation at $y < 0.5$, and carbamate reversion to bicarbonate at $y > 0.5$, are the only main reactions to take into account. However, in solutions of amines characterized by moderately stable carbamates, the situation is more complex since direct formation of the bicarbonate is also a main reaction at all values of y. In the limit of extremely unstable carbamates (e.g., with tertiary amines) direct formation of the bicarbonate is, in fact, the only main reaction that need be considered.

In conclusion, the following three main reactions need to be considered:

$$\text{CF: } CO_2 + 2RNH \longrightarrow RNH_2^+ + RNCOO^- \tag{9.4.3}$$

$$\text{BF: } CO_2 + RNH + H_2O \longrightarrow RNH_2^+ + HCO_3^- \tag{9.4.4}$$

$$\text{CR: } RNCOO^- + CO_2 + 2H_2O \longrightarrow RNH_2^+ + 2HCO_3^- \tag{9.4.5}$$

with CF taking place at $y < 0.5$, CR at $y > 0.5$, and BF at all values of y; the relative importance of the BF reaction increases with decreasing carbamate stability.

Carbamate formation (CF) is moderately well understood and has been studied extensively for monoethanolamine (MEA), where its forward rate has been established to be first-order with respect to both CO_2 and RNH:

$$r = k_{CF}[CO_2]\,[RNH] \tag{9.4.6}$$

The mechanism is believed to be made up of two steps, the rate-determining one being the formation of a zwitterion:

$$CO_2 + RNH \xrightarrow{\text{slow}} R\overset{+}{N}HCOO^- \tag{9.4.7}$$

$$R\overset{+}{N}HCOO^- + RNH \xrightarrow{\text{fast}} RNH_2^+ + RNCOO^- \tag{9.4.8}$$

Although a kinetic equation such as 9.4.6 has been well established for MEA and is believed to apply also for a variety of other amines, there is conflicting evidence regarding the next most studied amine, diethanolamine (DEA). Some data seem to support Eq. 9.4.6, but three sets of data indicate that the forward rate is second-order with respect to RNH. A suggested explanation for this may be that, in the case of DEA, reaction 9.4.8 is the rate-determining step. However, this appears unlikely: first, reaction 9.4.8 is a proton-transfer reaction; second, DEA and MEA have very similar chemical structures that make it hard to believe that the kinetic constants for reaction 9.4.8 are vastly different.

The zwitterion $R\overset{+}{N}HCOO^-$ is sometimes identified with its neutral structure (i.e., undissociated carbamic acid RNCOOH). Inasmuch as neither species can be isolated, the distinction may seem irrelevant but, in fact, it is not; tertiary amines could form a zwitterion $R_1R_2R_3\overset{+}{N}COO^-$ but, of course, they cannot form the neutral carbamic acid.

Bicarbonate formation (BF) can, of course, take place via the direct mechanism discussed in Section 9.2. If this were the only possible mechanism, the rate of BF could be easily calculated, provided a thermodynamic model were available for evaluating the hydroxyl ion concentration $[OH^-]$. However, it appears that BF in amine solutions takes place more rapidly than can be accounted for on the basis of reaction 9.2.1; this is briefly discussed in the following.

The best way of studying the BF reaction in amine solutions is to analyze the behavior of tertiary amine solutions where BF is the only overall reaction to take place. (In the case of tertiary alkanolamines, the alcoholic group has some reactivity with CO_2, but the alkylate $\rightarrow$ bicarbonate equilibrium is displaced to the right.) There is evidence in the literature that BF in tertiary amine solutions proceeds at a rate substantially higher than the rate of reaction 9.2.1; furthermore, the rate appears to be proportional to the amine concentration and possibly independent of the hydroxyl ion concentration.

A possible explanation of these results is that tertiary amines are catalysts for reaction 9.2.1 in the sense that they form an intermediate that can then more easily react with water to produce the bicarbonate ion. A possible intermediate is, of course, the zwitterion itself; in this case, the mechanism of the catalyzed reaction would be reaction 9.4.7 followed by

$$R\overset{+}{N}COO^- + H_2O \longrightarrow RNH^+ + HCO_3^- \tag{9.4.9}$$

The reaction of the zwitterion with water may, in fact, be rate-determining since it is not a proton-transfer reaction. Whichever is the rate-determining step, the rate would be first-order with respect to CO_2 and RNH:

$$r \propto [CO_2]\,[RNH] \tag{9.4.10}$$

The apparent kinetic constant, however, would be k_{CF} if the zwitterion formation is rate-determining, whereas it would have a different and lower value if reaction 9.4.9 is rate-determining.

Carbamate reversion to bicarbonate can also take place via parallel mechanisms. The first was discussed in the literature, and experimental evidence for it was obtained for CO_2 absorption into MEA solutions at y values in excess of 0.5. This mechanism simply consists in the reverse of the CF reaction, followed by direct formation of HCO_3^- via the CO_2 reaction with OH^-. The sequence of elementary steps is:

$$\left.\begin{array}{l} RNCOO^- + RNH_2^+ \longrightarrow R\overset{+}{N}HCOO^- + RNH \\ R\overset{+}{N}HCOO^- \longrightarrow RNH + CO_2 \\ 2CO_2 + 2OH^- \longrightarrow 2HCO_3^- \text{ (slow)} \\ 2RNH + 2H^+ \longrightarrow 2RNH_2^+ \\ 2H_2O \longrightarrow 2OH^- + 2H^+ \end{array}\right\} \tag{9.4.11}$$

The overall rate of CO_2 consumption by this route is one-half the rate of the CO_2 + $OH^- \rightarrow HCO_3^-$ reaction; the value is very low, since amine solutions where the CR reaction takes place have very low alkalinity.

The alternate mechanism starts again with formation of the zwitterion from the carbamate, and is followed by the zwitterion reaction with water:

$$\left.\begin{array}{l} RNCOO^- + RNH_2^+ \longrightarrow R\overset{+}{N}HCOO^- + RNH \\ CO_2 + RNH \longrightarrow R\overset{+}{N}HCOO^- \\ 2R\overset{+}{N}HCOO^- + 2H_2O \longrightarrow 2RNH_2^+ + 2HCO_3^- \end{array}\right\} \tag{9.4.12}$$

The overall rate via this mechanism is again given by Eq. 9.4.10, with the proportionality constant being k_{CF} if reaction 9.4.7 is slower than reaction 9.4.9, and having some lower value in the converse case.

With this background, we may classify amine solutions by the value of the parameter P, and consider three basic cases.

Case 1. $P \rightarrow 1$ (very stable carbamate). The MEA solution discussed in Section 2.4 is a good example.

At $y < 0.5$, CF is the only main reaction, whereas at $y > 0.5$, CR is the only main reaction. The vapor pressure of CO_2 at low temperature and low y is exceedingly low (see Figure 9.4.1); therefore, the residual content of CO_2 in the gas can be brought to very low levels (< 100 ppm) with MEA solutions.

However, K_c decreases with increasing temperature (see Figure 2.4.2); therefore, at the regeneration temperature, P is appreciably larger than unity even for MEA. This results in comparatively high vapor pressures (see Figure 9.4.2) and regeneration to very low values of y is feasible.

The rate behavior at $y < 0.5$ is governed by the CF kinetics. The free amine concentration is $b_1 = m(1 - 2y)$ (see Eq. 2.4.59); whereas the main reaction is 9.4.3. Therefore, the rate r is given by

$$r = k_{CF} b_1 (a - \bar{a}) = k_{CF} m(1 - 2y)(a - \bar{a}) \qquad (9.4.13)$$

where

$$\bar{a} = \frac{1}{K} \frac{b_2 b_3}{b_1^2} = \frac{1}{K} \left(\frac{y}{1 - 2y} \right)^2 \qquad (9.4.14)$$

with K the equilibrium constant of reaction 9.4.3 (we use the b_js as in Section 2.4). The kinetic constant k_{CF} has reasonably well-known values; an Arrhenius plot for k_{CF} is given in Figure 9.4.3. A Bronsted plot is given in Figure 9.4.4.

The reaction time (to be calculated at the bulk-liquid composition) is given by

$$t_r = \frac{1}{k_{CF} m_0 (1 - 2y_0)} \qquad (9.4.15)$$

From the data in Figure 9.4.3, values of Φ are estimated to be in the range 10–1000 and, therefore, significant rate enhancement is expected. This is well confirmed by experimental evidence obtained both with laboratory absorbers and with small packed columns.

The values of I_∞ should be calculated from the following equations (based on equal diffusivities for all species):

$$\psi \gg I, \qquad I_\infty = 1 + \frac{m_0(1 - 2y_0)}{a_i} \qquad (9.4.16)$$

$$\psi \ll I, \qquad I_\infty = 1 + m_0 K \frac{1 - 2y_0}{y_0} \qquad (9.4.17)$$

$$|\psi - 1| \ll 1, \qquad I_\infty = 1 + m_0 K \frac{1 + 2y_0^2 - 3y_0}{y_0(2 - y_0)} \qquad (9.4.18)$$

which are obtained from Eqs. 6.2.17, 6.2.18, and 6.2.20, with a_0 calculated from Eq. 2.4.41.

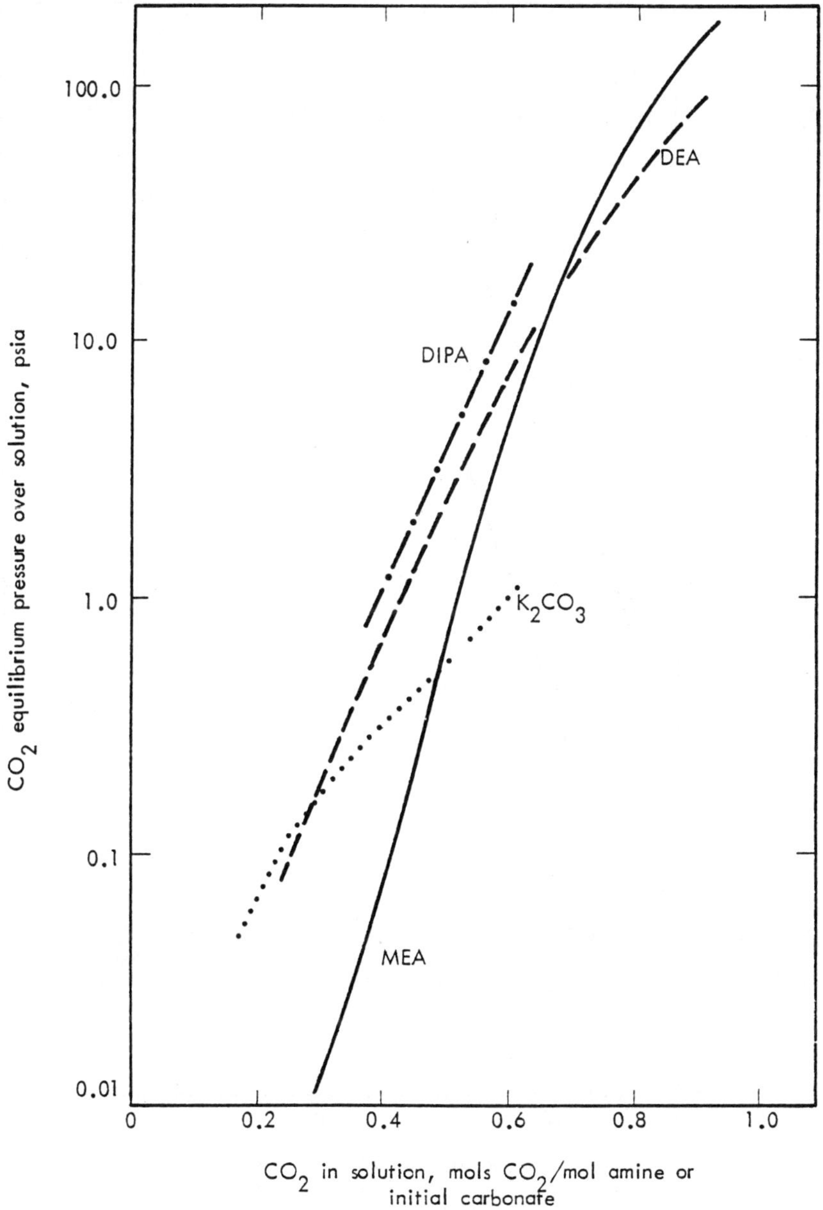

FIGURE 9.4.1. Equilibrium pressures of CO_2 over aqueous aminoalcohol and potassium carbonate solutions at absorber top conditions (40°C).

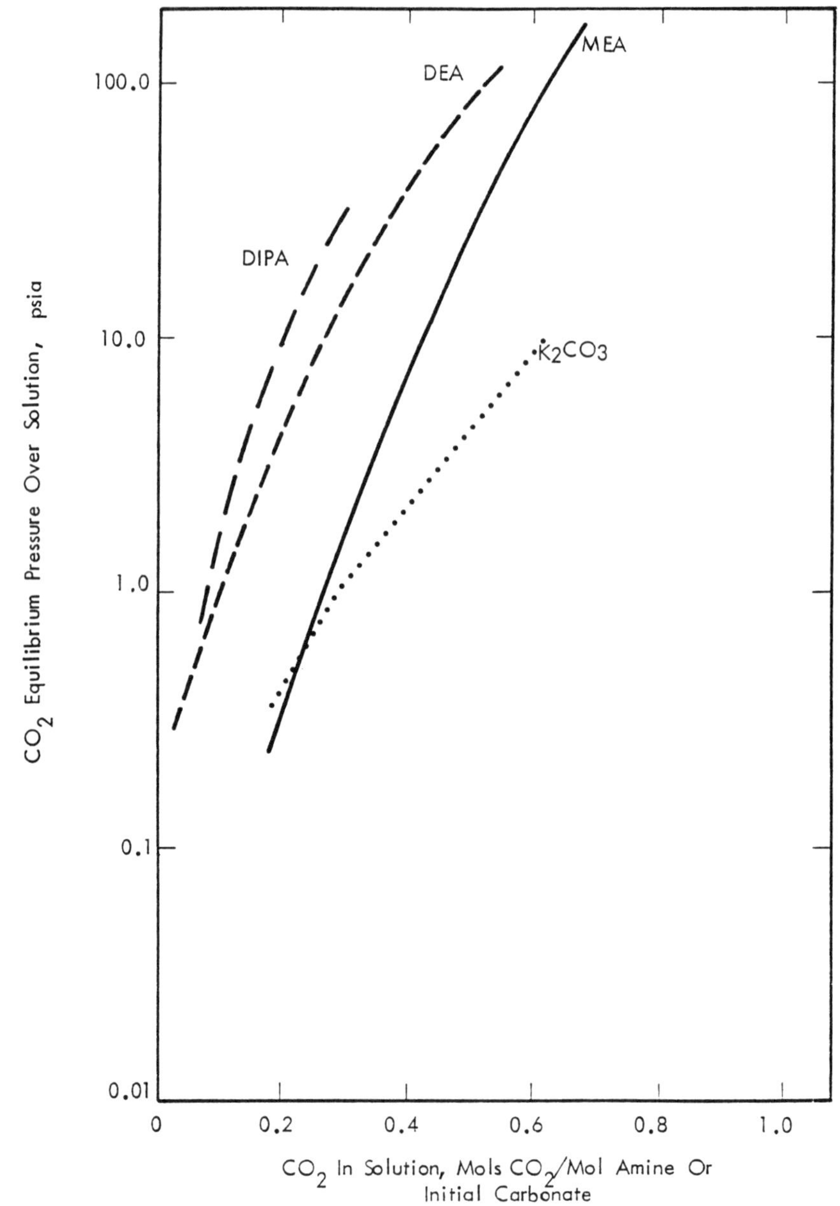

FIGURE 9.4.2. Equilibrium pressures of CO_2 over aqueous aminoalcohol and potassium carbonate solutions at regenerator conditions (100°C).

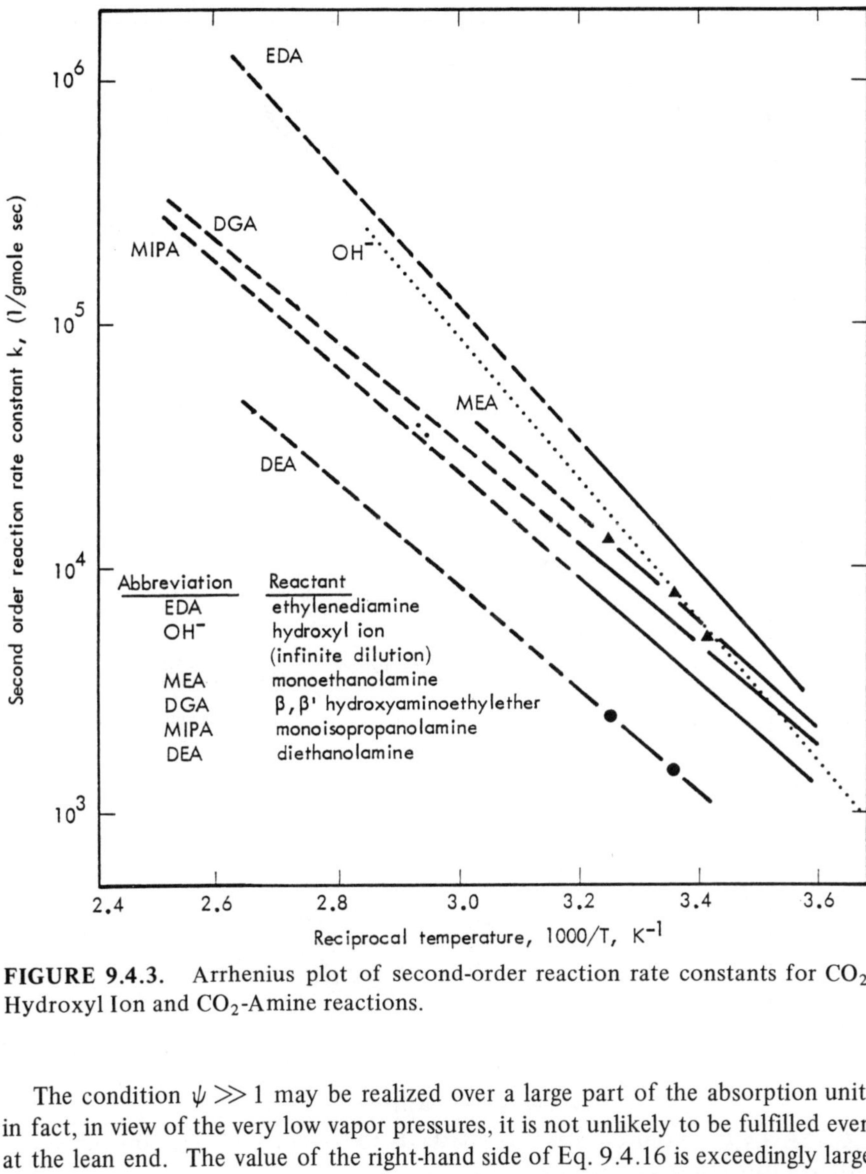

FIGURE 9.4.3. Arrhenius plot of second-order reaction rate constants for CO_2-Hydroxyl Ion and CO_2-Amine reactions.

The condition $\psi \gg 1$ may be realized over a large part of the absorption unit; in fact, in view of the very low vapor pressures, it is not unlikely to be fulfilled even at the lean end. The value of the right-hand side of Eq. 9.4.16 is exceedingly large at the lean end, say, as much as 10^5. Therefore, at the lean end, fast reaction conditions are expected to prevail. At the rich end of the absorber, the right-hand side of Eq. 9.4.16 may be small enough for instantaneous reaction conditions to be approached.*

*If such is the case, during simultaneous absorption of CO_2 and H_2S, H_2S may actually be *desorbed* in the lower part of the absorber via a mechanism of the type discussed in Section 8.4. There is some experimental evidence that this does, in fact, occur (see Figure 1.10.4).

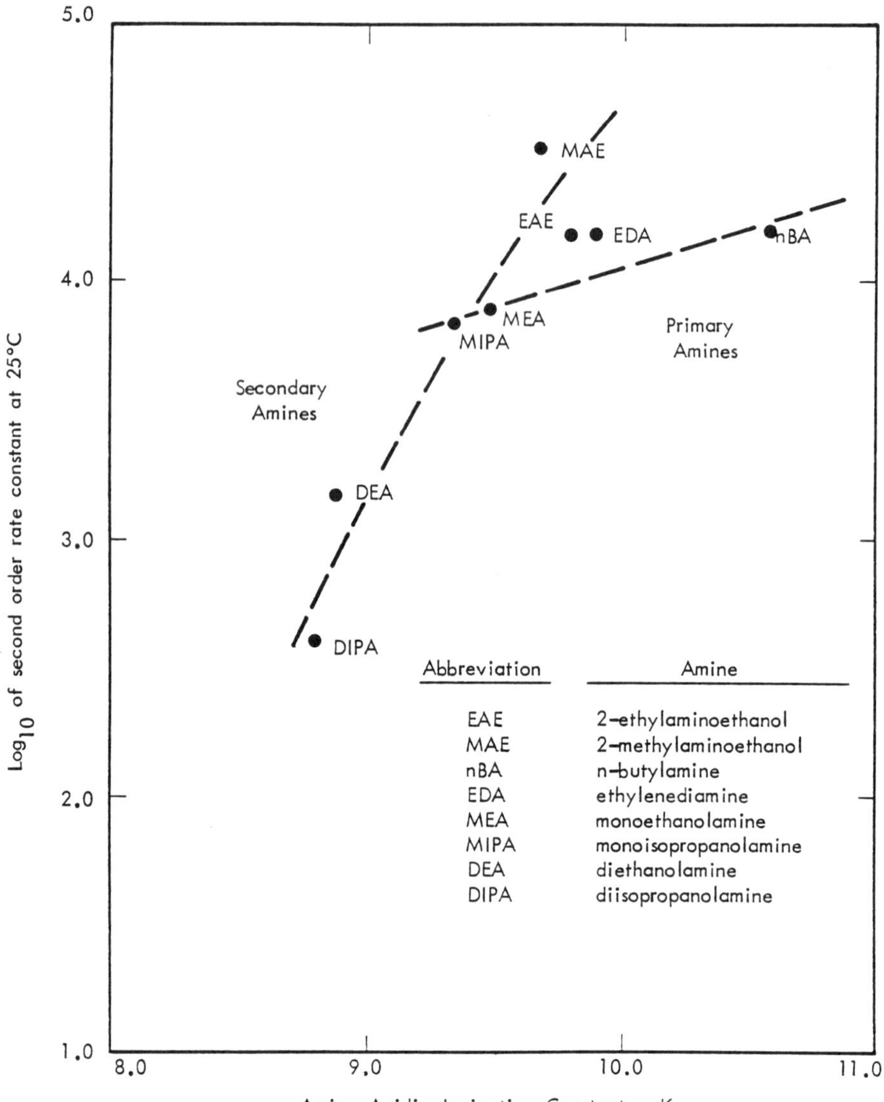

FIGURE 9.4.4. Second-order rate constants for reaction of CO_2 with amines of different base strength.

The condition $\psi \ll 1$ may be realized over some portion of the regeneration tower. The quantity m_0K is of the order of 10^3 at 100°C and, therefore, the right-hand side of Eq. 9.4.17 is very large unless y_0 approaches 0.5—which it does not in the regeneration unit if part of the desorption is carried out as a flash. Therefore, fast reaction conditions are expected.

Finally, the condition $|\psi - 1| \ll 1$ may be reached at the lean end of both absorption and desorption units and may, in fact, prevail over the whole desorption tower. The right-hand side of Eq. 9.4.18 is always very large (since y_0 is small at the lean end) and, therefore, again fast reaction conditions are expected.

In conclusion, fast reaction conditions are most likely, and the enhancement factor is given by

$$I = \sqrt{\frac{t_d}{t_r}} = \sqrt{\mathrm{k}_{\mathrm{CF}} m_0 (1 - 2y_0)\, t_D} \tag{9.4.19}$$

For absorption at $y > 0.5$, the carbamate must revert to bicarbonate. Both mechanisms for the CR reaction are very slow since both the OH^- and the RNH concentrations are very low. Inasmuch as k_{CF} is of the same order of magnitude as the kinetic constant k for reaction 9.2.1, the mechanism 9.4.12 is expected to be faster than 9.4.11, since the concentration of free amine, albeit small, is larger than the OH^- concentration. The only available experimental evidence is for MEA at room temperature; for this system, the fastest route appears to be the mechanism 9.4.11, but with the $CO_2 + OH^- \rightarrow HCO_3^-$ step substituted by the direct reaction of CO_2 with water, so that the rate is essentially given by Eq. 9.2.7.

In fact, it has been shown experimentally that the rate of CR in MEA solutions can be significantly enhanced by addition of homogeneous catalysts for reaction 9.2.1 such as arsenite ions. However, no industrial application of this is in existence.

A conventional single-stage flowsheet is generally used for CO_2 removal in ammonia plants using aqueous MEA. Two-stage and split-flow arrangements are not advantageous with MEA. The conventional flowsheet is shown in Figure 9.4.5. Water wash at the head of the tower is extremely important to prevent excessive carryover of the rather volatile MEA.

The traditional aqueous aminoalcohol process, typified by 3 molar MEA, fell from favor in the 1960s for bulk CO_2 removal owing to its high energy requirement. However, aqueous MEA appears to be making a comeback. The technical advance responsible for this is the discovery of corrosion inhibitors that allow MEA to be used up to concentrations of about 5 molar (30 wt%). The inhibited high concentration MEA process is competitive with organic-promoted hot carbonate processes for CO_2 removal in ammonia plants for gases containing no H_2S. The inhibitors break down in the presence of H_2S.

In the traditional MEA process, the "rich" CO_2 loading was limited to about 0.45 moles CO_2/mole MEA. The introduction of new inhibitors allows MEA concentration to be raised to $5M$, and provides an option to increase rich loading from 0.45 to over 0.6 moles CO_2/mole MEA. Raising rich loading above 0.45 moles/mole MEA will reduce reboiler heat duty, but it is apparently not economical to do so in ammonia plants because debits are incurred of higher CO_2 leakage, higher absorber pressure drop, and higher investment. Table 9.4.1 summarizes information on the traditional and high-concentration MEA processes for CO_2 removal.

The debits associated with "rich" loadings in excess of 0.5 moles CO_2/mole MEA are caused basically by the change in the mechanism of diffusion-reaction discussed before. At $y > 0.5$, the reactions are known to be slow and require a

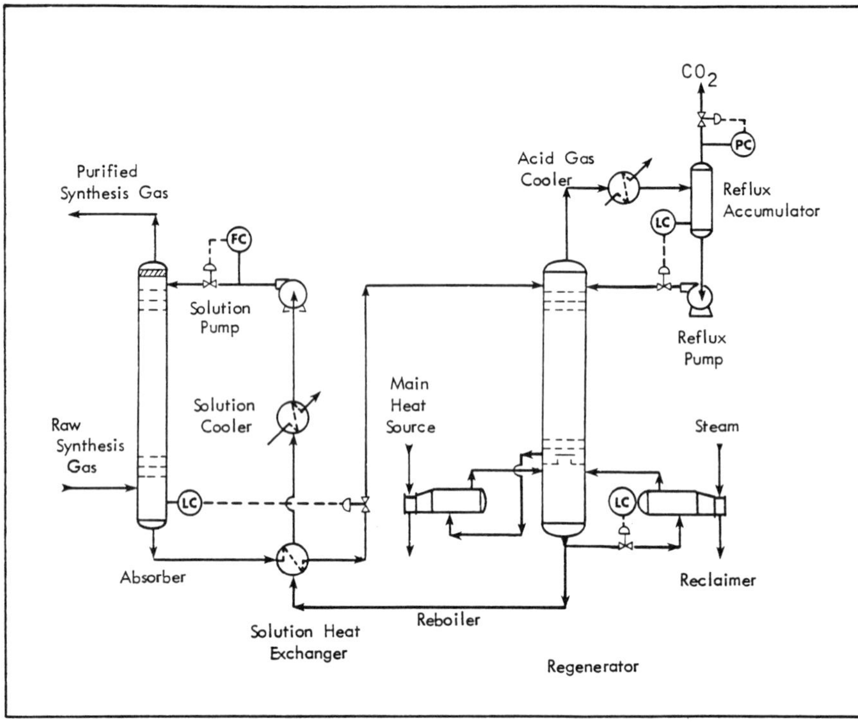

FIGURE 9.4.5. Flow diagram for a traditional MEA-CO_2 scrubbing unit.

considerable residence time of the liquid in the absorption apparatus to proceed to any significant extent. Thus achieving a rich loading of 0.6 moles CO_2/mole MEA requires the number of trays in the absorber to be doubled and high weirs to be employed on the bottom trays to give the required liquid residence time. These factors account for the higher absorber pressure drop and investment.

The desorption or stripping of CO_2 from aqueous MEA has been discussed in the literature. The quantity of stripping steam required depends importantly on the solution lean loading needed to produce the required product gas purity at the top of the absorber. The removal of CO_2 from MEA solutions is a difficult stripping operation owing to the high stability of the carbamate of MEA. The shape of the equilibrium curve sets the minimum stripping steam requirement for MEA. Increasing the number of trays is of little value. A representative desorber operating-equilibrium plot is shown in Figure 1.10.2, showing that a large fraction of the stripper tower operates under "pinch" conditions.

Case 2. $P \rightarrow \infty$ (very unstable carbamate; any tertiary monoamine is a good example). This is possibly the simplest case, inasmuch as BF is the only reaction to be considered at all values of y. The equilibrium vapor pressure of CO_2 is calcu-

TABLE 9.4.1. Improvements to Aqueous Monoethanolamine Process for CO_2 Removal

MEA process	MEA conc. Wt%	Lean loading (mols CO_2/ mol MEA)	Rich loading (mols CO_2/ mol MEA)	Reboiler heat duty (BTU/lb mole CO_2)	CO_2 leakage from absorber (VPPM)	Comments
Traditional	15–20	0.10	0.45	90,000[a]	<100	Expensive materials of construction
Inhibited high-concentration[b]						
Operating mode 1	30	0.10	0.45	53,000	~100	
Operating mode 2	30	0.10	0.65[c]	43,000	~1,000	Higher absorber pressure drop, higher CO_2 leakage, higher investment

[a]Including heat for reclaimer operation.
[b]A formulation containing a corrosion inhibitor including vanadium and antimony.
[c]Estimated by author from vapor–liquid equilibrium data.

lated as

$$p^* = \frac{HK_p}{K_{c1}} m \frac{y^2}{1 - y} \tag{9.4.20}$$

where K_p is the amine protonation constant and K_{c1} is the first dissociation constant of carbonic acid (values of K_p for several amines are given in Section 2.4). The right-hand side of Eq. 9.4.20 does not, or course, present any singularity at $y = 0.5$.

In spite of their attractive VLE characteristics, tertiary amines are not used widely for bulk CO_2 removal because of the low rates of reaction; BF is a rather slow reaction in tertiary amines, whichever of the two mechanisms prevails. The rate analysis is as follows.

At $y = 0$, the OH^- concentration is calculated as

$$[OH^-] = \sqrt{\frac{K_w}{K_p} m} \tag{9.4.21}$$

At sufficiently large values of y, the $[OH^-]$ concentration is

$$[OH^-] = \frac{K_w}{K_p} \frac{1 - y}{y} \tag{9.4.22}$$

Eq. 9.4.22 is correct provided it yields a value of $[OH^-]$ appreciably less than that given by Eq. 9.4.21, i.e., provided that

$$\frac{1 - y}{y} < \sqrt{m \frac{K_p}{K_w}} \tag{9.4.23}$$

The right-hand side of Eq. 9.4.23 is, for tertiary amines, of the order of 1000; therefore, Eq. 9.4.22 should be used at practically all values of y except when $y < 10^{-3}$. The pH of even a very slightly carbonated solution of a tertiary amine is very appreciably less than that of a solution of the pure amine.

Should the BF reaction take place via the $CO_2 + OH^- \rightarrow HCO_3^-$ mechanism, its rate would be independent of the molarity m (see Eq. 9.4.22). Experimental evidence seems, on the contrary, to suggest that Eq. 9.4.10 applies; therefore, the mechanism involving the zwitterion formation is more likely to be predominant. Furthermore, the OH^- concentrations calculated from Eq. 9.4.22 are low enough (for K_p values in the range 10^{-8}-10^{-9}) that no significant rate enhancement would be expected up to temperatures of about 60°C; in contrast with this, significant (but not very large) rate enhancement has been observed at 50°C.

Although BF via the zwitterion formation is faster than it would be if the amine provided only the alkalinity required for the $CO_2 + OH^- \rightarrow HCO_3^-$ reaction, it is still a rather slow reaction and enhancement factors are too low to make tertiary amines attractive for bulk CO_2 removal. However, for this very reason, solutions of tertiary amines are of interest for selective removal of H_2S from gases containing substantial amounts of CO_2.

Case 3. $P \simeq 2$ (moderately stable carbamate). Diisopropanolamine at moderately high temperature is a good example (see Figure 2.4.1). This is the most complex case, since the solution to Eq. 9.4.1 cannot be linearized.

At $y < 0.5$, both CF and BF are main reactions. We first introduce the concept of the thermodynamic progress ratio of two reactions. Given a solution with a given value of y, an additional intake of CO_2 (corresponding to an increase dy of the fractional saturation) will result, at equilibrium, in an increase $d\hat{b}_3$ of the carbamate concentration and an increase $d\hat{b}_4$ of the bicarbonate concentration (dimensionless concentrations are being considered). The values of $d\hat{b}_3$ and $d\hat{b}_4$ are calculated as

$$d\hat{b}_3 = \frac{db_3}{dy}\, dy = \frac{1 - 2y}{\sqrt{P^2 - 4y(1 - y)}}\, dy \tag{9.4.24}$$

$$d\hat{b}_4 = \left[1 - \frac{1 - 2y}{\sqrt{P^2 - 4y(1 - y)}}\right] dy \tag{9.4.25}$$

The increase $d\hat{b}_3$ is due to the progress of the CF reaction, whereas $d\hat{b}_4$ is due to the progress of the BF reaction. Therefore, from a thermodynamic viewpoint, the progress ratio for the two reactions would be

$$\left(\frac{\text{CF}}{\text{BF}}\right)_T = \frac{d\hat{b}_3}{d\hat{b}_4} = \frac{1}{\sqrt{[P^2 - 4y(1 - y)]}/(1 - 2y) - 1} \tag{9.4.26}$$

At $y > 0.5$, Eqs. 9.4.24 and 9.4.25 would still apply, with $d\hat{b}_3$ negative. In this case, $-d\hat{b}_3$ is the progress of the CR reaction, whereas $d\hat{b}_4$ is due to the progress of *both* the CR and the BF reactions; therefore, the progress of the BF reaction is measured by $d\hat{b}_4 + d\hat{b}_3$, and one has

$$\left(\frac{\text{CR}}{\text{BF}}\right)_T = \frac{-d\hat{b}_3}{d\hat{b}_4 + d\hat{b}_3} = \frac{2y - 1}{\sqrt{P^2 - 4y(1 - y)}} \tag{9.4.27}$$

(The suffix T on the left-hand side of Eqs. 9.4.26 and 9.4.27 is a reminder that the progress ratio is calculated from purely thermodynamic considerations.)

The ratio of the *rates* of the two main reactions, of course, need not coincide with the thermodynamic progress ratio. The progress ratio that will occur kinetically is, at very low times, simply the ratio of the rates of the two reactions; given a sufficiently long time, however, the thermodynamic progress ratio will be achieved. From the viewpoint of rate enhancement, only the short-time behavior is relevant, inasmuch as progress taking place over a time-scale exceeding the diffusion time has no effect on the mass transfer rates. Incidentally, experimental determination of the kinetics of two competing reactions, based on chemical analysis of the liquid phase or on any technique other than absorption rate measurements, will yield a progress ratio that depends on the time-scale over which the measurement is made. If the time-scale exceeds the two reaction times involved, the technique will yield simply the thermodynamic progress ratio, and no real information has been obtained on the ratio of the rates of reaction. Some of the data in Figures 9.4.3 and 9.4.4 may suffer from such uncertainty.

At $y < 0.5$, the two reactions are CF and BF. If the latter takes place via the zwitterion formation, and if step 9.4.9 is fast compared to 9.4.7, the overall rate is simply given by Eq. 9.4.6. Once the zwitterion is formed, it will undergo reactions 9.4.8 (leading to CF) and 9.4.9 (leading to BF) in a ratio that depends on their relative rates; presumably CF is faster, inasmuch as 9.4.8 is a proton-transfer reaction. However, since both 9.4.8 and 9.4.9 are fast compared to 9.4.7, the overall rate of CO_2 consumption is not influenced by the kinetic progress ratio.

The same argument applies also at $y > 0.5$ inasmuch as the two reactions (CR and BF) have the same slow step. It is, however, crucial that the parameter P be not too small so that a non-negligible concentration of amine is present, enabling CR to proceed via mechanism 9.4.12 rather than 9.4.11. The rate behavior of amines characterized by moderately stable carbamates is acceptable also at y values exceeding 0.5, in contrast to amines having very stable carbamates.

The parameter P also governs the equilibrium behavior of CO_2-amine systems. As was discussed in Section 2, except with very alkaline amines at $y \to 0$, formation of the carbonate ion may be neglected. With this approximation, the equilibrium partial pressure of CO_2 is calculated as

$$p^* = \frac{\mathrm{HK}_p}{\mathrm{K}_{c1}\mathrm{K}_c} \frac{f(y)\,y}{[1 - y - f(y)]^2} \qquad (9.4.28)$$

where $f(y)$ is the right-hand side of Eq. 9.4.1. Figure 9.4.6 shows a plot of Eq. 9.4.28 for three values of P: 1.01 ($\mathrm{K}_c m = 100$); 2 ($\mathrm{K}_c m = 1$); and 101 ($\mathrm{K}_c m = .01$). Different scales are used on the ordinate since different values of K_c are implied; the three scales would coincide if expressed as values of p^*. The figure shows the influence of P on the *shape* of the equilibrium curve; it should be compared with Figure 9.4.1. A direct comparison of the curve for $P = 1.01$ with CO_2-MEA data at low temperature is presented in Figure 9.4.7.

It is important to realize that the whole discussion given above, complex as it may appear, only applies to monoamines. In the case of diamines, the chemistry involved is remarkably more complex, particularly for unsymmetric diamines where

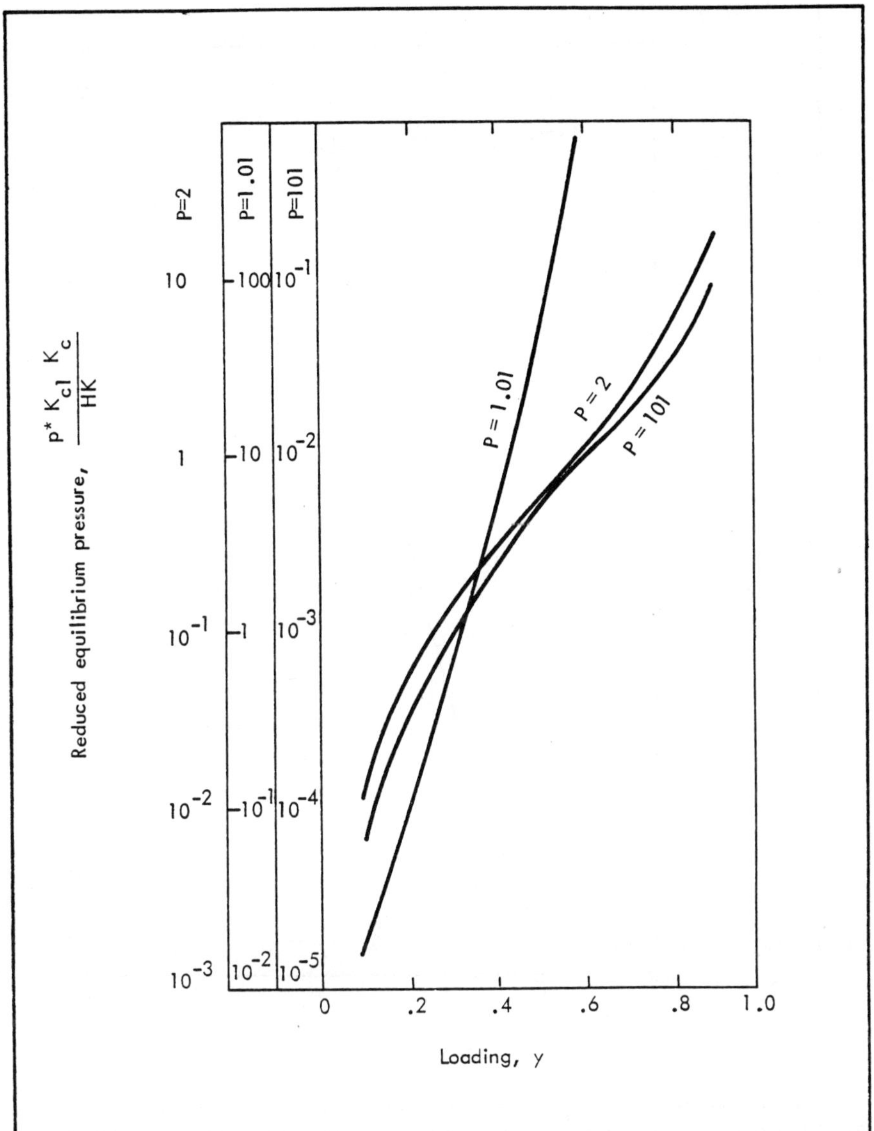

FIGURE 9.4.6. CO_2 vapor–liquid equilibrium curves for aqueous amines of different carbamate stability.

the two reactive nitrogen atoms may have different chemistries. In carbonated solutions of diamines, often a thermodynamically preferred form is the self-neutralizing carbamate; this, in turn, can easily transform to a zwitterion:

$$\overset{+}{H}NRNCOO^- \rightleftharpoons NR\overset{+}{N}HCOO^- \qquad (9.4.29)$$

so that the concentration of the zwitterion, though always very small, may be sub-

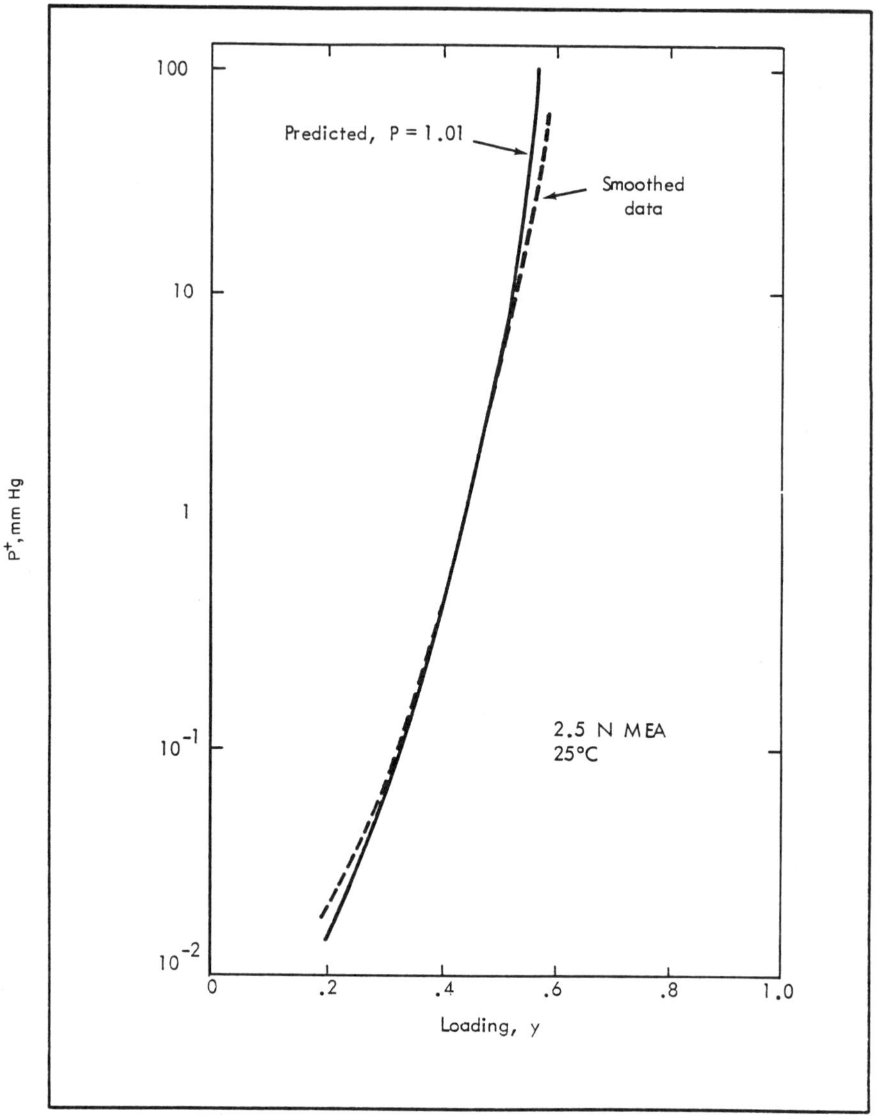

FIGURE 9.4.7. Comparison of CO_2 vapor–liquid equilibrium curve for aqueous MEA with predicted curve for amine having stable carbamate.

stantially larger than in solutions of monoamines. This may result in increased reactivity of diamines, particularly for the BF reaction.

9.5. CARBONATE SOLUTIONS

The thermodynamics of the system CO_2-aqueous carbonate solution have been discussed in Section 2.4. The main reaction that CO_2 undergoes is reaction 2.4.5. There are only two non-volatile components to be considered: $B_1 = CO_3^{2-}$ and $B_2 =$

HCO_3^-. Their concentrations are given in Eqs. 2.4.6 and 2.4.7 in terms of the molarity and the fractional saturation. The thermodynamic model developed in Section 2.4 leads to Eq. 2.4.13 which gives the equilibrium vapor pressure p^* in terms of the molarity m and the fractional saturation y.

Although Eq. 2.4.13 does very well in predicting the influence of y on p^* (see Figure 2.4.1), it does poorly in predicting the influence of molarity m: actual data show that p^* is proportional to the 0.4 power of the molarity. The activation energy for p^* appears to be independent of both m and y. All the available data are well correlated by the following equation:

$$p^* = 1.95 \times 10^9 \, m^{0.4} \left(\frac{y^2}{1 - y}\right) \exp\left(\frac{-8160}{T}\right) \tag{9.5.1}$$

with p^* in atmospheres, m in gmol/liter, and T in °K. The value of Henry's Law constant has also been determined for this system and it is well correlated by the following equation:

$$\log \frac{\mathrm{H}}{\mathrm{H}^0} = 0.125 \, m \tag{9.5.2}$$

where H^0 is the value of H in water at the same temperature, and m is the molarity in gmol/liter.

The chemical capacity of carbonate solutions is of the order of the molarity m, which can be made as high as 40% by weight; the upper limit is set by bicarbonate solubility. Gas treating processes using carbonate solutions are particularly efficient (in terms of pounds of steam required/pound of CO_2 removed) when both absorption and desorption are carried out at rather high temperatures, typically about 90°C for absorption and about 110°C for desorption.

The main reaction for CO_2 absorption in carbonate solution is reaction 2.4.5, which proceeds via reaction 9.2.1 followed by the reverse of reaction 2.4.14; the latter is instantaneous. The hydroxyl ion concentration is determined by the carbonate-bicarbonate equilibrium (see Eq. 2.4.16) which is reported below:

$$[\mathrm{OH}^-] = \mathrm{K}_1 \frac{b_1}{b_2} = \mathrm{K}_1 \frac{1 - y}{2y} \tag{9.5.3}$$

where K_1 is the ratio of the water dissociation constant and the constant plotted in the lower part of Figure 9.2.2. An Arrhenius plot for K_1 is reported in Figure 9.5.1.

Substitution of Eq. 9.5.3 into 9.2.2 gives the rate of the main reaction as

$$r = \mathrm{k} \, \mathrm{K}_1 \frac{1 - y}{2y} (a - \bar{a}) \tag{9.5.4}$$

where

$$\bar{a} = \frac{2y}{1 - y} \frac{1}{\mathrm{K}_2} \tag{9.5.5}$$

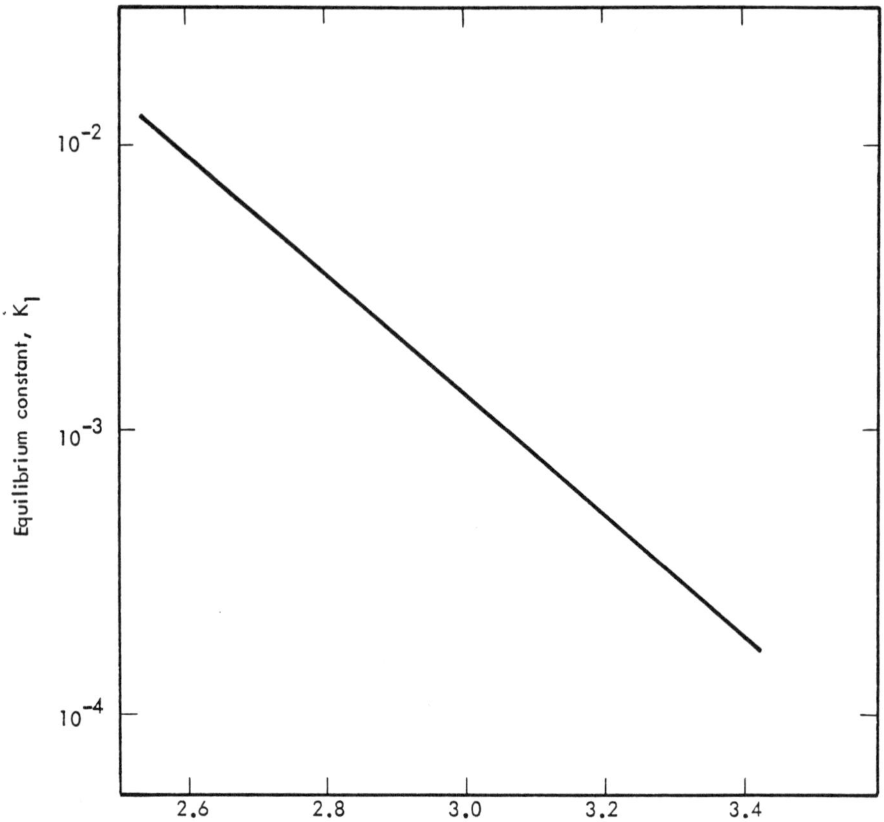

FIGURE 9.5.1. Equilibrium constant K_1.

with K_2 the equilibrium constant for reaction 9.2.1. The product $k\,K_1$ may be regarded as a pseudo-first-order kinetic constant for the main reaction; Figure 9.5.2 is an Arrhenius plot for $k\,K_1$ corresponding to an ionic strength of 7.75 gion/liter (which is typical of concentrations in industrial use).

The reaction time is given by

$$t_r = \frac{1}{k\,K_1}\,\frac{2y}{1-y} \tag{9.5.6}$$

With reference to Figure 9.5.2, it is seen that no significant rate enhancement is to be expected at temperatures below 40°C. This point has been widely confirmed experimentally; indeed, laboratory experiments carried out at low temperature have allowed determination of the Henry's Law constant H based on the equation

$$N = \frac{k_L^0\,(p_i - p^*)}{H} \tag{9.5.7}$$

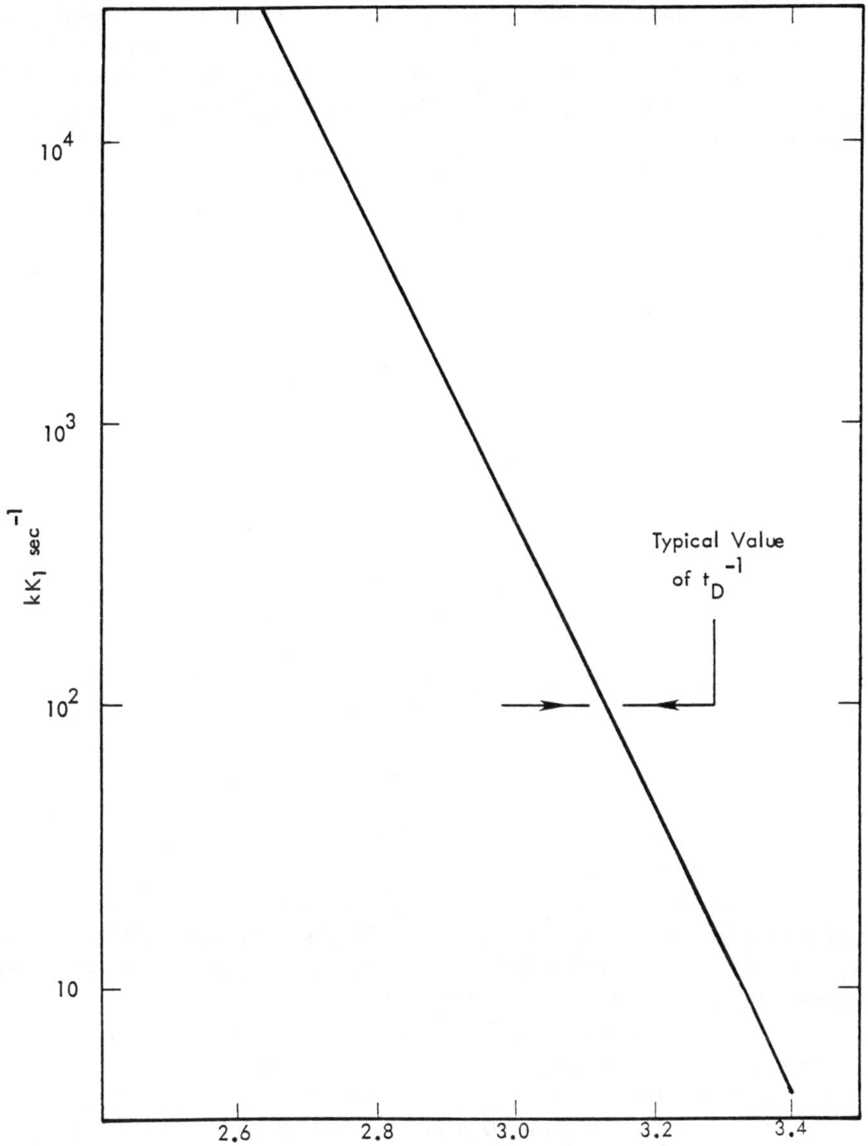

FIGURE 9.5.2. Pseudo-first-order kinetic constant for bicarbonate formation.

which, of course, only holds if the enhancement factor I is unity. Such experiments have allowed the establishment of the relationship for the dependency of H on molarity given in Eq. 9.5.2.

At the temperatures of interest in the hot carbonate process (90–110°C), Φ is of the order of 100 and, therefore, significant rate enhancement is expected.

The diffusivity D' of the two non-volatile components CO_3^{2-} and HCO_3^-, is very nearly the same and, therefore, there is no need to consider diffusive coupling of ions; furthermore, the interface molarity may be assumed to be the same as the bulk-liquid molarity (see Section 2.6). However, the value of D' is about one-half the diffusivity of CO_2 and, therefore, $D'/D = 0.5$ is significantly different from unity.

The value of I_∞ is to be calculated from the following equations (see Chapter 6):

$\psi \gg 1$: large driving force absorption

$$I_\infty = 1 + \sqrt{\frac{D'}{D}}\,\frac{m(1-y_0)}{a_i} \tag{9.5.8}$$

$\psi \ll 1$: large driving force desorption

$$I_\infty = 1 + \sqrt{\frac{D'}{D}}\,\frac{my_0}{a_0} \tag{9.5.9}$$

$|\psi - 1| \ll 1$: pinched mass transfer

$$I_\infty = 1 + \sqrt{\frac{D'}{DK}}\,\frac{(1-y_0)^2}{4y_0(2-y_0)} \tag{9.5.10}$$

The values of I_∞ calculated from Eqs. 9.5.8–9.5.10 are invariably very large, even at the absorber rich end. With ϕ values of the order of 100, fast reaction conditions will apply almost everywhere and, in fact, the effectiveness of rate enhancement is rather low so that very long columns are required. In actual modern industrial practice, additives to the carbonate solution (which act as rate promoters) are nearly always used. The chemistry of CO_2 reactions in promoted carbonate solutions is discussed in the following.

Rate promoters are components that are able to enhance the rate of bicarbonate formation over the value given by Eq. 9.5.4. They act, at least in a global sense, as homogeneous catalysts. Industrially, a rate promoter that has been used extensively is arsenious acid. In fact, most acids with a pK comparable to that of arsenious acid are known to have a promotional effect; boric acid is used in one industrial process. Another category of rate promoters is that of amines, which may be used by themselves or in conjunction with acid-type rate promoters.

The vapor–liquid equilibrium behavior of promoted carbonate solutions is, in general, somewhat different from that of unpromoted solutions. The whole $p^*(y)$ curve may be shifted either to lower or higher values of y and its shape may also be changed so that the cyclic capacity of the promoted solution is different from that of the unpromoted one. Not surprisingly, acid promoters tend to slightly decrease the cyclic capacity, whereas certain alkaline promoters such as amines may tend to increase it. Complete thermodynamic modeling of amine-promoted carbonate solutions is rather complex but not infeasible; that of acid-promoted solutions is relatively straightforward. A brief discussion is given in the following.

First consider the case of arsenious acid, H_3AsO_3. Consider a solution of m

moles/liter of K_2CO_3 to which an amount of mc moles/liter of arsenious acid is added. The following reaction may take place:

$$H_3AsO_3 + CO_3^{2-} \rightleftharpoons H_2AsO_3^- + HCO_3^- \qquad (9.5.11)$$

The equlibrium constant for reaction 9.5.11 is the ratio of the first dissociation constant of arsenious acid (room temperature $pK = 9.23$) to the second dissociation constant of carbonic acid (room temperature $pK = 10.25$). Its value is very large and, therefore, arsenious acid is present in carbonate solutions essentially in the form of the $H_2AsO_3^-$ ion (the pK for the second dissociation constant of arsenious acid is so large that the $HAsO_3^{2-}$ form can be neglected). The chemistry of arsenic-promoted carbonate solutions is complicated by the tendency of the arsenious acid to dimerize or trimerize. The influence of polymerization is ignored in the analysis given here.

It follows from the argument above that, in a solution of only K_2CO_3 and H_3AsO_3 to which a value $y = 0$ would be assigned, the concentrations of CO_3^{2-} and HCO_3^- are given by

$$[CO_3^{2-}] = m(1 - c) \qquad (9.5.12)$$

$$[HCO_3^-] = mc \qquad (9.5.13)$$

The equilibrium vapor pressure of CO_2 over such a solution, however, is non-negligible. In fact, from the equilibrium condition for reaction 2.4.5, one has

$$p^* = \frac{H}{K} \frac{[HCO_3^-]^2}{[CO_3^{2-}]} \qquad (9.5.14)$$

Equation 9.5.14 gives a finite value of p^* when Eqs. 9.5.12 and 9.5.13 are substituted into it. Therefore, additional CO_2 can be stripped from a solution at $y = 0$, and the lean solution exiting the bottom of the desorber may be characterized by a negative value of y.

Inasmuch as reaction 9.5.11 is strongly displaced to the right, its extent does not change appreciably with changing y and, therefore, the main reaction occurring is still reaction 2.4.5. The concentrations of CO_3^{2-} and HCO_3^- are given in general by

$$[CO_3^{2-}] = m(1 - c - y) \qquad (9.5.15)$$

$$[HCO_3^-] = m(c + 2y) \qquad (9.5.16)$$

so that the equilibrium pressure is given by

$$p^* = \frac{Hm}{K} \frac{(c + 2y)^2}{1 - c - y} \qquad (9.5.17)$$

which, of course, degenerates into Eq. 2.4.13 when $c = 0$.

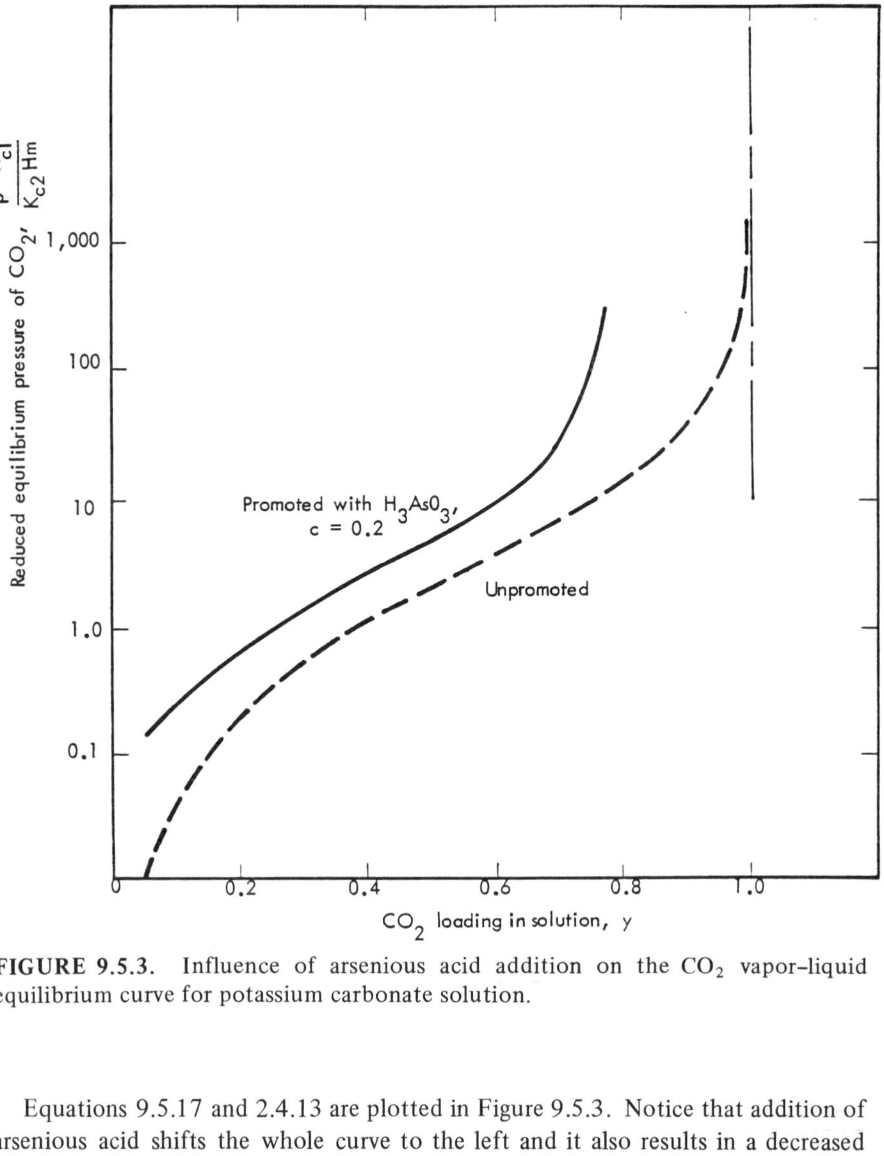

FIGURE 9.5.3. Influence of arsenious acid addition on the CO_2 vapor–liquid equilibrium curve for potassium carbonate solution.

Equations 9.5.17 and 2.4.13 are plotted in Figure 9.5.3. Notice that addition of arsenious acid shifts the whole curve to the left and it also results in a decreased cyclic capacity since the high pressure branch of the curve is shifted more than the low pressure one.

It is also of interest to note that, over the whole range of y values, the main reaction is still reaction 2.4.5 since reaction 9.5.11 remains displaced to the right over the whole range. Therefore, in a global sense, arsenious acid acts as a catalyst, inasmuch as it increases the rate of the main reaction without participating in it.

The thermodynamics of amine-promoted solutions is remarkably more complex. First of all, the amine can act as a base. Certain amine promoters provide cyclic capacity for CO_2; these base-promoted carbonate solutions will have a higher cyclic capacity than an unpromoted one. Amines can also form carbamates and, therefore, the chemistry of amine-promoted solutions is more complex than that of a base-promoted solution. However, in the following, a brief discussion is given for the case of a base-promoted solution and carbamate formation is neglected. The discussion can be regarded as applying to an amine that forms a very unstable carbamate such as, in the limit, a tertiary amine.

Let B be the base and let K_p be its protonation constant:

$$K_p = \frac{[B]\,[H^+]}{[BH^+]} \tag{9.5.18}$$

As was discussed in Section 2.4, for most amines, K_p is larger than the second dissociation constant of carbonic acid K_{c2} (i.e., the amine is a weaker base than the CO_3^{2-} ion). It follows that the reaction

$$B + HCO_3^- \rightleftharpoons BH^+ + CO_3^{2-} \tag{9.5.19}$$

is displaced to the left. Therefore, as long as y does not approach the value of unity, the base will remain essentially in its free form B, reaction 2.4.15 is still the main reaction, and the value of p^* will be essentially the same as in an unpromoted solution. However, as y approaches unity, the $[HCO_3^-]/[CO_3^{2-}]$ ratio becomes very large and some BH^+ will start to form; i.e., the base B will exhibit its ability to react with CO_2 according to

$$B + CO_2 + H_2O = BH^+ + HCO_3^- \tag{9.5.20}$$

Values of y in excess of unity can therefore be achieved; at $y > 1$, the following equation for the equilibrium vapor pressure is easily established by assuming that no CO_3^{2-} is left in the solution and only reaction 9.5.20 is taking place:

$$p^* = \frac{HK_p}{K_{c1}}\, m\, \frac{(y - 1)(1 + y)}{c + 1 - y} \tag{9.5.21}$$

with mc being the total amount of amine added to the solution. Notice that, since $K_p > K_{c2}$, Eq. 9.5.21 tends to yield p^* values considerably larger than those exhibited by an unpromoted solution.

Figure 9.5.4 is a plot of p^* versus y for a base-promoted solution for the case where $K_p = 300K_{c2}$. The cyclic capacity is increased provided the partial pressure in the feed gas is large enough for the high-pressure branch of the curve to be utilized. As a rough rule of thumb, one may assume that at $y < 1$, the main reaction is reaction 2.4.15 and the base acts simply as a catalyst; at $y > 1$, the main reaction is reaction 9.5.20 and the base acts as a chemical sink for CO_2.

The discussion above is oversimplified insofar as it neglects the possibility of carbamate formation. A general analysis of all possible cases would be very complex. It may be stated, in general, that amine promoters tend to increase the cyclic capac-

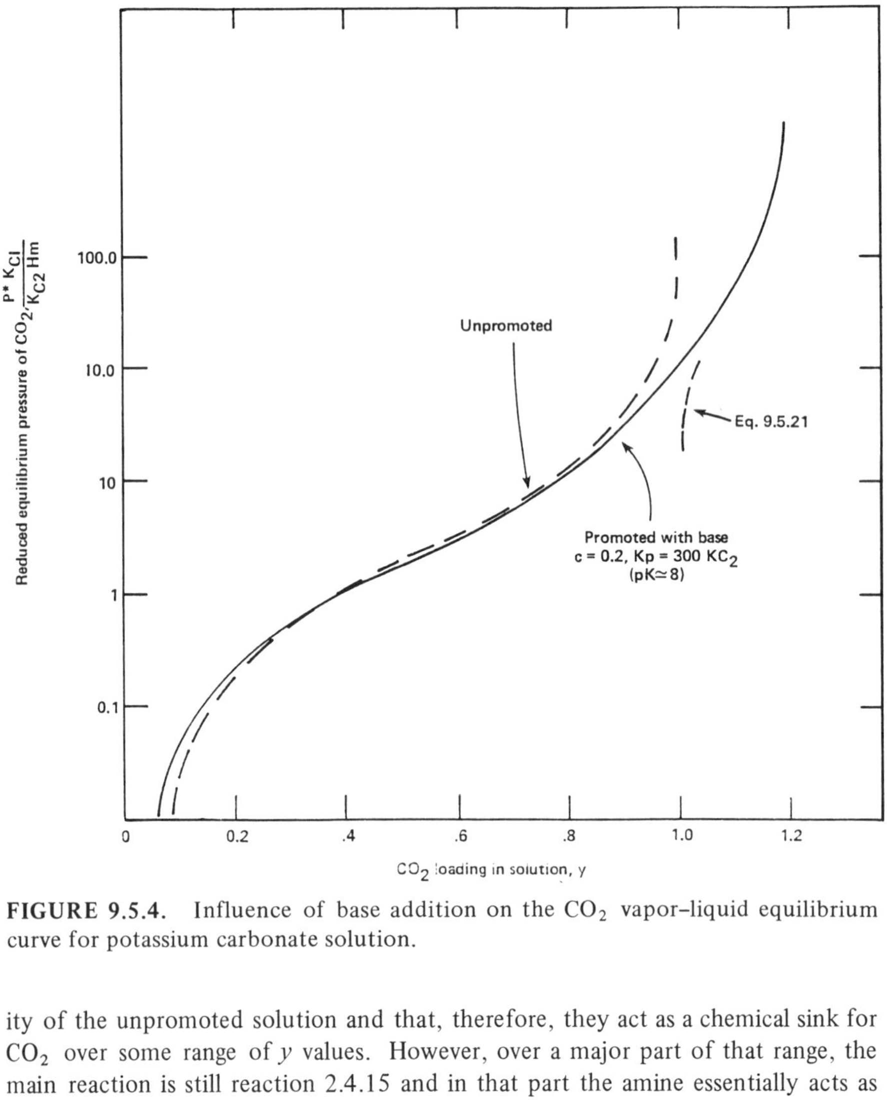

FIGURE 9.5.4. Influence of base addition on the CO_2 vapor–liquid equilibrium curve for potassium carbonate solution.

ity of the unpromoted solution and that, therefore, they act as a chemical sink for CO_2 over some range of y values. However, over a major part of that range, the main reaction is still reaction 2.4.15 and in that part the amine essentially acts as a catalyst. In the following, the rate promotion effect is discussed on the basis of the assumption that the main reaction is reaction 2.4.15.

Let mc^0 be the total amount of promoter added to the solution. The promoter may undergo reactions in the solution and, therefore, be present in different chemical forms with mc^0 being the sum of the concentrations of all such forms. Let mc be the concentration of the form that is the actual rate promoter; in general, $c < c^0$ and the value of c can be calculated from an appropriate thermodynamic model.

Consider the case of arsenious acid. At low temperatures and comparatively large pH values, the forward rate of bicarbonate formation is given by

$$r = k_S a m c^0 \tag{9.5.22}$$

At room temperature, $k_S = 164$ 1/gmol sec, so that if mc^0 is larger than 0.01 gmol/liter, the rate given by Eq. 9.5.22 by far exceeds that given by Eq. 9.5.4, i.e., all bicarbonate is formed via the catalytic mechanism.

At the pH values at which Eq. 9.5.22 has been established, arsenious acid is completely dissociated. This suggests that the real catalyst is the ion $H_2AsO_3^-$, which has a concentration of the order of mc^0 in carbonate solutions. Indeed, this has been confirmed experimentally by working with monoethanolamine solutions at $y > 0.5$ where the pH is low enough for c to be appreciably less than c^0; thus the kinetic expression for the forward rate should be written as

$$r = k_S amc \tag{9.5.23}$$

There is, however, some difficulty with the reverse rate; desorption data exhibit rates substantially lower than would be predicted from the kinetic constant k_S as measured in absorption. Figure 9.5.5 is a plot of absorption rate versus fractional saturation obtained in a laboratory unit; it clearly shows that although very significant rate promotion is observed for lean solutions, the rate promotion effect almost dies out at fractional saturations of 0.5 or more. The reasons for this behavior may be related to the polymerization of arsenious acid in solution.

Rate promotion by amines has been discussed in the literature on the basis of a special mechanism called the shuttle mechanism. This mechanism assumes that, during the absorption, amine reacts rapidly with CO_2 near the interface, thus increasing the rate of mass transfer. The product of the reaction, presumed to be the carbamate, diffuses to the bulk of the liquid, where it reverts to the bicarbonate (presumably through reaction 9.4.9) liberating free amine. The latter diffuses back to the interface region where it reacts again with CO_2, and so on. Low-temperature mass transfer rates have been well correlated by equations based on this physical model.

Several observations are in order concerning the shuttle mechanism; these are developed concisely below. The first observation is that the shuttle mechanism could not be efficient in desorption, since for desorption the reverse of reaction 9.4.9 should take place near the interface and this reaction is not any faster than the reaction taking place in unpromoted solutions.

The second observation is that the rate promotion effect observed at high temperatures is larger than the maximum possible via a shuttle mechanism. In fact, even if the promoter-CO_2 reaction were instantaneous and irreversible in the interface region, the rate promotion would be

$$E = \frac{I_{\text{promoted}}}{I_{\text{unpromoted}}} = \sqrt{1 + \frac{c^2 k_L^{0 2}}{D k_0 a_i^2}} \tag{9.5.24}$$

Equation 9.5.24 is based on the sketch of concentration profiles in the film in Figure 9.5.6. In the region $0 \leq x \leq \lambda$, CO_2 reacts with OH^- following Eq. 9.2.2, since this region is unpromoted. Rate promotion effects larger than predicted by Eq. 9.5.24 have been observed; these, of course, cannot be explained via a shuttle mechanism hypothesis.

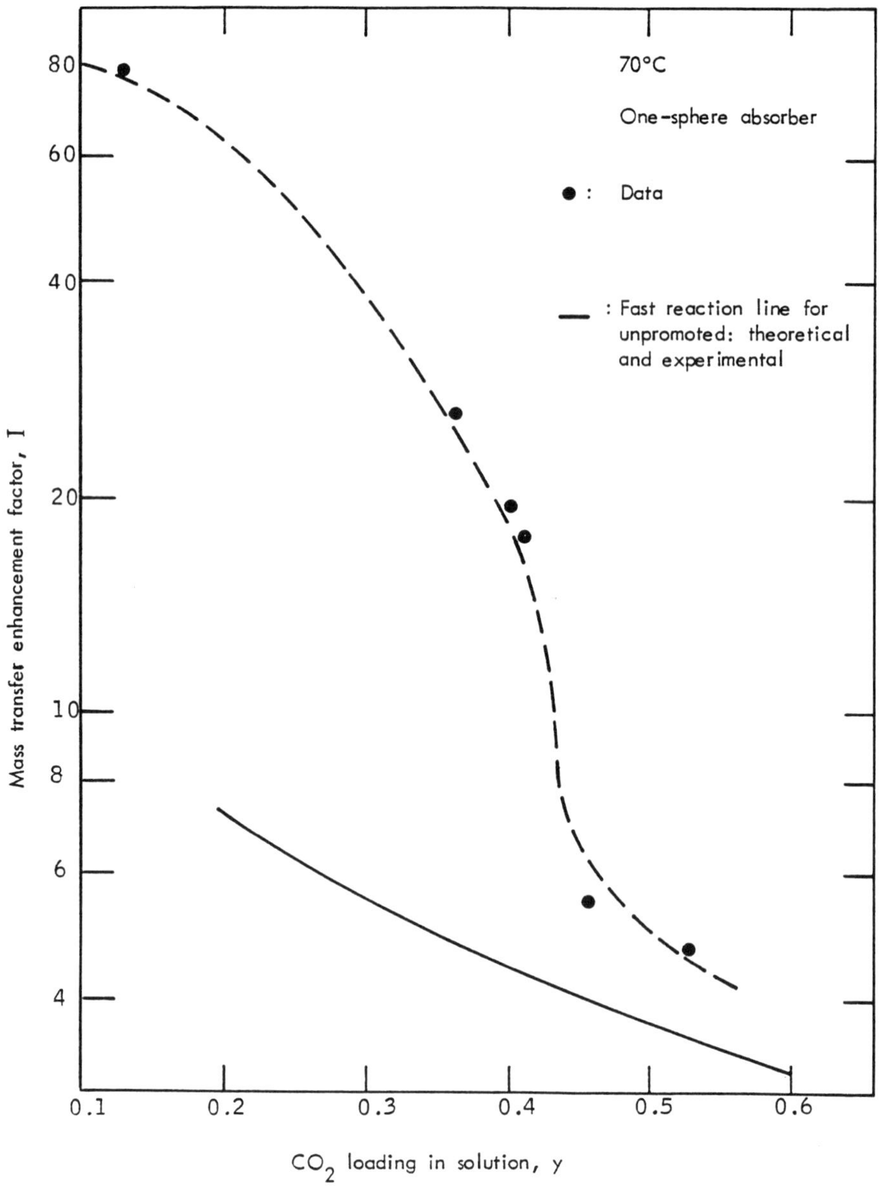

FIGURE 9.5.5. Dependence of I on y for CO_2 absorption in arsenite-promoted K_2CO_3.

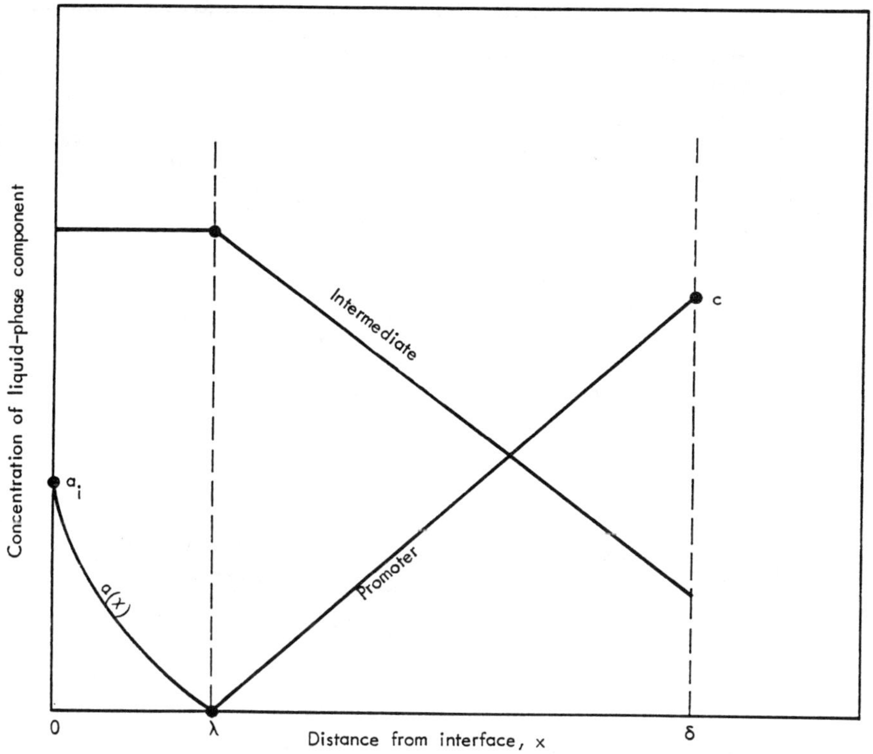

FIGURE 9.5.6. Liquid-phase concentration profiles for promoted carbonate solutions instantaneous regime shuttle mechanism.

The third observation is related to the fact that the shuttle mechanism model is based on the idea that the two chemical steps, which may be indicated concisely as

$$CO_2 + \text{Promoter} \longrightarrow \text{Intermediate} \tag{9.5.25}$$

$$\text{Intermediate} + OH^- \longrightarrow HCO_3^- + \text{Promoter} \tag{9.5.26}$$

take place at different locations (interface and bulk) and are, therefore, separated by diffusion processes; the reason for this is that reaction 9.4.9 is slow. However, should no diffusion processes be involved, the sequence 9.5.25-9.5.26 is a classical sequence of homogeneous catalysis. Therefore, a rate promoter that acts via a shuttle mechanism at low temperature may well act as a homogeneous catalyst at higher temperature, where reaction 9.5.26 may be fast enough to take place in the interface region.

If a rate promoter is regarded as a homogeneous catalyst, its largest possible

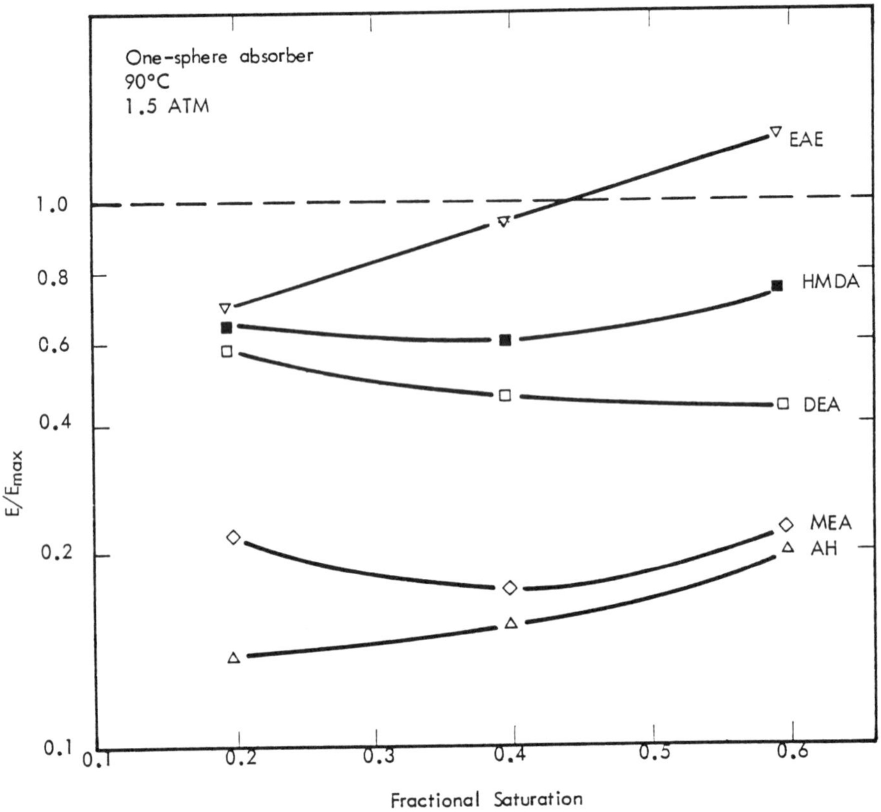

FIGURE 9.5.7. Normalized CO_2 absorption rates for amine-promoted potassium carbonate based on absolute upper limit for homogeneous catalysis hypothesis.

effect is to speed up the rate of reaction 9.2.1 to the point where mass transfer takes place in the instantaneous reaction regime; hence the largest possible rate promotion factor is

$$E_{max} = \frac{I_\infty}{I_{unpromoted}} \tag{9.5.27}$$

where I_∞ is to be calculated from whichever of Eqs. 9.5.8–9.5.10 is appropriate.

Figure 9.5.7 is a plot of E/E_{max} versus y, obtained on a laboratory unit with carbonate solutions promoted with several different amines and aminoalcohols; it is seen that the most efficient amines result in the largest possible promotion effect. The data in Figure 9.5.7 correspond to I_∞ values of the order of 300; this implies that Φ values are certainly well in excess of 10^5, but it cannot be established by how much. However, under conditions where I_∞ is appreciably larger than 300, even if $E < E_{max}$, the overall rate could not be increased significantly anyhow, because at such high values of I, gas-phase mass transfer control becomes important.

It should be pointed out that, if a rate promoter is efficient enough for mass transfer to take place in the instantaneous reaction regime, the mass transfer rate will be proportional to the chemical driving force. It follows that of two such promoters, the one that also increases the capacity of the solution and, therefore, the chemical driving force, will also result in the larger absolute rates of mass transfer.

The present understanding of the chemical mechanism of rate promotion is very poor, particularly for the case of amine promoters. The rate promotion effect is much larger than one would expect and differences observed with different amines are hard to explain. However, from a practical viewpoint, rate promoters are very useful and they are in common industrial use. Industrial processes of CO_2 removal with unpromoted and promoted carbonate solutions are discussed in the following.

Three types of liquid flow arrangement have been used, depending on the degree of CO_2 removal required. These are shown in Figure 9.5.8, and briefly discussed below.

1. Single-stream flow in absorber and regenerator. For carbon dioxide removal to a residual partial pressure of about 5 psia or higher, this scheme is generally used. Lean carbonate solution flows from the bottom of the regenerator, without cooling, to the top of the absorber. The rich solution leaving the bottom of the absorber flows to the top of the regenerator after pressure letdown. The temperature of the solution to the absorber varies from about 220–245°F, depending on the regeneration pressure.

2. Split-stream flow in the absorber and single-stream flow in the regenerator. Carbonate solution from the regenerator is divided into two portions. The major portion, 60–70% of the total, flows to an intermediate level of the absorber without cooling, and the minor portion is cooled by 20–70°F before being pumped to the top inlet of the absorber. The equilibrium back pressure of carbon dioxide over the minor stream is decreased to permit attaining low residual partial pressure, down to about 2 psia, in the purified gas.

3. Split-stream flow in the absorber and split-stream flow in the regenerator, with two stages of regeneration of the split stream. Two regenerated streams leave the regenerator; the major portion (60–85% of the total) is removed from an upper level in a partially regenerated state for return to an intermediate level of the absorber. The minor stream, 15–40% of the total, is regenerated further in the bottom section of the regenerator by the total boilup steam, removed from the sump, and cooled before being pumped to the top of the absorber. Typical conversions of the two streams are 40–45% conversion to bicarbonate for the semi-lean and 15–30% conversion to bicarbonate for the lean stream. Details of a two-stage promoted hot carbonate CO_2 removal design are shown in Figure 9.5.9.

The CO_2 cleanup attainable with the various flowsheet configurations described

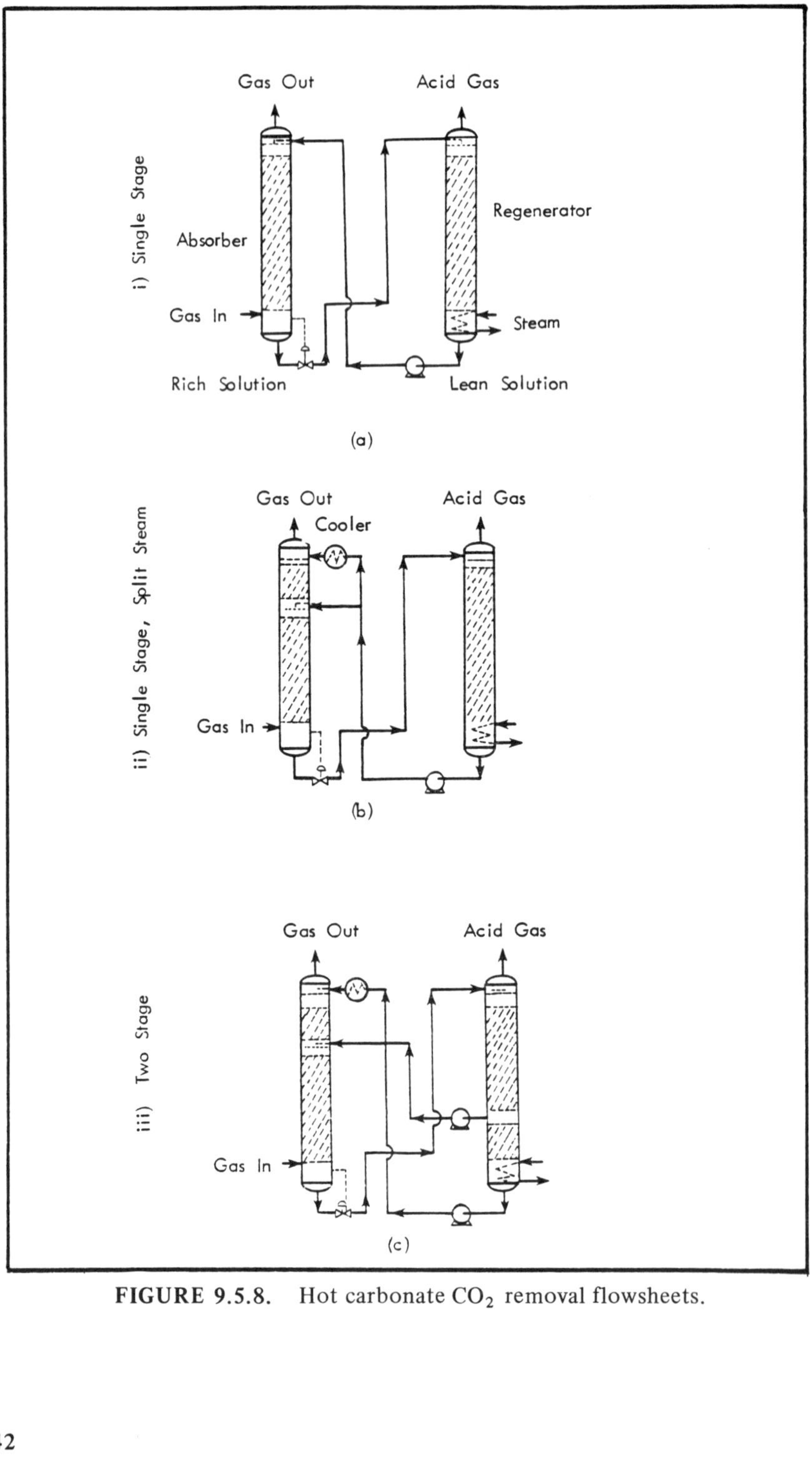

FIGURE 9.5.8. Hot carbonate CO_2 removal flowsheets.

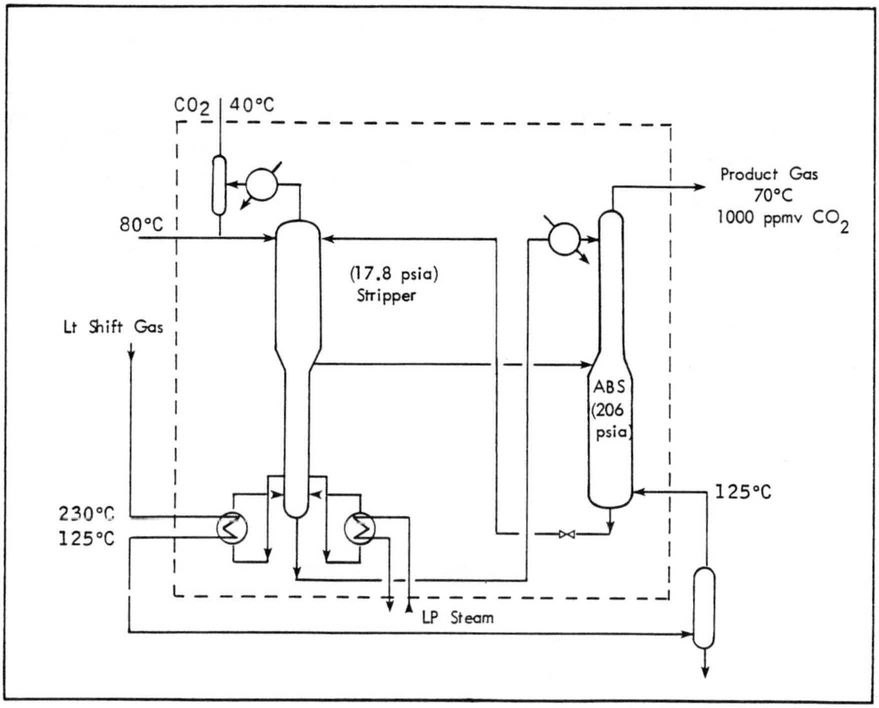

FIGURE 9.5.9. Two-stage promoted hot carbonate CO_2 removal design.

above is summarized in Table 9.5.1. The regenerator efficiencies that can be obtained with these flow arrangements are shown in Figure 9.5.10, which refers to unpromoted hot carbonate solutions.

Recently, other flowsheet configurations have been proposed for reducing energy consumption further. One approach involves flashing at least a portion of the partially stripped solution from the regenerator and using the steam raised in this manner in an auxiliary stripper. Figure 15.7.2 shows this arrangement.

Another approach makes use of a hydraulic turbine in the high-pressure rich solution line to drive the lean solution pumps. This can be attractive with pro-

TABLE 9.5.1. CO_2 Cleanup Attainable with Promoted Hot Carbonate Processes

Configuration	Final CO_2 content
Single stage	1% or greater
Cooled split flow absorber	0.1–1%
Cooled split flow absorber, two-stage regenerator	500–1000 ppm

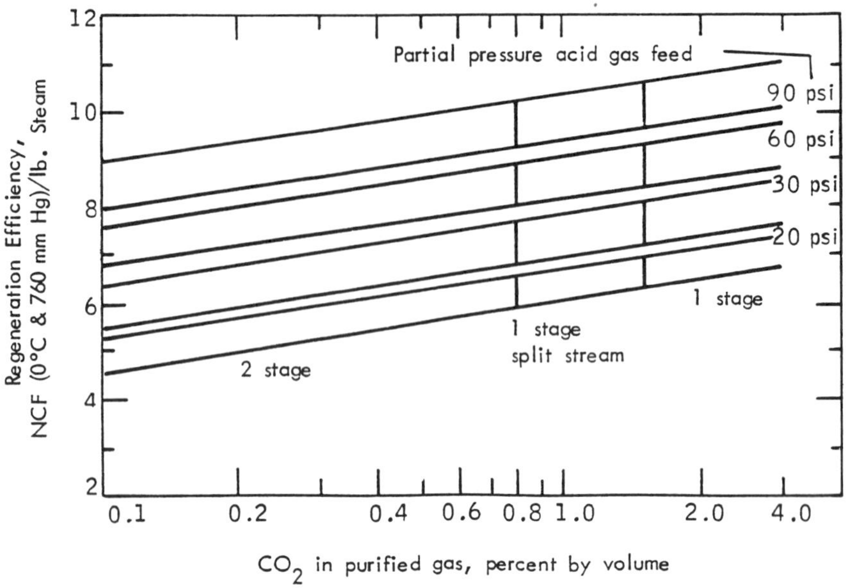

FIGURE 9.5.10. Regeneration efficiencies associated with hot carbonate CO_2 removal flow arrangements.

moted hot carbonate; it may be relatively more attractive with physical solvents that have a high pumping energy requirement owing to the large solution circulation required.

Information on the energy requirements for removing carbon dioxide with organic-promoted hot carbonate solutions is available. Solution circulation rate and regeneration heat requirement are dependent mainly on the partial pressure of CO_2 in the feed gas. To purify a 400 psig gas containing 30% CO_2 normally requires 75-90 gal of solution circulation and 30,000-40,000 BTU heat input/mole of CO_2 removed. If the feed gas contained only 10% CO_2, 100-140 gal of solution circulation and 50,000-60,000 BTU heat input/mole of CO_2 removed would be required.

10 Removal of Hydrogen Sulfide and Other Sulfur Contaminants

"What do boiled eggs stink of?"
"We know that; they stink of . . ."

N. GINZBURG

Removal of sulfur contaminants from gas streams is a common industrial requirement and will become increasingly important in the future with the expanded utilization of low-grade high sulfur content fuels. The most important sulfur contaminant is H_2S.

Removal of H_2S from gases is required for a variety of reasons such as health, safety, corrosion limitation during transport and distribution, pollution prevention, and avoidance of catalyst poisoning. Specifications are in general severe, with 4 ppm by volume being typical for pipeline gas, and as low as 0.01 ppm being required to limit catalyst poisoning when the gas is to be used in the chemical industry.

Other common sulfur contaminants, which generally occur at much lower concentrations than H_2S, are carbonyl sulfide (COS), carbon disulfide (CS_2), and mercaptans (RSH). Atomic sulfur is chemically similar to atomic oxygen and, therefore, the first two have a chemical behavior somewhat analogous to that of CO_2, whereas mercaptans are the sulfur analogues of alcohols. Specifications for these sulfur contaminants are similar to those for H_2S.

10.1. PROCESS SELECTION

An important constraint on process selection for H_2S removal is the fact that the separated H_2S must be disposed of in an environmentally acceptable way. In most cases, conversion to elemental sulfur is the preferred way, with the separated gas containing H_2S being sent to a Claus-type plant. Claus plants have been in operation for 80 years and they are still unsurpassed today. A brief description of a Claus plant is given in the following.

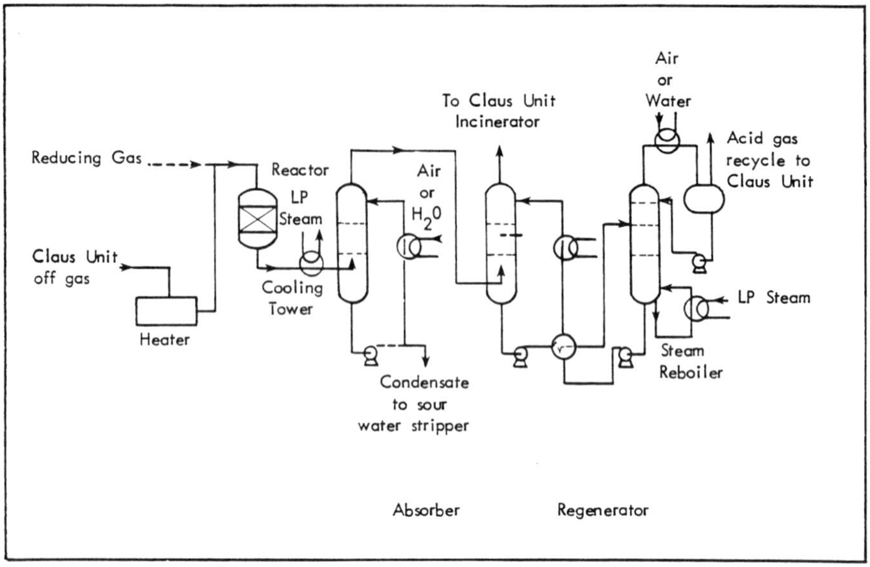

FIGURE 10.1.1. Claus off-gas treating process.

The gas stream is split into two streams. The first one, consisting of one-third of the total, undergoes oxidation:

$$H_2S + \frac{3}{2} O_2 = SO_2 + H_2O \qquad (10.1.1)$$

whereas the second stream is reacted catalytically with the SO_2 from the first step:

$$2H_2S + SO_2 = \frac{3}{N} S_N + 2H_2O \qquad (10.1.2)$$

There are thermodynamic limitations to the H_2S conversion obtainable in a Claus plant and, therefore, the tail gas must be treated for removal of the residual H_2S. Claus off-gas treating processes are scaled down replicas of the mainstream acid gas separation unit (see Figure 10.1.1). They are needed in order to meet the increasingly tight air pollution control requirements. Integration of the Claus off-gas and mainstream treating units is possible, and some options are shown in Figure 10.1.2.

Often CO_2 and H_2S are removed simultaneously from a gas stream (these processes will be discussed in Chapter 11). When this is so, the acid gases exiting from the regenerator unit must be sent to a Claus plant. The feed to a Claus plant should generally contain 15% or more of H_2S for acceptable operation; therefore, if the H_2S-CO_2 ratio in the feed gas is too low, simultaneous removal may be unacceptable and selective removal of H_2S may be required. Selective removal is also discussed in Chapter 11.

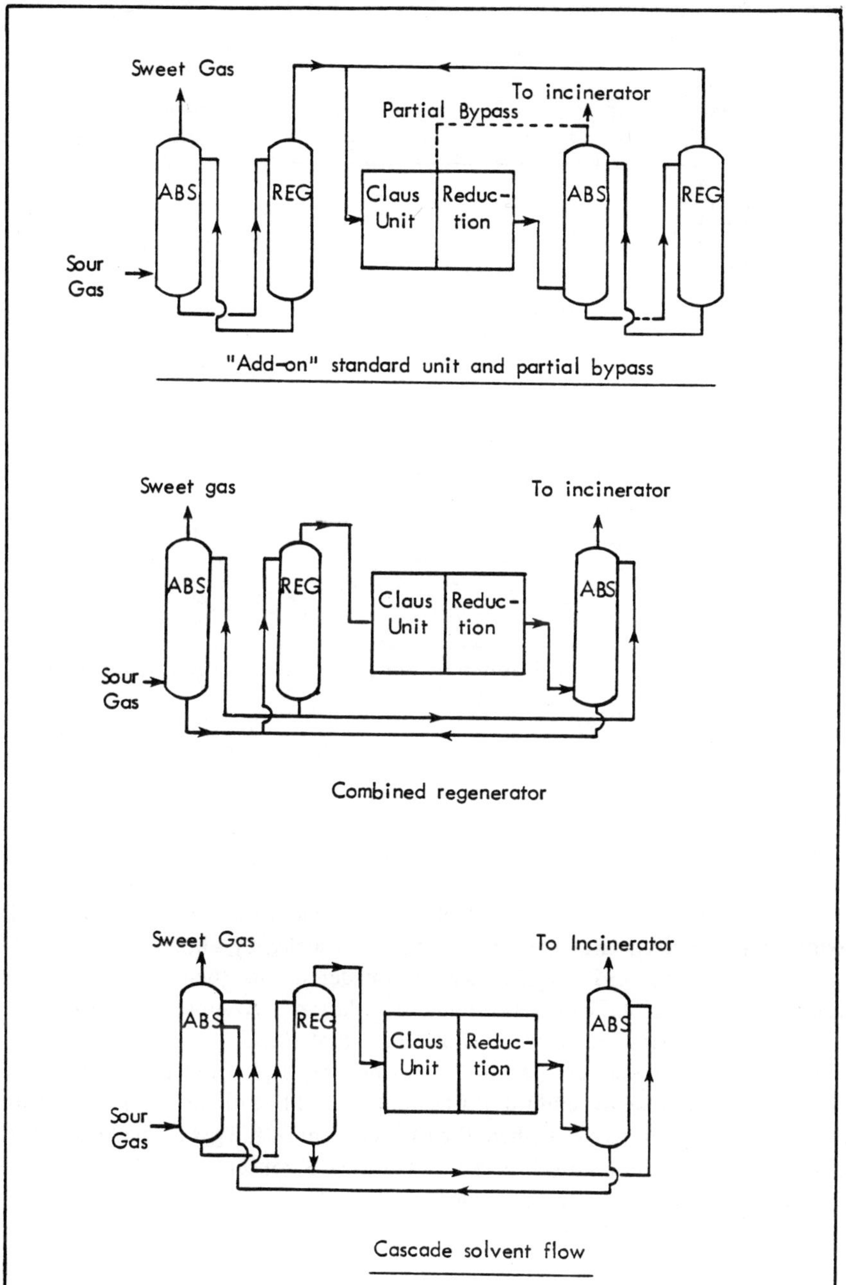

FIGURE 10.1.2. Integration of Claus tail gas treating with mainstream acid gas removal.

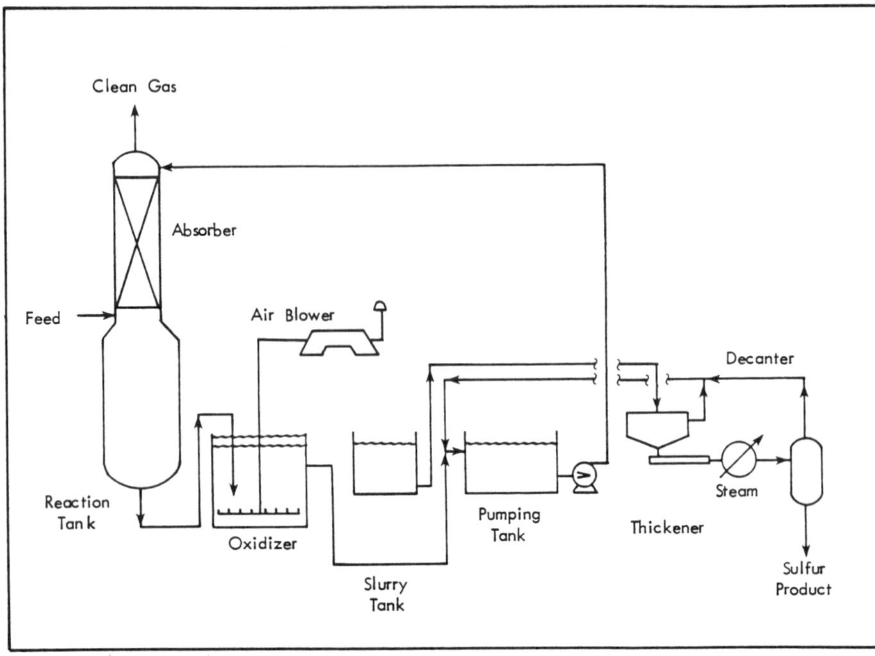

FIGURE 10.1.3. Simplified flowsheet of direct conversion H_2S removal process.

An alternative to H_2S separation from feed gas combined with Claus plant sulfur recovery is the use of a special oxidative treating process that produces elemental sulfur directly. These direct conversion processes employ aqueous alkaline solutions to absorb H_2S from the gas. The solutions also contain chemical oxidants such as vanadate or arsenate ions which react with the dissolved H_2S, converting it to elemental sulfur which can be removed as a separate phase. The oxidants are reduced and must be reoxidized, usually by air blowing, for reuse. The solutions may also contain oxygen transfer agents such as anthraquinone-disulfonic acid (ADA) which facilitate reoxidation. A simplified flow plan of one of these processes, Stretford, is shown in Figure 10.1.3.

Consider the simple problem of removing only H_2S from a gas stream. (In Chapter 11 simultaneous removal of CO_2 and H_2S will be discussed. In this chapter we restrict attention to cases where the CO_2 content of the gas is small enough that it does not affect the removal of H_2S.) If the H_2S concentration in the gas is extremely low, aqueous hydroxide solutions may be used. These cannot easily be regenerated and can, therefore, not be used if substantial amounts of H_2S need to be removed.

At higher H_2S concentrations, up to about 3 psi in the sour gas, alkaline salt solutions with chemical oxidants are preferred. At still higher concentrations, aqueous aminoalcohol processes, principally monoethanolamine, diethanolamine, and diisopropanolamine, are preferred. Finally, at H_2S partial pressures exceeding about 60 psi, physical solvents should be considered; these, however, cannot easily reduce H_2S contents to the required specifications and a second removal stage may be required. Figure 10.1.4 summarizes the ranges of application of different solvents.

The most important instance of a process where essentially only H_2S must be removed occurs as part of hydrodesulfurization processes. These processes, which are of growing importance because the supply of sweet crude oil in the world is shrinking, remove organic sulfur compounds from petroleum fractions and coal-derived liquids by catalytic reaction with hydrogen at high temperature and pressure. The sulfur contained in the organic compound is removed and transforms to H_2S. The gas exiting from the hydrodesulfurization reactor is hydrogen containing H_2S and a certain amount of light hydrocarbons formed in the reactor. H_2S must be removed from the gas so that H_2 may be recycled, and the solvent used in the removal should exhibit low hydrocarbon solubility. The specifications are typically 50–100 ppm H_2S; the process must be integrated with a sulfur recovery plant. The most common solvents are aqueous monoethanolamine, diethanolamine, and diisopropanolamine, which are to be preferred because the partial pressure in the feed gas is not high enough to make physical solvents attractive. Figure 10.1.5 is a schematic flowsheet of a typical hydrodesulfurization process; Figure 10.1.6 is a flow diagram of the H_2S removal section.

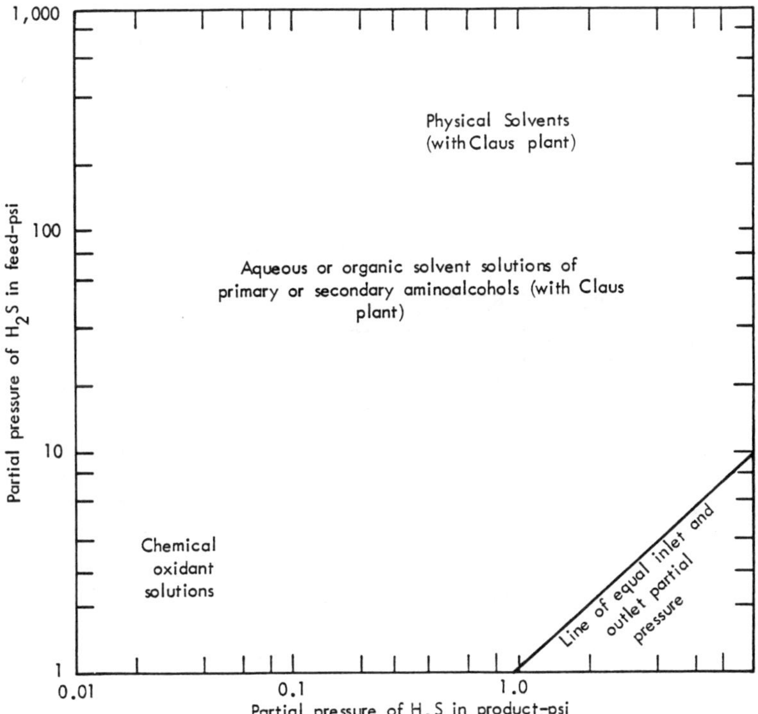

FIGURE 10.1.4. Range of gas treating process application for H_2S removal with no CO_2 or other impurities present.

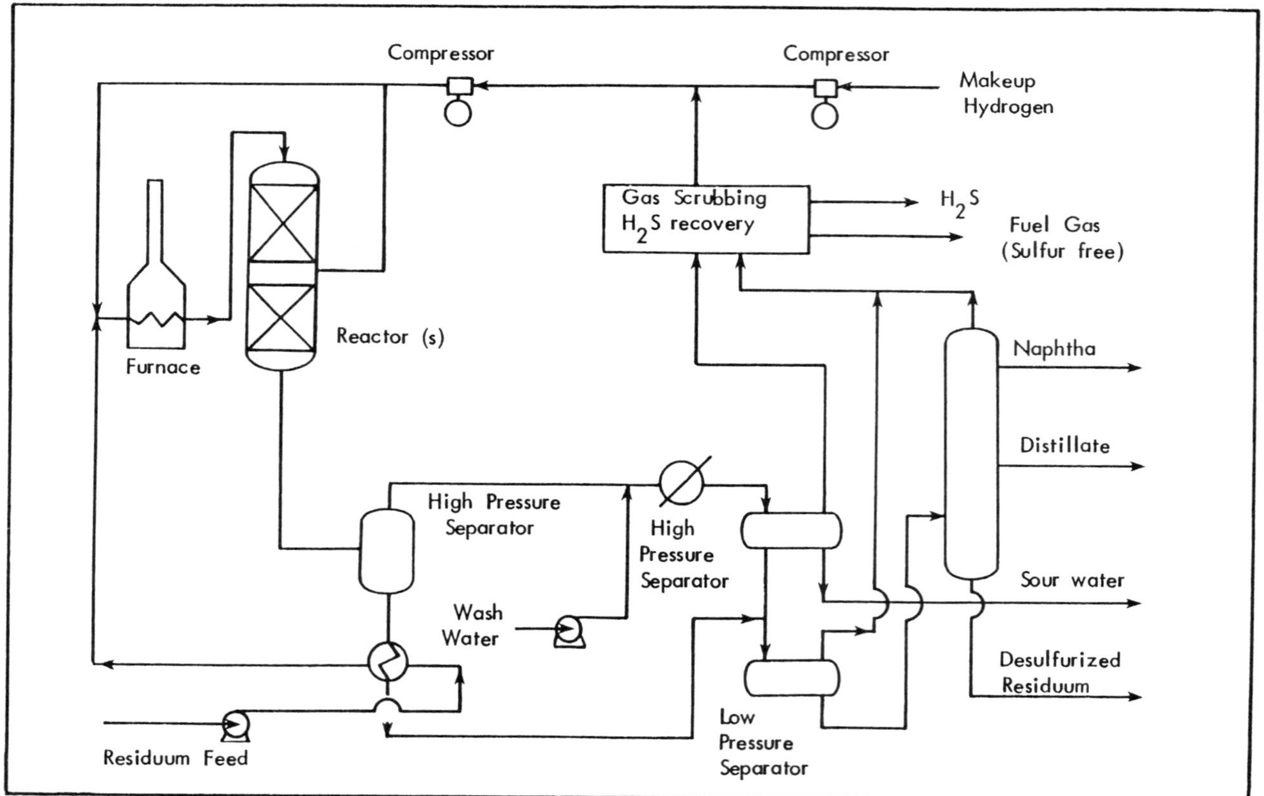

FIGURE 10.1.5. Residual oil hydrodesulfurization process.

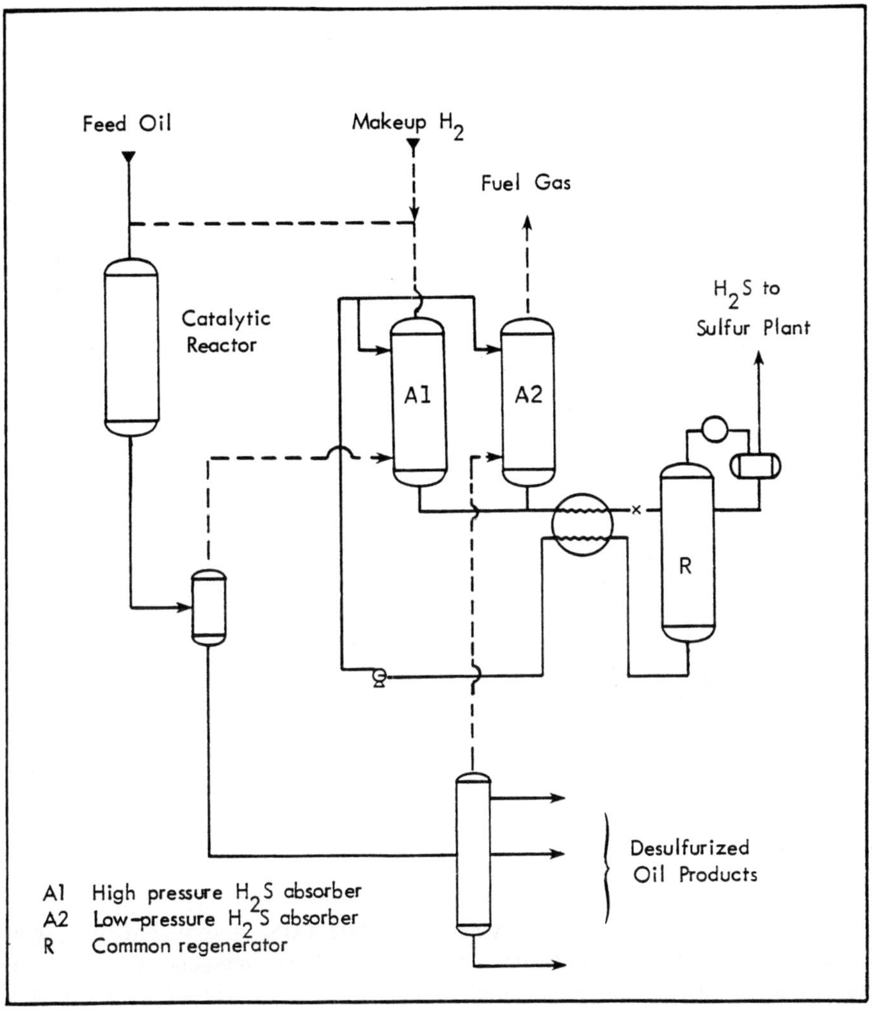

FIGURE 10.1.6. Gas scrubbing and H_2S recovery facilities associated with oil hydrodesulfurization.

Increasingly, in the future, treating processes will have to be able to remove small but significant amounts of carbonyl sulfide along with large amounts of CO_2 and H_2S. For example, gas generated by partial oxidation of heavy petroleum oils contains typically 300 ppm COS, along with 3.9% CO_2 and 0.7 H_2S. The variation of acid gas levels with weight percent sulfur in the feed oil is shown in Figure 10.1.7.

Co-removal of H_2S, COS, CS_2, and mercaptans presents some special problems. The reactivity of these impurities in alkaline solutions is low, so that their removal must rely on physical solubility. Their solubility in water, and hence in aqueous solutions, is rather low and, furthermore, COS may cause degradation of primary amines though not of secondary ones.

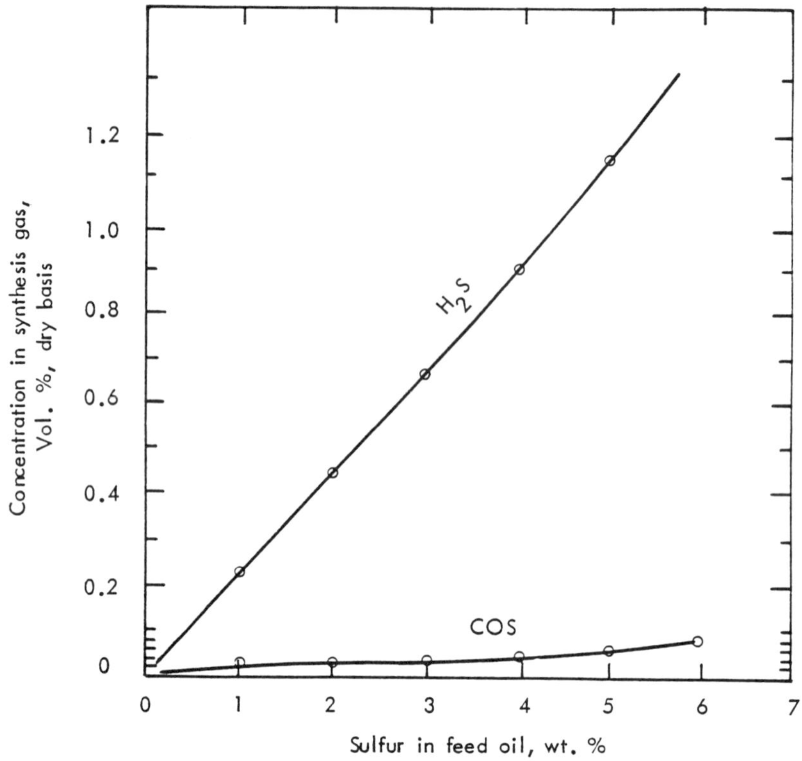

FIGURE 10.1.7. Process effects of sulfur content of feed oil.

The classical approach to COS removal has been catalytic hydrolysis as a pretreatment ahead of the acid gas removal unit. Absorption in organic (physical) solvents is another approach for nearly complete COS removal, but many contacting stages or high solvent rates are needed. Chilled methanol ($<$273K) and Sulfolane (tetrahydrothiophene dioxide) are reported to give virtually complete COS removal.

Carbon disulfide is usually at least as difficult as COS to remove from gases. It also occurs at low concentration relative to H_2S and CO_2, and its low aqueous solvent solubility is a limiting factor. Organic solvents are more effective in removing it. Organic solvents, particularly Sulfolane, are reported to be very effective also for mercaptan absorption.

10.2. CHEMISTRY AND THERMODYNAMICS

Hydrogen sulfide is essentially a hard acid but it also has soft base character. Acting as a hard acid, it reacts readily with all hard bases. It acts as a soft base in donating electrons from the sulfur atom, as in oxidation processes involving metals.

In contrast with carbon dioxide, H_2S does not need to undergo a chemical

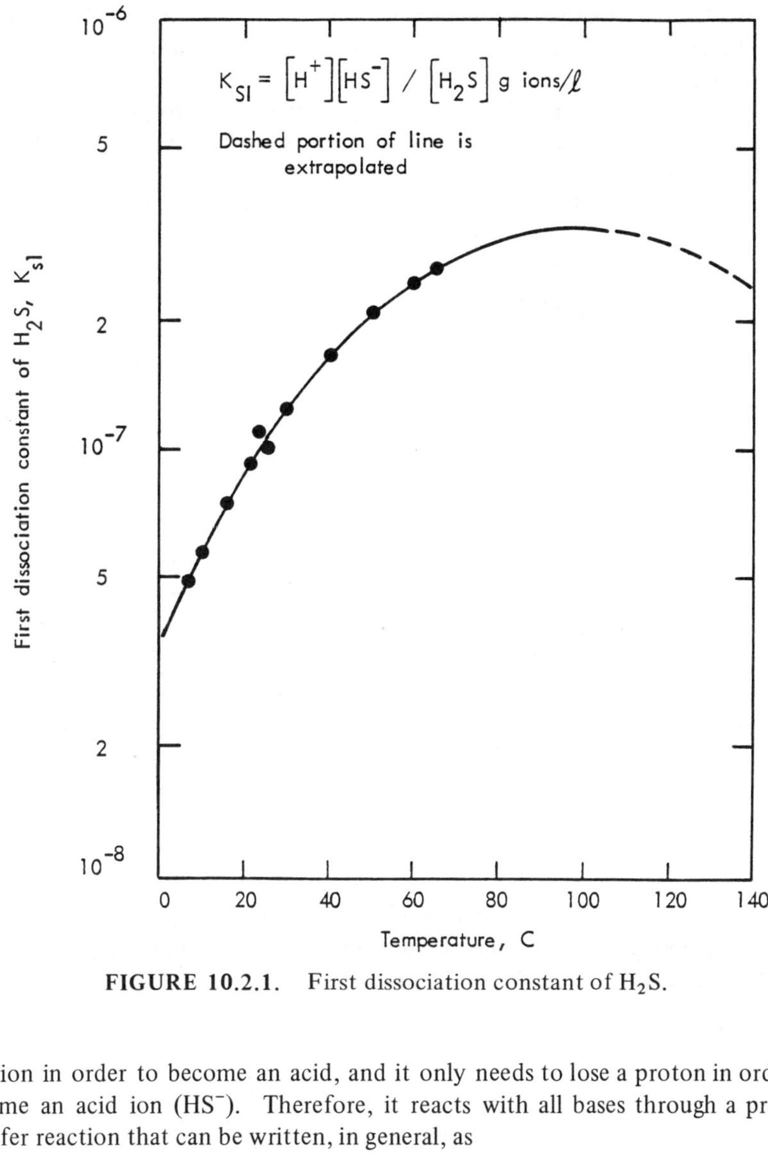

FIGURE 10.2.1. First dissociation constant of H_2S.

reaction in order to become an acid, and it only needs to lose a proton in order to become an acid ion (HS^-). Therefore, it reacts with all bases through a proton-transfer reaction that can be written, in general, as

$$H_2S + B = BH^+ + HS^- \tag{10.2.1}$$

Proton-transfer reactions are extremely fast (reaction times are typically of the order of 10^{-13} sec) so that, to all practical purposes, reactions of the type 10.2.1 can be regarded as instantaneous and, therefore, at equilibrium everywhere in the liquid.

Figures 10.2.1 and 2.7.2 give the first and second dissociation constants of H_2S, K_{s1}, and K_{s2}. Several important observations are in order.

Figure 2.7.2 shows that the second dissociation constant of H_2S is of the same order of magnitude as the water dissociation constant. This implies that practically no S^{2-} will form in any but extremely strong alkaline solutions. This observation considerably simplifies the chemistry of H_2S, since the only product to be considered is the acid sulfide ion, HS^-. (In contrast with CO_2, H_2S does not react to chemical forms, such as the carbamic ion, different from its own ion.)

Inasmuch as only HS^- ions need be considered as a possible chemically combined form of H_2S, reaction 10.2.1 is invariably the main (and only) reaction to be considered in alkaline solutions. Let K be the equilibrium constant for reaction 10.2.1; the value of K is given by

$$K = \frac{K_{s1}}{K_p} \tag{10.2.2}$$

where K_p is the protonation constant of the generic base B:

$$K_p = \frac{[B]\ [H^+]}{[BH^+]} \tag{10.2.3}$$

In order that an aqueous solution of base B be an acceptable chemical solvent for H_2S, the value of K should be rather large, inasmuch as K is a measure of the ratio of the chemical capacity of the solvent to the physical capacity of water. [See the discussion in Section 6.4. For reaction 10.2.1, $1 + \Sigma\nu_j = 0$ and, therefore, K in Eq. 10.2.2 equals $K(m_0)^{1+\Sigma\nu_j}$ in Eq. 6.4.6.] This sets an upper limit on the value of K_p, which needs to be at least, say, 10 times smaller than K_{s1}; in other words, an acceptable chemical solvent for H_2S must be a base with a room temperature pK of at least about 8.

As long as reaction 10.2.1 is the only one to be considered (this is not true when CO_2 and H_2S are both present in appreciable amounts in the gas phase; see Chapter 11), the thermodynamic modeling is very simple. Let the undissociated base B be B_1, the protonated base BH^+ be B_2, and the acid sulfide ion HS^- be B_3. One gets immediately

$$b_1 = m(1 - y) \tag{10.2.4}$$

$$b_2 = b_3 = my \tag{10.2.5}$$

where y is fractional saturation. The equilibrium vapor pressure is given by

$$p^* = \frac{HK_p}{K_{s1}}\, m \frac{y^2}{1 - y} \tag{10.2.6}$$

which shows that p^* increases with increasing K_p—the stronger the base B, the lower the H_2S equilibrium vapor pressure.

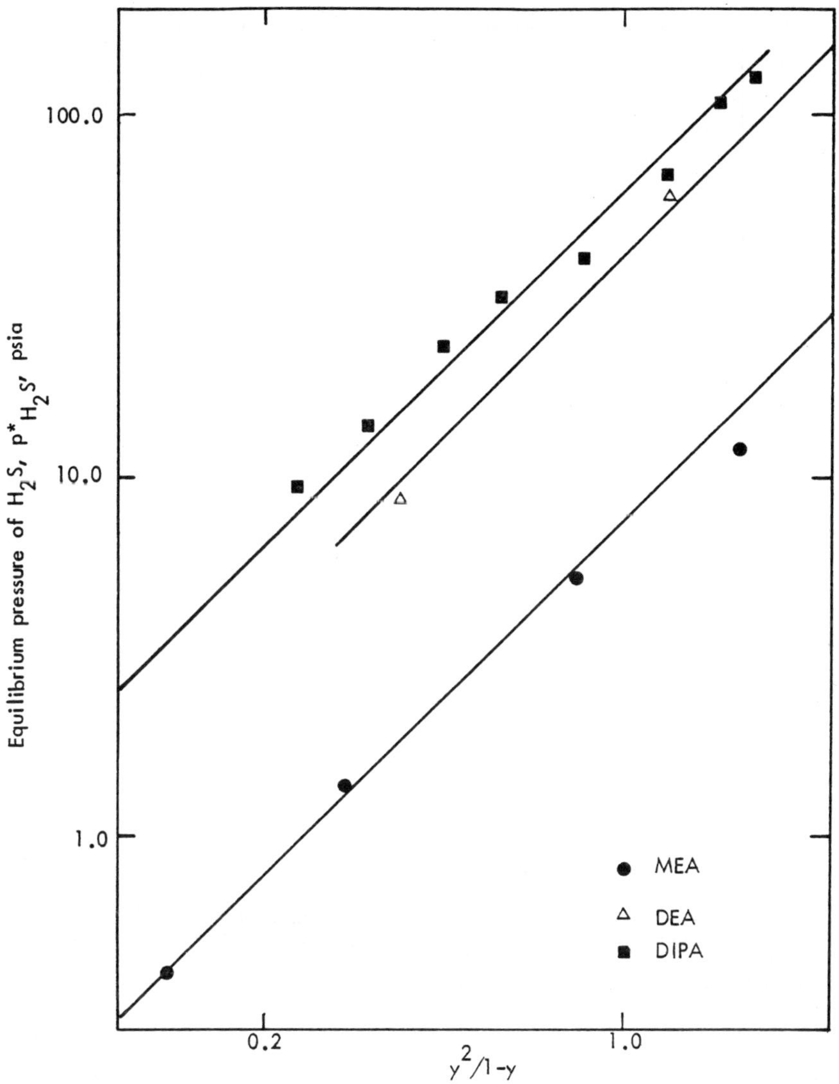

FIGURE 10.2.2. Correlation of vapor-liquid equilibrium data for H_2S in aqueous aminoalcohols.

Equation 10.2.6 is well confirmed by available vapor-liquid equilibrium data; Figure 10.2.2, which shows this, is a plot of p^* versus $y^2/(1 - y)$ for three different aminoalcohols. The fact that p^* increases with increasing m is also well established experimentally. However, as usual, the simple thermodynamic model does well in predicting the dependency of p^* on y, but less well in predicting the dependency of p^* on m (see Section 2.9).

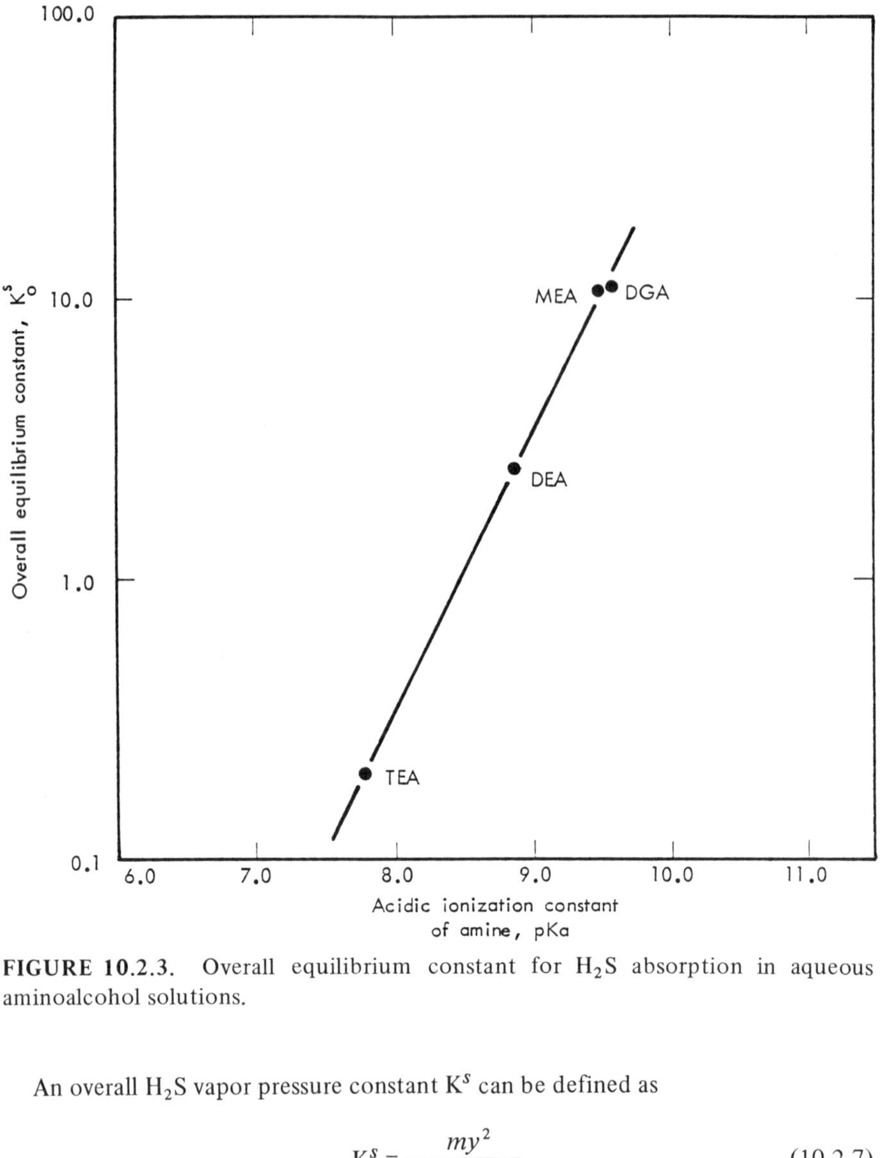

FIGURE 10.2.3. Overall equilibrium constant for H_2S absorption in aqueous aminoalcohol solutions.

An overall H_2S vapor pressure constant K^s can be defined as

$$K^s = \frac{my^2}{(1 - y)p^*} \qquad (10.2.7)$$

which, on the basis of Eq. 10.2.6, is expected to be inversely proportional to K_p. This is well confirmed by experimental data (see Figure 10.2.3). However, measured values of K^s are not proportional to m, and a slight residual dependency on y is also observed. The following empirical equation has been proposed:

$$K^s = K_0^s \exp\left[\frac{-\Delta H}{R}\left(\frac{1}{T} - \frac{1}{313}\right) + 2.4\sqrt{I_i} - 0.72 I_i\right] \qquad (10.2.8)$$

where K_0^s is the value of K^s at 40°C, ΔH is the heat of reaction, and I_i is the ionic strength.

A lower limit to K_p is imposed by the necessity to regenerate the solution; if K_p is extremely low, even at high temperature the value of p^* is too low for regeneration (see Eq. 10.2.6). This is why hydroxide solutions with K_p values of the order of the water dissociation constant cannot be regenerated. In practice, the base B is an acceptable chemical solvent for H_2S if its pK is in the range 8–12 at room temperature.

This range includes practically all aminoalcohols, including tertiary amines (which are of particular interest for selective H_2S removal), and all salts of a strong base with a moderately weak acid. Inorganic salt solutions that have been used in industry include tripotassium phosphate, sodium phenolate, sodium alanine, and the potassium salts of diethyl glycine and dimethyl glycine (K_2CO_3 will be discussed in Chapter 11).

Organic physical solvents can also be used for H_2S removal; the H_2S solubility is generally larger than that of CO_2 because of the specific nonbonding interactions of H_2S with solvents of industrial interest such as propylene carbonate and the dimethyl ether of polyethylene glycol.

The chemistry of the other sulfur contaminants is briefly discussed below. Both COS and CS_2 may react with amines to give thiocarbamates:

$$COS + 2RNH = RNCOS^- + RNH_2^+ \qquad (10.2.9)$$

$$CS_2 + 2RNH = RNCSS^- + RNH_2^+ \qquad (10.2.10)$$

However, these reactions are slow for COS and even slower for CS_2. Both gases can also hydrolyze to a significant degree in hot alkaline solutions:

$$COS + H_2O = CO_2 + H_2S \qquad (10.2.11)$$

$$CS_2 + H_2O = COS + H_2S \qquad (10.2.12)$$

with again the hydrolysis of CS_2 being much slower than that of COS.

The thiocarbamic ion formed from primary amines may degrade the amine according to one of the following two reactions:

$$RNHCOS^- + RNH_2 = (RNH)_2CO + H_2S \qquad (10.2.13)$$

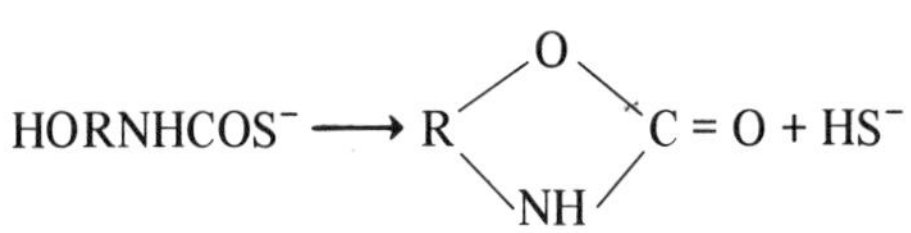

(10.2.14)

with, of course, the second reaction taking place only with the thiocarbamic ion of a primary aminoalcohol. The degradation reactions can be at least partially reversed by thermally hydrolyzing (reclaiming) the solution at high temperature.

Mercaptans are very weakly acidic and, therefore, only strong hydroxide solutions will act as appropriate chemical solvents; e.g., the *p*K at room temperature of ethyl mercaptan is 10.6 and, therefore, even the carbonic ion CO_3^{2-} is not a strong enough base to react to a significant extent with it.

10.3. KINETICS OF HYDROGEN SULFIDE REMOVAL

The basic reaction that H_2S undergoes in alkaline chemical solvents is reaction 10.2.1, a proton-transfer reaction. Proton transfer reactions are extremely fast with reaction times typically of the order of 10^{-13} sec and can, therefore, be regarded as infinitely fast in comparison with diffusional processes (i.e., $\sqrt{\Phi} \to \infty$). Absorption and desorption of H_2S take place invariably in the instantaneous reaction regime. The only exception is the case of highly carbonated solutions of amines with very stable carbamates (e.g., MEA), where dissociation of H_2S forces the reversion of carbamate to bicarbonate, so that the slow step is again reaction 9.2.1. However, a highly carbonated MEA solution has such a low *p*H that it cannot be regarded as an acceptable chemical solvent for H_2S.

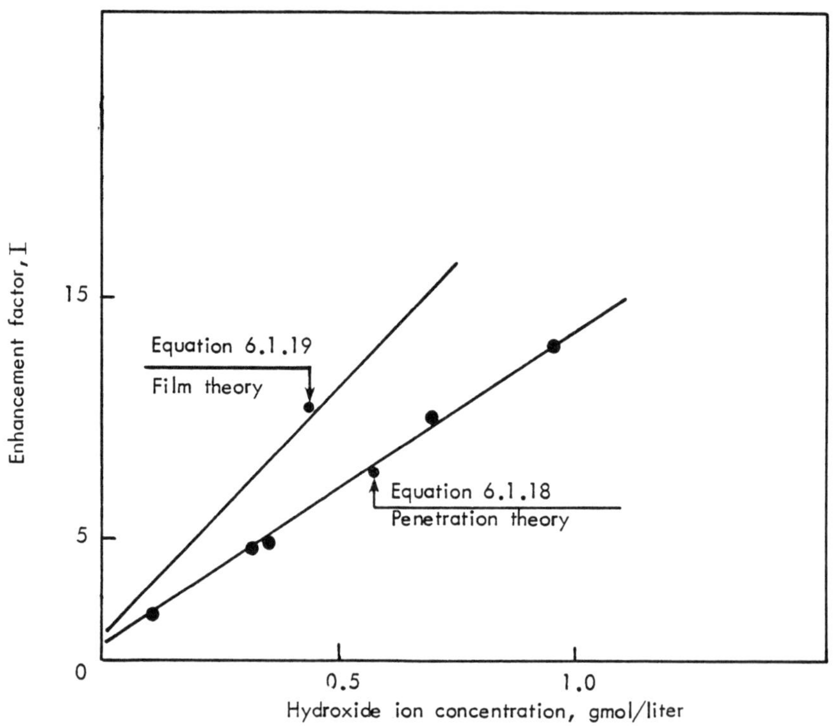

FIGURE 10.3.1. H_2S absorption in hydroxide solutions.

Laboratory verifications of the fact that H_2S mass transfer takes place in the instantaneous reaction regime are abundant in the literature. However, all these experiments are for absorption in the "irreversible" limit, $\psi \gg 1$. In this limit, Eq. 6.1.18 applies, with b_{10} the bulk-liquid concentration of free base. This has been well confirmed experimentally for sodium hydroxide, sodium carbonate, sodium triphosphate, and monoethanolamine solutions.

In the case of hydroxide solutions, the diffusivity of the base (OH^-) is sufficiently different from that of H_2S to make the difference between Eqs. 6.1.18 and 6.1.19 significant. As can be seen from Figure 10.3.1, the data strongly support the penetration-theory equation.

For desorption under conditions of large driving force ($\psi \ll 1$), Eq. 6.2.18 should be used, with $\sqrt{D_2/D}$ substituted for D_2/D:

$$\Psi \ll 1, \qquad I_\infty = 1 + \sqrt{\frac{D_2}{D}}\,\frac{[HS^-]_0}{a_0} \tag{10.3.1}$$

In Eq. 10.3.1, HS^- has been regarded as the DLRD; this is reasonable in view of the following argument. If H_2S is the only component that combines with the base B, the concentrations of BH^+ and HS^- in the bulk will be equal, and so either one could be regarded as the DLRD (the equivalent diffusivities of BH^+ and HS^- would necessarily need to be equal; see Section 10.4). If other components in addition to H_2S neutralize the base B, then in the bulk $[HS^-] < [BH^+]$ and, therefore, HS^- is the DLRD.

The ratio $[HS^-]_0/a_0$ in Eq. 10.3.1 can be calculated from the equilibrium of reaction 10.2.1:

$$\frac{[HS^-]_0}{a_0} = \mathrm{K}\frac{[B]_0}{[BH^+]_0} = \frac{\mathrm{K}_{s1}}{\mathrm{K}_p}\,\frac{[B]_0}{[BH^+]_0} \tag{10.3.2}$$

which shows that, at a given level of saturation of the base B, the value of I_∞ is almost proportional to the thermodynamic constant K.

Values of I_∞ calculated from Eq. 6.1.18 or 10.3.1 invariably turn out to be extremely large for the conditions prevailing in industrial operation. This, in turn, implies that liquid-side mass transfer resistance is often depressed to the point where the gas-phase resistance is non-negligible and may, in fact, be controlling. It is, therefore, important to consider also the asymptote $\psi \simeq 1$, where Eq. 6.3.10 should be used (or, almost equivalently, Eq. 6.2.20 with $\sqrt{D/D_j}$ substituted for D/D_j). The right-hand side of Eq. 6.3.10 is best written in the form of Eq. 6.4.6; the final result for the case at hand is

$$\Psi \simeq 1: \qquad I_\infty = I_\infty^S = 1 + \frac{\mathrm{K}}{([BH^+]_0\,[HS^-]_0/[B]_0)\,(1/[BH^+]_0 + 1/[HS^-]_0 + 1/[B]_0)} \tag{10.3.3}$$

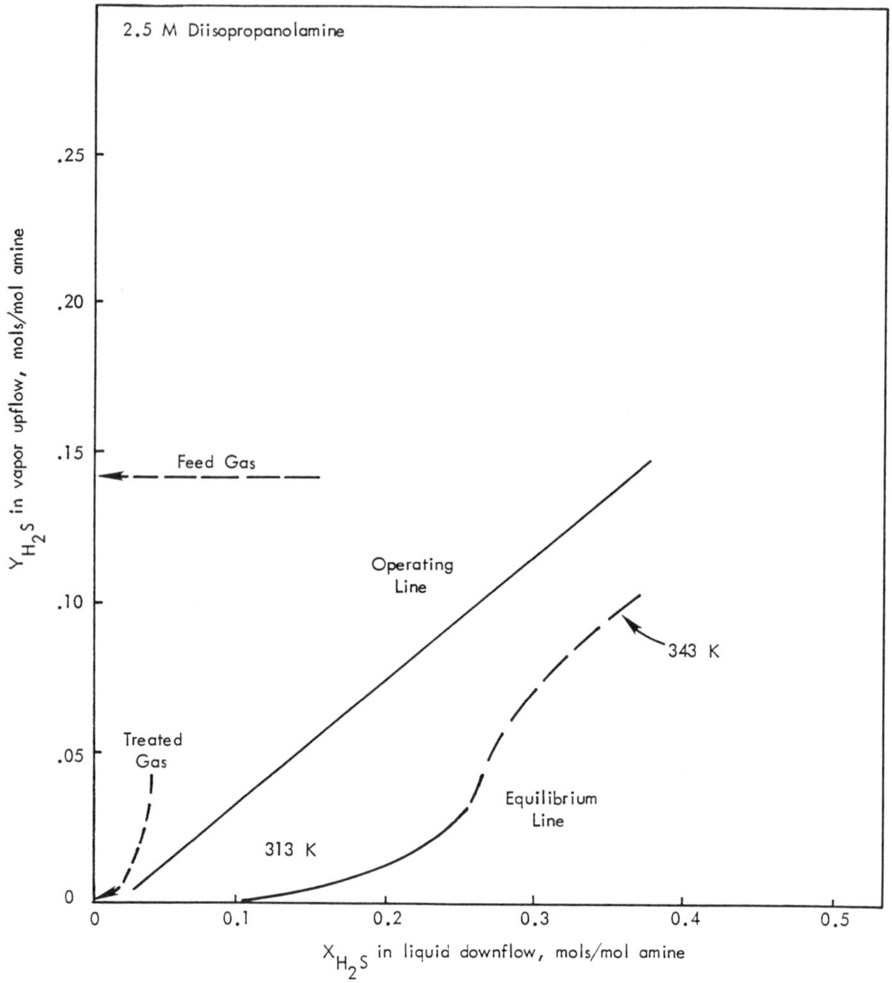

FIGURE 10.3.2. H_2S absorber operating diagram.

An analysis of a process of H_2S removal by means of an aqueous diisopropanolamine solution based on Eqs. 6.1.18, 10.3.1, and 10.3.3 has recently been published. The operating plots for the absorption and the regeneration units are reported in Figures 10.3.2 and 10.3.3. Values of I_∞ calculated from the three equations are given in Table 10.3.1. It may be seen that I_∞^S is always less than I_∞, as calculated from the "irreversible" equations; with reference to Figure 6.4.2, this shows that I_∞^S is a much better estimate of the actual value of I_∞. Gas-phase and liquid-phase resistances are compared in Table 10.3.2. It is believed that this kind of behavior is rather typical of H_2S removal units.

Since H_2S invariably undergoes mass transfer in the instantaneous reaction

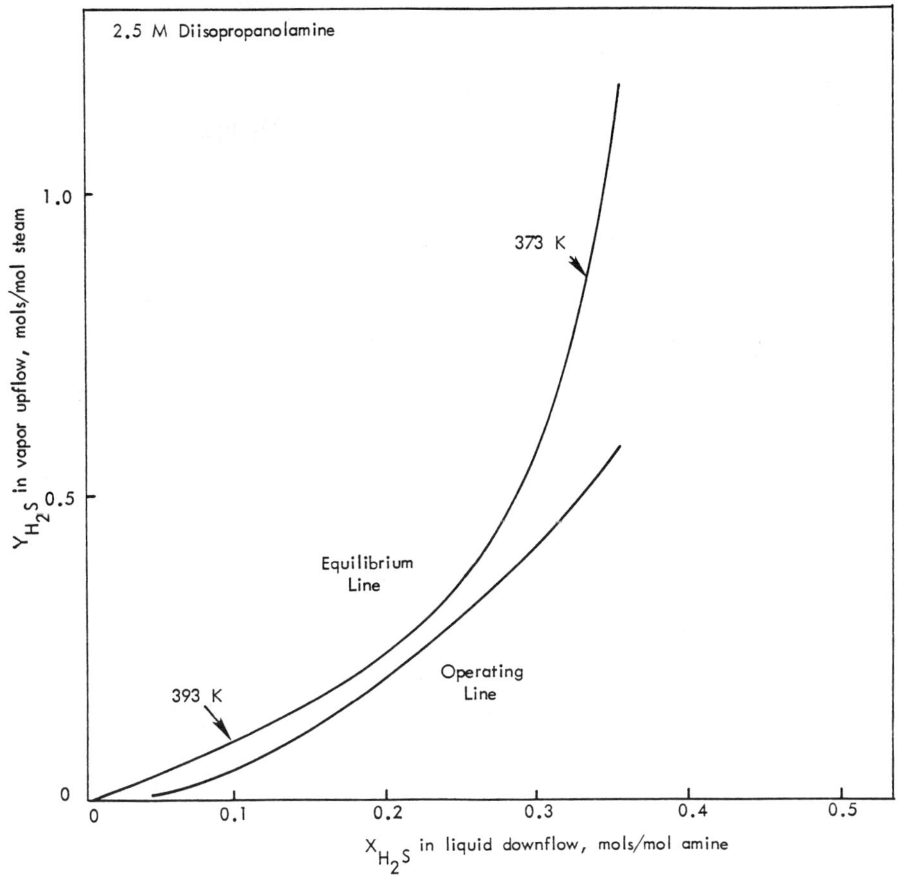

FIGURE 10.3.3. H_2S regenerator operating diagram.

TABLE 10.3.1. Enhancement Factors for $\psi \approx 1$ and "Irreversible" Cases

	$(I_\infty)_{irr.}$ Eqs. 6.1.18, 10.3.1	I_∞^S Eq. 10.3.3
Absorber		
Top	28,146	936
Bottom	25.4	11.1
Regenerator		
Top	8.5	2.2
Bottom	1,058	15.2

TABLE 10.3.2. Comparison of Gas-Phase and Liquid-Phase Resistances

	$\frac{K_G a R T}{H}$	$(I_\infty^s)\, k_L^0 a$
Absorber		
Top	8.6	131
Bottom	5.7	1.9
Regenerator		
Top	10.0	0.5
Bottom	9.6	3.7

regime, the mass transfer rate is determined not only by its own diffusivity but, in fact, predominantly by the diffusivity of other species in solution. The problem of estimating diffusivities in reacting electrolyte solutions is discussed briefly in the next section.

10.4. DIFFUSIVITIES IN ELECTROLYTE SOLUTIONS

As shown in Chapters 3-7, for reaction rates up to conditions of the fast reaction regime, the mass transfer rate depends on the diffusivity of only the volatile component in the liquid phase. In the instantaneous reaction regime, however, the mass transfer rate depends on the diffusivity of all components in the liquid phase and is, in fact, dominated by that of the non-volatile components. Inasmuch as H_2S invariably undergoes mass transfer in the instantaneous reaction regime in aqueous chemical solvents, prediction of its mass transfer rate needs an ability to estimate the diffusivity of the nonvolatile components.

For non-ionic species, correlations for liquid-phase diffusivity are available in the literature. The situation is remarkably more complex when, as invariably turns out to be the case, the diffusivity of ionic species needs to be estimated. The reason for this is that diffusion of different ions is coupled, in the sense that the diffusive flux of any one ion depends on the concentration gradients of all ions in solution. In principle, one would need to set up the diffusion-reaction equations taking this into effect; the equations would then be coupled not only through the rate-of-reaction term but also through the diffusion term (in other words, one would have both chemical and ionic coupling). Such a problem is hopelessly complex and some approximation scheme is in order. A brief discussion of the techniques that can be followed is given below.

Ions may move in a solution not only under the influence of a concentration gradient, but also under the influence of a gradient of electrical potential with negative ions tending to move towards the higher potential and vice versa. Let Z_j be the electrical valence of the ion B_j, and let ϕ be the product of the electrical potential and Faraday's constant divided by RT (ϕ is a dimensionless electrical potential). The flux of the B_j ion is given by:

$$N_j = -D_j[\nabla b_j + Z_j b_j \nabla \phi] \tag{10.4.1}$$

where the first term on the right-hand side describes the concentration driven flux and the second term, the electrical potential driven flux. Values of D_j are available in the literature for common ions and can be calculated from measurements of electrical conductance.

The electrical capacity of liquids is so low that even a very minor imbalance in the local charge concentration would result in very large peaks of electrical potential. To all practical purposes, the solution is electrically neutral everywhere, say,

$$\sum_j Z_j b_j = 0 \tag{10.4.2}$$

The flux of the B_j ion carries a current proportional to Z_jN_j. Although in an electrochemical cell the total net current is non-zero, in solutions of interest in gas treating there is no net electrical current and, therefore,

$$\sum_j Z_j N_j = 0 \tag{10.4.3}$$

Substitution of Eq. 10.4.1 into 10.4.3 and algebraic rearrangement yields the electrical potential gradient

$$\nabla \phi = -\frac{\sum Z_j D_j \nabla b_j}{\sum Z_j^2 D_j b_j} \tag{10.4.4}$$

Notice that, in view of the electroneutrality condition 10.4.2, if all the D_js are equal, the gradient of electrical potential is zero, and Eq. 10.4.1 reduces to the ordinary diffusion equation for non-ionic species. The difficult case is the one where the ionic diffusivities are sufficiently different from each other to generate a non-negligible electrical potential gradient; in this case, the second term on the right-hand side of Eq. 10.4.1 may provide strong ionic coupling of diffusion. The approximate procedure that one tries to develop in such cases is to estimate an "apparent" diffusivity of ionic species $\tilde{D}_j$ defined by

$$N_j = -\tilde{D}_j \nabla b_j \tag{10.4.5}$$

If the procedure is successful, the apparent diffusivities can be used in the equations developed in Chapters 6 and 7. Two specific examples are discussed in the following.

Consider absorption of H_2S in a monoamine RNH solution. The species to be considered are $B_1 \equiv$ RNH, $B_2 \equiv RNH_2^+$, and $B_3 \equiv HS^-$. Notice that H^+ and OH^-

are not taken into consideration since their concentrations are known to be negligible as compared to b_2 and b_3.

The electroneutrality condition becomes, trivially,

$$b_2 = b_3 \tag{10.4.6}$$

and the electrical potential is calculated from Eqs. 10.4.4 and 10.4.6 as

$$\nabla\phi = \frac{(D_3 - D_2)\nabla b_2}{(D_3 + D_2)\, b_2} \tag{10.4.7}$$

Substitution of Eq. 10.4.7 into the appropriate form of Eq. 10.4.1 and rearrangement yields the apparent diffusivities

$$\tilde{D}_2 = \tilde{D}_3 = \frac{2D_2D_3}{D_2 + D_3} \tag{10.4.8}$$

This is a particularly simple system and the apparent diffusivities are independent of the local concentrations. Furthermore, even if $D_3 \neq D_2, \tilde{D}_3 = \tilde{D}_2$ and, therefore, the simplifications discussed in Section 3.5 for the case where the diffusivities of all non-volatile species are equal are likely to apply (though not necessarily, since D_1 may well be different from $\tilde{D}_2$).

A physical interpretation of the result is obtained by considering the sketch in Figure 10.4.1, referring to absorption of H_2S. Inasmuch as RNH_2^+ is considerably more bulky than HS^-, it is reasonable to assume that $D_3 > D_2$. Therefore, Eq. 10.4.7 shows that the electrical potential ϕ decreases with decreasing b_2.

The electroneutrality condition requires b_2 and b_3, as well as their gradients, to be equal everywhere. Should there be no electrical potential gradient, HS^- would diffuse faster than RNH_2^+ and thus there would be a net current towards the interface. Since this is impossible, an electrical potential gradient develops that slows down the flux of HS^- and accelerates that of RNH_2^+ to the point where they are equal and, therefore, there is no net current. As both the fluxes and the concentration gradients of B_2 and B_3 are equal everywhere, their equivalent diffusivity is the same.

The case considered above is extremely simple and things don't work out so easily in general. The next example to be discussed is that of absorption of H_2S into a salt MX, where the species to be considered are $B_1 \equiv X^-, B_2 = HX, B_3 = HS^-$ and $B_4 = M^+$; we assume that $Z_1 = -1$ and $Z_4 = 1$.

The electroneutrality condition is

$$b_4 = b_1 + b_3 \tag{10.4.9}$$

so that the electrical potential gradient is calculated from Eq. 10.4.4 as

$$\nabla\phi = \frac{(D_3 - D_1)\nabla b_3 - (D_4 - D_1)\nabla b_4}{(D_3 - D_1)\, b_3 + (D_1 + D_4)\, b_4} \tag{10.4.10}$$

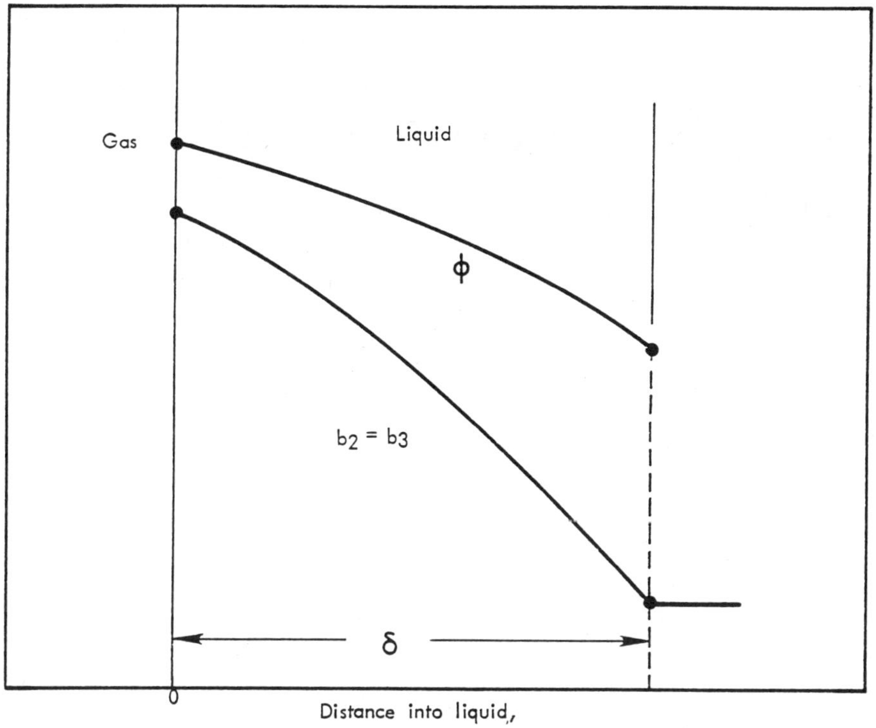

FIGURE 10.4.1. Absorption of H_2S into amine solution; concentration and electrical potential distribution sketch.

Notice that in this case $\nabla\phi$ is not expressed in terms of only one concentration gradient as in Eq. 10.4.7, but rather in terms of both ∇b_3 and ∇b_4. However, an additional condition is imposed by the physical fact that the flux of $M^+ \equiv B_4$ needs necessarily to be zero. Therefore, from the appropriate form of Eq. 10.4.1,

$$\nabla\phi = -\frac{\nabla b_4}{b_4} \tag{10.4.11}$$

Algebraic rearrangement of the equations yields the apparent diffusivities

$$\tilde{D}_1 = D_1 \frac{2(D_4 + D_3 - D_1)\, b_4}{(D_3 - D_1)\, b_3 + (2D_4 + D_3 - D_1)\, b_4} \tag{10.4.12}$$

$$\tilde{D}_3 = D_3 \frac{2D_4 b_4}{(D_3 - D_1)\, b_3 + 2D_4 b_4} \tag{10.4.13}$$

with the equivalent diffusivity $\tilde{D}_4$ being zero, since N_4 is zero although ∇b_4 is not.

Notice that, if $D_1 = D_3$, then the apparent and actual diffusivities coincide, no matter what D_4 may be. The complex case is when $D_1 \neq D_3$ since in that case the apparent diffusivities, as given by Eqs. 10.4.12 and 10.4.13, will depend on the local concentrations and, therefore, will be non-constant in the liquid phase.

11 Simultaneous and Selective Removal of Hydrogen Sulfide and Carbon Dioxide

The law is equal for everybody. For some people it is just a little bit more equal.

OLD ITALIAN PROVERB

In many industrial gas treating processes the sour gas contains substantial amounts of both CO_2 and H_2S. Two main cases are to be considered. The first is the case where the aim of the gas treating process is the simultaneous removal of most of both CO_2 and H_2S; this will be referred to as Case I in the following. Case II is the situation where the aim of the process is to remove only H_2S, and whatever simultaneous removal of CO_2 occurs is simply a largely undesirable side effect. An example of a Case II process is a low-joule content fuel gas obtained from coal or oil gasification, where removal of H_2S is required for the environmental reasons discussed in Chapter 10, but removal of CO_2 only marginally increases the heating value of the gas. Another example of a Case II process is where the H_2S to CO_2 ratio in the gas is low, e.g., 1-10 or 1-30. Selective removal of H_2S is desirable in this case to provide a high concentration H_2S stream for conversion to elemental sulfur in a Claus plant.

For Case I processes, the specifications on the sweet gas are, in general, much stricter for H_2S than they are for CO_2; therefore, selectivity of the solvent for H_2S is desirable but not crucial to the process. For Case II processes, high selectivity for H_2S removal is of primary importance.

Physical solvents may be used for simultaneous removal of CO_2 and H_2S. Figure 1.7.1 and Table 1.7.2 give solubility data for CO_2, H_2S, and light hydrocarbons that are often present in sour gases. The H_2S solubility in polar organic solvents is always larger than the CO_2 solubility, though never by more than an order of magnitude.

11.1. PROCESS SELECTION

For the purpose of discussing the industrial chemistry of simultaneous H_2S and CO_2 removal, a subdivision of Case I is useful:

IA. High H_2S cleanup, e.g., $<$4-100 ppm
Modest CO_2 cleanup, e.g., $<$2%

IB. High H_2S cleanup, e.g., 4 ppm
High CO_2 cleanup, e.g., $<$ 100 ppm

Further subclassification in terms of H_2S/CO_2 ratio in the raw gas to treating is helpful as this determines the *type* of Claus plant needed to convert the separated H_2S to elemental sulfur:

$H_2S/(CO_2 + H_2S) > 35\%$	Once-through Claus
$10\text{-}20 < H_2S/(CO_2 + H_2S) < 35\%$	Split-flow Claus

Case IA processes tend to occur often in treating natural gas to pipeline specification (<4 ppm H_2S, $<$ 2% CO_2). The gas sweetening (treating) technology chosen for a recent grassroots sour gas plant in Wyoming is an example. The inlet gas contains 17.6% CO_2, 3.8% H_2S, trace hydrocarbons up to C_7, and unspecified trace impurities that could cause degradation of some chemical solvents. A two-stage process was chosen. In the first sweetening stage, a promoted hot potassium carbonate solution is used for bulk acid-gas removal, followed by a second stage using an aqueous aminoalcohol to produce pipeline specification gas. A flowsheet is shown in Figure 11.1.1. The process incorporates some interesting features such as common overhead condenser for the two separate regeneration towers, use of flash cooling of the lean carbonate solution before it is charged to the absorber, and a power recovery turbine on the rich carbonate stream. The rationale for these features in minimizing plant investment and energy costs is discussed in detail in Chapter 15.

The sulfur plant is a three converter, split-flow modified Claus unit processing the lean ($H_2S/(CO_2 + H_2S) = 18\%$) acid gas from the treating section. With H_2S lean acid gas, the Claus plant requires special design considerations such as hot gas bypass and inline reheat (see Figure 11.1.2). Both the amine regenerator and the carbonate regenerator use low-pressure (55 psig) steam generated in the sulfur plant.

Case IB processes, in which a high degree of cleanup of both H_2S and CO_2 is needed, occur in the chemical industry and in the manufacture of liquefied natural gas (LNG). Chemical synthesis gas consisting of hydrogen and carbon monoxide can be produced by partial oxidation of a wide variety of fluid hydrocarbons, including sour heavy liquids. A typical composition of the raw synthesis gas made from heavy high sulfur (3.5 wt%) fuel oil is shown in Table 11.1.1. The gas must be treated to produce a 50/50 H_2/CO syngas containing <30 ppmv CO_2 and <5 ppmv H_2S plus COS. Acid gas ratio in the raw untreated synthesis gas is $H_2S/(H_2S + CO_2) = 17\%$. The acid gas must be processed in a sulfur recovery plant. The pro-

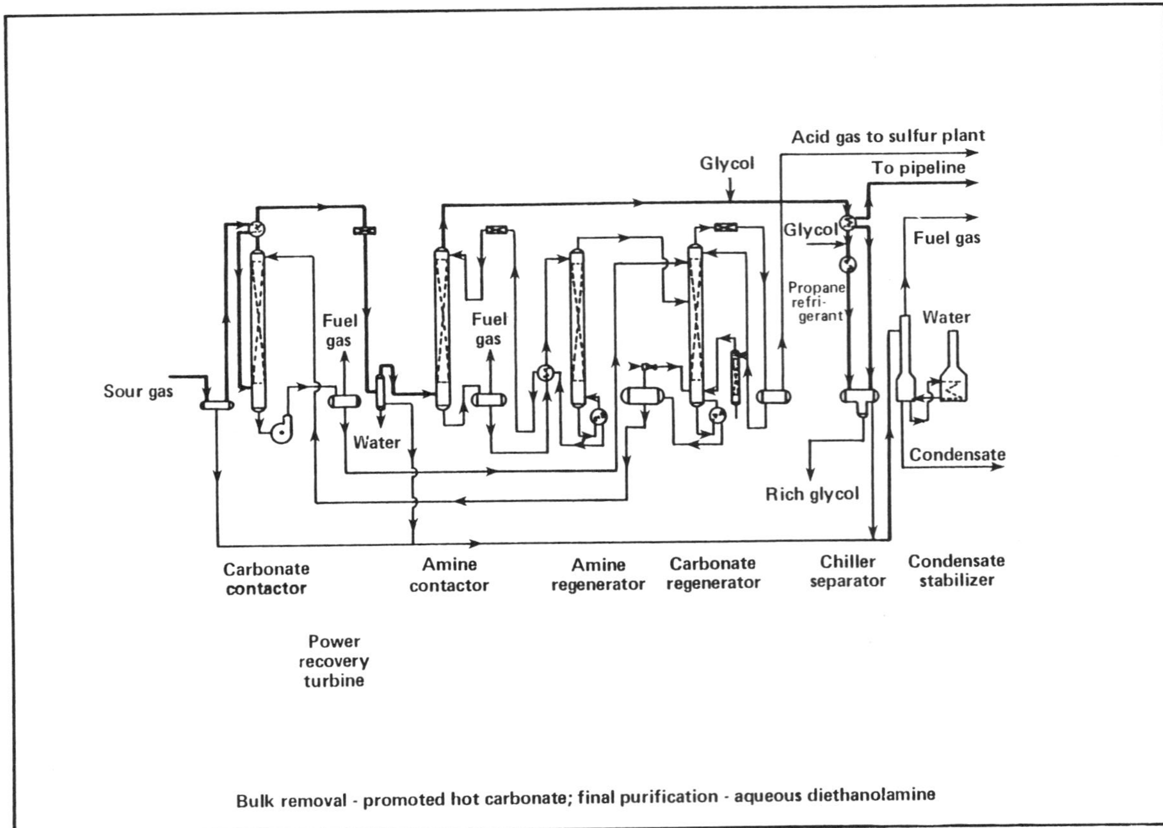

FIGURE 11.1.1. Natural gas treatment for simultaneous removal of CO_2 and H_2S to pipeline specification.

TABLE 11.1.1. Compositions of Raw Synthesis Gas, Desulfurized Gas and Acid Gas from Oil Gasification Process (mole %, dry basis)

	Feed	Desulfurized gas	Acid gas
H_2	47	49	4
CO	48	50.4	4
CO_2	3.7	<30 ppm	76
N_2	0.2	0.2	–
CH_4	0.4	0.4	–
H_2S	0.7 }	<5 ppm	16
COS	0.03 }		

cessing scheme is outlined in Figure 11.1.3. Simultaneous removal of H_2S, COS, and CO_2 is carried out by absorption in a chemical solvent comprising diisopropanolamine in an organic solvent, Sulfolane, with some water using a conventional absorber-regenerator flowsheet.

The acid gas from the regenerator may be treated to selectively separate the H_2S (Case II process) to provide a concentrated H_2S/CO_2 acid gas which is sent to a Claus sulfur plant as shown in Figure 11.1.3. The CO_2 that passes through the selective H_2S separation unit may be vented or used as feedstock for another process. Gas treating solvents of the type aminoalcohol-organic solvent-water have the unique capability of simultaneously removing all acid gas constituents, including the trace sulfur contaminants.

A different approach to selective treating involves simultaneous absorption of H_2S and CO_2 from the feed gas followed by selective regeneration to produce one acid gas stream rich in H_2S and another containing mainly CO_2. Figure 11.1.4 shows a physical solvent process using methanol with selective regeneration. Selective regeneration using chemical solvents is also possible. Whereas in the case of a physical solvent the CO_2 may be stripped out in preference to H_2S (owing to the higher solubility of H_2S), in the case of a chemical solvent such as monoethanolamine which forms a relatively stable carbamate with CO_2, the H_2S can be stripped out in preference to the CO_2. CO_2 can be stripped from the solution in a subsequent step at higher temperature.

Simultaneous removal of H_2S and CO_2 from natural gas to make it suitable for cryogenic processing to produce LNG is another demanding treating operation. For LNG production, feed conditions, process design considerations, and LNG specifications usually dictate the necessary separation processes. Table 11.1.2 shows the allowable ranges of impurities in LNG product.

Acid gas removal units used in some baseload LNG plants are listed in Table 11.1.3. Three main processes have been used: promoted hot carbonate followed by caustic wash or aqueous DEA treat for final purification, aqueous MEA, and DIPA-Sulfolane-water.

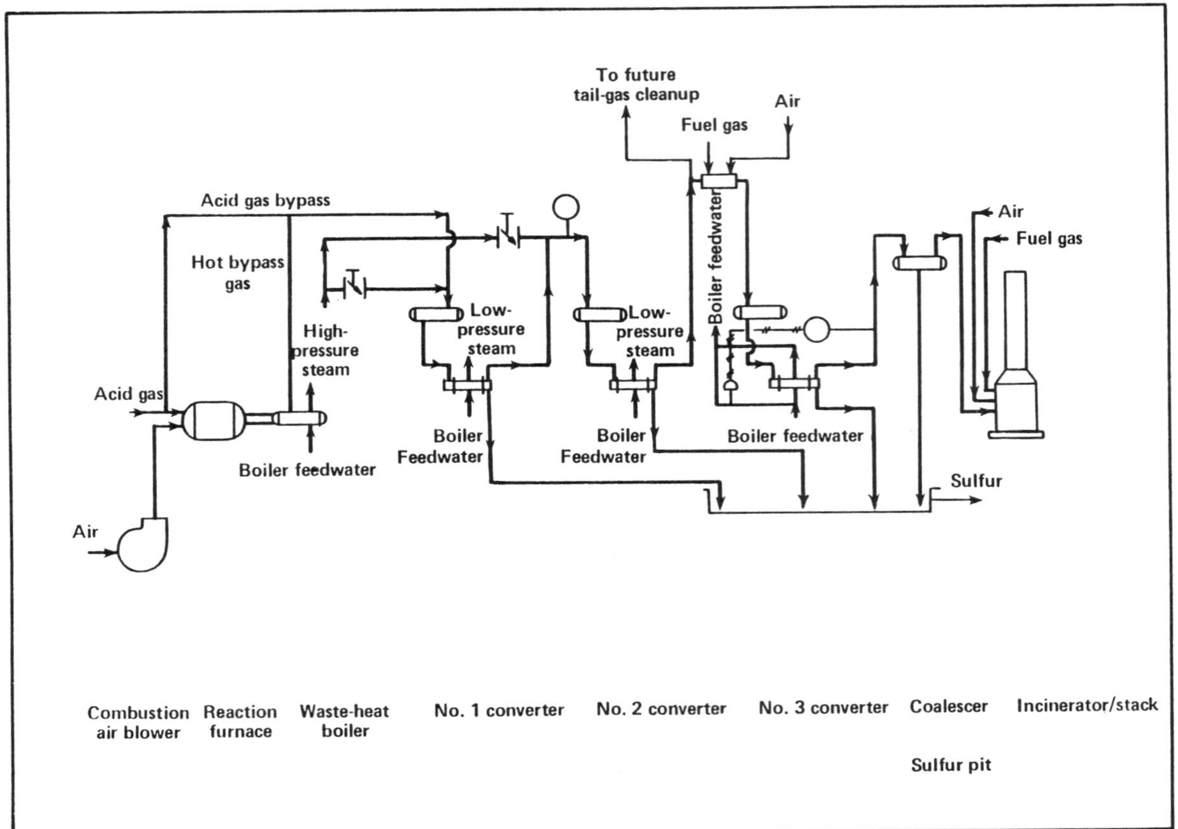

FIGURE 11.1.2. Split-flow Claus sulfur plant.

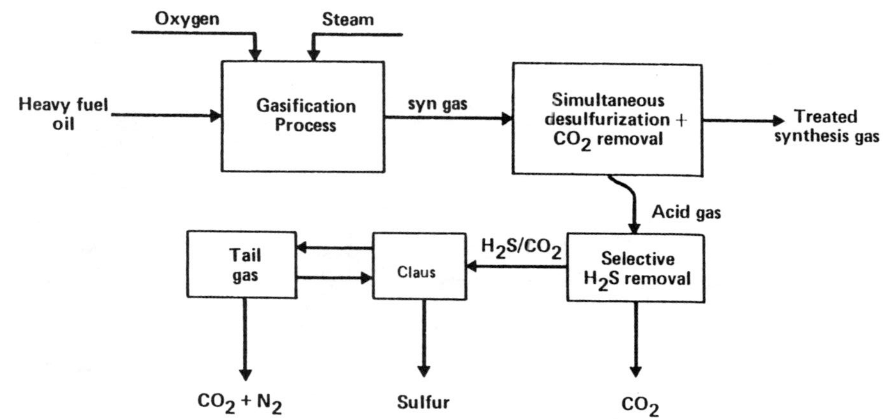

FIGURE 11.1.3. Processing scheme for synthesis gas manufacture using both simultaneous and selective gas treating.

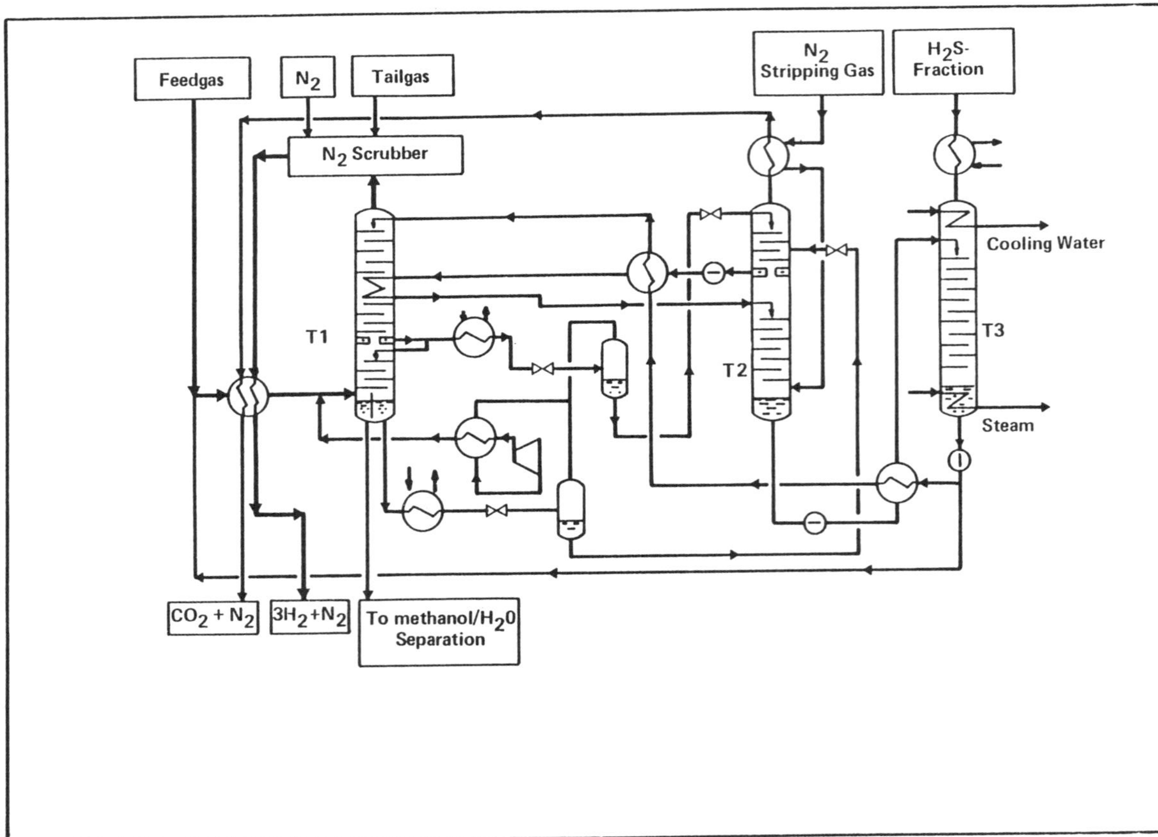

FIGURE 11.1.4. Physical solvent (chilled methanol) process for simultaneous absorption of H_2S and CO_2 with selective regeneration.

TABLE 11.1.2. Allowable Ranges of Impurities in LNG Product

Impurity	Allowable level in LNG product	Limit set by[a]
Water	0.5 ppm	A
H_2S	0.25 grains/100 SCF (3.5.mg/Nm^3)	C
CO_2	50–125 ppm	B
Organic sulfur (mercaptans, disulfides, thioethers, CS_2)	150 mg total sulfur/Nm^3 of regasified gas	C
COS[b]		
Nitrogen	0.5–1.5 mole%	C
Aromatics	1–10 ppm by volume	A or B

[a]A = quantity allowed to accumulate at the cold end of the plant above the solubility limit without restricting production, B = solubility limit, and C = product specifications.
[b]Hydrolysis of COS, $COS + H_2O \rightleftharpoons H_2S + CO_2$, will add to the concentration of H_2S.

If rational process selection is followed, the choice of gas treating process should correlate with feed gas composition. However, there are other factors which should be, and are, taken into account. For example, most LNG liquefaction plants are fed by gases from numerous wells and several different fields. Gas quality can vary widely from one field to another and the composite feed to a plant often changes. In addition, composition from given wells can change over time.

TABLE 11.1.3. Acid Gas Removal Units Used in Some Baseload LNG Plants

Company and plant site	Liquefaction plant capacity (MCFD)	Type of acid gas removal unit
Sonatrack (LNG-2) Argen, Algeria	1100	MEA
Esso Libya Marsa el Brega, Libya	385	PHC[a] followed by caustic wash and water wash
Sonatrack, Phase I Skikda, Algeria	430	MEA
Brunei LNG Lumut, Brunei, Borneo	750	DIPA-Sulfolane-H_2O
Phillips-Marathon Kenai, Alaska	185	MEA
Abu Dhabi Gas Liquefaction Co., Ltd. Das Island, Abu Dhabi	350	PHC followed by DEA
Pertamina/Mobil Arun, N. Sumatra, Indonesia	1200	PHC followed by DEA
Pertamina/Huffro Badak, E. Kalimantas, Indonesia	550	MEA

[a]Promoted hot carbonate.

The purified gas specification for LNG feedstock quality is usually <3 ppm H_2S, <5 ppm total sulfur (as H_2S), and <150 ppm CO_2. Raw natural gas is usually supplied by pipeline to the liquefaction plant from fields located within about 100 miles of the plant. Only minimal purification is carried out at the wellhead, the gas being sent to the LNG plant as a two-phase mixture with liquid hydrocarbons. After condensate separation and mercury removal (if needed), the gas is treated for acid gas removal. The gas is then dried and, finally, liquefied. A schematic flow diagram of an LNG plant is shown in Figure 11.1.5.

Case II processes for selective separation of H_2S from gas containing both H_2S and CO_2 represent a more specialized class of treating processes than simultaneous treating, but they are expected to experience rapid growth as industry switches to high-sulfur oil and coal feedstocks. The chemical oxidation approach to selective H_2S separation was described briefly in Chapter 10, and it will not be discussed further. The processes based on kinetic selectivity have high industrial potential and will be reviewed in the following paragraphs.

A commercial plant using aqueous methyldiethanolamine (MDEA) for selective H_2S removal in ethane purification has been described in the literature. The MDEA unit is a converted refinery DEA treater. Feed to the unit during a test period was 32% H_2, 21.5% CH_4, 12% C_2H_6, 8.5% H_2S, and 1.4% CO_2. (Notice the very high ratio of H_2S to CO_2.) Plant data show that at temperatures below 110°F, the H_2S content of the absorber overhead product is less than 10 ppm and that the H_2S content of the treated gas rises steeply as the acid gas loading of the MDEA solution increases above about 3.0 SCF/gal. At 3.3 SCF/gal loading and 25 ppm H_2S in the overhead product, approximately 70% of the CO_2 is rejected (passes through the absorber). No other descriptions of commercial selective tertiary amine units have appeared in the literature, though pilot plant tests have been described.

Selective H_2S removal has been practiced for many decades for treating coke oven gas. Some processes are based on aqueous ammonia scrubbing, using ammonia absorbed from the coke oven gas. Others are based on the use of cold sodium carbonate regenerated under vacuum (vacuum carbonate process). Available technologies are summarized in Table 11.1.4.

TABLE 11.1.4. Potential Coke Oven Gas Desulfurization Technologies

Liquid absorption processes
Vacuum carbonate
Aqueous MEA (Sulfiban)
Aqueous NH_3
Wet oxidative processes
Vanadate (Stretford, Takahax)
Ammonia/Picric Acid (Fumaks)
Arsenite/Arsenate (Giammarco-Vetrocoke)
Dry oxidative processes
Iron oxide boxes

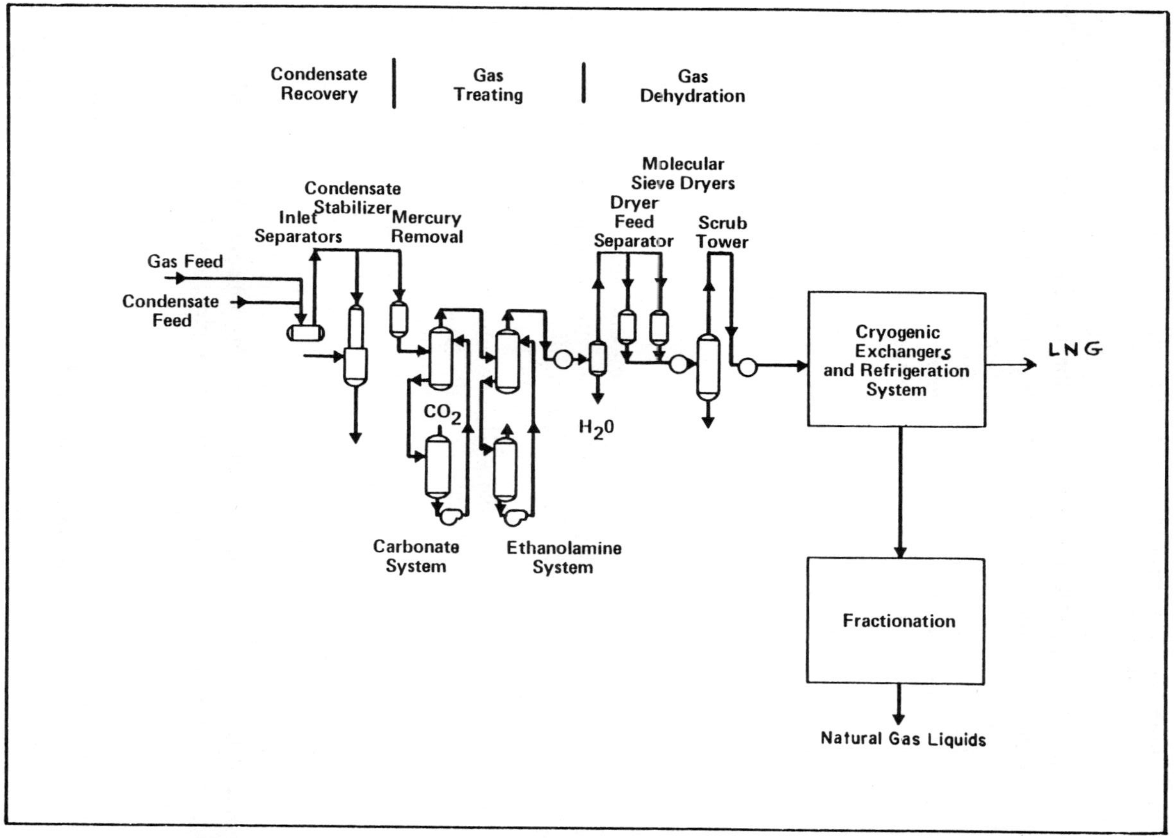

FIGURE 11.1.5. Schematic flow diagram of a liquified natural gas (LNG) plant.

The vacuum carbonate process, which is the one most widely used in the United States, employs a Na_2CO_3 absorbing solution. Raw gas countercurrently contacts an aqueous solution of Na_2CO_3 into which H_2S, HCN, and some CO_2 are absorbed. The solution then passes to a regenerator (actifier) where acid gases are removed by steam stripping. Stripping is carried out at a partial vacuum (4.4 in. Hg absolute, 54°C) to minimize high-pressure steam consumption. Under these conditions, all required regenerator steam can normally be raised using low grade heat available in coke plant flushing liquor. Vacuum is maintained in the regenerator using two steam-operated ejector jets. Recovered acid gases are usually sent to a Claus plant for sulfur recovery.

Table 11.1.5 lists H_2S removal efficiencies and H_2S/CO_2 selectivity information for the vacuum carbonate process, the aqueous alkanolamine (MEA) process, and the Stretford (oxidative) process. Flowsheets of these processes are shown in Figure 11.1.6. HCN pickup is an important consideration in all cases as it results in contamination of the acid gas stream that can cause plugging and corrosion problems in the treating plant. The vacuum carbonate process offers good selectivity for H_2S but H_2S cleanup below about 30 gr/100 SCF cannot be achieved without a large increase in steam consumption. The aqueous ammonia scrubbing process, widely used in Europe, suffers from the disadvantage of ammonia volatilization losses. All of these processes use short gas-liquid contact times to maximize the selective removal of H_2S over CO_2. The data in Table 11.1.5 demonstrate the complex considerations of H_2S selectivity, H_2S cleanup, and steam consumption that must be weighed in selecting the most attractive technology for a given application.

Increasingly, in the future many gases that require treating for acid gas removal will contain CO_2, H_2S, COS, and possibly other trace sulfur contaminants. These gases include natural gases and synthetic gases generated by gasification of sour heavy oils and sulfur-containing coal. Typical gas compositions are shown in Table 11.1.6. These gases are characterized by (1) relatively high content of CO_2; (2) low ratio of H_2S to CO_2; and (3) small but significant content of COS. The problem of treating these gases is a challenging one that will probably require a combination of gas treating technologies. Some possible examples from the future manufacture of hydrogen are given in Figure 11.1.7.

TABLE 11.1.5. Comparative Performance of Potential Coke Oven Gas (COG) Desulfurization Technologies[a]

Liquid absorption process	H_2S in clean gas (Gr H_2S/100 SCF)	Acid gas to sulfur plant, %			Equivalent energy (10^6 BTU/day)
		H_2S	CO_2	HCN	
Vacuum carbonate	30	55–70	15–25	5–15	585
Aqueous MEA	10	50	44	4	437
Chemical oxidation (Stretford)	5	–	–	–	499

[a]Basis: 60 million ft^3/day COG: sour COG contains 500 gr H_2S/100 SCF and 60 gr HCN/100 SCF. Ninety percent H_2S removal efficiency.

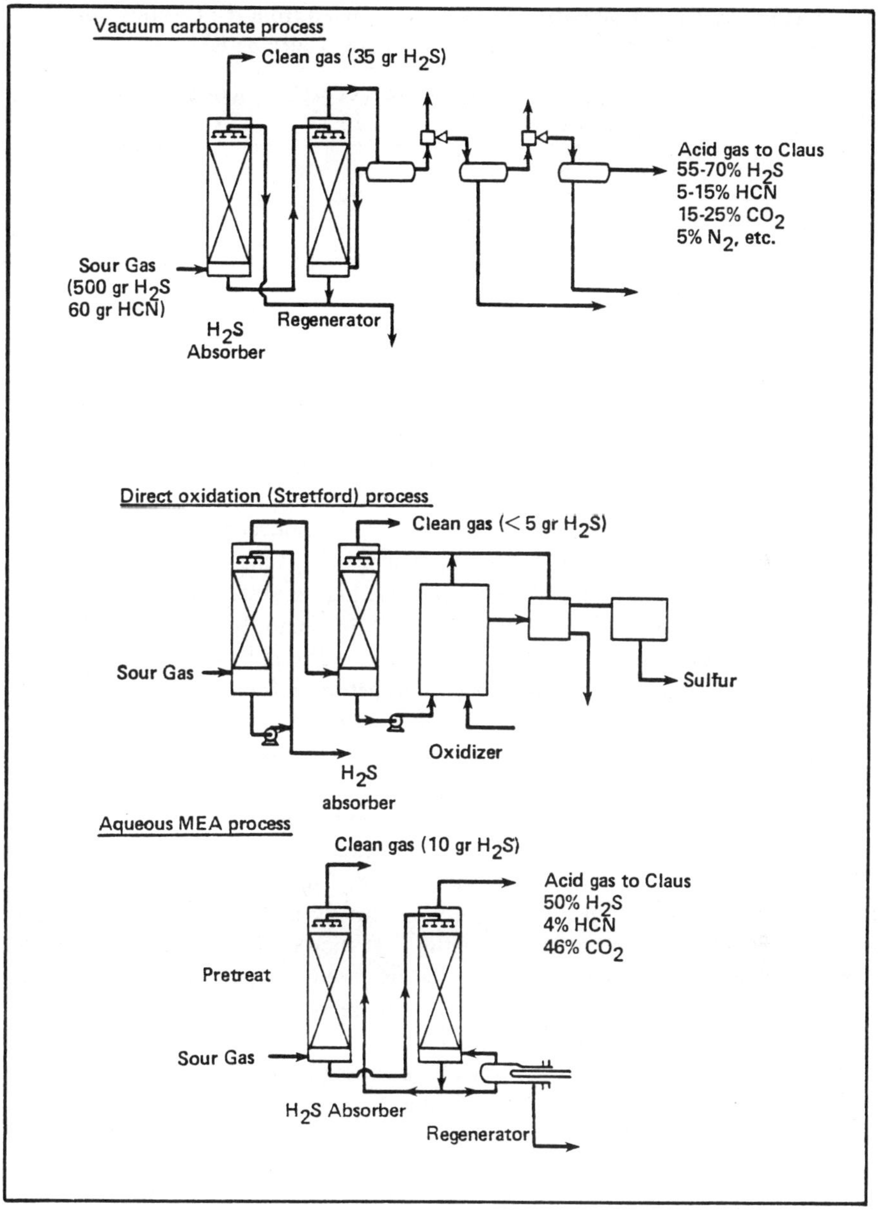

FIGURE 11.1.6. Process flowsheets for selective desulfurization of coke oven gas.

TABLE 11.1.6. Composition of Gases Obtained from the Gasification of Heavy Oil and Coal (mole %, dry basis)

Component	Heavy oil partial oxidation	Lurgi coal gasifier	Koppers–Totzek coal gasifier
H_2	46.1	38.3	35
CO	46.9	17.4	55
CO_2	4.3	32.0	7.3
H_2S	0.8	1.2	1.1
COS	0.1	Trace	Trace

In scheme (A), the gas, after CO removal by shift conversion using a sulfur-tolerant shift catalyst, is treated for simultaneous acidic gas ($CO_2 + H_2S +$ COS) removal from the H_2 by absorption in a physical solvent. The solvent is then *regenerated selectively* to produce a high concentration stream of H_2S (and COS) and an H_2S-free CO_2 stream (see Figure 11.1.4).

In scheme (B), the first step is again simultaneous removal of all acid gases. Conventional regeneration is used producing an acid gas stream containing H_2S, CO_2, and COS in the same ratio in which they appeared in the shift effluent. In a second selective absorber, the acid gas stream is treated to selectively remove the H_2S (and COS) in a concentration suitable for conversion to sulfur in a Claus plant.

In scheme (C), the raw gas from the gasifier, before shift conversion, is treated to remove H_2S (and COS) selectively. CO is then removed by the water gas shift reaction over a classical shift catalyst and, finally, CO_2 is removed from the product H_2. Insufficient information is available today to decide which of these three schemes is likely to be preferred. Conceivably, all three could become important.

The usual range of gas treating process application for simultaneous removal of CO_2 and H_2S from gases is shown in Figure 11.1.8. The same kind of plot for selective H_2S removal is given in Figure 11.1.9. It should be borne in mind that plots such as these describe the existing situation but they do not necessarily yield the optimal choice; actual choices of processes are often related to historical reasons, with individual companies tending to select those processes for which they have more technological expertise.

An important element in the choice of a process for selective H_2S removal is the *process selectivity* S_p which may be defined as

$$S_p = \frac{(H_2S/CO_2) \text{ removed gas}}{(H_2S/CO_2) \text{ raw gas}} \tag{11.1.1}$$

Inasmuch as removal of H_2S is invariably required to be almost complete, due to the severe H_2S specifications, the amount of H_2S in the removed gas is practically the same as in the feed gas; therefore, the process selectivity S_p can be expressed as

$$S_p = \frac{(CO_2) \text{ raw gas}}{(CO_2) \text{ removed}} \tag{11.1.2}$$

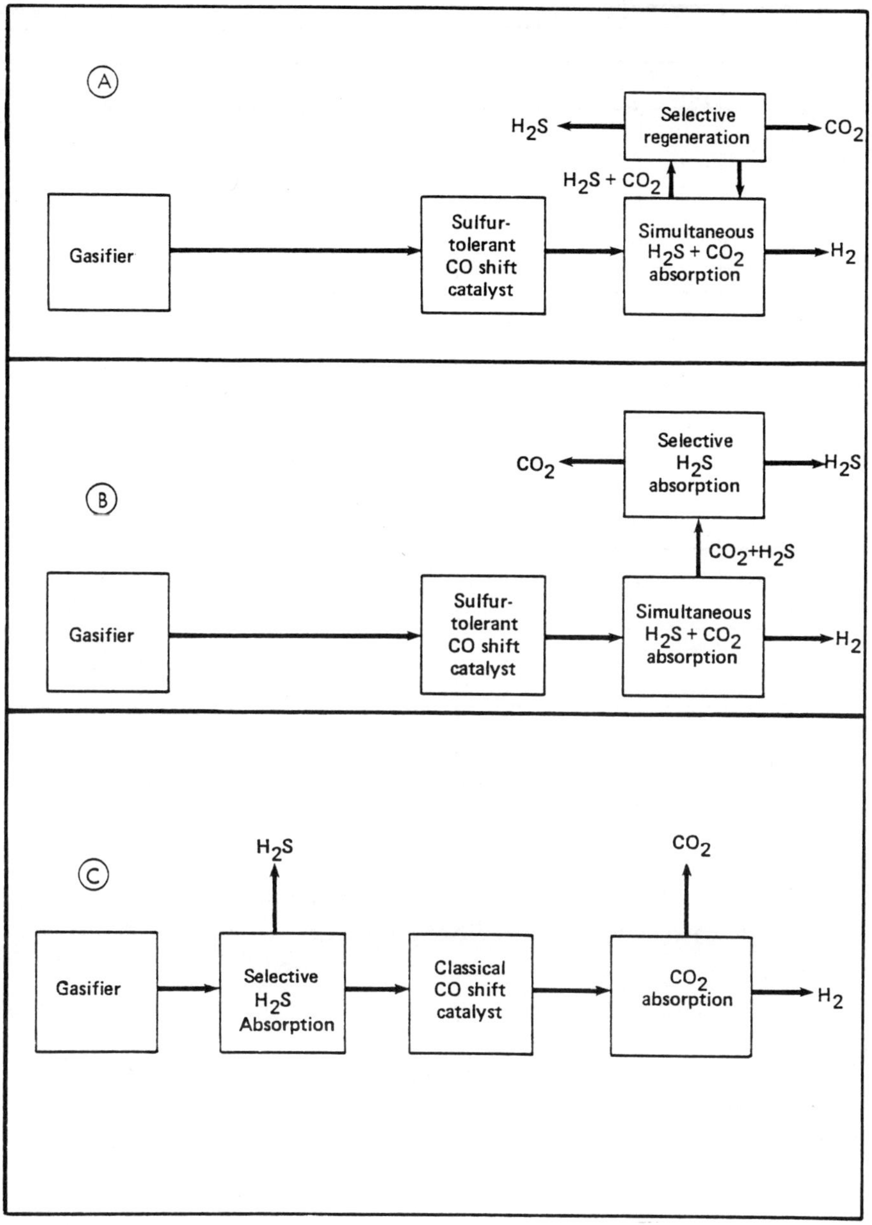

FIGURE 11.1.7. Approaches for treating gases containing CO_2 and H_2S and having high CO_2-to-H_2S ratio.

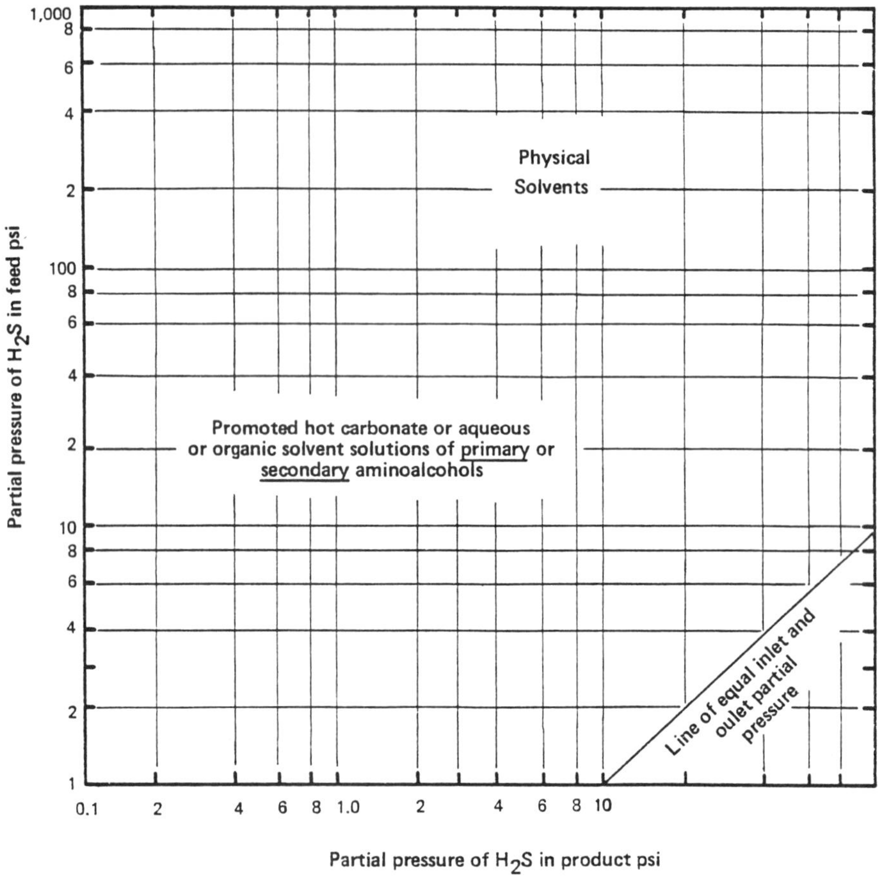

FIGURE 11.1.8. Range of gas treating process application for simultaneous H_2S plus CO_2 removal with no other impurities present.

Typically, S_p values of 10 or better are desirable; i.e., only 10% of the CO_2 should be removed.

In the case of physical solvents, the process selectivity is simply related to the thermodynamic selectivity, as shown below. The thermodynamic selectivity of a physical solvent is given by (see Section 2.7)

$$S_T = \frac{H_{CO_2}}{H_{H_2S}} \tag{11.1.3}$$

where H is the Henry's Law constant.

Figure 11.1.10 is a partial pressure versus acid gas load in the liquid plot for simultaneous absorption of H_2S and CO_2. The two equilibrium lines are independent of each other (the solubility of CO_2 is independent of the H_2S load in the

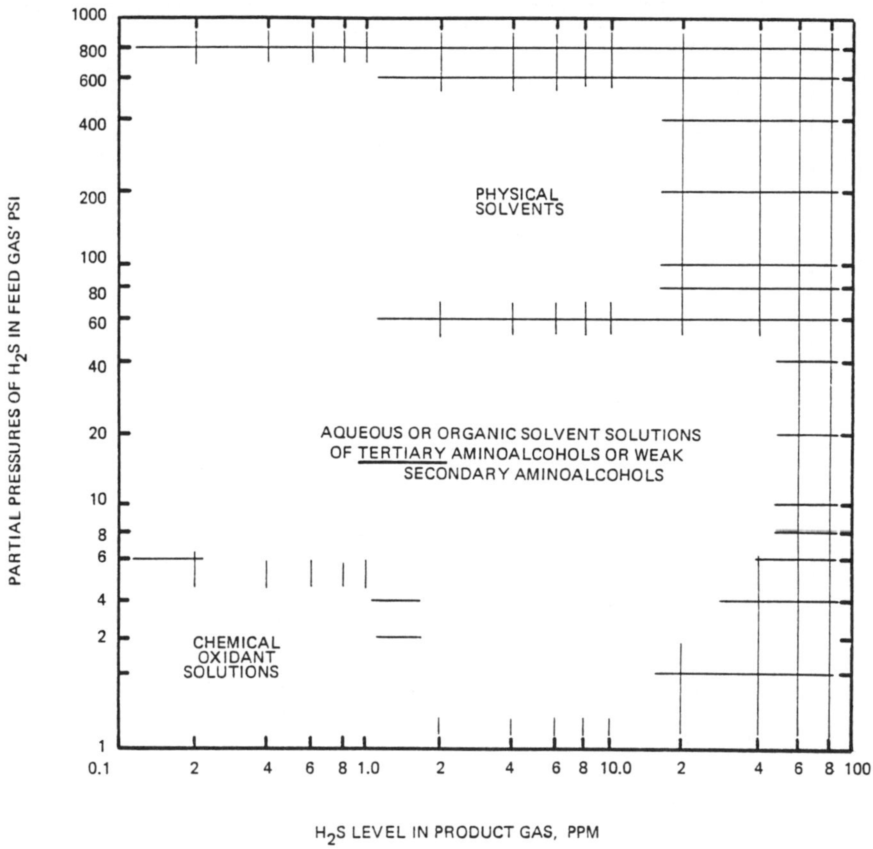

FIGURE 11.1.9. Range of gas treating process application for selective H_2S removal in the presence of CO_2.

liquid and vice versa), have slopes equal to the Henry's Law constants, and, therefore, the ratio of their slopes equals the thermodynamic selectivity.

In designing for selective H_2S removal, one would choose the liquid flowrate so as to achieve complete H_2S removal; for the simple dilute case where the operating lines are straight, their slope would be slightly larger than the slope of the equilibrium line for H_2S, say,

$$\text{slope} = \beta H_{H_2S} \tag{11.1.4}$$

with β being typically about 1.3.

Since almost all the H_2S must be removed, a large packing height or a large number of trays is required and the absorber will be H_2S-pinched at the lean (top) end. If $S_T < \beta$, the absorber will be CO_2-pinched also at the lean end and, therefore, almost all the CO_2 will be removed; i.e., $S_p = 1$. However, if $S_T > \beta$ (which is prac-

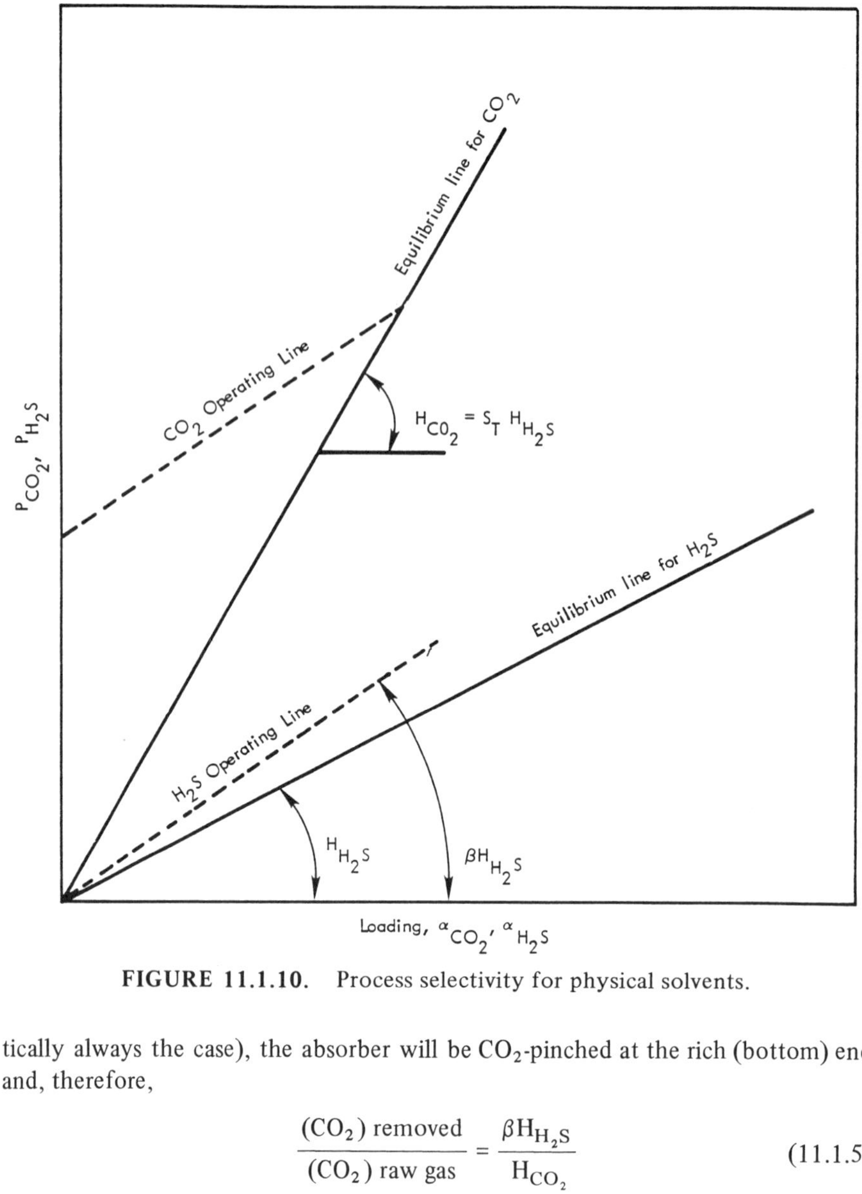

FIGURE 11.1.10. Process selectivity for physical solvents.

tically always the case), the absorber will be CO_2-pinched at the rich (bottom) end and, therefore,

$$\frac{(CO_2)\text{ removed}}{(CO_2)\text{ raw gas}} = \frac{\beta H_{H_2S}}{H_{CO_2}} \tag{11.1.5}$$

Combinations of Eqs. 11.1.2, 11.1.3, and 11.1.5 gives

$$S_p = \frac{S_T}{\beta} \tag{11.1.6}$$

i.e., *the process selectivity of a physical solvent is marginally less than the thermodynamic selectivity.* This conclusion is true also for the more complex case where the operating lines are curved. Inasmuch as the thermodynamic selectivity of physical solvents practically never exceeds 10, an upper bound is set on the process selectivity obtainable with physical solvents.

In the case of chemical solvents that are generally aqueous solutions, the thermodynamic selectivity can never exceed the value given by Eq. 2.7.9, which is of the order of 3. However, the process selectivity that can be achieved may be remarkably larger than the thermodynamic selectivity. The reasons for this are briefly reviewed in the following.

First of all, the equilibrium curves for a chemical solvent are not uncoupled, in the sense that the equilibrium vapor pressure of CO_2 depends on the H_2S load in the liquid and vice versa (see Section 2.6). This implies that in a chemical solvent with a low load of CO_2, H_2S could be absorbed up to a load *exceeding* that which would correspond to the liquid being in equilibrium with the gas. Furthermore, complex kinetic coupling, of the type discussed in Chapter 8, may result in the absorption rate of H_2S being remarkably faster than that of CO_2.

It should be borne in mind that a process selectivity S_p well in excess of the thermodynamic selectivity can only be achieved by making use of the different rates of mass transfer of H_2S and CO_2. Therefore, not only process analysis but also process selection in this area requires a thorough understanding of the mass transfer mechanisms.

11.2. INORGANIC SALT SOLUTIONS

In this section we consider the simultaneous removal of CO_2 and H_2S by means of aqueous solutions of inorganic salts of a strong base with a moderately weak acid; we shall indicate such salts by the notation *MX*. We do not imply that the acid involved is monobasic. Consider, for example, the case of tripotassium phosphate, K_3PO_4. By *HX* we identify the acid H_3PO_4, and by X^- the monovalent ion $H_2PO_4^-$. Salts of this type have already been considered as generic bases in Chapter 10, and their solutions have been shown to be acceptable chemical solvents for H_2S if their room temperature pK is in the range 8–12. Since H_2S and carbonic acid have comparable strengths, such solutions are also acceptable chemical solvents for CO_2 *from the thermodynamic viewpoint.*

The thermodynamics of the system considered depend essentially on the value of the first dissociation constant K_x of the acid *HX*. K_x is equal to the protonation constant for the base *MX*. The second dissociation constant of carbonic acid K_{c2} (see Figure 10.2.1) is about three orders of magnitude larger than the second dissociation constant of H_2S (see Figure 2.7.2). The value of K_x, as discussed in Chapter 10, must lie in the interval

$$K_{s1} \gg K_x \gg K_{s2} \tag{11.2.1}$$

However, the value of K_{c2} also lies in this interval and, therefore, K_x may be much larger than, of the same order of magnitude as, or smaller than K_{c2}. Different types of behavior are encountered in these three cases, as discussed below.

The reactions to be considered are

$$CO_2 + H_2O + X^- = HCO_3^- + HX \tag{11.2.2}$$

$$H_2S + X^- = HS^- + HX \tag{11.2.3}$$

as well as a CO_3^{2-}–HCO_3^- shift equilibrium which can be written as

$$HCO_3^- + X^- = CO_3^{2-} + HX \tag{11.2.4}$$

Let $B_1 \equiv X^-$, $B_2 \equiv HX$, $B_3 \equiv HS^-$, $B_4 \equiv HCO_3^-$, and $B_5 \equiv CO_3^{2-}$. The molarity m is given by

$$m = b_1 + b_2 \tag{11.2.5}$$

Let $\hat{b}_j = b_j/m$ be the dimensionless concentrations, and let y and y' be the CO_2 and H_2S fractional saturations. An admissible composition, corresponding to the condition where all the chemically combined CO_2 is in the HCO_3^- form is

$$\left.\begin{aligned} \hat{b}_1^0 &= 1 - y - y' \\ \hat{b}_2^0 &= y + y' \\ \hat{b}_3^0 &= y' \\ \hat{b}_4^0 &= y \\ \hat{b}_5^0 &= 0 \end{aligned}\right\} \tag{11.2.6}$$

Let $\hat{\xi}$ be the dimensionless extent of reaction 11.2.4. The actual dimensionless concentrations will be

$$\left.\begin{aligned} \hat{b}_1 &= 1 - y - y' - \hat{\xi} \\ \hat{b}_2 &= y + y' + \hat{\xi} \\ \hat{b}_3 &= y' \\ \hat{b}_4 &= y - \hat{\xi} \\ \hat{b}_5 &= \hat{\xi} \end{aligned}\right\} \tag{11.2.7}$$

The equilibrium of reaction 11.2.4 is written as

$$\frac{K_{c2}}{K_x} = \frac{\hat{\xi}(y + y' + \hat{\xi})}{(y - \hat{\xi})(1 - y - y' - \hat{\xi})} \tag{11.2.8}$$

Equation 11.2.8 can be solved in general for $\hat{\xi}$ and the liquid-phase composition can then be calculated from Eq. 11.2.7. However, asymptotic conditions are very easily established, namely:

1. $K_{c2} \ll K_x$ (i.e., HX is a stronger acid than HCO_3^-). The solution of Eq. 11.2.8 is, quite obviously, $\hat{\xi} \simeq 0$; practically no CO_3^{2-} ion is formed. Eq. 11.2.6 represents the *actual* concentrations.

2. $K_{c2} \gg K_x$ (i.e., HX is a weaker acid than HCO_3^-). The solution of Eq. 11.2.8 is

$$\hat{\xi} = \min\left[y, 1 - y - y'\right] \tag{11.2.9}$$

i.e., as much CO_3^{2-} ion is formed as is stoichiometrically possible. Both results are, in a sense, intuitive since how much CO_3^{2-} is formed depends simply on the relative strengths of the HCO_3^- and HX acids.

Of course, K_{c2} may be of the same order of magnitude as K_x and, in that case, solution of the quadratic Eq. 11.2.8 cannot be avoided. The particularly relevant case in this category will be discussed in Section 11.3. In this section we focus attention on salts falling into Category 1, say, characterized by a room temperature pK in the interval 8–10.

For such systems, reaction 11.2.4 can be neglected altogether. The equilibrium constants for reactions 11.2.2 and 11.2.3 are

$$K = \frac{K_{c1}}{K_x} \tag{11.2.10}$$

$$K' = \frac{K_{s1}}{K_x} \tag{11.2.11}$$

The equilibrium vapor pressures are calculated from the equilibrium conditions for reactions 11.2.2 and 11.2.3, and from Eqs. 11.2.6 which give the *actual* concentrations in the liquid phase:

$$p^* = Hm \frac{K_x}{K_{c1}} \frac{y(y + y')}{1 - y - y'} \tag{11.2.12}$$

$$p^{*\prime} = H'm \frac{K_x}{K_{s1}} \frac{y'(y + y')}{1 - y - y'} \tag{11.2.13}$$

$$\frac{p^*}{p^{*\prime}} = \frac{H}{H'} K_{cs} \frac{y}{y'} \tag{11.2.14}$$

It is important to realize that, although both p^* and $p^{*\prime}$ depend on the value of K_x, the ratio $p^*/p^{*\prime}$ does not. The same ratio is also independent of the molarity m, whereas both p^* and $p^{*\prime}$ are proportional to it.

From Eq. 11.2.4, the thermodynamic selectivity of inorganic salt solutions is calculated as

$$S_T = \frac{H}{H'} K_{cs} \tag{11.2.15}$$

Comparison with Eq. 2.7.8 shows that inorganic salt solutions of type 1 have the highest possible thermodynamic selectivity for H_2S; furthermore, the value of S_T is independent of both y and y'.

Let us now turn attention to the kinetics of the process considered. Reaction 11.2.3 is a simple proton-transfer reaction and as such may be regarded as instantaneous. In contrast with this, reaction 11.2.2 takes place at a rate determined by the slow step, i.e., reaction 9.1.1 (the other two steps are water dissociation and HX reassociation, both of which are instantaneous). Therefore, if the X^- ions of the acid HX are not catalysts for reaction 9.1.1, the rate of reaction 11.2.2 is given by Eq. 9.1.2. Correspondingly, the reaction time is

$$t_r = \frac{1}{k[OH^-]_0} \tag{11.2.16}$$

where k is the kinetic constant for reaction 9.1.1, values of which are plotted in Figure 6.9.1. The hydroxyl ion concentration in the bulk of the liquid is given by

$$[OH^-]_0 = \frac{K_w}{K_x} \frac{1 - y - y'}{y + y'} \tag{11.2.17}$$

unless y and y' are vanishingly small (K_w is the water dissociation constant). If this is compared with Eq. 9.3.2, one sees that the hydroxyl ion concentration in inorganic salt solutions of the type considered here is *less* than in carbonate solutions, since K_x is *larger* than the second dissociation constant for carbonic acid. It follows that reaction 11.2.2 proceeds (in the absence of H_2S) at most in the fast reaction regime, and possibly in the slow reaction regime. As discussed in Chapter 8, this will result in a very strong kinetic selectivity for H_2S and, in fact, in the bottom part of an absorption tower, CO_2 may actually be *desorbed* from the liquid phase by a mechanism of the type discussed in Section 8.5. Very high kinetic selectivity for H_2S has been reported in the literature; the data were obtained at room temperature in laboratory-type absorbers. (With reference to the discussion in Chapter 8, the shift reaction in inorganic salt solutions is reaction 11.1.5, and it is kinetically driven to the right.)

The possibility that reaction 9.1.1 may be catalyzed by the X^- ion, analogous to the $H_2AsO_3^-$ ion catalysis discussed in Chapter 9, should not be overlooked. The kinetic selectivity that is achievable depends critically on the rate of CO_2 reaction, as discussed below.

Absorption of H_2S takes place in the instantaneous reaction regime, say,

$$N' = k_L^0(\alpha_i' - \alpha_0') \tag{11.2.18}$$

Absorption of CO_2 takes place at most in the fast reaction regime and, therefore, the discussion in Section 8.5 applies. The value of α_i' is to be calculated from Eq. 11.2.13, say,

$$p_i' = H'm \frac{K_x}{K_{s1}} \frac{y_i'(y_0 + y_i')}{1 - y_0 - y_i} \tag{11.2.19}$$

which yields

$$y_i' = \frac{Ap_i' + y_0}{2}\left\{-1 + \sqrt{1 + \frac{4Ap_i'(1 - y_0)}{(Ap_i' + y_0)^2}}\right\} \tag{11.2.20}$$

where

$$A = \frac{K_{s1}}{H'mK_x} \tag{11.2.21}$$

Alternatively, α_i' can be calculated as the H_2S load which, combined with the CO_2 bulk-liquid load, yields an equilibrium H_2S pressure equal to its interface partial pressure; this can be read off a reliable VLE curve. The important point is that, at given loads y_0 and y_0', the rate of H_2S absorption depends only on the value of K_x, with the rate increasing as K_x decreases. At a given alkalinity of the salt, the rate of H_2S absorption is uniquely determined by y_0 and y_0'.

Should reaction 9.1.1 not be catalyzed, the same conclusion would be drawn also for CO_2, as can be seen from Eqs. 11.2.16 and 11.2.17. However, salts with equal alkalinity may have different catalytic effects on reaction 9.1.1 and, therefore, the kinetic selectivity obtainable in practice may not be the same for two such salts.

A straightforward calculation shows that the thermodynamic selectivity for solutions of salts of type 2 is much less than given by Eq. 11.2.15, by at least a factor K_x/K_{c2} (which is much less than unity in such solutions). Furthermore, the kinetic selectivity is also expected to be much less, since reaction 9.1.1 is the slow step also for the formation of CO_3^{2-} (reaction 11.2.4 is a proton-transfer reaction), but solutions of salts of type 2 have $[OH^-]_0$ values appreciably *larger* than carbonate solutions. For very strong alkalis, a kinetic selectivity mechanism of the type discussed in Section 8.6 for hydroxide solutions may set in.

11.3. CARBONATE SOLUTIONS

The thermodynamics of the system H_2S-CO_2-K_2CO_3 solution, have been discussed in Section 2.6; the equilibrium vapor pressures are given by

$$p^* = \frac{Hm}{K}\,\frac{(2y + y')^2}{1 - y - y'} \tag{11.3.1}$$

$$p^{*\prime} = \frac{H'm}{K'}\,\frac{(2y + y')\,y'}{1 - y - y'} \tag{11.3.2}$$

where $K = K_{c1}/K_{c2}$ and $K' = K_{s1}'/K_{c2}$. The vapor-pressure *ratio* is given by

$$\frac{p^*}{p^{*\prime}} = \frac{H}{H'}K_{cs}\left(2\frac{y}{y'} + 1\right) \tag{11.3.3}$$

The case of Na_2CO_3 solutions is analogous.

Equation 11.3.3 shows that *K_2CO_3 solutions cannot be used for H_2S removal unless a substantial amount of CO_2 is simultaneously removed.* This can easily be seen from the following argument.

Should a K_2CO_3 solution be used for removal of *only* H_2S, the value of y should be a constant throughout the absorption and regeneration units. In the regeneration unit, the ratio $p^*/p^{*\prime}$ would increase continuously from top to bottom and would become very large indeed towards the lean end as $y' \to 0$ (unless $y = 0$). Therefore, there would be a strong driving force for CO_2 desorption and, of course, if CO_2 is desorbed in the regeneration unit, it must be absorbed in the absorption unit. Even if the constant value of y were zero, the ratio $p^*/p^{*\prime}$ would be finite and desorption of CO_2 would take place in the regeneration unit.

If CO_2 and H_2S are simultaneously removed from the gas (in comparable amounts), both y and y' will decrease as the liquid proceeds from the top to the bottom of the regeneration unit and their ratio would, therefore, stay finite even at very low y' values. This allows the solution to be regenerated to a very low value of y', as required in order to obtain a lean solution with a low enough value of $p^{*\prime}$ (see Eq. 11.3.2). Such a lean solution can then be used to meet the severe H_2S specification on the clean gas.

Carbonate solutions exhibit a thermodynamic selectivity which is calculated from Eq. 11.3.3 as

$$S = \frac{\mathrm{H}}{\mathrm{H}'} \mathrm{K}_{cs} \left(2 + \frac{y'}{y}\right) \tag{11.3.4}$$

and, therefore, even if $y'/y \to 0$ (as is likely at the lean end of the absorption unit), the thermodynamic selectivity is twice the maximum value given by Eq. 11.1.6. The reason for this is a simple stoichiometric one; carbonate solutions are the only ones for which chemical absorption of one mole of CO_2 results in the formation of *two* moles of HCO_3^-.

From the kinetic viewpoint, absorption of H_2S proceeds in the instantaneous reaction regime, with a rate given again by Eq. 11.2.18; the value of y_i' is to be calculated from Eq. 11.3.2, leading to

$$y_i' = \frac{Ap_i' + 2y_0}{2} \left[-1 + \sqrt{1 + \frac{Ap_i'(1 - y_0)}{(Ap_i' + 2y_0)^2}} \right] \tag{11.3.5}$$

where

$$A = \frac{\mathrm{K}'}{\mathrm{H}'m} \tag{11.3.6}$$

For CO_2, there is, of course, no possibility of catalysis inasmuch as the X^- ion is, in this case, the CO_3^{2-} ion itself. Therefore, the regime will be either FI or SI according to whether the temperature is higher or lower than about 40°C (see Figure 9.5.2). This is the rationale for the low temperature chosen for operation of the selective vacuum carbonate process.

The shift reaction is

$$H_2S + HCO_3^- = HS^- + CO_2 + H_2O \tag{11.3.7}$$

and it is kinetically driven to the right.

With *promoted* carbonate solutions, particularly at high temperatures, simultaneous absorption of CO_2 and H_2S may approach the conditions of I-I behavior, and a lower kinetic selectivity is expected. Indeed, if I-I behavior is attained, as is possible at the rich end of the absorber, the shift reaction may proceed in the direction of H_2S formation.

11.4. ORGANIC SOLUTIONS

Aqueous amine solutions are commonly used in industry for simultaneous removal of CO_2 and H_2S. Different amines exhibit, of course, different behavior but certain general features of this class of processes can be identified.

A very important role is played by the carbamate stability constant K_c, values of which for three different amines are reported in Figure 2.4.1. The constant K_c is the equilibrium constant for reaction 2.4.27; it follows that Eq. 2.6.30 can be written in the following form:

$$\frac{b_3}{m} = \frac{1 - y' + 1/K_c m}{2}\left[1 - \sqrt{1 - \frac{4y(1 - y - y')}{(1 - y' + 1/K_c m)^2}}\right] \tag{11.4.1}$$

where the b_js have the same meaning as in Section 2.6, and hence b_3 is the carbamate concentration. The simplified results in Table 2.6.31 apply to amines (such as monoethanolamine at low temperature) characterized by very low values of $1/K_c m$.

Now consider an amine solution for which $1/K_c m$ is not very much less than unity. We are interested in the situation prevailing at the lean end of the absorption tower where $y \ll 1$, $y' \ll 1$, and therefore Eq. 11.4.1 can be expanded to $0(y)$, $0(y')$ to yield

$$b_3 = my \frac{K_c m}{1 + K_c m} \tag{11.4.2}$$

which shows that even at very low values of y not all the chemically combined CO_2 is in the carbamate form. To within the same approximation, the other b_js are calculated as

$$\left.\begin{aligned} b_1 &= m \\ b_2 &= m(y + y') \\ b_4 &= my \frac{1}{1 + K_c m} \\ b_5 &= 0 \\ b_6 &= my' \end{aligned}\right\} \tag{11.4.3}$$

The equilibrium vapor pressures are to be calculated from the equilibrium equations for reactions 2.4.25 (constant K) and 2.6.23 (constant K′). One has

$$p^* = \frac{H}{K} \frac{y(y+y')}{1 + 1/K_c m} \tag{11.4.4}$$

$$p^{*\prime} = \frac{H'}{K'} \frac{y'(y+y')}{m} \tag{11.4.5}$$

$$\frac{p^*}{p^{*\prime}} = \frac{H}{H'} K_{sc} \frac{y}{y'} \frac{1}{1 + K_c m} \tag{11.4.6}$$

since the following identity holds true:

$$\frac{K'}{K} = \frac{K_{sc}}{K_c} \tag{11.4.7}$$

From Eq. 11.4.7, the thermodynamic selectivity of a lean amine solution (i.e., at $y \ll 1, y' \ll 1$) is calculated as

$$S = \frac{H}{H'} K_{sc} \frac{1}{1 + K_c m} \tag{11.4.8}$$

Equation 11.4.8 is very important and should be discussed in detail. Consider the case of amines characterized by very stable carbamates ($K_c m \gg 1$); Eq. 11.4.8 shows that such amines will exhibit thermodynamic selectivities much smaller than the ones achievable with inorganic salt solutions (see Eq. 11.2.15). Monoethanolamine solutions at low temperatures exhibit thermodynamic selectivities as low as 10^{-2} (see, e.g., Figure 2.6.3). Fortunately, however, these solutions may exhibit very high kinetic selectivities.

At the other extreme, consider the case of amines characterized by extremely unstable carbamates ($K_c m \to 0$). This is unlikely to be the case with primary and secondary amines, but is exactly the case with tertiary amines, which cannot form the carbamate and can, therefore, be described mathematically by $K_c \to 0$. For such amines, the thermodynamic selectivity becomes as high as it can possibly be; i.e., it is given again by Eq. 11.2.15.

We now turn attention to the *kinetics* of simultaneous absorption of H_2S and CO_2 in amine solutions. Consider the case of monoethanolamine, where $K_c m \gg 1$ and, therefore (as long as $2y + y' < 1$; see Table 2.6.31), the main reactions are

$$CO_2 + 2RNH = RNCOO^- + RNH_2^+ \tag{11.4.9}$$

$$H_2S + RNH = RNH_2^+ + HS^- \tag{11.4.10}$$

and the shift reaction is

$$CO_2 + RNH_2^+ + 2HS^- = RNCOO^- + 2H_2S \tag{11.4.11}$$

Now consider this system in an absorber unit. At the lean end, both I_∞ and I'_∞ are very large and, therefore, whereas H_2S is absorbed in the instantaneous reaction regime, CO_2 is absorbed in the fast reaction regime. This implies that the H_2S-to-CO_2 partial pressure ratio in the gas phase may be reduced to a value considerably less than that which would be calculated from the thermodynamic selectivity given by Eq. 11.4.8. Indeed, as can be seen from Figure 2.6.3, thermodynamically, the ratio p/p' would be, at the lean end, about 50 times less than the y/y' ratio. In contrast to this, actual p/p' values at the lean end are, in industrial practice, typically of the order of 10^2 and y/y' values of the order of 10.

The situation is quite different at the rich end, where both I_∞ and I'_∞ are much lower and conditions may, in fact, be such that both gases undergo mass transfer in the instantaneous reaction regime. If that is the case, an I-I regime mechanism of the type discussed in Chapter 8 may result in *desorption* of H_2S; there is one documented case where this has been observed to occur (see Figure 1.10.4).

Next consider the case of a tertiary amine R_3N. Inasmuch as tertiary amines cannot form a carbamate, the two main reactions are

$$CO_2 + R_3N + H_2O = R_3NH^+ + HCO_3^- \tag{11.4.12}$$

$$H_2S + R_3N = R_3NH^+ + HS^- \tag{11.4.13}$$

The shift reaction is reaction 11.3.7.

The thermodynamic modeling of the system is straightforward and leads to the following equations (the modeling is strictly analogous to that of inorganic salt solutions, since as far as no carbamate can form, the amine simply provides the required alkalinity).

$$p^* = \frac{m\text{HK}_p}{\text{K}_{c1}} \frac{y(y+y')}{1-y-y'} \tag{11.4.14}$$

$$p^{*\prime} = \frac{m\text{H}'\text{K}_p}{\text{K}_{s_1}} \frac{y'(y+y')}{1-y-y'} \tag{11.4.15}$$

where K_p is the amine's protonation constant:

$$\text{K}_p = \frac{[R_3N]\,[H^+]}{[R_3NH^+]} \tag{11.4.16}$$

Notice that the ratio $p^*/p^{*\prime}$ is given again by Eq. 11.2.14, and the thermodynamic selectivity by Eq. 11.2.15. The latter is, of course, the limit of Eq. 11.4.8 when

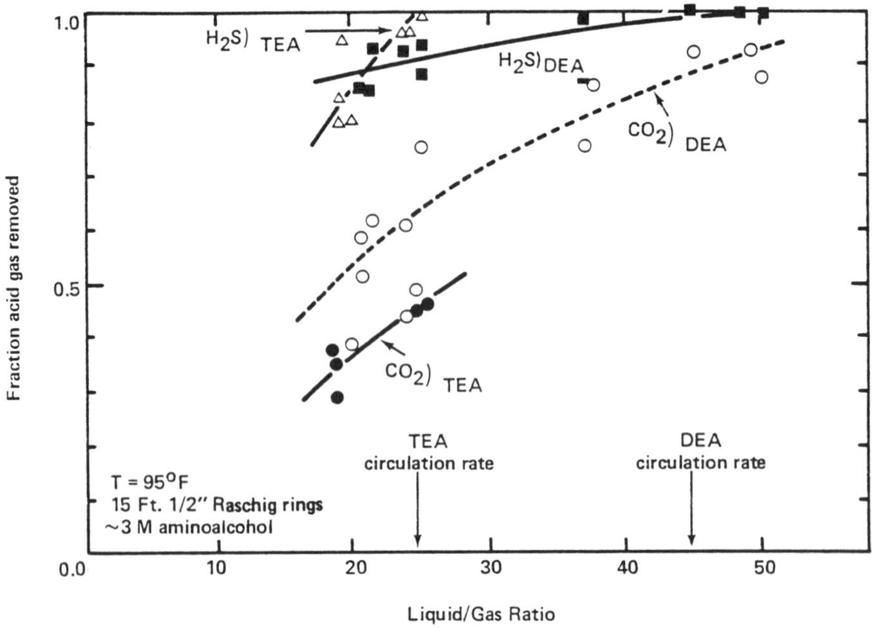

FIGURE 11.4.1. Kinetically selective separation of H_2S from gas containing H_2S plus CO_2 with aqueous aminoalcohols.

$K_c \rightarrow 0$. The protonation constant K_p plays exactly the same role as the dissociation constant K_x plays in the case of inorganic salts of type 1. Tertiary amines corresponding to inorganic salts of type 2, i.e., characterized by room temperature *p*Ks in excess of 11, are not known.

Also the kinetics of mass transfer in tertiary amine solutions are of the same type as in inorganic salt solutions. There is convincing evidence in the literature that tertiary amines may catalyze reaction 9.1.1, so that the kinetic selectivity exhibited by them is less than that which could be achieved if the reaction time for CO_2 were given by Eq. 11.2.16. For the tertiary amine in actual industrial use, methyldiethanolamine, experimental evidence is available to support Eq. 11.2.18 for the H_2S absorption rate.

A comparison of the selectivity obtainable with a secondary and a tertiary amine is presented in Figure 11.4.1. With 15 ft of packing, H_2S cleanup is attainable at an L/G ratio of 24 with the tertiary amine (triethanolamine), where only 40% of the CO_2 is removed; i.e., the process selectivity S_p is 2.5. With the secondary amine (diethanolamine), H_2S cleanup is only attained at an L/G ratio of about 45, when 85% of the CO_2 is removed; i.e., the process selectivity S_p is 1.15.

Understanding of the catalytic effect of tertiary amines on reaction 9.1.1 is poor. Without such understanding, it is impossible to predict the process selectivity obtainable with a given amine, and to rationally design the equipment required for optimal selective removal of H_2S. Process selectivities achieved in existing commercial units may be far less than optimal.

12 Removal of Sulfur Dioxide

. . . quod fortasse facile intellectu est quia Carnotus auxilium analyseos mathematicae in ratiocinationibus suis haud facile adhibuit.

C. A. TRUESDELL

The problem of removing sulfur dioxide from gases arises in two main areas: flue gases from combustion, which normally contain 0.1-0.4 mole percent SO_2, depending on the type of fuel, oil, or coal being fired; and stack gas from smelters handling sulfide ores, where the concentration of SO_2 may be as high as 8%. Other sources include catalytic cracking regenerators, sulfuric acid plant tail gases, and Claus sulfur recovery plants.

The reason for requiring SO_2 removal is environmental protection. Sulfur dioxide is an air pollutant and may result in acid rains far away from the pollutant source. Accelerated coal usage will require technology to adequately guard environmental quality.

The U.S. Environmental Protection Agency's standards for flue gas desulfurization at large new coal-fired power plants limit emissions of particulates and sulfur dioxide. The 1980 standards set the ceiling for particulates at 0.03 lb/10^6 BTU (13 ng/J). SO_2 emissions must satisfy the tighter of two requirements. The absolute emission is limited to 1.2 lbs/10^6 BTU (520 ng/J); in addition, SO_2 in the flue gas must be reduced by 70-90%. If controlled emissions are less than 0.6 lb/10^6 BTU (260 ng/J), the percentage reduction can be less than 90% but in no case less than 70%. If controlled emissions are above 0.6 lb/10^6 BTU (260 ng/J), the 90% limitation applies. Compliance is determined on a 30-day rolling average basis. The impact of the 1980 regulations is to require scrubbing of low sulfur coal that was not required previously.

Flue gas desulfurization is still considered to be an infant technology. At the end of 1979 there were 57 systems, mainly of the throwaway type, in operation on industrial and utility boilers in the United States, but many problems remain to be solved before they can be considered reliable trouble-free technology. Many more systems are in operation in Japan.

Flue gas desulfurization is a process where an acidic impurity is removed from a gas phase. This would seem to imply that the technology should be analogous to that employed in the removal of H_2S and CO_2. However, this is often not the case for a variety of reasons, not all of which are of a technical nature.

The most important non-technical reason for the difference in technology is that flue gas desulfurization is a gas treating process that is often run not by a chemical company but by an electrical utility. It follows that simple processes that do not require sophisticated tail units and are not integrated into a complex chemical plant are preferred. Throwaway scrubbing has been selected by utility companies because it is the most economical process available. It also happens to be a relatively simple process.

Technical reasons for the difference in technology are abundant. First of all, SO_2 is often present together with some oxygen in the gas (see Table 12.1.1) and, therefore, some oxidation to sulphates always takes place in the process. The sulphate products are often not regenerable and must be purged and discarded: this means a loss of reagent in any regenerable solvent system. Such a loss may be intolerable with any reagent that is not a very cheap base, so processes that would be technically attractive need to be discarded if they involve a moderately expensive basic reagent.

Another important point is that SO_2 occurs at rather low concentrations in the low-pressure influent gas (see Table 12.1.1). This means that the range of partial pressures of the absorbing gas in FGD processes is markedly lower than that encountered in CO_2 and H_2S removal.

Finally, there are differences in the physico-chemical behavior of SO_2 in aqueous systems as compared to that of other acidic gases; these will be discussed in Sections 12.2 and 12.3.

TABLE 12.1.1. Representative Influent Flue Gas to Desulfurization

Fuel		SO_2 in flue gas (%)
Coal	4% sulfur	0.35
Fuel oil	2% sulfur	0.12
Fuel oil	5% sulfur	0.31
Refinery acid sludge	40% sulfuric acid	2.0

Representative flue gas composition

Component	Mole%
CO_2	12.60
SO_2	0.22
SO_3	0.001
NO	0.045
NO_2	0.005
O_2	1.78
N_2	73.44
H_2O	11.90

12.1. PROCESS SELECTION

Representative flue gas compositions that need to undergo desulfurization are given in Table 12.1.1. Available processes can be classified in two basic categories: throwaway and regenerable solvents. In throwaway processes, a cheap base is used that reacts with SO_2 to form a solid sulphite, which is simply disposed of as waste. Regenerable processes use either thermal regeneration to recover the acid gas as a concentrated SO_2 stream (sodium sulfite, magnesium oxide) or reaction with H_2S to produce elemental sulfur (citrate and phosphate processes).

In throwaway processes there is a strong economic incentive to use the cheapest available chemical bases, lime or limestone. Most of the flue gas desulfurization systems being operated or planned are based on the limestone-water slurry process. Many different scrubber types and process systems are offered by several vendors. A process configuration using "double loop" scrubbing is shown in Figure 12.1.1. The most general slurry scrubbing process employs a single recycle loop for both quenching and SO_2 removal.

After particulate collection, the flue gas enters the quencher section where it is cooled adiabatically. The 10-15% limestone slurry removes about 40% of the SO_2 in this stage. The flue gas then passes to the main absorbing stages, where countercurrent scrubbing with the 10% limestone slurry from the absorber feed tank removes most of the remaining SO_2. The flue gas then passes through a demister where entrained droplets are removed, is reheated to provide buoyancy, and sent to the stack.

The solubility of $CaCO_3$ in water is a limitation on absorption rates in this system. High stoichiometric ratios of 1.3-1.5 moles reagent/mole SO_2 removed are used because of the limited reactivity of ground natural limestones. Calcium sulfate scaling is minimized by high slurry solids concentration, a high slurry-to-gas ratio, and a large holding tank. The equilibrium pH of the scrubber loop is 5.5-6.5. The spent solid, a mixture of $CaSO_3$, $CaSO_4$, and unreacted limestone, is discharged as 50% solid after thickening or 80% solid after thickening and filtration. Spent solids may be used for landfill. Despite the extensive commercial experience with the limestone slurry process, problems still occur with the chemistry, controls, and materials of construction; the maintenance requirements of the process are high.

Another limestone slurry process uses limestone to which dolomite lime has been added in a horizontal spray absorber. Addition of dolomite lime, which contains magnesium oxide, increases the amount of dissolved alkalinity, and allows higher SO_2 removal at lower liquid-to-gas ratios than a limestone only process.

Use of a horizontal absorber without internals, such as shown in Figure 12.1.2, results in a low pressure drop and simplifies gas ducting. It has no advantage over vertical spray scrubbers in minimizing scaling problems. Gas-liquid reactions occur in the absorber itself, whereas liquid-solid reactions occur mainly in an external slurry reactor before recycling. Flue gas flows horizontally at 20-25 ft/sec in the absorber where several spray stages are directed downwards; the slurry flows to a compartmentalized reactor vessel underneath the absorber. The scrubbed flue gas then flows through a vertical demister and is reheated before the stack. The par-

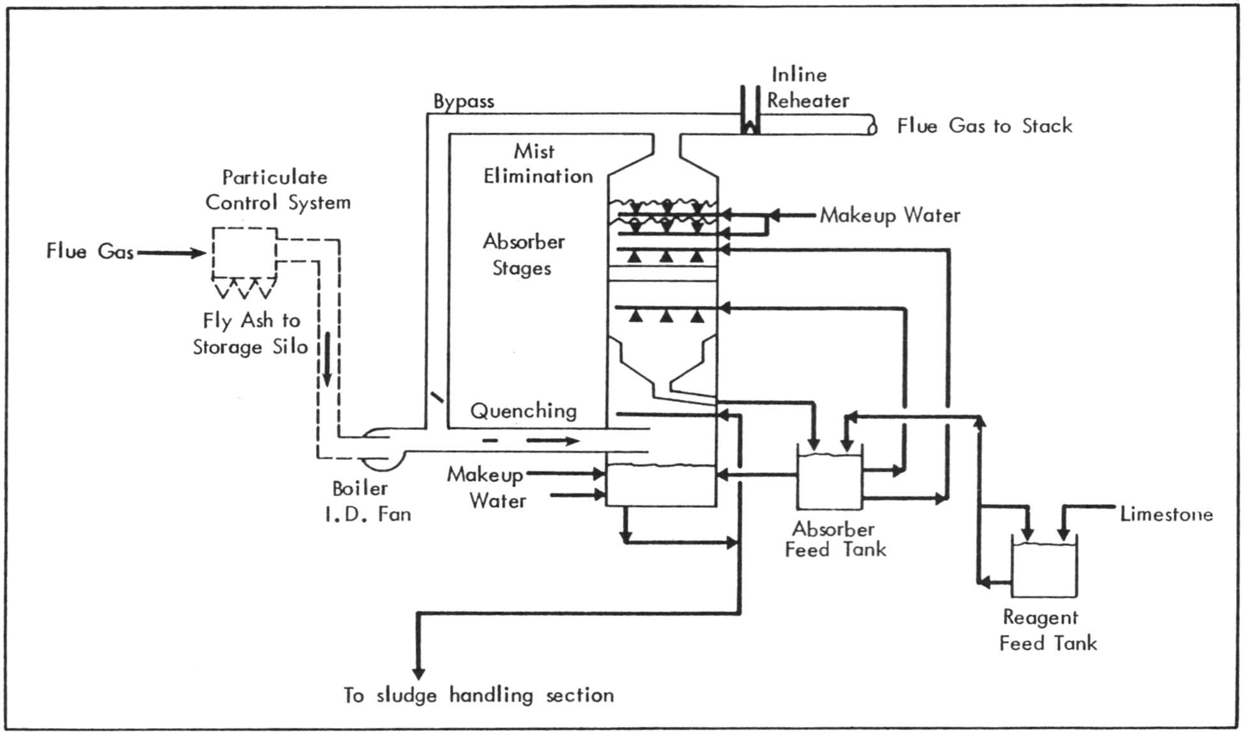

FIGURE 12.1.1. Limestone-water slurry flue gas desulfurization process.

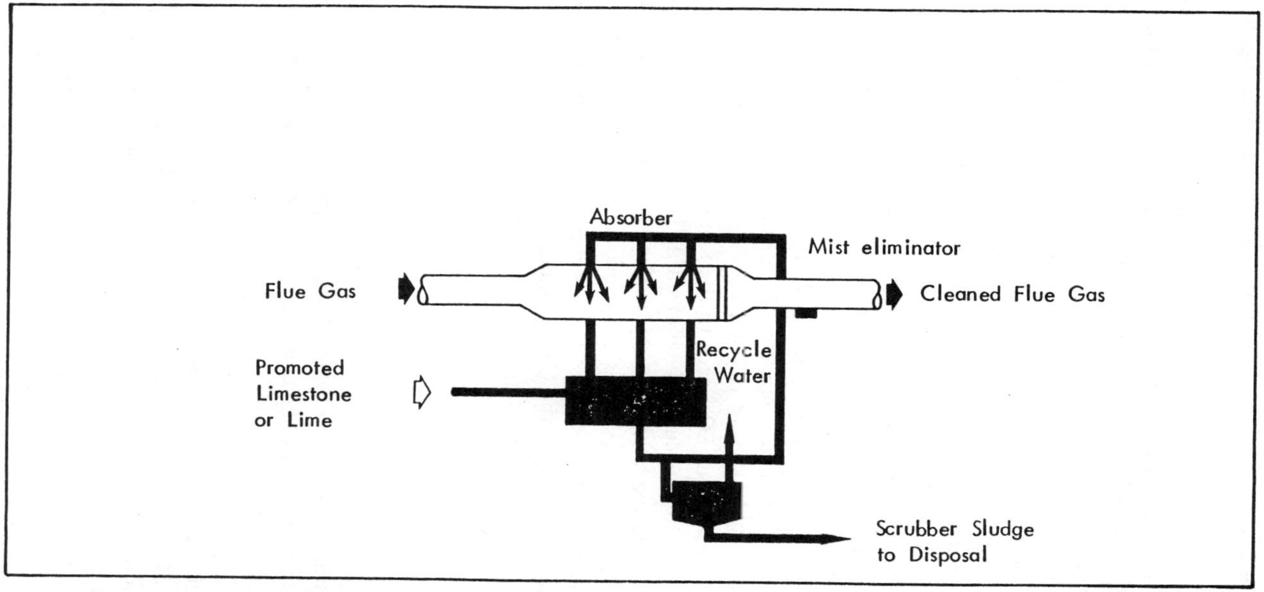

FIGURE 12.1.2. Promoted limestone-water slurry process with horizontal spray absorber.

ticulate collection system and induced draft fans are located upstream of the absorber.

Clear solution scrubbing may avoid most of the problems associated with slurry scrubbing. One process, dual alkali, absorbs SO_2 in a clean solution of sodium sulfite to produce bisulfite. The scrubbing solution is regenerated by lime, with the resulting crystallization of calcium sulfite solids.

Another lime-based process is the dry alkali, once through reagent process. A concentrated lime-water slurry is dispersed in the boiler flue gas through nozzles in a spray dryer. The SO_2 is absorbed in the less than 10-micron size droplets and neutralized by the alkali. The small quantity of water in the slurry is completely evaporated by the heat released by the hot flue gas so that a dry powder of calcium sulfites and sulfates admixed with the boiler fly ash is obtained. The solids are removed in a baghouse which in this service is more economical than electrostatic precipitators. A flowsheet is given in Figure 12.1.3.

Another process uses a clear (i.e., not a slurry) alkaline solution based on lime. It is possible to solubilize lime completely by using an alkaline solution of calcium hydroxide, calcium chloride, and calcium formate; the latter acts as a buffer for the solution. Absorption of SO_2 leads to the formation of water soluble calcium bisulfite, rather than the insoluble calcium sulfite hemihydrate. Calcium sulfite is precipitated by forced oxidation in a separate integrated process step. One version of the solubilized lime process uses a special cocurrent spray contactor with a separator that eliminates the need for a conventional mist eliminator. This process has not been commercialized in the United States.

Throwaway systems generate large volumes of waste sludge and are not always feasible because in certain regions there is a shortage of landfill area required for disposal of the sludge. Therefore, regenerable processes are of great interest as they offer utilities and other users an alternative to conventional lime/limestone-based scrubbing.

The solubility of SO_2 in water, discussed in Chapter 1, is large enough to make a regenerable process based on water feasible. However, the amount of water to be circulated would be rather large and the steam requirements for desorbing SO_2 would be unacceptably high. Therefore, processes using alkaline solutions of reasonably large capacity are needed. There are basically two such processes available, one using sulfite solutions and one using citrate solutions.

A flowsheet of a sodium sulfite process is shown in Figure 12.1.4. The scrubber effluent is regenerated by evaporative crystallization of sodium sulfite and evolution of SO_2 and water. A purge stream is required in order to remove sodium sulfate formed by oxidation. The separated SO_2 may be recovered as such, transformed to sulfur, or converted to sulfuric acid.

The sodium citrate process can accept a wide range of SO_2 concentrations, from the 0.3% produced from coal-fired boilers to over 3% discharged from a non-ferrous smelter. Some oxidation to sodium sulfate does occur, but this is claimed to be a minimum owing to the formation of citrate-metal ion complexes; if not blocked, the transition metal ions from the flue gas enhance oxidation. The citrate process uses a conventional absorber-stripper flowsheet.

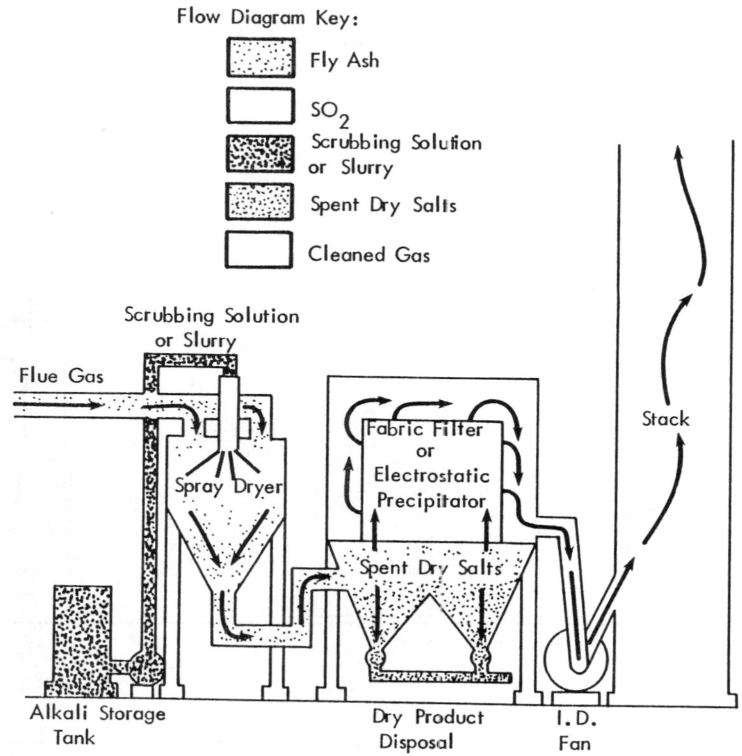

FIGURE 12.1.3. Dry alkali flue gas desulfurization process.

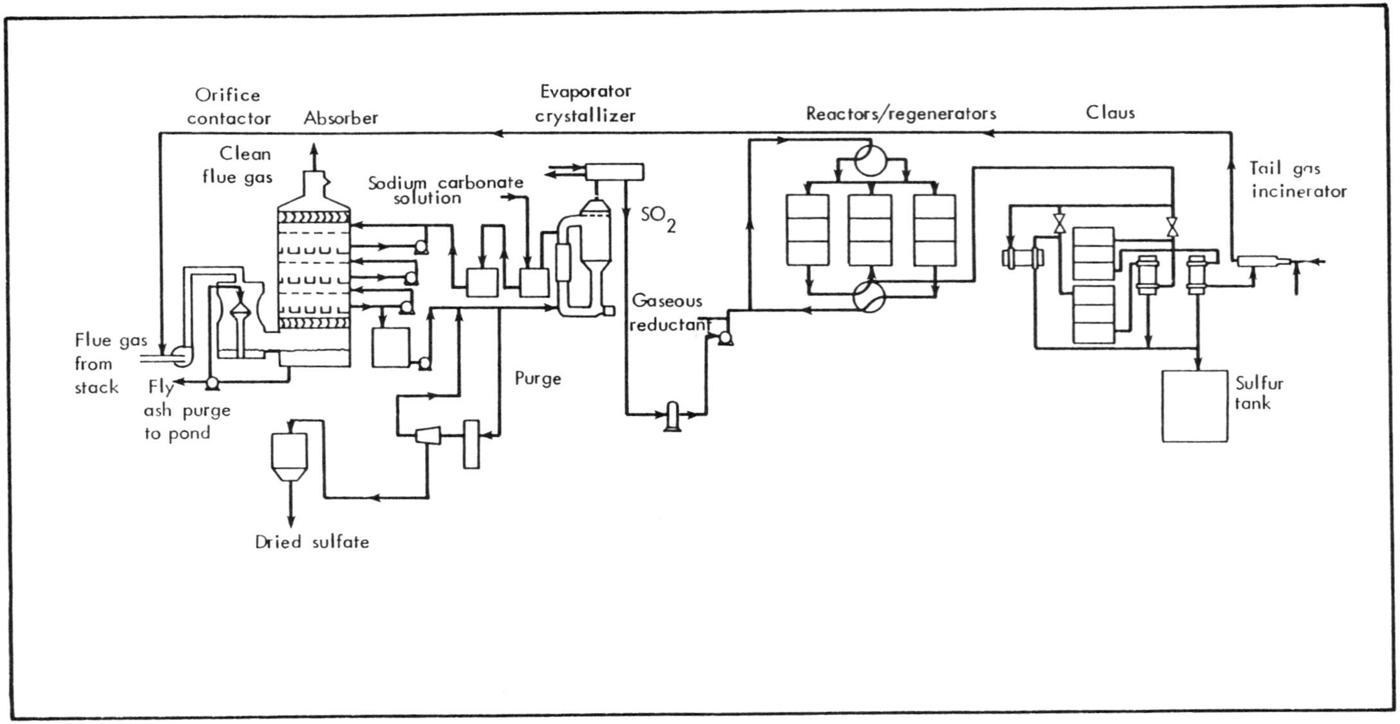

FIGURE 12.1.4. Regenerable flue gas desulfurization process using sodium sulfite solution.

In regenerable processes, the stripped SO_2 may be reduced to elemental sulfur. This requires addition of a chemical reductant which may be H_2S, a CO/H_2 mixture, or a hydrocarbon. In the case of H_2S, the overall chemistry can be regarded as an SO_2/H_2S redox reaction in solution and is, therefore, closely related to the classical Claus process. An interesting combination process has been proposed for petroleum refineries that need both hydrodesulfurization (which liberates H_2S) and flue gas desulfurization. In one scheme, SO_2 is scrubbed from the flue gas with a sodium phosphate solution. The SO_2-laden solution is then regenerated with hydrogen sulfide from the amine regenerator associated with the hydrotreating unit, to precipitate elemental sulfur. A flowsheet of this scheme is given in Figure 12.1.5. A similar scheme based on aqueous ammonia scrubbing of SO_2 has also been suggested.

12.2. CHEMISTRY AND THERMODYNAMICS

Although both SO_2 and CO_2 are Lewis acids, listed below are several important differences in the chemistry of these two species.

1. H_2SO_3 is a much stronger acid than H_2CO_3.
2. The physical solubility of SO_2 in water and in aqueous solutions is much larger than that of CO_2.
3. The hydrolysis reaction of SO_2 in aqueous solutions is much faster than the analogous reaction of CO_2.
4. SO_2 does not react with amines to form the equivalent of a carbamate and, therefore, its chemistry in all aqueous solutions of commercial interest is essentially of acid-base nature.
5. SO_2 can oxidize to SO_3 and, therefore, irreversible transformation to sulfates needs to be taken into account.

At 25°C, the dissociation constants of sulfurous acid are

$$K_1 = \frac{[HSO_3^-]\,[H^+]}{[SO_2]} = 1.7 \times 10^{-2} \tag{12.2.1}$$

$$K_2 = \frac{[SO_3^{2-}]\,[H^+]}{[HSO_3^-]} = 6.2 \times 10^{-8} \tag{12.2.2}$$

where $[SO_2]$ is to be understood in the sense discussed in Section 1.8, i.e., $[SO_2]$ + $[H_2SO_3]$. The units of K_1 and K_2 are gmol/liter. The temperature dependence of K_1 can be taken to be

$$K_1 = \exp\left(\frac{1972.5}{T} - 10.6819\right) \tag{12.2.3}$$

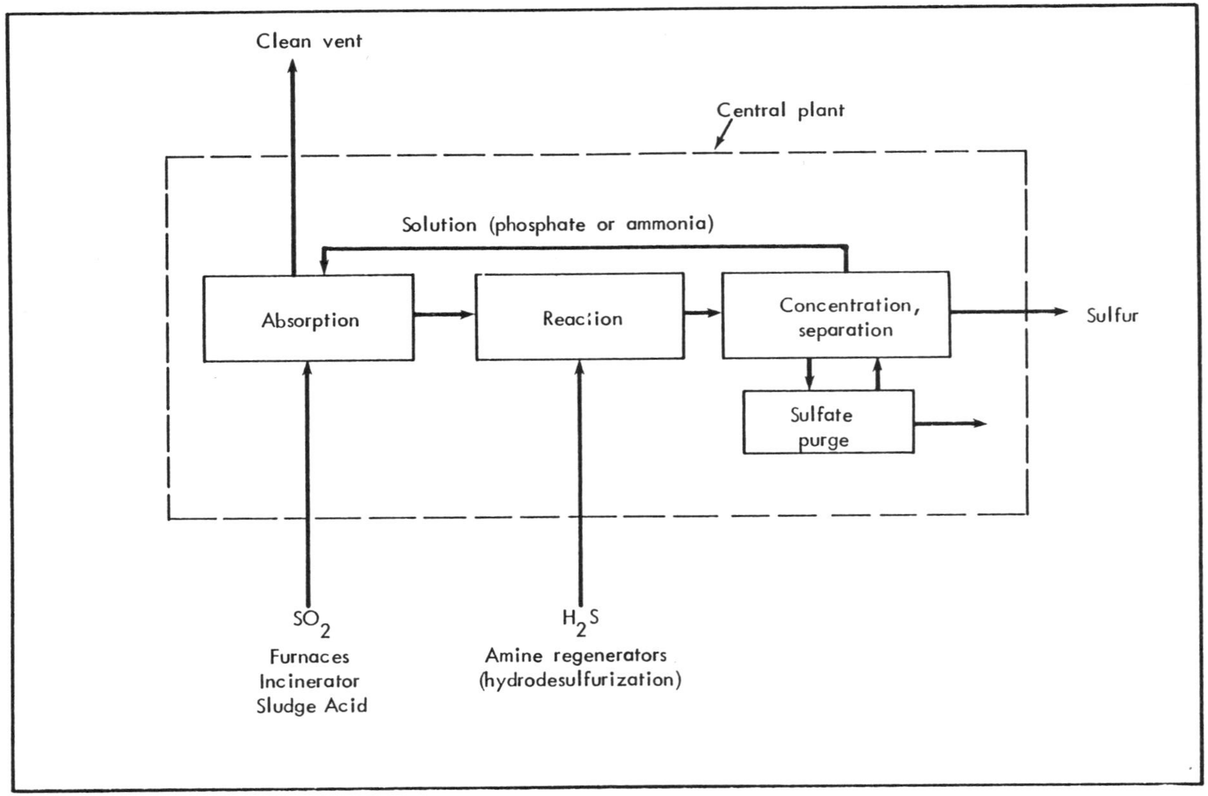

FIGURE 12.1.5. Combined SO_2 removal/H_2S removal/sulfur recovery scheme for refineries.

We begin by considering the thermodynamics of regenerable solutions for SO_2 removal. Consider an aqueous solution of a base B characterized by a protonation constant K_p. The following two reactions need to be considered:

$$SO_2 + H_2O + B = BH^+ + HSO_3^- \tag{12.2.4}$$

$$HSO_3^- + B = BH^+ + SO_3^{2-} \tag{12.2.5}$$

A thermodynamic calculation following the methods discussed in Chapter 2 is straightforward. Let $B_1 \equiv B$, $B_2 \equiv BH^+$, $B_3 \equiv HSO_3^-$, and $B_4 \equiv SO_3^{2-}$. The total base molarity is m and the fractional saturation is y. The result is

$$b_1 = m(1 - y) - \xi \tag{12.2.6}$$

$$b_2 = my + \xi \tag{12.2.7}$$

$$b_3 = my - \xi \tag{12.2.8}$$

$$b_4 = \xi \tag{12.2.9}$$

where ξ is the extent of reaction 12.2.5, and is given by the solution of the following equation:

$$\frac{(my + \xi)}{(my - \xi)\,[m(1 - y) - \xi]} = \frac{K_2}{K_p} \tag{12.2.10}$$

Two special cases need to be considered. The first is the case where the base B is moderately strong, say, $K_p \ll K_2$. In view of Eq. 12.2.2, any base with a room temperature pK in excess of 8 falls into this category. The solution to Eq. 12.2.10 becomes

$$\xi \simeq \min\,[my, m(1 - y)] \tag{12.2.11}$$

which implies that the chemically combined SO_2 will be as much in the sulphite (SO_3^{2-}) form as is stoichiometrically possible.

In a base solution of the type discussed above, at $y < 0.5$ practically all chemically combined SO_2 is in the SO_3^{2-} form. A standard calculation yields

$$P^* = \frac{HK_p^2}{K_1K_2}\, m\, \frac{4y^3}{(1 - 2y)^2} \tag{12.2.12}$$

If K_p is no more than 10^{-8} (which it needs to be if $K_p \ll K_2$), the quantity mHK_p^2/K_1K_2 is, at room temperature, no more than 4×10^{-7} atm; the values of p^* are so low that the solution is not regenerable. Therefore, solutions of this type cannot be used in a conventional regenerable cycle at $y < 0.5$.

At $y > 0.5$, a solution of a base with $pK > 8$ will contain as much sulphite as stoichiometrically possible, with the excess of chemically combined SO_2 being in the bisulphite form. Since it is impossible to regenerate the sulphite, one may as well regard the sulphite itself as the base (i.e., assign the value $y = 0$ to a solution of B_2SO_3). Sodium and ammonium sulphite solutions are used industrially as regenerable chemical solvents for SO_2. The chemistry and thermodynamics of SO_2 dissolution in sulphite solutions is exactly analogous to that of CO_2 dissolution in carbonate solutions. The main reaction to be considered is

$$SO_2 + SO_3^{2-} + H_2O \longrightarrow 2HSO_3^- \qquad (12.2.13)$$

and the equilibrium vapor pressure is given by

$$p^* = \frac{HK_2}{K_1} m \frac{4y^2}{1 - y} \qquad (12.2.14)$$

(notice that reaction 12.2.13 is the *difference* between reactions 12.2.4 and 12.2.5; that is why K_p, the protonation constant of the alkaline ion, does not enter the equilibrium equation). Equation 12.2.14 is well confirmed by the experimental data.

Now consider the converse case where B is a weak base, say, $K_p \gg K_2$. Inasmuch as K_2 is of the order of 10^{-7} (see Eq. 12.2.2), a base in this category may, in fact, also be a weak acid. Aromatic amines, which have lower pKs than amino-alcohols, have been used as regenerable solvents for SO_2 removal. The solution to Eq. 12.2.10 is, in this case, $\xi \simeq 0$ and a standard calculation shows that

$$p^* = \frac{HK_p}{K_1} m \frac{y^2}{1 - y} \qquad (12.2.15)$$

Solutions of this type will result in higher equilibrium partial pressures than sulphite solutions, as can be seen by comparing Eqs. 12.2.14 and 12.2.15.

The capacity of solutions of weak bases for SO_2 can be analyzed in general terms. Since no SO_3^{2-} will form in solutions of bases with $pK < 7$, the total amount of SO_2 that can be dissolved is

$$\alpha = [SO_2] + [HSO_3^-] \qquad (12.2.16)$$

and, in view of the equilibrium condition for the hydrolysis reaction,

$$\alpha = [SO_2] \left\{1 + \frac{K_1}{[H^+]}\right\} \qquad (12.2.17)$$

or, equivalently,

$$\alpha_{max} = \frac{p_B}{H} \left\{1 + \frac{K_1}{[H^+]}\right\} \qquad (12.2.18)$$

where P_B is the partial pressure of SO_2 in the raw gas entering the absorber. Equation 12.2.18 shows that the capacity will be highest for solutions where, even at comparatively large $[HSO_3^-]$ levels, the pH is moderately high, say, in the range 3–6 (if it were any higher, formation of SO_3^{2-} should be taken into account, and that would lead to the problem of regeneration being impossible).

The pH range 3–6 is an interesting one since it includes the pKs of several polyfunctional organic acids. Consider a solution of a salt M_nX of a strong base MOH and a polyfunctional acid H_nX. The protonation constant of such a salt equals the dissociation constant of the acid H_nX. If H_nX is an inorganic acid, its dissociation constants differ by several orders of magnitude and, therefore, at most one of them will be in an acceptable range. With a polyfunctional organic acid, however, the successive dissociation constants differ only by about one order of magnitude, so that more than one M^+ ion per mole of salt can be liberated to neutralize the HSO_3^- in an acceptable pH range.

The point above is illustrated in Fig. 12.2.1, where the pH of a solution of a salt M_2X is plotted versus the fractional saturation y (i.e., the ratio of $[HSO_3^-]$ to the molarity of the salt). The upper curve refers to an acid H_2X where the two dissociation constants are 10^{-3} and 10^{-6}, which is typical of polyfunctional inorganic acids. At $y < 1$, practically no H_2X is found, and the pH is governed by the X^{2-}/HX^- equilibrium. The pH is so large that the solution could not be regenerated. At $y > 1$, practically no X^{2-} is left over, and the pH is governed by the HX^-/H_2X equilibrium. The solution could be loaded up to values of y approaching 2, and regenerated down to values of y approaching 1. Therefore, the capacity is of the order of the molarity.

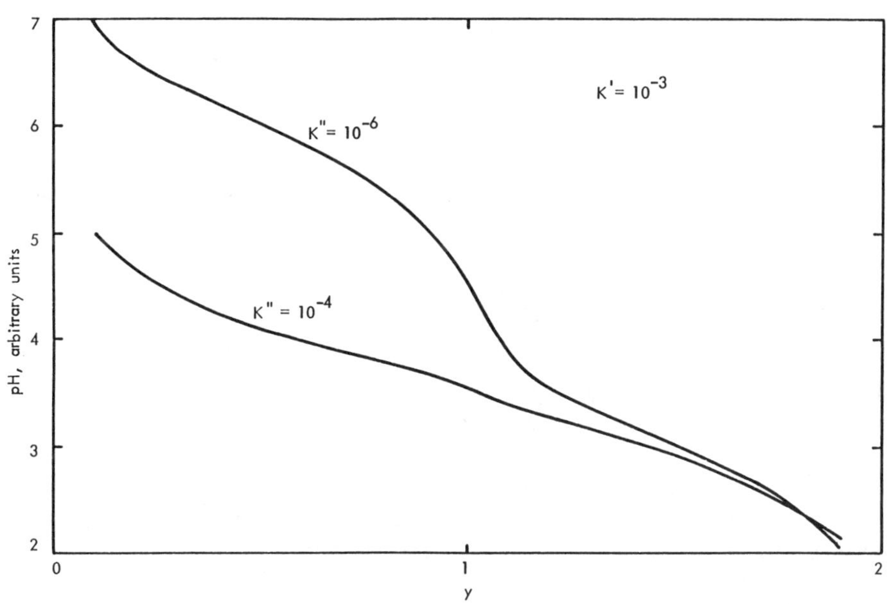

FIGURE 12.2.1. Typical pH versus loading curves for inorganic acid salt and polyfunctional organic acid salt solutions.

The lower curve refers to an acid where the two dissociation constants are 10^{-3} and 10^{-4}, which is typical of polyfunctional organic acids. The pH remains in an acceptable range over the whole range of y values so that the capacity of the solution is of the order of twice the molarity. A process that has been developed up to the demonstration stage uses aqueous solutions of sodium citrate; citric acid is a tri-functional organic acid, and capacities of the order of three times the molarity of sodium citrate are attainable.

We now turn our attention to the thermodynamics of slurried systems. For the sake of simplicity, we consider the case of a lime slurry and we neglect the possibility of the formation of sulphate.

As long as there is both solid lime and solid calcium sulphite suspended in the system, the following equilibria should be considered in addition to the two dissociation constants of sulfurous acid:

$$\text{water dissociation:}\quad [\mathrm{H}^+]\,[\mathrm{OH}^-] = \mathrm{K}_w \tag{12.2.19}$$

$$\text{lime dissolution:}\quad [\mathrm{Ca}^{2+}]\,[\mathrm{OH}^-]^2 = \mathrm{K}_L \tag{12.2.20}$$

$$\text{calcium sulfite dissolution:}\quad [\mathrm{Ca}^{2+}]\,[\mathrm{SO}_3^=] = \mathrm{K}_s \tag{12.2.21}$$

From Eqs. 12.2.19 and 12.2.20 one obtains

$$[\mathrm{Ca}^{2+}] = \frac{\mathrm{K}_L\,[\mathrm{H}^+]^2}{\mathrm{K}_w^2} \tag{12.2.22}$$

whereas from Eqs. 12.2.1, 12.2.2, and 12.2.21 one obtains

$$[\mathrm{Ca}^{2+}] = \frac{\mathrm{K}_s\,[\mathrm{H}^+]^2}{[\mathrm{SO}_2]\,\mathrm{K}_1\mathrm{K}_2} \tag{12.2.23}$$

The last two equations can be simultaneously satisfied at only one particular value of $[\mathrm{SO}_2]$; i.e.,

$$[\mathrm{SO}_2] = \frac{\mathrm{K}_s\mathrm{K}_w^2}{\mathrm{K}_L\mathrm{K}_2\mathrm{K}_1} \tag{12.2.24}$$

If p/H is larger than the right-hand side of Eq. 12.2.24, equilibrium will be reached only when solid lime has been completely converted to calcium sulphite; on the other hand, if p/H is less than the right-hand side of Eq. 12.2.24, no calcium sulphite at all will form. (These on-off equilibrium compositions are typical of heterogeneous systems where one or more of the components are present as pure separate phases. The reason lies in the fact that, as long as there is solid lime present, its activity is independent of the amount of lime.)

The right-hand side of Eq. 12.2.24 has, at conditions typical of industrial operations, a value of the order of 10^{-7} gmol/liter. This implies that even at very low partial pressures complete conversion to calcium sulfite will be attained at equilibrium.

Note that, once $[SO_2]$ has been established by Eq. 12.2.24, the set of five equations 12.2.1, 12.2.2, 12.2.14, 12.2.20, and 12.2.21, together with the electroneutrality condition, completely determine the equilibrium liquid-phase composition. At typical industrial operating conditions, the bulk-liquid pH is of the order of 5.

The analysis above can be generalized to the case of limestone, where the carbonic acid equilibria also need to be taken into account. Calculation of the sulfite-sulfate equilibrium is irrelevant, inasmuch as oxidation to sulfate is a slow reaction and equilibrium is never approached under industrial conditions.

12.3. MASS TRANSFER–GENERAL ANALYSIS

Inasmuch as H_2SO_3 is a sufficiently strong acid that even water is a chemical solvent for it, there are some peculiarities in the diffusion-reaction behavior of SO_2 in any aqueous solution. The hydrolysis reaction

$$SO_2 + H_2O = HSO_3^- + H^+ \tag{12.3.1}$$

must always be taken into account, though in some cases its influence on the overall mass transfer rate is negligible. The peculiarity of reaction 12.3.1 lies in the fact that the liquid-phase reactant, H_2O, is the solvent itself and, therefore, it can never get depleted near the interface, even when instantaneous reaction conditions are approached.

The reaction time for reaction 12.3.1 is of the order of 10^{-7} sec. Though this is not extremely small, for practical purposes it can be regarded as small enough for instantaneous reaction conditions to be achieved. The reasons for this are briefly discussed in the following.

With diffusion times typically of the order of 10^{-2} sec, the value of ϕ for reaction 12.3.1 is of the order of 10^5. Therefore, the condition for the instantaneous reaction regime equations to apply,

$$\sqrt{\phi} \gg I_\infty \tag{12.3.2}$$

is satisfied provided I_∞ does not exceed the value 50. It turns out that in practice one can regard I_∞ as never exceeding 50, for the following reason.

Typical values of $k_G a/k_L^0 a$ are of the order of 10 gmol/liter atm. Therefore, for SO_2 absorption in aqueous solutions, where the Henry's Law constant is of the order of 1 liter atm/gmol, the ratio $Hk_G a/k_L^0 a$ is of the order of 10; without any rate enhancement in the liquid phase, the SO_2 mass transfer rate would be controlled by the liquid-state resistance (see the discussion following Eq. 1.6.8). However, when the rate enhancement factor in the liquid phase exceeds 50, the rate of mass transfer becomes controlled by the gas-side resistance and, therefore, accurate modeling of the liquid-side resistance to mass transfer becomes irrelevant.

Mass transfer rates of SO_2 in water are easily calculated. If all diffusivities are equal, I_∞ is simply calculated from Eq. 3.5.19:

$$I_\infty = \frac{\alpha_i^* - \alpha_0}{a_i - a_0} \tag{12.3.3}$$

The value of α is given by Eq. 1.8.12, which can be written as

$$\alpha = a + \sqrt{\mathrm{K}_1 a} \tag{12.3.4}$$

Substitution of Eq. 12.3.4 into 12.3.3 and algebraic rearrangement gives

$$I_\infty = 1 + \frac{\sqrt{\mathrm{K}_1}}{\sqrt{a_i} + \sqrt{a_0}} \tag{12.3.5}$$

The question of unequal diffusivities should, however, be considered. Although the ions H^+ and HSO_3^- have considerably different real diffusivities, with H^+ being about six times more mobile, their equivalent diffusivities must be equal, as can easily be seen by applying the procedure discussed in Section 10.4 which yields

$$\tilde{D}_{H^+} = \tilde{D}_{HSO_3^-} = \tilde{D} = \frac{2D_{H^+}D_{HSO_3^-}}{D_{H^+} + D_{HSO_3^-}} \tag{12.3.6}$$

A sufficiently large electrical potential gradient will develop and slow down the flux of H^+ and accelerate that of HSO_3^-. The value of $\tilde{D}$ is appreciably larger than the diffusivity D of SO_2 itself, so that I should be calculated with the procedure discussed in Section 6.2, which yields

$$I_\infty = 1 + \sqrt{\frac{\tilde{D}}{D}}\,\frac{\sqrt{\mathrm{K}_1}}{\sqrt{a_i} + \sqrt{a_0}} \tag{12.3.7}$$

With K_1 of order 10^{-2}, Eq. 12.3.7 will result in I values appreciably larger than unity only if a_i is less than 10^{-3} gmol/liter, which would correspond to interface partial pressures of SO_2 lower than about 10^{-3} atm. Therefore, significant rate enhancement will be observed in water only at very low partial pressures and, in fact, none had been reported in the literature until 1979 since all experiments had been performed at high SO_2 partial pressures. At such low partial pressures, addition of neutral salts such as NaCl will result in a much lower gradient of electrical potential, and hence in a larger value of $\tilde{D}$, so that mass transfer is enhanced by such additions.

We next consider the case of alkaline solutions, which may be either regenerable ones or throwaway process slurries. In alkaline solutions, the possibility of SO_3^{2-} formation should be taken into account:

$$HSO_3^- = H^+ + SO_3^{2-} \tag{12.3.8}$$

Reaction 12.3.8, as well as any other reaction involving additional ions in the solution, is a proton-transfer reaction that can be regarded as instantaneous. Therefore, the mass transfer analysis should be carried out following the procedure discussed in Section 6.2. The procedure needs to be generalized since, instead of only one instantaneous reaction, a network of such reactions must be taken into account.

Should all diffusivities be equal, the enhancement factor would still be given by Eq. 12.3.3. The only sulfur-carrying species are $A \equiv SO_2$, $B_1 \equiv HSO_3^-$, $B_2 \equiv SO_3^{2-}$, since SO_2 does not undergo any chemical reaction other than hydrolysis (oxidation to the sulphate ion may take place, but its rate is too slow to influence the mass transfer rate). Therefore, the total SO_2 content α is given by

$$\alpha = [SO_2] + [HSO_3^-] + [SO_3^{2-}] \tag{12.3.9}$$

When Eq. 12.3.9 is substituted into Eq. 12.3.3, one obtains

$$I_\infty = 1 + \frac{(b_{1i} - b_{10}) + (b_{2i} - b_{20})}{a_i - a_0} \tag{12.3.10}$$

Notice that Eq. 12.3.5 is a special case of Eq. 12.3.10, since in the case of water $b_2 \ll b_1$ and $b_1^2 = K_1 a$. In alkaline solutions, the interface concentrations b_{1i} and b_{2i} are not determined solely by SO_2 hydrolysis but also by the coupling of the hydrolysis reactions with other ionic reactions taking place in the liquid phase.

Should one wish to take into account the effect of unequal diffusivities, a rather complex diffusion-reaction problem needs to be solved; the general procedure for doing so is discussed in the Appendix to this section. In the following sections, some general conclusions about mass transfer rates in both regenerable and throwaway systems are obtained on the basis of the simplifying assumption that all diffusivities are equal.

Appendix to Section 12.3

When the effect of unequal diffusivities is taken into account, Eq. 12.3.10 takes the following form:

$$I_\infty = 1 + \frac{\sqrt{(\tilde{D}_1/D)}\,(b_{1i} - b_{10}) + \sqrt{(\tilde{D}_2/D)}\,(b_{2i} - b_{20})}{a_i - a_0} \tag{12.A.1}$$

where the equivalent diffusivities $\tilde{D}_1$ and $\tilde{D}_2$ need to be estimated with the procedures discussed in Section 10.4. Equation 12.A.1 is obtained with the "rule of thumb" of substituting square roots of diffusivity ratios for the ratios themselves in the equations obtained from the film theory approach.

The interface values b_{1i} and b_{2i} cannot be obtained simply from the thermodynamic model, since the interface molarity may not be the same as that in the bulk-liquid. Therefore, the complete set of diffusion-reaction equations needs to be solved. The general procedure for doing so, for an arbitrary set of ionic reactions, is briefly reviewed in the following.

Inasmuch as one wishes to substitute the square root of diffusivity ratios to the ratios themselves in the final result of the calculation, it is useful to write the diffusion-reaction equations in such a form that the ratios appear directly in them:

$$\frac{\tilde{D}_j}{D}\frac{d^2 b_j}{dx^2} = \frac{1}{D}\sum_k \nu_{jk} r_k \tag{12.A.2}$$

By introducing the following definitions:

$$b_j' = \sqrt{\frac{\tilde{D}_j}{\tilde{D}}}\, b_j \tag{12.A.3}$$

$$\hat{r}_k = \frac{r_k}{D} \tag{12.A.4}$$

Eq. 12.A.2 is written as

$$\frac{d^2 b_j'}{dx^2} = \sum_k \nu_{jk} \hat{r}_k \tag{12.A.5}$$

Let r and r' be the local rates of the two hydrolysis reactions, i.e., reactions 12.3.1 and 12.3.8. The rate of consumption of the hydrogen ion $H^+ \equiv B_3$ is given by

$$-r - r' + \sum_{R=1}^{M} \nu_{3k} r_k \tag{12.A.6}$$

where ν_{3k} is the stoichiometric coefficient of H^+ in the kth ionic reaction. It is always possible to write the ionic reactions in such a way as not to include the SO_3^{2-} and HSO_3^- species, say, to set $\nu_{1K} = \nu_{2K} = 0$, by appropriate transformation of the ν_{jk} matrix with the procedure discussed in Section 2.2. Therefore, the local rate of consumption of HSO_3^- is $r' - r$, and that of SO_3^{2-} is $-r'$. Elimination of r and r' between the diffusion equations for H^+, SO_3^{2-} and HSO_3^- yields

$$\frac{d^2}{dx^2}[b_3' - b_1' - 2b_2'] = \sum_{k=1}^{M} \nu_{3k} \hat{r}_k \tag{12.A.7}$$

Let there be N ionic species to be considered, including SO_3^{2-} and HSO_3^-. Equations 12.A.5 and 12.A.7 represent a set of $N - 2$ equations; the number of independent reactions M is certainly less than $N - 2$. The matrix ν_{jk} has rank M and thus at least one of its square submatrices is non-singular and can be inverted. It follows that the $\hat{r}_k$s can be eliminated from the system 12.A.5-12.A.7 (this elimination makes the difference between r_k and $\hat{r}_k$ irrelevant) to yield $N - M - 2$ equations of the following form:

$$\frac{d^2}{dx^2}\left[\sum_{j=1}^{N} \sigma_{jp} b'_j\right] = 0 \tag{12.A.8}$$

where the σ_{jp}s are uniquely determined by the ν_{jk}s. The boundary conditions at the gas–liquid interface have exactly the same structure as the diffusion-reaction equations and can be written as

$$x = 0, \qquad \frac{d}{dx}[b'_3 - b'_1 - 2b'_2] = \sum_{k=1}^{M} \nu_{3k}\hat{r}_{k0} \tag{12.A.9}$$

$$x = 0, \qquad \frac{db'_j}{dx} = \sum_{k=1}^{M} \nu_{jk}\hat{r}_{k0} \tag{12.A.10}$$

with $\hat{r}_{k0}$ being the rate of the kth reaction at the interface per unit area divided by the SO_2 diffusivity. The $\hat{r}_{k0}$s can again be eliminated and since the same matrix ν_{jk} is involved, the result is

$$x = 0, \qquad \frac{d}{dx}\left[\sum_{j=1}^{N} \sigma_{jp} b'_j\right] = 0 \tag{12.A.11}$$

Equation 12.A.8 and 12.A.11 imply that the quantities $\Sigma_{j=1}\ \sigma_{jp} b'_j$ are constant throughout the liquid phase, say,

$$\sum_{j=1}^{N} \sigma_{jp} b'_{ji} = \sum_{j=1}^{N} \sigma_{jp} b'_{j0} \tag{12.A.12}$$

This completes the solution of the general problem. The N interface values b_{ji} can be calculated from the $N - M - 2$ equations 12.A.12, the M equilibrium equations for the ionic reactions, and the two equilibrium equations for the hydrolysis reactions. The value of I_∞ is then calculated from Eq. 12.3.11. The procedure is somewhat cumbersome, and numerical techniques are required for the calculation of the interface concentrations. Special examples of this procedure have been presented in the literature for solutions of NaOH, Na_2SO_3, and for lime slurries to which polyfunctional organic acids are added as rate promoters; the theoretical predictions are well confirmed by experimental data.

12.4. THROWAWAY PROCESSES

With the exception of one process which uses a clear alkaline solution based on lime, throwaway processes are all based on slurries. The thermodynamics of slurries have been discussed in Section 12.2; as long as there is some solid reagent still present in the slurry (which is always the case since the reagent is fed in excess of the stoichiometric requirement), as well as some solid calcium sulfite (which forms as soon as some SO_2 is absorbed in the slurry), the composition of the bulk-liquid is constant, irrespective of the reagent/product ratio. Therefore, the relative circulation pattern of slurry and gas is irrelevant, with no advantage whatsoever for the countercurrent flow configuration. This allows a great flexibility of design, of which use has been made. Countercurrent flow configurations are in use (see Figure 12.1.1) but so are configurations where the slurry is essentially in a perfectly mixed stirred tank configuration (see Figure 12.1.2), and even cocurrent flow is in commercial use.

Analysis of slurried processes is made particularly easy by the fact that the liquid-phase composition is constant; in fact, only the gas-phase composition variations along the absorber need be taken into account. Furthermore, the SO_2 concentration in the liquid bulk is negligibly small, so that the theory of mass transfer accompanied by irreversible instantaneous chemical reaction may be used. (In principle, when the SO_2 partial pressure in the gas phase is reduced to extremely low levels, the reversibility should be taken into account. However, in this limit, the enhancement factor becomes large enough for the resistance to mass transfer to become gas-phase controlled. Therefore, the approximate procedure discussed at the end of Section 6.4 can be used.)

With the notation introduced in the Appendix to Section 12.3, let b_1' be defined as

$$b_1' = \sqrt{\frac{\tilde{D}_1}{D}}\, b_1 \tag{12.4.1}$$

where b_1 is the liquid-phase DLRA (i.e., the dissolved alkalinity). At typical values of bulk-liquid alkalinities (pH = 5.5), formation of SO_3^{2-} can be neglected in the film near the interface and thus the stoichiometric coefficient is unity. Therefore, Eqs. 6.4.14 to 6.4.16 can be used with $\nu_1 = 1$ and b_{10}' substituted for b_{10}. Let $\tilde{a}$ be the interface area/unit volume of the absorber, π the total pressure, Z the coordinate along the gas-flow path, p the partial pressure in the gas phase, and G the superficial molar velocity of the gas. A mass balance on the gas phase yields

$$\frac{G}{\pi}\frac{dp}{dz} = N\tilde{a} \tag{12.4.2}$$

and the flux N is given by Eqs. 6.4.14–6.4.17. These can be simplified by introducing the parameter P defined as follows (see also Section 15.3):

$$P = \frac{\mathrm{k}_G\tilde{a}\mathrm{H}}{\mathrm{k}_L^0\tilde{a}} \tag{12.4.3}$$

so that the equations for N become

$$p \leqslant \frac{\mathrm{H}b_{10}'}{P}, \qquad N = \mathrm{k}_g p \tag{12.4.4}$$

$$p > \frac{Hb'_{10}}{P}, \qquad N = k_G \frac{p + Hb'_{10}}{1 + P} \tag{12.4.5}$$

Three cases are possible. The first is the one where the partial pressure of SO_2 in the inlet gas p_B is so low that Eq. 12.4.4 applies even at the gas inlet and, therefore, throughout the absorber. The whole process is entirely gas-phase controlled. Substitution of Eq. 12.4.4 into Eq. 12.4.2 and integration yields

$$N_G = ln \left(\frac{p_B}{p_T}\right) \tag{12.4.6}$$

where p_T is the partial pressure in the exit gas and N_G is the gas-side number of transfer units:

$$N_G = \frac{Z k_G \tilde{a} \pi}{G} \tag{12.4.7}$$

Z being the total length of the absorber along the gas flow path.

The second case is where p_T is large enough that Eq. 12.4.5 applies even at the gas exit. In this case, the liquid-side resistance to mass transfer can never be neglected. Substitution of Eq. 12.4.5 into 12.4.2 and integration yields

$$N_G = (1 + P)\, ln \frac{p_B + Hb'_{10}}{p_T + Hb'_{10}} \tag{12.4.8}$$

The third case is where the partial pressure in the gas phase takes the value Hb'_{10}/P somewhere within the absorber, say,

$$p_B > \frac{Hb'_{10}}{P} > p_T \tag{12.4.9}$$

In this case, there is a section of the absorber near the gas inlet where Eq. 12.4.4 applies; this extends up to the point where $p = Hb'_{10}/P$. From that point to the gas exit, Eq. 12.4.5 applies. The number of transfer units required is

$$N_G = (1 + P)\, ln \left[P \, \frac{1 + p_B/Hb'_{10}}{1 + P}\right] + ln \frac{Hb'_{10}}{P p_T} \tag{12.4.10}$$

where the two terms on the right-hand side represent the number of transfer units required in the two sections of the absorber.

The dissolved alkalinity of lime or limestone slurries is rather low and, therefore, the rate enhancement of mass transfer is low. Several schemes have been proposed and tested for increasing the dissolved alkalinity. These include addition of magnesium oxide, sodium sulphite, and ammonium sulphite. The effect of such additives is essentially to accumulate as soluble sulfites in the bulk-liquid, thus increasing the dissolved alkalinity; magnesium sulfite is about 600 times more soluble then calcium sulfite.

An interesting possibility for enhancing the mass transfer rate is the use of additives that act as buffers in the appropriate range of pH values. Additives of this type should not be too alkaline, because if they were, they would simply be neutralized by dissolved HSO_3^- and, therefore, would rapidly become depleted. On the other side, as can be seen from Eq. 12.3.10, additives should have the ability to increase the HSO_3^- concentration at the interface b_{1i}. The equilibrium condition for reaction 12.3.1 shows that the latter requirement is accomplished if the additive keeps the interface pH at a sufficiently high value. These considerations suggest the use of additives that have buffering capacity in a pH range between the value in the liquid bulk (typically 5.5) and the pH at the interface (typically 3.5). Weak organic acids, particularly polyfunctional ones (see the discussion in Section 12.2), are obvious candidates. Increased mass transfer rates have been observed experimentally with addition of such acids. The concentrations required are rather low, since above a few millimoles/liter the rate enhancement is already large enough to make the gas-phase resistance to mass transfer the controlling one. Quite obviously, additional increases of additive concentration have no further effect on mass transfer rates.

Modeling mass transfer rates in slurried liquids, particularly when additives are present, becomes rather complex due to the large number of ionic species and reactions that need to be taken into account. The procedures discussed in the Appendix to Section 12.3 should be used. The predicted mass transfer rates are in good agreement with experimental data. However, it should be borne in mind that an excess of mathematical complexity is not particularly useful, inasmuch as enhancement factors in excess of about 50 will invariable result in gas-phase mass transfer control.

In practice, it is comparatively easy to enhance the mass transfer rate in the liquid to the point where gas-phase mass transfer control is reached. Therefore, the simple Eq. 12.4.6 is often appropriate for design.

12.5. REGENERABLE PROCESSES

Regenerable processes have been less studied than throwaway processes and very few mass transfer studies have been carried out under conditions approaching those of industrial operation (i.e., low to very low SO_2 partial pressures in the gas phase).

Since all reactions in the liquid phase are instantaneous, the enhancement of the mass transfer rate is approximately equal to the local ratio of the chemical capacity of the liquid to its physical capacity (see Eq. 12.3.3). With both sulfite and citrate solutions, the right-hand side of Eq. 12.3.3 is a rather large number, except perhaps near the gas inlet where the partial pressure in the gas is comparatively high and the concentration of alkaline solute in the liquid is comparatively low. Therefore, gas-phase control of mass transfer is very likely in these processes;

however, design cannot be based on the simple formulation in Eq. 12.4.6, inasmuch as the equilibrium vapor pressure corresponding to the bulk-liquid composition is not negligible.

Formation of the sulfate is always a problem and from this viewpoint, processes based on ammonium sulfite are preferable as the ammonium sulfate is a more desirable by-product. However, in ammonium sulfite processes the cation is volatile, which poses some problems in the regeneration step.

The analysis of SO_2 mass transfer rates in sulfite solutions has been carried out successfully with the procedure discussed in the Appendix to Section 12.3; the only ionic reaction that need be considered in addition to the two hydrolysis reactions is the water dissociation. With citrate solutions, the analysis would be considerably more complex since the three successive dissociations of citric acid should be taken into account. There is some doubt that such analyses have any more than academic interest.

PART FOUR: INDUSTRIAL PROBLEMS

13 Laboratory Experiments

Vitruvius says that small models are of no avail for ascertaining the behavior of big ones. I here propose to show that this conclusion is a false one.

LEONARDO DA VINCI

Process selection and design in gas treating as in any other area requires the availability of reliable physico-chemical data. Often the data that are needed cannot be found in the literature and, therefore, they must be obtained with appropriately designed laboratory-scale experiments.

We believe that data obtained in the laboratory should be of a fundamental nature and not purely empirical information that cannot be extrapolated confidently to conditions other than the ones prevailing when the experiment was made. The temptation of planning experiments that resemble as closely as possible the foreseeable industrial unit should be resisted; such experiments rarely, if ever, provide the fundamental understanding that is to be sought in laboratory-scale experiments and often they do not yield reliable predictions of the behavior of an industrial-scale unit.

Experiments can be classified in two fundamental categories: differential and integral. In a differential experiment, the operating variables (such as temperature, pressure, composition, etc.) are fixed at some *unique* predetermined value; the value of some dependent variable (e.g., absorption rate) which is presumed to depend on the operating variables and only on them, is measured. By repeating the experiment over a range of values of the operating variables, the functional dependency of the dependent variable is determined.

In an integral experiment, the operating variables are not constant, but their values vary from point to point or in time within a single experiment. A pilot plant experiment is a typical example; composition, temperature, and pressure have different values at different points in the pilot plant column. When some dependent variable is measured in an integral experiment, one cannot deduce any fundamental data from it, unless a theoretical model is available that relates the overall behavior of the experiment to the actual local values of physico-chemical parameters. However, if such a model is available, it would be better to determine the values of the parameters in differential experiments and then use the model to *predict* the result of the integral experiment. If this is done, the integral experiment becomes unnecessary. Unfortunately, often a reliable model is not available and, therefore, integral experiments are needed; this is the main reason for performing pilot plant work.

In this chapter we focus attention on differential experiments aimed at the measurement of values of physico-chemical parameters such as solubility, heat of reaction, mass transfer coefficient, and so on. Pilot plant operation will be discussed in Chapter 15.

13.1. VAPOR-LIQUID EQUILIBRIUM DATA

Vapor-liquid equilibrium (VLE) data are gathered with the purpose of determining the partial pressure(s) of the gas components to be removed from a gas stream as a function of the relevant independent variables, namely, temperature, liquid phase molarity m, and liquid phase conversions y, y'. It is useful to also obtain information on the equilibrium partial pressure of the solvent (e.g., water in the case of aqueous chemical solvents) as a function of the same variables. Vapor-liquid equilibrium data can be correlated, and often successfully extrapolated, by appropriate modeling of the kind discussed in Chapter 2.

Three different types of VLE cells have been used to obtain high quality data for pure and mixed acid gases in chemical and physical gas treating solvents. The cells all provide means for rapidly bringing vapor and liquid phases into thermal and composition equilibrium at known temperature and pressure. The compositions of the equilibrium phases are then determined by withdrawing samples and analyzing them. This requires the use of specialized analytical techniques such as mass spectrometry and gas chromatography. Although many types of cells are possible, only three have been used extensively: the rocking autoclave cell, the vapor recirculation cell, and the gas flow through cell. These are all sophisticated units requiring careful design, extensive instrumentation, and control equipment. Great care and patience are required on the part of the operator to obtain reliable data.

The rocking autoclave cell has been used to determine the VLE data for CO_2 in aqueous potassium carbonate/bicarbonate solutions (without promoters) which are reported in Figure 2.4.1. The data cover 20–40 wt% K_2CO_3 solutions, temperatures of 70–140°C, and pressures up to 150 psia. These are the conditions of industrial interest. The method also gives equilibrium water vapor pressures.

The rocking autoclave unit is shown schematically in Figure 13.1.1. The autoclave is a 3 in. dia pipe, 3 ft long with a heating jacket. The autoclave is charged through the flanged end with weighed amounts of high-purity potassium carbonate and bicarbonate chemicals. After evacuating air, the quantity of water necessary to yield the desired concentration is drawn into the autoclave. The autoclave is rocked (through an arc of approximately 30° at a rate of 24 cycles/min) to insure uniform temperature in the contained charge, thorough dissolution of the solids, and inti-

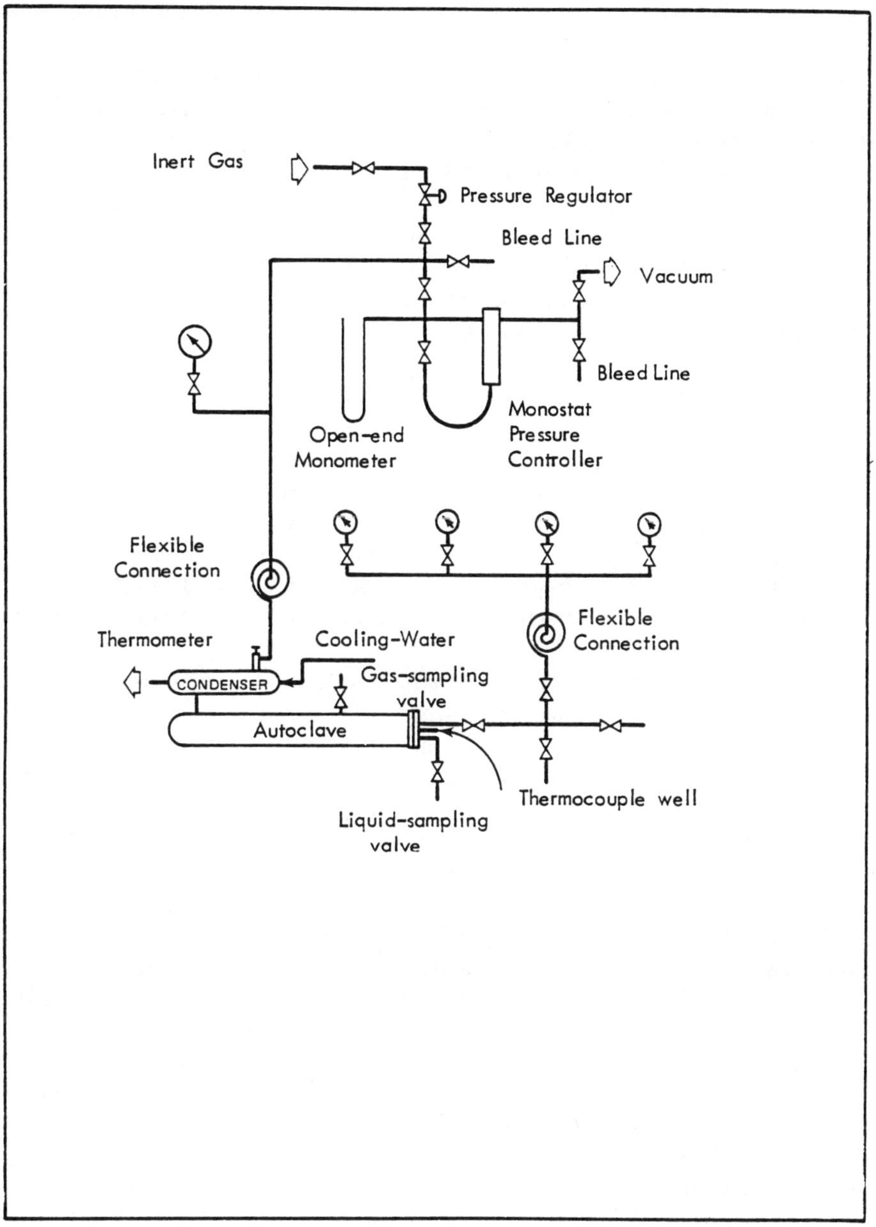

FIGURE 13.1.1. Rocking autoclave vapor–liquid equilibrium cell.

mate contact between liquid and gas phases. These factors assist in reaching equilibrium in a reasonable time.

Heat is applied through an electric winding on the outside of the jacket. Desired temperatures are obtained by boiling water at appropriate pressures in the jacket. Heat in excess of that required to maintain boiling is removed in the condenser. The temperature of the solution in the autoclave can be maintained to $\pm 0.1°C$. Internal pressures are read on a range of gauges. The autoclave is attached to a vacuum system needed for evacuating air from the unit at the start of a run and for withdrawing gas samples for analysis after equilibrium has been reached.

The experimental techniques used to determine the compositions of the liquid and vapor phases at equilibrium are extremely important. Obtaining accurate analysis on the liquid presents special difficulties.

No direct liquid-phase analysis was done in determining the data. Instead, the equilibrium composition (bicarbonate-carbonate ratio) was estimated from the weights of ingredients charged (H_2O, K_2CO_3, $KHCO_3$) making an allowance, based on gas analysis, for the fraction of carbonate converted to bicarbonate under final equilibrium conditions. This approach is open to the criticism that no independent check on equilibrium solution composition (K^+ ion concentration, HCO_3^-/CO_3^{2-} ratio) was made. However, it appears to have given high-quality data. Recently, an X-ray fluorescent analyzer for rapid determination of K^+ concentration in aqueous solutions has become available.

Gas sampling and analysis techniques also require careful attention. To sample the gas, the sampling apparatus was attached to the sampling point and evacuated to about 1 mm Hg. Inasmuch as the gas in the autoclave contains water vapor, the pressure of the gas sample during sampling must be kept below the saturated water vapor pressure to prevent condensation. After evacuation, the vacuum line is closed off and the gas sampling valve is throttled, permitting vapor from the autoclave to flow into the sampling apparatus. Sampling is stopped when the manifold pressure reaches 20 mm Hg. Prior to taking the vapor sample for analysis, the 10-liter sample bottle is purged three times with vapor from the autoclave. The purging and sampling procedure uses about 200 cc (standard) of gas. Duplicate samples of gas are taken for each test and analyzed immediately to minimize air contamination. The precision of the mass spectrometer for a particular analysis is $\pm 0.2\%$.

Mass spectrometry has the advantage over gas chromatography that both acid gas and water vapor pressures can be obtained readily. The accuracy of mass spectrometry measurements is about $\pm 1\%$ or better at the 50% mole fraction level.

The data obtained with the technique discussed above have been used as a basis for the correlation discussed in Section 2.4. Equation 9.5.1 correlates all the data within a few percent; it reflects the fact that the dependency of p^* on T is linear in an Arrhenius plot (see also Section 13.2) and that the slope in such a plot is independent of molarity m and fractional saturation y. This, of course, implies that the heat of reaction is constant.

The water vapor pressure data exhibit some scatter and they seem to indicate that $p^*_{H_2O}$ goes through a maximum when plotted against y. There is no known theoretical basis for this and, in fact, one would expect $p^*_{H_2O}$ to decrease monotonically as y is increased.

Another version of the rocking autoclave-type of VLE cell was used to measure equilibrium pressures of CO_2, H_2S, and CO_2/H_2S mixtures over aqueous MEA and DEA solutions. The cell body was a nominal one-liter stainless steel cylinder with liquid sample ports at each end and a vapor sample line at the top center. It was suspended in an oil bath (±0.2°F at 100°F; ±1.0°F at 300°F) and rocked at 14 cycles/min.

After evacuation, the cell was charged with about 600 g of amine solution and lowered into the oil bath. Predetermined amounts of H_2S and/or CO_2 were added; the cell was then pressured up to the desired total pressure with methane. The mixture was rocked for at least 16 hours to achieve equilibrium.

Liquid samples were analyzed for H_2S content by electrometric titration with silver nitrate. CO_2 in the liquid was measured by a method in which CO_2 is stripped from the solution and the evolved CO_2 trapped on an adsorbent and weighed. When CO_2 and H_2S are both present, the evolved H_2S was trapped ahead of the CO_2 absorption tube in a scrubber containing sulfuric acid and silver nitrate.

Vapor-phase samples were taken by different methods depending on acid gas concentration. Glass sample containers were felt to be preferable to stainless steel as H_2S loss by adsorption on stainless steel was suspected. At 1 mole% concentration and above, analyses were done routinely by mass spectrometer. At the 0.1% concentration level, mass spec became inaccurate and the methylene blue color method was used. This involves taking the vapor sample into a bottle containing zinc acetate and sodium hydroxide.

The vapor recirculation closed cell design has been used with aqueous amine and organic solvent-amine solutions with CO_2, H_2S, and mixtures of CO_2 and H_2S for partial pressures of 0.1-1000 psia and temperatures of 40-130°C. Below 0.1 psia partial pressure, the method becomes quite inaccurate owing to limitations on the size of vapor samples that can be withdrawn without disturbing equilibrium significantly. A different experimental approach is needed in the very low partial pressure region.

The vapor recirculation closed cell is shown schematically in Figure 13.1.2. The term closed refers to the fact that the cell is charged with a batch of solution and gas that is then allowed to come to equilibrium by recirculating the vapor through the liquid. Other designs, of the open type, use a batch of liquid but gas of constant acid gas and moisture content is allowed to flow once through the liquid and out of the cell.

A batch of solution is contained in a pressure vessel with sight glass (Jerguson gauge). CO_2 and H_2S are introduced in desired amounts and N_2 is added to bring total system pressure up to at least 100 psig at operating temperature. Vapor is drawn from the top of the Jerguson and recirculated by a magnetic circulating pump back to the base of the Jerguson. The sight glass gives a visual check that vapor is actually circulating. The vapor circulation loop contains a 250 cc vapor reservoir to increase vapor volume. This allows many vapor samples to be withdrawn for analysis without disturbing equilibrium significantly. The Jerguson, reservoir, and magnetic pump are housed in an air bath for temperature control. At equilibrium, vapor and liquid samples are withdrawn for analysis.

Vapor samples are taken directly off the vapor reservoir into a gas chromatograph

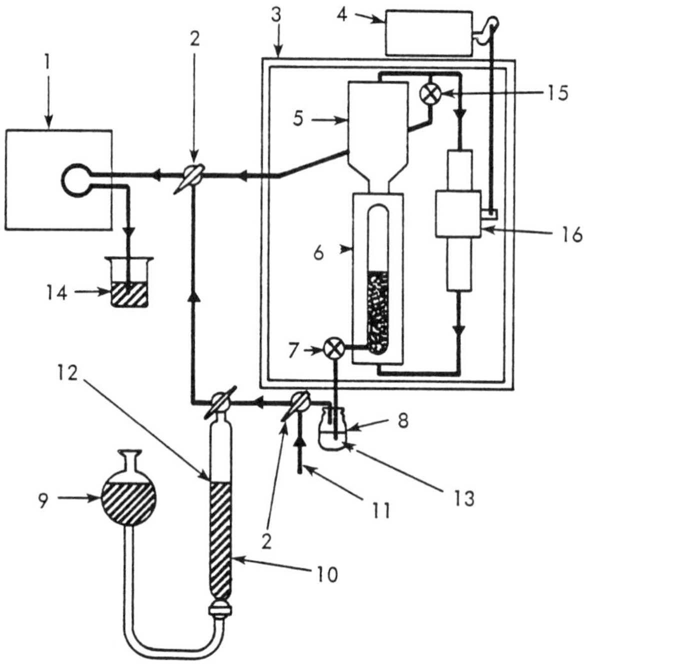

LEGEND

1. Gas Chromatograph
2. 3-Way Ball Valves
3. Temperature Controlled Air Bath
4. Pump Motor
5. Vapor Reservoir (250 ml.)
6. High Pressure Jerguson Gauge (200 ml.)
7. Liquid Sampling Valve
8. Liquid Sampler (with 5N H_2SO_4)
9. Mercury Leveller
10. 200 ml. Gas Collecting Burette
11. Air Line
12. Silicone Oil Layer
13. Needle (.01 inch I.D.)
14. Flow Indicator
15. Vapor Sampling Valve
16. Magnetic Pump

FIGURE 13.1.2. Vapor recirculation closed vapor–liquid equilibrium cell.

for analysis for CO_2 and H_2S contents. Use of a 10 ft column of Chromosorb W gives well shaped, well spaced peaks for N_2, CO_2, H_2S, and H_2O. Calibration gas mixtures must be used as standards. The water vapor peak is not very reproducible and quantitative analysis for H_2O vapor is not generally possible.

Vapor composition is determined as follows. The acid gas to N_2 ratio is determined on a dry basis from the gas chromatograph. This is combined with known water vapor pressure data for the aqueous amine system under study. It is usually assumed that the vapor pressure of water over an aqueous amine solution changes by a negligible amount as the acid gas loading in solution increases.

Liquid samples are taken from the Jerguson into a liquid sample vessel containing 5N sulfuric acid. All of the acid gases are evolved into a gas collecting burette that measures the volume of evolved acid gas. The evolved gases are then analyzed by gas chromatography to get the CO_2-to-H_2S ratio.

The closed design of VLE cell described in the last paragraphs cannot be used for measuring very low acid gas partial pressures because the vapor samples that would have to be withdrawn for accurate analysis would disturb the equilibrium. The gas flow-through cell type gives ample gas available for analysis and gives accurate data in the acid gas partial pressure range of about 0.001–0.3 psia. This range, and below, is of great industrial importance for determining thermodynamic limits to attainable acid gas cleanup in countercurrent adsorbers.

For partial pressures below 0.001 psia, where measurements prove very difficult, VLE models are very useful for extrapolating data to very low pressures.

A schematic of the gas flow-through cell is shown in Figure 13.1.3. Nitrogen gas presaturated with water vapor is passed at a steady controlled rate through a series of bubblers containing acid gas-loaded solution. Acid gas desorbs from solution into the nitrogen, the vapor leaving the third bubbler being in equilibrium with the solution. The off-gas comprising N_2, H_2O, CO_2, H_2S is passed directly into a gas chromatograph.

Despite its superficial simplicity, this apparatus is quite difficult to run. This is partly owing to the long times needed to reach equilibrium at very low conversions. Flow control and solution depletion can be problem areas if the apparatus is not designed and operated carefully. An alternative to the gas flow-through type for very low partial pressure measurements is the batch autoclave type with large enough vapor volume to provide large vapor samples for accurate GC analysis.

Figures 13.1.4 and 13.1.5 show comparisons of data measured by the vapor recirculation method (high p_{CO_2} range) and the gas flow-through method (low p_{CO_2} range). The good correspondence attests to the high quality of the data.

13.2. MEASUREMENT OF HENRY'S LAW CONSTANT

When considering the problem of determining the value of Henry's Law constant (see Equation 1.6.2) for a gaseous component, a fundamental distinction should be drawn between two cases: when the liquid phase is non-reactive (i.e., it is a physical solvent); and a reactive liquid (i.e., a chemical solvent).

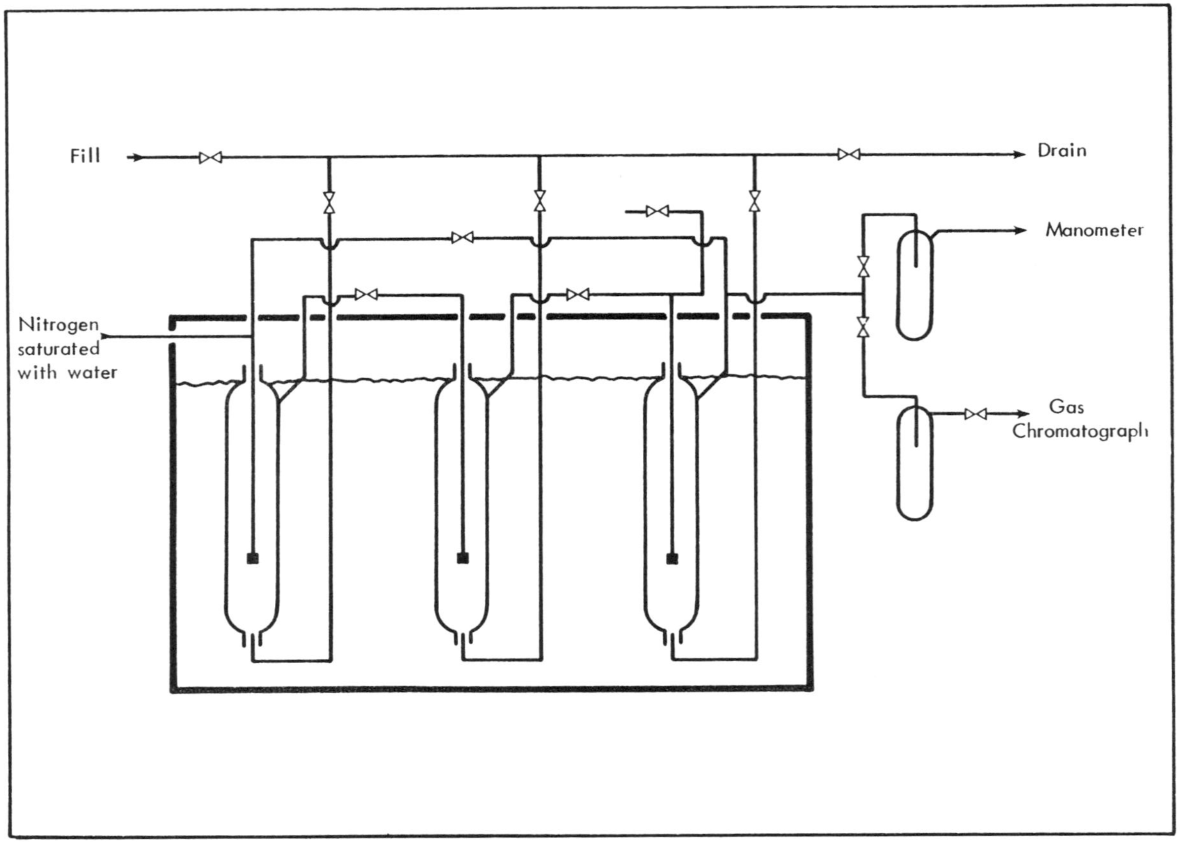

FIGURE 13.1.3. Gas flow-through vapor–liquid equilibrium cell.

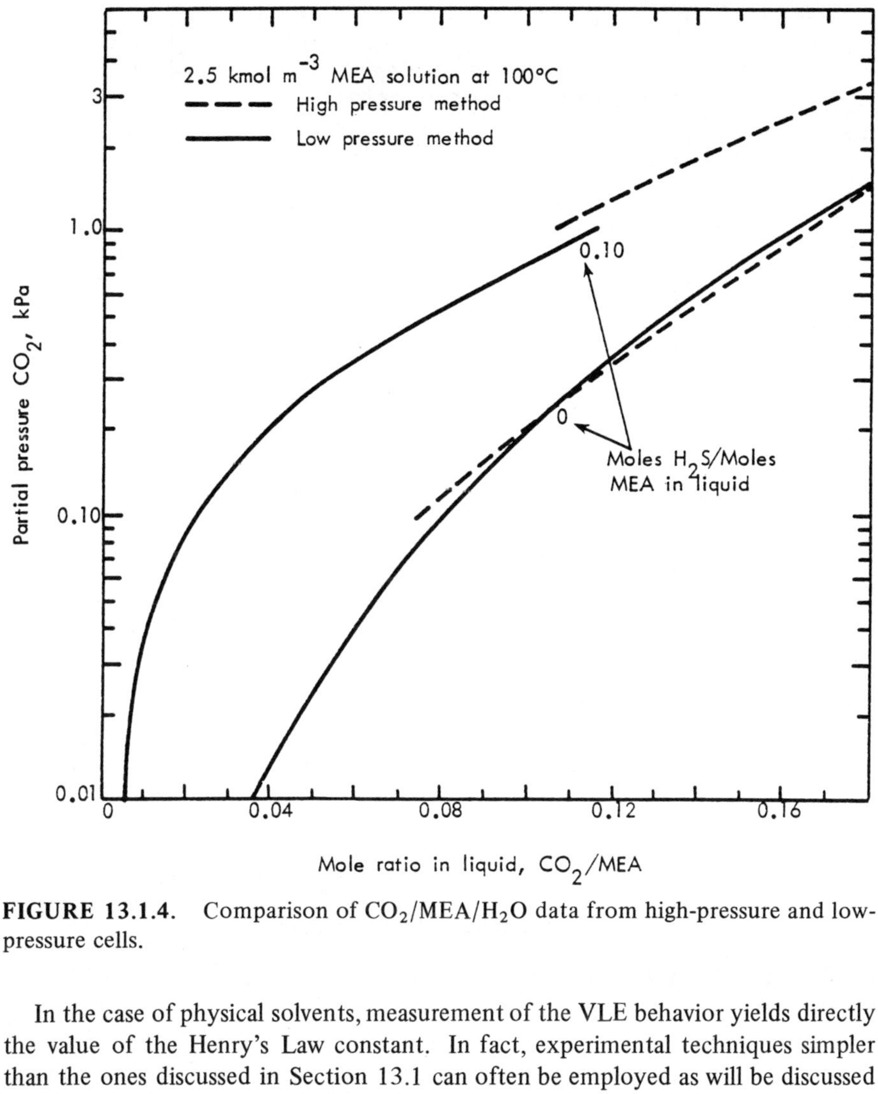

FIGURE 13.1.4. Comparison of CO_2/MEA/H_2O data from high-pressure and low-pressure cells.

In the case of physical solvents, measurement of the VLE behavior yields directly the value of the Henry's Law constant. In fact, experimental techniques simpler than the ones discussed in Section 13.1 can often be employed as will be discussed below.

In the case of chemical solvents, the value of the Henry's Law constant H cannot be calculated from overall vapor–liquid equilibrium data. Special experimental techniques are needed and the results obtained often require some theoretical modeling in order to yield the value of H. A simple example of this was discussed in Chapter 1 (Equation 1.8.1) concerning the value of H for the SO_2-water system (SO_2 is an acid strong enough for water to be regarded as a chemical solvent for it).

We begin by illustrating the experimental techniques that may be used for physical solvents, in addition to those that have already been discussed in Section 13.1. These techniques can be classified in two basic categories; volumetric methods and pressure methods.

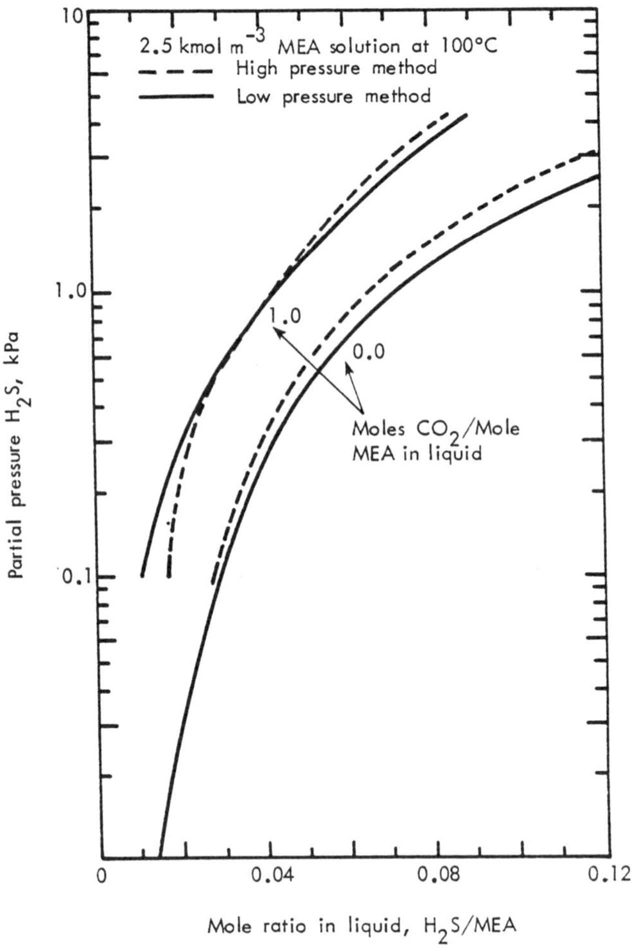

FIGURE 13.1.5. Comparison of $H_2S/MEA/H_2O$ data from high-pressure and low-pressure cells.

Volumetric methods are based on the principle of bringing a measured volume of gas in contact with a known quantity of gas-free liquid, establishing equilibrium by agitation, and measuring the volume of gas remaining. Various designs of apparatus have been used; these differ mainly in the method of providing a gas–liquid interface and the means of agitation of the vessel.

Pressure methods are based on injecting a known amount of degassed liquid into a vessel of known volume filled with gas. Equilibrium is established by agitation and the pressure decrease in the vessel is measured. From this measurement, together with knowledge of the pressure-volume-temperature behavior of the gas, the amount of gas dissolved in the liquid can be calculated.

Volumetric and pressure methods do not require chemical analysis, inasmuch as the gas phase is known to comprise only the absorbing gas, preferably saturated with solvent vapor. Gas circulation methods of the type described in Section 13.1 are also used, but these require chemical analysis of the liquid phase.

We now turn attention to the problem of determining the value of H in a reactive (chemical) solvent. These may be classified in two categories; rate methods and methods by analogy.

Rate methods rely on the measurement of the rate of absorption of the gas into the liquid under carefully controlled conditions. The following conditions need to be fulfilled:

1. The experiment must be carried out in an apparatus where the value of $k_L^0 A$ is known, where k_L^0 is the physical mass transfer coefficient and A is the total interface area.
2. The experiment must be carried out under conditions such that the absorption process is known to proceed in the slow reaction regime.

The fulfillment of condition 1 is always possible by appropriate choice of apparatus, theory, and calibration. Laboratory-scale differential absorbers will be discussed in Section 13.4. Of course, calibration must be carried out with non-reactive liquids, which may not match the physical properties of the reactive liquid to be studied. This problem can be circumvented by using non-reactive liquids having the appropriate physical properties. Since the most important property to match is viscosity, aqueous chemical solvents are well imitated by aqueous sucrose solutions which span a wide range of viscosities with increasing sucrose concentration.

Condition 2 is more difficult to fulfill and, in fact, is strictly impossible to fulfill unless the reactive solvent is expected to exhibit slow reaction behavior also in actual operation. This is hardly likely to be the case, as was discussed in Chapter 4. In fact, some form of method by analogy is always required in order to obtain useful results from experiments performed under conditions such that 2 is fulfilled.

Before discussing the methods by analogy, it is useful to recall that data on H values in reactive solutions (e.g., aqueous solutions) are best correlated in terms of the ratio of H to the value of H^0, the Henry's Law constant in the pure solvent (water for aqueous solutions). The correlations are usually cast in the general form:

$$\frac{H}{H^0} = f(\text{solution composition}) \tag{13.2.1}$$

with different correlations varying in the choice of parameters required for identifying the solution composition, and in the analytical form of the function $f(\ )$. The Setchenov equation given in Section 2.8, Eq. 2.8.1, as well as the alternate form in Eq. 2.8.3, are examples of such correlations.

Methods by analogy are based on the assumption that the form of the function $f(\ )$ in Eq. 13.2.1 is independent of at least one of the following variables: temperature, nature of the gaseous component, and nature of the liquid-phase solutes. These assumptions will be discussed in some detail below.

The solubility of O_2, N_2, and H_2 in aqueous sucrose solutions at 25°C is well correlated by the following equation:

$$\log \frac{H}{H^0} = 0.205\ C_s \tag{13.2.2}$$

where C_s is the sucrose concentration. This seems to suggest that $f(\ )$ is independent of the nature of the absorbing gas. If that is true, the value of H/H^0 for a gas in a reactive liquid can be obtained by measuring the solubility in the same reactive liquid of an inert gas. As an example, measuring the value of H/H^0 for N_2O in alkaline solution could be used to infer the value for CO_2 in the same solutions.

If the Setchenov equation (Eq. 2.8.1) is considered as acceptable and if the constant K_i is regarded as independent of the nature of the ions in solution, the value of H/H^0 for an acid gas in an alkaline solution could be inferred from measurements of H/H^0 for the same gas in a non-reactive solution of equal ionic strength (typically, a sodium chloride solution). The ionic contributions to the value of K_i listed in Table 2.8.1 have been obtained by this procedure.

Finally, suppose that the form of the $f(\ )$ function in Equation 13.2.1 is independent of temperature. It is then possible to determine the values of H/H^0 directly on the system of interest by performing a rate method experiment at a temperature low enough for the absorption process to proceed in the slow reaction regime. These values would then be assumed to be the same also at the higher tem-

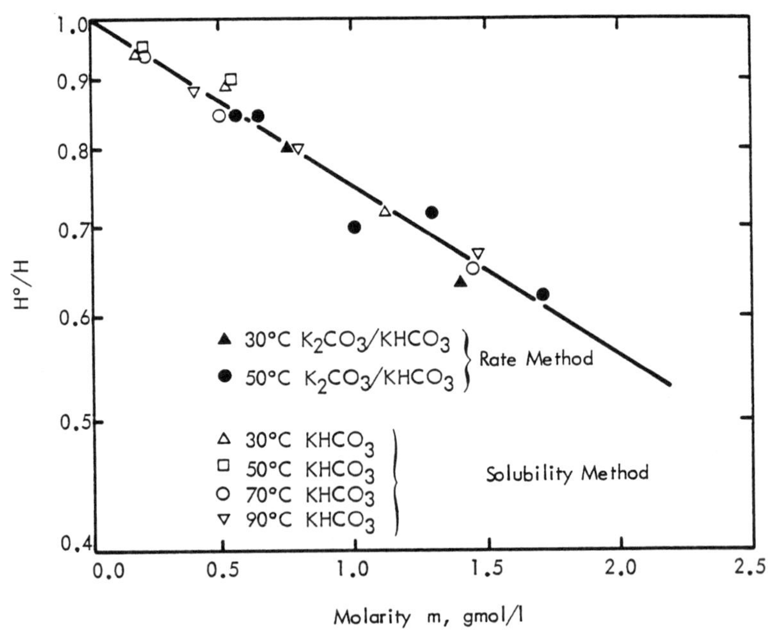

FIGURE 13.2.1. Solubility of CO_2 in aqueous solutions of potassium carbonate/bicarbonate.

perature of interest where the absorption process would *not* proceed in the slow reaction regime.

Figure 13.2.1 is a plot of data on H values for CO_2 in aqueous K_2CO_3 solutions. The full symbols refer to data obtained at temperatures low enough for a rate method to be used. The open symbols refer to data obtained with solutions of pure $KHCO_3$ (with no K_2CO_3), i.e., non-reactive solutions. Several observations can be made.

The fact that all the open symbols are correlated by the same line supports the idea that $f(\ )$ is independent of temperature. Further support for this is given by the fact that the full circles and triangles are also correlated by the same line. Finally, the fact that open and full symbols are correlated by the same line supports the idea that $f(\ )$ is independent of the nature of the liquid-phase solutes. The correlating line (Eq. 2.8.3) is interpreted as giving the value of H/H^0 for CO_2 in partly converted K_2CO_3 solutions at temperatures exceeding 50°C by a combination of rate-method data and method-by-analogy data.

13.3. OTHER THERMODYNAMIC DATA

An important thermodynamic parameter in gas treating systems is the heat of acid gas absorption and reaction in the solvent. The value of the heat of reaction determines the magnitude of thermal effects in gas treating units; it also plays a role in setting the heat load required for solvent regeneration.

Equation 1.7.12 given earlier for physical absorption applies also to dissolution of a gas in a chemical solvent. Therefore, the value of ΔH^{diss} can be obtained by plotting VLE data in an Arrhenius plot (i.e., ln p^* versus $1/T$) and measuring the slope. However, there are problems associated with this procedure.

The first problem is that values of p^* in an Arrhenius plot must correspond to the same values of m and y. Since it is unlikely that p^* data at exactly the same values of m and y but at different temperatures are available, one needs to plot values read off smoothed curves of p^* versus y. This procedure, of course, induces some loss of precision.

The second problem is that the heat of dissolution is not exactly equal to the heat of reaction inasmuch as some of the dissolved gas remains in its physically dissolved form rather than in the chemically combined one. This problem becomes relatively more important at high gas partial pressures. If Henry's Law constant is available, a correction can be made, but again this results in loss of precision.

In spite of the difficulties discussed above, most of the available thermodynamic information on enthalpy effects in gas/liquid solutions has been calculated from the temperature coefficients of measured gas solubilities; only two direct calorimetric studies of heat effects associated with acid gas absorption in chemical solvents of interest for gas treating have been published.

An isothermal gas solution calorimeter that yields results under conditions approaching constant volume and temperature has been developed. Comparison of literature data of the enthalpy of solution of CO_2 in 0.1 mol/1 aqueous NaOH

(ΔH^0 approximately 110 kJ/mol) indicated that the accuracy was in the 1-2% range.

The calorimeter is of a convenient easy to use design that provides enthalpies of solution of gases at a useful level of precision. The vessel is constructed with vacuum-tight O-ring seals that allow for the insertion of a variety of pressure transducers, control devices, and so on, into the cell (see Figure 13.3.1). During a run, the vessel is completely filled with solvent (45 cc); the gas that is injected is assumed to dissolve completely. Complicated corrections and fluctuations connected with solvent evaporation or condensation are avoided. This feature is an important part of the design. Automatic control circuitry responds to maintain a constant temperature in the vessel. The solutes (gas) are injected from a gas-tight syringe through a rubber septum that is protected from the solvent by a Teflon gasket. Injection of 1-5 cc gas is used. Solvents must be degassed on a vacuum line before use. The experiment measures the partial molar enthalpy of solution rather directly, only minor corrections being needed to the measured instrument response.

In an adiabatic calorimeter with gas bubbling through the solvent, results for CO_2 absorption in aqueous DEA were in good agreement with those obtained by differentiation of partial pressure-solubility data; both showed ΔH decreasing with increasing y. However, for H_2S absorption in aqueous DEA, the enthalpy of solution was found to increase slightly with increasing loading, in contrast with the results for CO_2. Based on these somewhat variable results, the mode of operation with gas bubbling through must be considered suspect, especially if volatile solvents are being used.

Another important parameter is the equilibrium constant for the chemical reactions of interest in gas treating. As was discussed in Chapter 2, no thermodynamic equilibrium constant can be extracted directly from VLE data. VLE data yield a combination of Henry's Law constant and one particular equilibrium constant; the value of the latter can, of course, be calculated if the Henry's Law constant is measured independently.

The technique of extracting values of equilibrium constants from VLE data can be extended as follows. A thermodynamic model is developed with at most few simplifying assumptions; this will yield p^* as a complex function of m, y, and a number of equilibrium constants. A nonlinear regression analysis on the VLE data can then be used to deduce the values of the equilibrium constants. The procedure is unlikely to yield anything better than reasonable estimates.

Measurement of pH as a function of liquid composition can also be used to infer values of equilibrium constants. It should be borne in mind, however, that pH meters may give spurious readings in concentrated solutions.

The best technique for measuring equilibrium constants is, of course, that of actually measuring the liquid-phase composition at equilibrium. Standard analytical techniques are generally not applicable to the systems of interest in gas treating but nuclear magnetic resonance (NMR) techniques are often useful.

Carbon-13 NMR spectroscopy utilizes a signal from the naturally occuring C^{13} atoms (about 1% of total carbon) to identify quantitatively the species in solution: bicarbonate ion, carbonate ion, free amine, carbamate ion, and protonated amine.

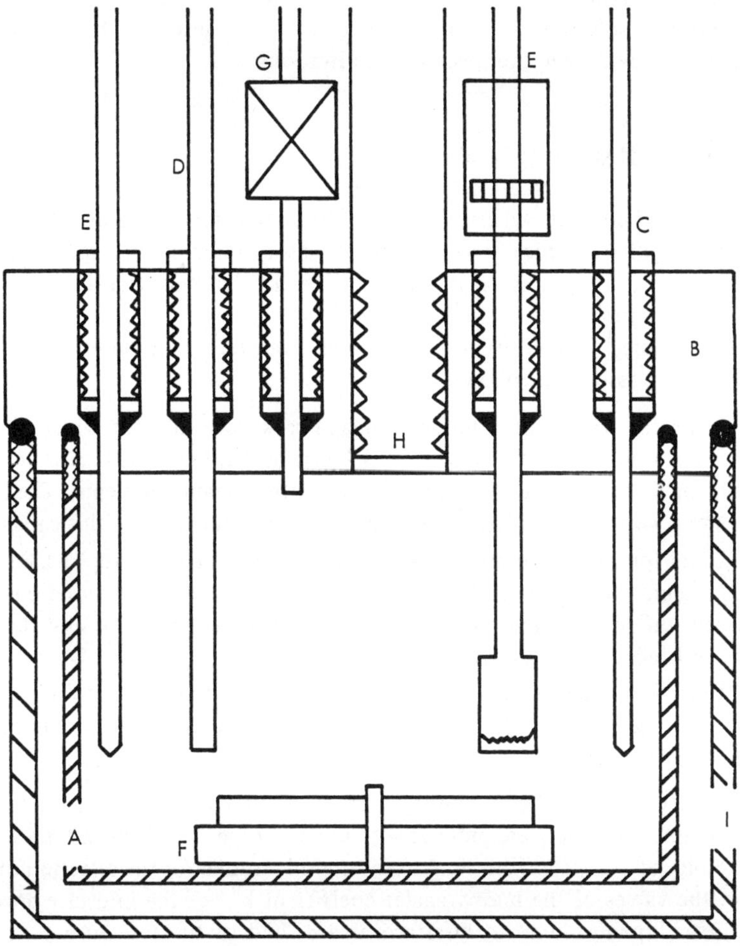

A. Reaction vessel; B. Cap with vacuum feed-throughs; C. Peltier cooler-resistance assembly; D. Calibrating resistor; E. Thermistors; F. Magnetic stirrer; G. Valve for solvent introduction; H. Septum and holder; I. Outer nylon jacket.

FIGURE 13.3.1. Isothermal gas–liquid calorimeter.

C^{13} atoms in the sample are first excited and the response is analyzed by Fourier Transform techniques. The method has the advantages that it can be applied easily to concentrated solutions at temperatures of industrial interest. Equilibrium constants are calculated from the species concentrations.

13.4. MASS TRANSFER DATA

The goal of experiments where the rate of mass transfer is measured is to understand the physico-chemical mechanism of absorption-reaction and desorption-reaction. For such a goal to be attained, it is crucial that mass transfer experiments be of the differential type.

Several classical laboratory units for mass transfer experiments are available. These may be classified as follows:

1. Units for which the interfacial area A is known and for which the value of the physical mass transfer coefficient k_L^0 can be estimated with satisfactory accuracy from the solution of the appropriate hydrodynamic problems. The laminar jet, the short wetted-wall column, and the one-sphere unit fall into this category.

2. Units for which the interfacial area is known, but the mass transfer coefficient k_L^0 cannot be predicted from first principles. The several variations of the stirred cell, the long wetted-wall column, the string-of-spheres unit and the string-of-disks unit fall into this category.

3. Units for which neither the interfacial area nor the mass transfer coefficient are known. The sparged cell, the single sieve tray, and the single bubble cap plate fall into this category.

The choice of a specific laboratory unit is based on a number of considerations; in general, units of type 1 are preferable to units of type 2, and the latter are preferable to units of type 3. Other important considerations for the selection of a unit include: the values of the mass transfer coefficient k_L^0, of the interface area A, of the ratio of liquid volume to interface area ϕ, and of the residence time of the liquid t_R. In some units it is possible to vary these quantities over a certain range by varying the operating conditions.

It is always sound policy to calibrate the unit, i.e., to measure the absorption rate of a gas in a non-reactive liquid in which the solubility of the gas is known. The calibration yields the value of the product $k_L^0 A$ for the unit. With units of type 1 and 2, the value of k_L^0 is obtained from the calibration since A is known.

The calibration should be carried out with a system that matches as closely as possible the physical properties of the reactive system one wishes to investigate. Usually the most important property to match is the viscosity in the case of units of type 1 and 2.

For units of type 1, the calibration is required to check the validity of the theoretical equation for k_L^0. A check within a factor as large as two is often acceptable. The value obtained in the calibration should be used in the subsequent analysis. For these units, calibration does not need to be carried out on a system with closely matching physical properties inasmuch as the influence of the latter on the value of k_L^0 can be predicted from the theoretical equation for k_L^0. For units of type 2, the calibration is indispensable in order to know the value of k_L^0, which is often needed in the subsequent analysis.

For units of type 3, calibration is not very meaningful as it does not yield the value of k_L^0, but only of the product $k_L^0 A$. The interfacial area A may change very significantly with minor changes of the interfacial properties of the gas–liquid pair (not only the surface tension, but also the derivative of the latter with respect to concentration) that strongly influence the value of A. Therefore, for units of type 3, the calibration at best offers an estimate for the actual value of $k_L^0 A$ for the reactive system one wishes to investigate.

The principal characteristics of the units quoted above are reviewed briefly in the following paragraphs.

The laminar jet absorber sketched in Figure 13.4.1 is possibly the most flexible of the units of type 1. By varying the jet length Z_j (which is easily accomplished with a movable receiver) and the liquid flowrate L, both the physical absorption coefficient and the interfacial area can be changed. Furthermore, k_L^0 is independent of the physical properties of the liquid and, therefore, calibration is very easy.

The hydrodynamics involved are extremely simple when an idealized rod-like cylindrical jet is considered. The distribution function discussed in Section 3.2 has the form given in Eq. 3.2.25 and t^* is given by

$$t^* = \frac{\pi d_j^2 Z_j}{4L} \tag{13.4.1}$$

with d_j the jet diameter. The interface area and physical mass transfer coefficient are given by

$$A = \pi d_j Z_j \tag{13 4.2}$$

$$k_L^0 = \frac{4}{\pi d_j} \sqrt{\frac{DL}{Z_j}} \tag{13.4.3}$$

Equations 13.4.2 and 13.4.3 show that the product $k_L^0 A$ and hence the physical absorption rate is independent of the jet diameter.

Departures from ideal behavior are due to entrance effects (the velocity profile at the nozzle is not flat), to gravity effects (the vertical jet accelerates and shrinks under the influence of gravity), and to exit effects (ripples and surface stagnancy induced by the jet receiver). The interface area of laminar jets ranges from 0.3–10 cm^2 and the diffusion time from 0.01–0.1 sec. The theoretical predictions of physical-absorption rate in laminar jets are well confirmed experimentally when well designed jets are used.

When a wetted-wall column such as sketched in Figure 13.4.2 is used, the interface area cannot be varied, but a range of interface areas can be realized with dif-

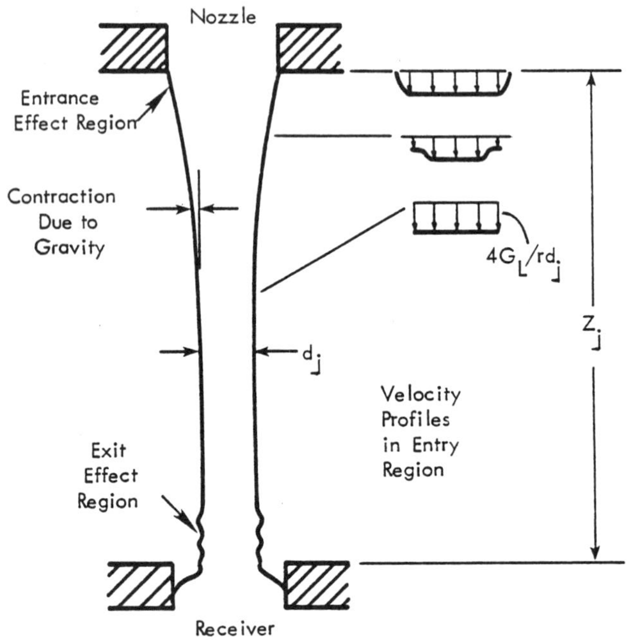

FIGURE 13.4.1. Sketch of the laminar jet absorber.

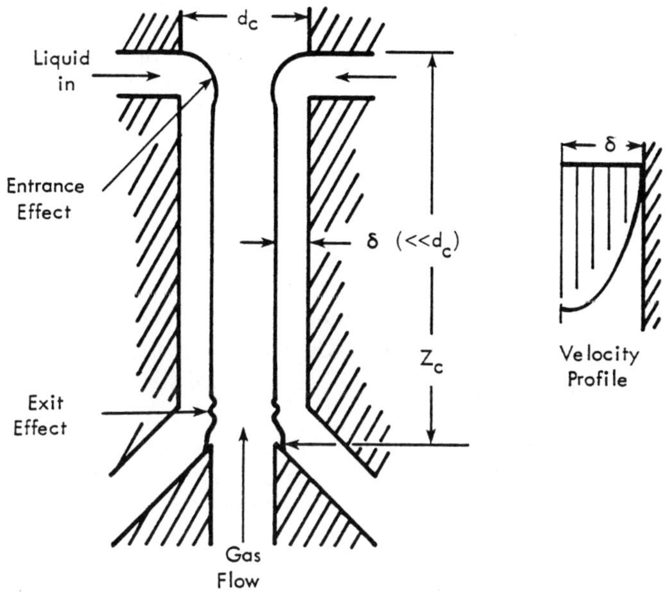

FIGURE 13.4.2. Wetted-wall column absorber.

ferent columns. The physical-absorption coefficient can be varied by varying the liquid flowrate.

The hydrodynamics involved in the ideal case are rather simple. If d_c is the column diameter and z_c the column length, the width of the liquid film is given by

$$\delta = \left[\frac{3v_L G_L}{\pi d_c g}\right]^{1/3} \tag{13.4.4}$$

The surface velocity of the liquid is $\frac{3}{2}$ the average velocity; the distribution function is again given by Eq. 3.2.25, with

$$t^* = \frac{2}{3} Z_c \left[\left(\frac{\pi d_c}{L}\right)^2 \frac{3\mu}{\rho g}\right]^{1/3} \tag{13.4.5}$$

where μ and ρ are the liquid viscosity and density.

The absorption coefficient and interface area are thus very easily calculated. It is important to point out that k_L^0 is proportional to the cube root of the liquid flowrate, in contrast with the laminar jet where it is proportional to the square root of L. The proportionality of k_L^0 with $L^{1/3}$ is expected whenever a laminar falling film of liquid is the absorbing medium.

Although (as shown above) the hydrodynamics involved are, in principle, rather simple, entrance effects, ripples, surface renewal, and exit effects may be very important. The interface area for wetted-wall columns ranges from 10–100 cm^2 and diffusion times from 0.1–1 sec. The theoretical predictions of physical-absorption rate are well confirmed only when short and extremely well designed wetted-wall columns are used. In long wetted-wall columns, measured absorption rates are always significantly larger than predicted. That is why long wetted-wall columns should be regarded as type 2 units.

For the one-sphere absorber sketched in Figure 13.4.3, an exact solution of the hydrodynamic equations is not available. However, an approximate solution is available that results in satisfactory accuracy. The mass transfer coefficient is given by

$$k_L^0 = 0.583\sqrt{D}\left(\frac{gL^2\rho}{\mu d_s^5}\right)^{1/6} \tag{13.4.6}$$

with d_s the sphere diameter, whereas the interface area is given by

$$A = \pi d_s^2 \tag{13.4.7}$$

The one-sphere absorber is easy to operate and measured values of k_L^0 are, in general, close to the theoretical value given in Eq. 13.4.6. The interface area ranges from 10–40 cm^2 and the diffusion time from 0.1–1 sec. A one-sphere absorber/desorber unit suitable for experiments at high temperature (up to 110°C) and high pressure (up to 3 atm) has been developed.

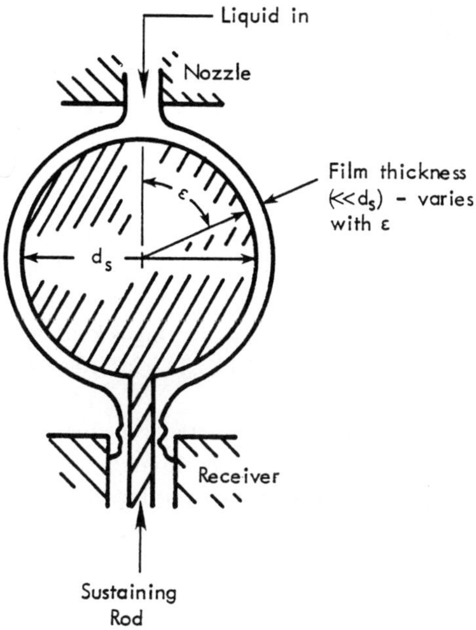

FIGURE 13.4.3. One-sphere absorber.

The units of type 1 discussed above have certain disadvantages. First of all, they have little flexibility. The interface area can be varied only with the laminar jet, and not over a wide range. If the jet is too short, end effects become too important; if it is too long, operation of the receiver becomes difficult. The mass transfer coefficient k_L^0 can be changed only over a limited range, due to the restricted range of possible values of L; in the wetted-wall and one-sphere absorbers, a lower bound on L is imposed by the requirement of completely wetting the entire solid surface, whereas an upper bound is imposed by the need to avoid excessive rippling and turbulence. In the laminar jet, a lower bound is imposed by the requirement of avoiding dripping and an upper bound again by the need to avoid turbulence. The residence time of the liquid and the liquid volume-to-interface area ratio, again can be varied only over a very limited range; the latter, in fact, not at all in the laminar jet. These shortcomings occasionally force the selection of a type 2 unit. Type 2 units are briefly described in the following paragraphs.

Basically, three types of stirred cells have been used as absorber/desorber laboratory units. The standard configuration is one where the liquid phase impeller is totally immersed and the rate of stirring is kept below the limit at which gas bubbles would be drawn into the liquid. The interface area is simply the geometrical area of the exposed liquid surface.

In another version of the stirred cell, mechanical stirrers break the gas–liquid interface. In a third version, a perforated plate sits right at the gas–liquid interface; interfacial areas may be varied by varying the hole area of the perforated plate.

The stirred cells exhibit three advantages over type 1 cells. The mass transfer coefficient k_L^0 can be varied over a fairly wide range by varying the rate of stirring. The residence time of the liquid can be set at will by regulating the liquid flowrate. Finally, raising the level of the gas–liquid interface allows the liquid volume-to-interface area ratio to be varied.

The long wetted-wall column, the string-of-spheres unit sketched in Figure 13.4.4, and the string-of-disks unit sketched in Figure 13.4.5 have very similar overall characteristics. The interface area is known, can be fixed at will by fixing the length of the unit, and is rather large; however, it cannot be varied on any given unit. The mass transfer coefficient k_L^0 is not known; it can be varied over a fairly wide range by varying the liquid flowrate, since it is proportional to L raised to a power larger than $\frac{1}{3}$. This is because as L is increased, not only does the interfacial velocity of the liquid increase (as in a short wetted-wall column or a one-sphere absorber), but also the remixing of the liquid at sphere or disk junctions or at random spots in the long wetted-wall unit becomes more vigorous. For the string-of-spheres and string-of-disks units, an upper limit is reached when mixing at the sphere or disk junctions is complete; in that case, the value of k_L^0 equals that on a single sphere or single disk. The residence time of the liquid is large in these units but can be varied only over a very limited range. The liquid volume-to-interface area ratio is small and can be varied only over a very limited range.

Before discussing units of type 3, which are used for different purposes, we now turn attention to the mode of operation of laboratory units of type 1 or 2. We first

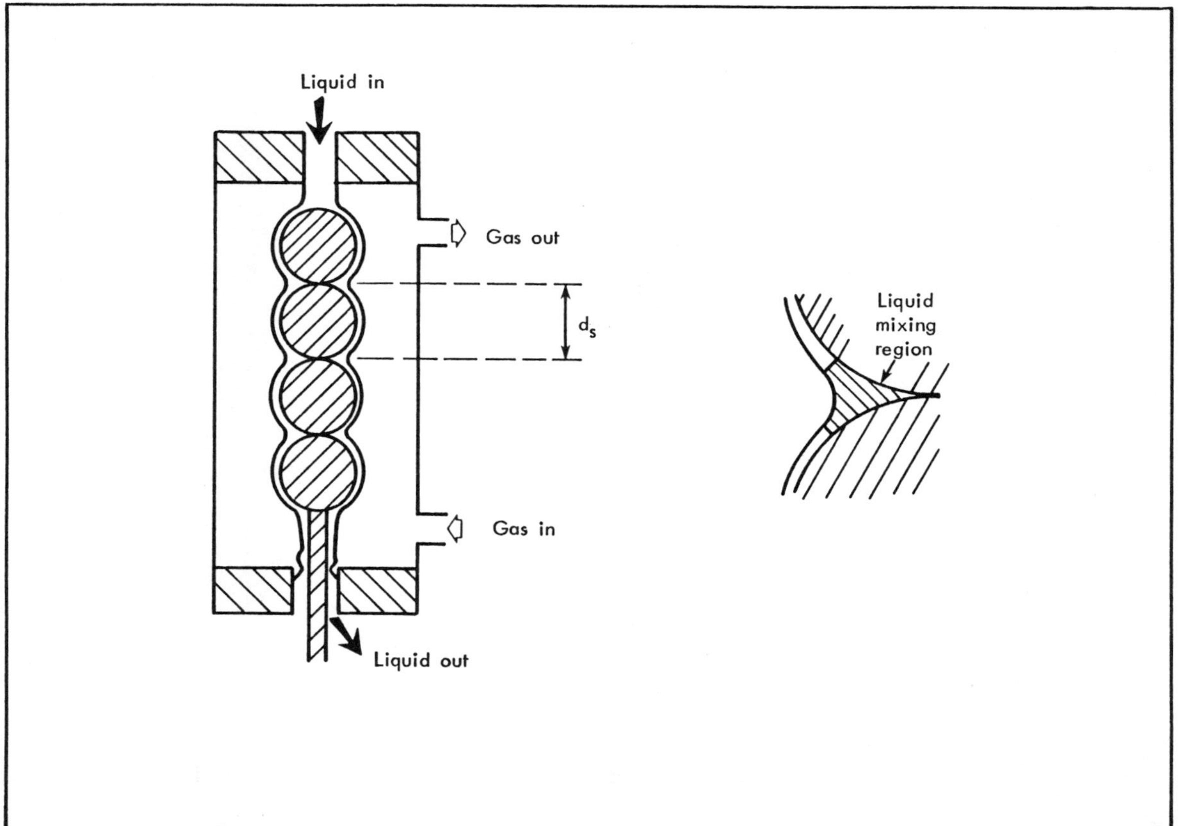

FIGURE 13.4.4. Multiple-sphere absorber.

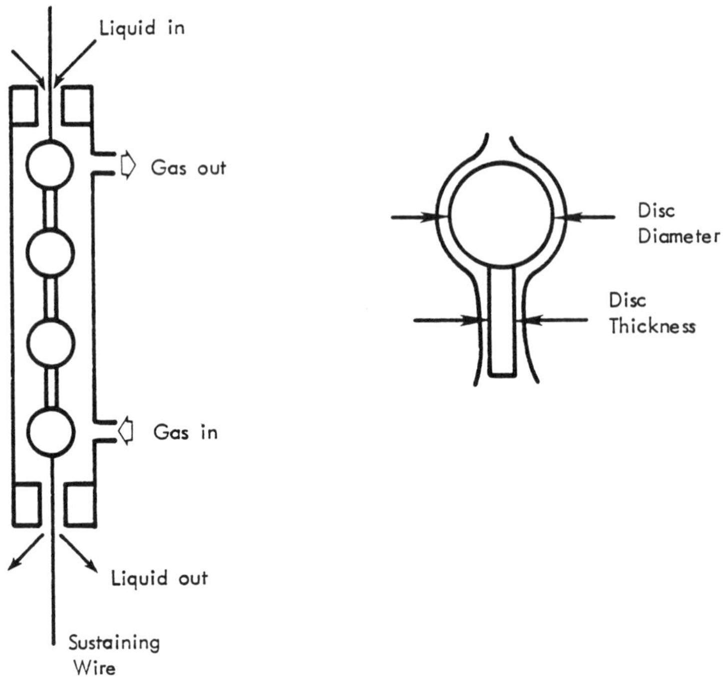

FIGURE 13.4.5. Disc-column absorber.

consider the simple case of absorption of a single gas in a reactive liquid. For this case, the best technique for measuring the mass transfer rate is a volumetric one. The technique is illustrated schematically in Figure 13.4.6.

A gaseous stream consisting of only the gas to be absorbed is forced to flow through the unit with the gas purge valve left open and the release valve closed. After sufficient time has elapsed to make sure that any inert gas originally in the unit has been flushed out, the gas purge valve is closed and the release valve is opened. At this point, the flowrate of gas at the VFM point equals the volumetric rate of absorption and it can easily be measured. In addition to standard measuring techniques of volumetric gas flowrate such as the wet test meter, a very precise and easy technique is the soap film meter sketched in Figure 13.4.7.

In the soap film meter, the gas is forced to flow through the graduated glass cylinder. The rubber container filled with a soap solution is squeezed until the soap solution reaches point A and then released. A soap film will form at point A and move along the glass cylinder forced by the gas flow. Measurement of the time taken by the soap film to travel between two marks on the glass cylinder will yield the volumetric flowrate. At the top of the glass cylinder, point B, the soap film will break (because it is suddenly accelerated) and the soap solution will fall back along the walls of the glass cylinder keeping them wet for successful operation of the next measurement. By choosing the diameter of the glass cylinder appropriately, a very wide range of gas flowrates can be measured with great accuracy.

If the operating pressure is atmospheric and the gas is a non-toxic one (say CO_2) both the purge and the release valve can be simply vented to the laboratory. In all other cases, they need to be vented to a gas container that is kept at the constant pressure of the experiment.

The analysis of the data obtained is straightforward. The measured absorption rate $\overline{V}$ is related to the mass transfer coefficient k_L by the following equation:

$$\overline{V} = v k_L A(a_i - a_0) \tag{13.4.8}$$

where v is the molar volume of the gas and A is the interface area. The driving force $a_i - a_0$ can be expressed as

$$a_i - a_0 = \frac{p_i - p^*}{\mathrm{H}} \tag{13.4.9}$$

where p^* is the equilibrium vapor pressure corresponding to the (known) bulk-liquid composition and p_i is the interface partial pressure of the absorbing gas. The value of p_i is calculated as the difference between the total pressure and the solvent vapor pressure. Therefore, the value of k_L can be calculated provided enough thermodynamic information–values of p^*, H, and of the solvent vapor pressure–is available. It is important to realize that the analysis as sketched above does not

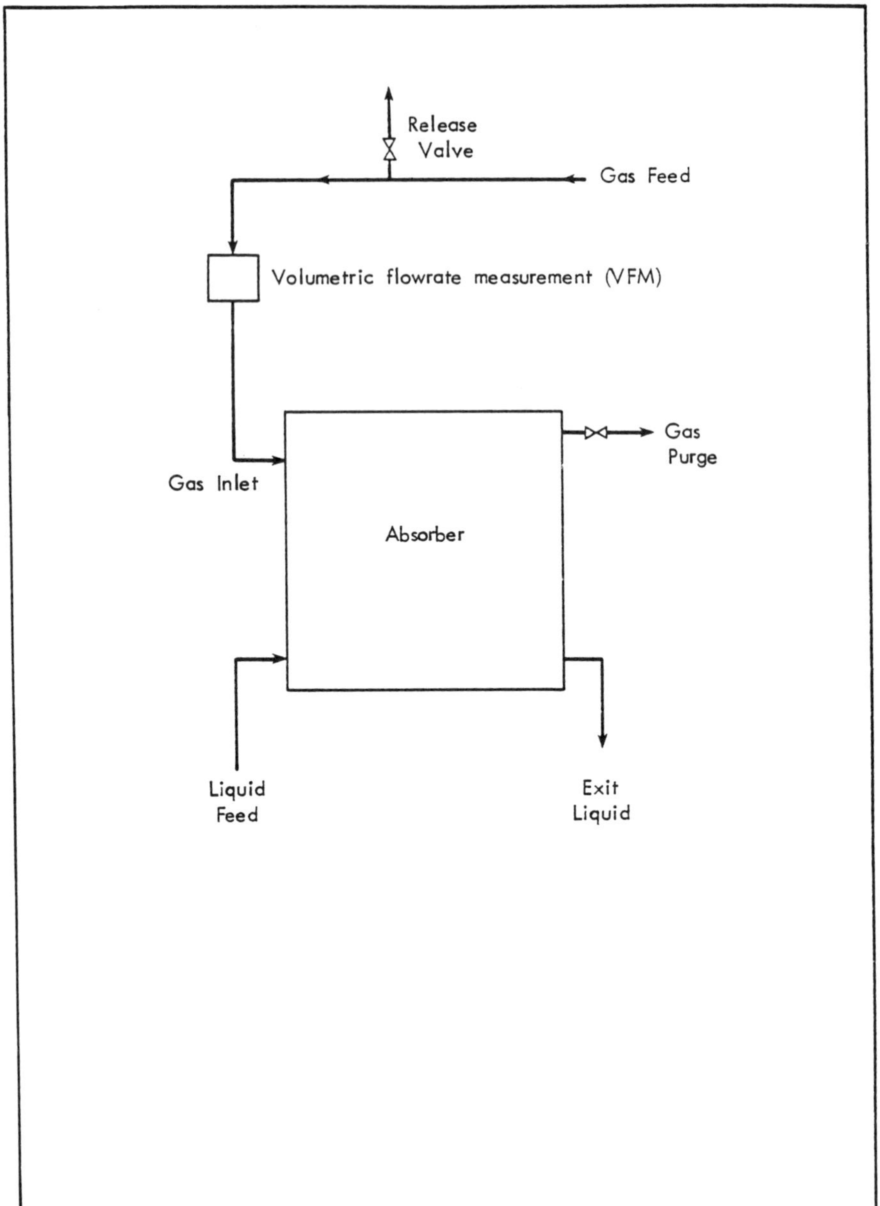

FIGURE 13.4.6. Schematic of operation for absorption of a single gas in a reactive liquid.

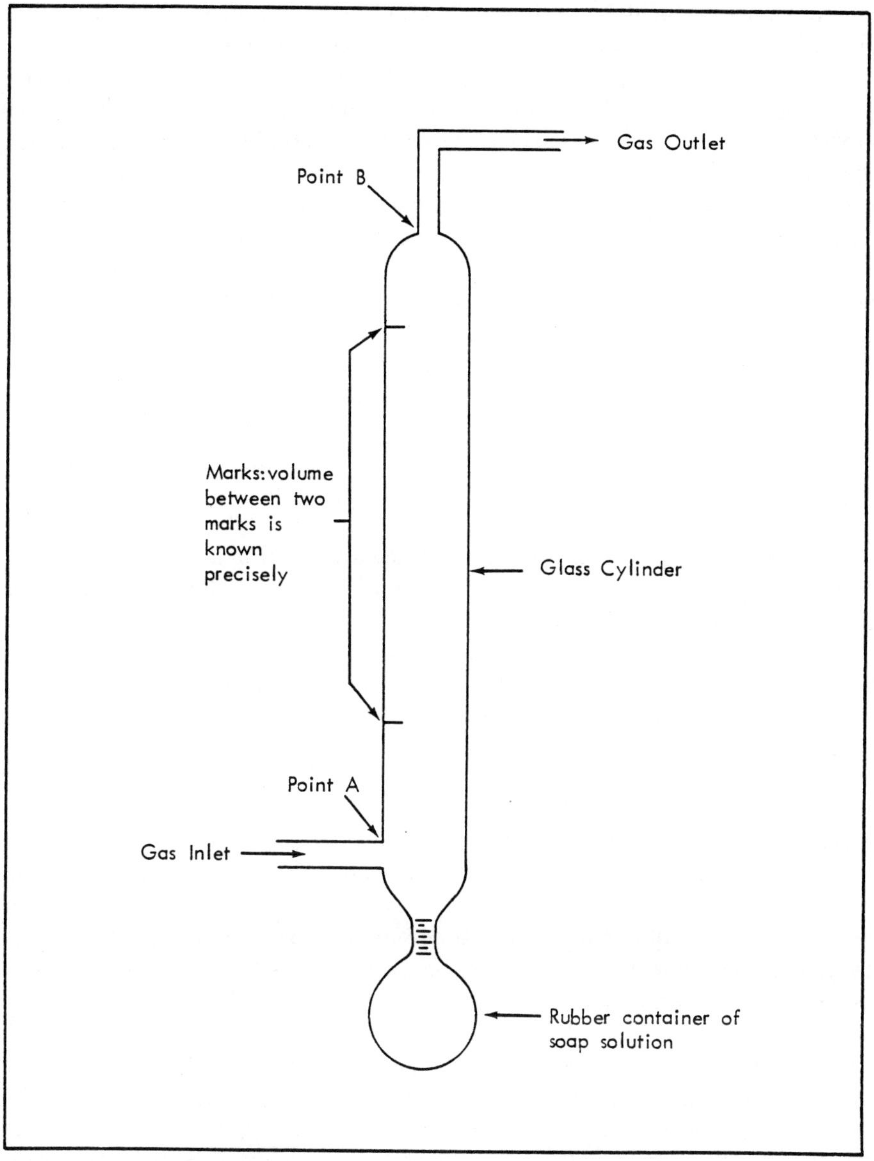

FIGURE 13.4.7. Schematic of soap film meter.

require knowledge of the mass transfer coefficient in the gas phase. This offers the great advantage that control of the gas-phase mixing in the unit is not required.

Once the value of k_L has been determined, it can be compared with the value predicted by a theoretical model of the type discussed in Chapters 3-8. Since such models yield a prediction for the enhancement factor I (rather than for k_L directly), the value of k_L^0 needs to be known. In this regard, a word of caution is needed. It has been reported in the literature that occasionally interfacial turbulence develops during absorption of a gas in a reactive solvent. When such is the case, the actual value of k_L^0 for the reactive solvent may be appreciably larger than that which has been measured during calibration of the unit. Interfacial turbulence during chemical absorption of CO_2 in MEA solutions has been detected by simultaneously desorbing an inert tracer gas, propylene. The high value of k_L^0 based on the measured propylene desorption rate was used in the analysis of chemical absorption and gave a good fit with the theoretical model for rate enhancement although the fit was bad when the calibration value of k_L^0 was used. The technique of simultaneous desorption of an inert tracer gas may be regarded as simultaneous calibration of the unit.

For single gas absorption experiments using a volumetric measuring technique, units of type 1 are clearly preferable. The only limitation is that experiments become very difficult, if not impossible, at low partial pressures of the absorbing gas, inasmuch as the total pressure in the unit cannot be too low.

Desorption experiments, even of a single gas, present many more difficulties than absorption experiments. First the temperature of a desorption experiment is generally significantly higher than that of an absorption experiment. A volumetric technique of measuring the rate of mass transfer is not possible as it would require the total pressure of the unit to be less than the desorbing gas equilibrium vapor pressure corresponding to the bulk-liquid composition. If such a low pressure were imposed on the system, the liquid phase would boil, making the experiment impossible.

It follows from this argument that a desorption experiment must be carried out by continuously flowing an inert stripping gas through the unit. The measurement of the rate of mass transfer must then be based on chemical analysis of either the gas or the liquid phase. A flowsheet of a desorption unit is given in Figure 13.4.8.

Consider the analysis of the exit gas phase. Assume that the partial pressure p of the desorbing gas is measured in the exit gas; in the inlet gas, p is zero.

Ideally, one would like the mass transfer resistance in the gas phase to be negligible and the gas phase in the unit to be perfectly well mixed. If these two conditions are satisfied, the measured value of p is also the value in the bulk of the gas everywhere in the unit (since the gas is perfectly well mixed) as well as at the gas liquid interface (since there is no resistance to mass transfer in the gas phase).

The volumetric desorption rate is calculated as

$$\overline{V} = G\frac{p}{\pi} \tag{13.4.10}$$

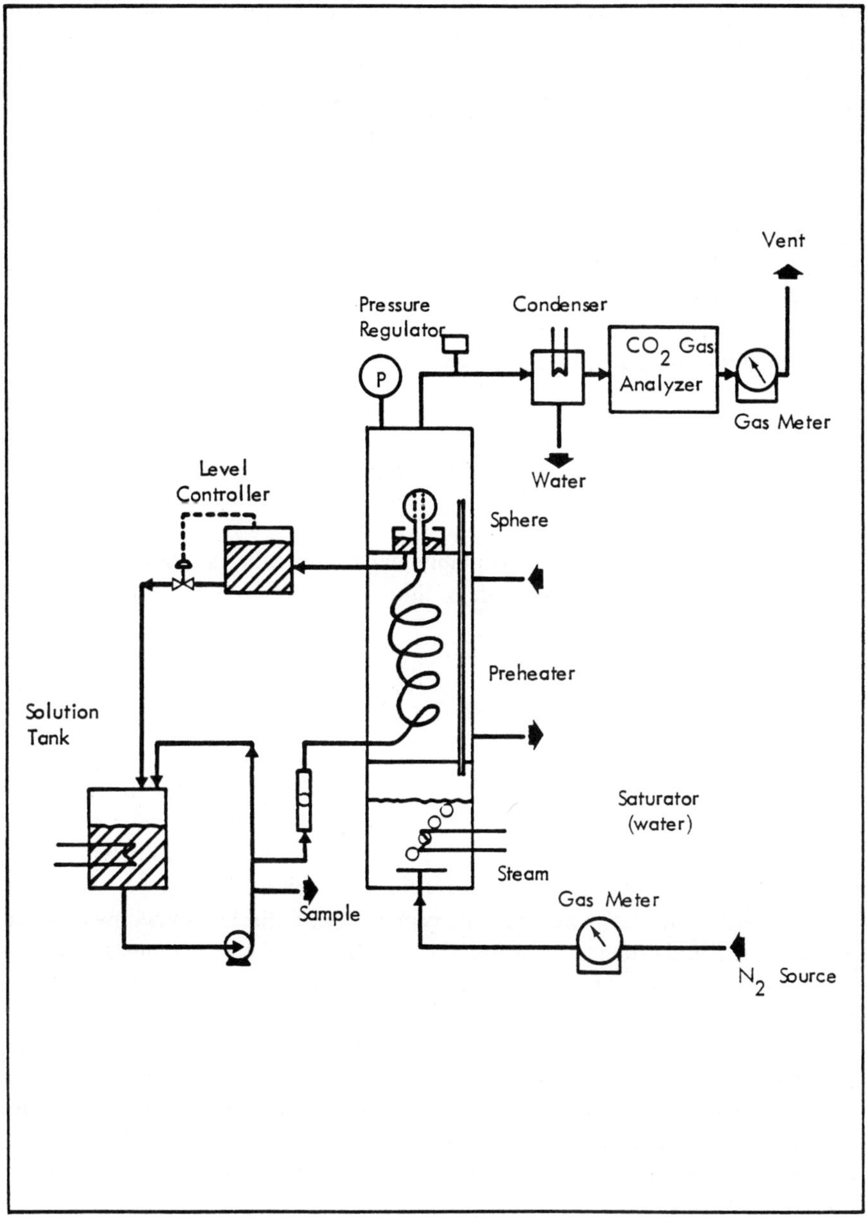

FIGURE 13.4.8. Apparatus for studying desorption at high temperature on a one-sphere unit.

where G is the exit gas volumetric flowrate and π the total pressure in the unit. Since $p = p_i$, combination of equations analogous to 13.4.8 and 13.4.9 yields

$$\frac{Gp}{\pi} = \upsilon k_L A \frac{(p^* - p)}{H} \tag{13.4.11}$$

Notice that the structure of Eq. 13.4.11 is such that any error in the measurement of p will result in a larger error in the evaluation of k_L.

If the mass transfer resistance in the gas phase is not negligible, the comparatively simple analysis discussed above is not applicable. Equation 13.4.10 still applies but $p \neq p_i$ and, therefore, Eq. 13.4.11 does not hold. If the value of k_G is known, the following equation can be used to calculate k_L:

$$G\frac{p}{\pi} = \upsilon k_L A \frac{p^* - p}{H + \dfrac{k_L A}{k_G A}} \tag{13.4.12}$$

Equation 13.4.12 leads to unacceptably high errors if the gas-phase resistance to mass transfer is predominant; as a rule of thumb, the following condition should be satisfied:

$$k_G > \frac{k_L}{H} \tag{13.4.13}$$

Very large rate enhancement factors in desorption are, therefore, difficult to measure.

If the gas phase is not perfectly well mixed, meaningful data can be obtained only if $p \ll p^*$, so that the actual value of p does not need to be known in order to evaluate the right-hand side of either Eq. 13.4.11 or Eq. 13.4.12.

Now consider the case where the rate of mass transfer is determined by chemical analysis of the liquid phase. A differential unit is one where the liquid-phase composition is practically constant throughout the unit. Therefore, unless the liquid phase is perfectly well mixed, the composition of the exit liquid will be almost equal to that of the inlet liquid and chemical analysis will not be a feasible technique for determining the mass transfer rate. This difficulty does not arise with a stirred cell where the liquid phase is perfectly well mixed and has everywhere the same composition as the exit liquid. The composition of the inlet liquid to a stirred cell may be made to be quite different from that in the exit liquid by keeping the liquid flowrate low.

If a differential unit other than a stirred cell is used, chemical analysis of the liquid phase can be adopted with the flowsheet design sketched in Figure 13.4.9. Although the difference in composition between point B and point A may be too small to be measured with accuracy, the composition at point C will slowly change in time and if steady operation is maintained, the mass transfer rate can be determined by periodic sampling at point C. This closed loop flowsheet configuration

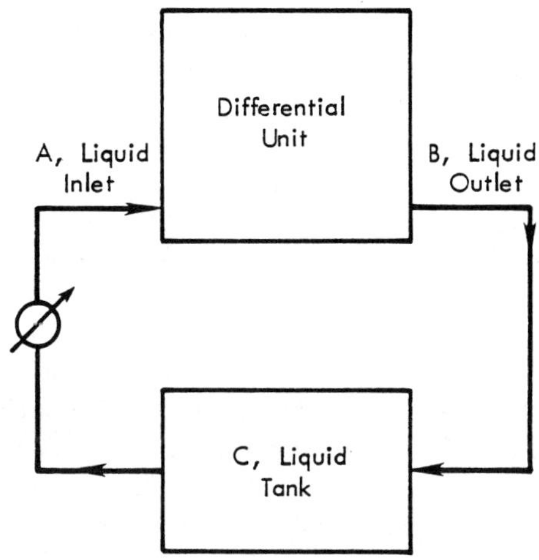

FIGURE 13.4.9. Differential unit operating in the batch recycle mode.

has the additional advantage of allowing a range of liquid-phase compositions to be scanned easily.

The most difficult mass transfer data to obtain are those of simultaneous mass transfer of two gases. All the difficulties encountered in desorption experiments (except the high temperature) are encountered again in simultaneous mass transfer. In addition to these, other problems arise. The first is simply related to the fact that chemical analysis for two species is more difficult than for only one species. The second problem is that analysis of the gas phase runs into the difficulties discussed for liquid-phase analysis in the case of desorption inasmuch as a *difference* in composition between inlet and outlet gas would need to be determined. The only satisfactory approach is to have a well mixed gas phase in the unit, since closed-loop operation on the gas phase, though conceivable, would lead to a variety of problems. The third problem that arises is that, in the liquid-phase closed loop flowsheet configuration, the composition of the liquid phase will, in general, *not* follow the path of main interest and, therefore, scanning of the appropriate composition path requires a large number of separate experiments. A flowsheet of a simultaneous absorption unit is given in Figure 13.4.10.

We now turn attention to laboratory mass transfer units of type 3. Since the value of k_L^0 for such units cannot be determined, they can rarely be used for testing theoretical models of the type discussed in Chapters 3–8. The exception is the case where the mass transfer process is known to proceed in the instantaneous reaction regime, inasmuch as the analysis in this case does not require that the values of k_L^0 and A be known separately. Units of type 3 are useful.for quick comparative studies of different chemical solvents.

A simple batch absorber, which gives comparative measurements of absorption rate, absorption capacity, and desorption rate is sketched in Figure 13.4.11. In this unit, neither k_L^0 nor A are known. However, results with known solvents show that the apparatus gives valid comparisons for CO_2 absorption in aqueous aminoalcohol and aminoalcohol-promoted carbonate solutions.

The mode of operation is as follows. CO_2 is absorbed and desorbed quickly (15–60 min) by providing a large interfacial area for gas–liquid contact. This is achieved by continuously drawing off liquid and recirculating it through a sparger in the vapor space of the absorber, thus creating a spray and froth. In addition to this, CO_2 gas is admitted to the vessel through a second sparger located beneath the liquid surface, thus generating bubbles of gas in the liquid. This highly agitated setup has many features of a plate contactor.

At the beginning of an experiment, a batch of fresh solution is fed to the unit and a first test of absorption is carried out. The gas phase consists of only the gas to be absorbed; the pressure is of the order of the partial pressure of the absorbing gas in the gas feed to be treated industrially and the temperature is the one that one expects at the bottom section of the industrial absorber. This first absorption cycle is carried out until the absorption rate drops to zero; from the total amount absorbed, one can calculate the highest possible conversion y_{max} of the solvent in the industrial unit.

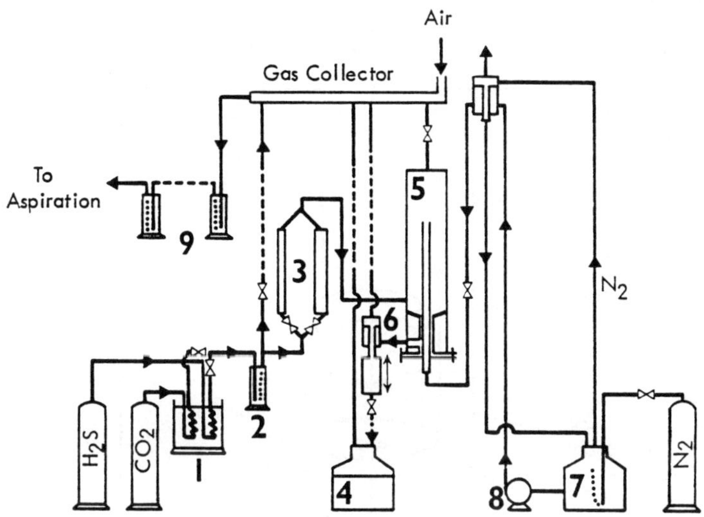

1-Preheater (to balance Joule-Thomson effect); 2-Presaturator; 3-Soap-film meters; 4-Exit liquid tank; 5-Absorber; 6-Lute vessel and liquid flow rate measure; 7-Liquid feed reservoir; 8-Pump; 9-H_2S scrubbers.

FIGURE 13.4.10. Wetted-wall column apparatus for studying the simultaneous absorption of H_2S and CO_2 in alkaline liquids.

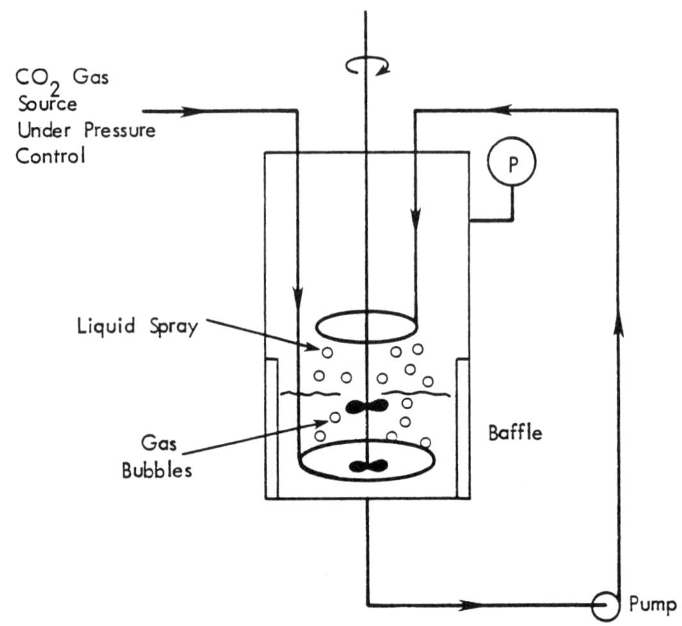

FIGURE 13.4.11. Batch absorber suitable for rapid comparisons.

Once the first cycle is completed, a desorption test is run by bringing the temperature to the value corresponding to the reboiler for the commercial unit and stripping the solution with inert gas. The test is stopped when the desorption rate drops to zero. At this point, the liquid has the composition expected for the lean solution in the industrial unit. A new cycle of absorption is now performed and the total amount absorbed is a measure of the cyclic capacity of the solvent. The lengths of the desorption and reabsorption tests give an indication of the rate behavior of the solvent.

Other units of type 3 that have been discussed in the literature are not designed for solvent screening purposes but as actual laboratory-scale mockups of industrial units. Occasionally, type 2 units have also been used as mockups for industrial-size equipment. This type of experiment is perhaps best considered as a pilot-plant type of experiment, performed on a scale small enough to fit into the laboratory. Indeed, laboratory-scale experiments on packed towers of sufficiently small size have been performed.

13.5. ANALYSIS OF RATE DATA

The main purpose of gathering mass transfer rate data in the laboratory is to understand the chemical mechanism and the corresponding kinetics well enough to allow the development of a reliable model of the rate enhancement phenomenon. From such a model, the enhancement factor I can be calculated as a function of temperature and composition of both gas and liquid phases. This function forms the basis for the design procedures to be discussed in Chapter 15.

When a set of rate data is obtained, the first step in the analysis is to ascertain which regime of mass transfer applies to the data. The quantity measured is the total mass transfer rate $\overline{V}$; the chemical mass transfer coefficient k_L can be extracted from the data by the procedures discussed in Section 13.4. The key to the identification of the regime is the dependency of $\overline{V}$ on the operating variables. The latter are: the degree of mixing as represented by k_L^0, the interface area A, the liquid volume V, temperature, and the physical driving force $a_i - a_0$.

The dependency of $\overline{V}$ on the operating variables has been discussed in Chapters 3-8; the results are summarized in Table 13.5.1. The entries in the table are sufficiently different from one regime to another to permit identification of the regime of the available set of data (of course, data may fall in a transition region).

Once the first step of the analysis has been completed and the regime has been identified, the data can be scrutinized to gain insight into the chemical mechanisms involved. The kind of information that can be obtained is briefly reviewed in the following paragraphs.

Consider the case where the data indicate fast reaction regime behavior. Substitution of Eq. 5.2.13 and 13.4.9 into 13.4.8 yields

$$\overline{V} = v\sqrt{Dk_0}\, A\, \frac{p_i - p^*}{H} \tag{13.5.1}$$

TABLE 13.5.1. Dependency of $\overline{V}$ on Operating Variables (Rate of Reaction First-Order with Respect to Transferring Component)

Regimes:	Slow			Instantaneous	
Operating Variables	Kinetic	Diffusional	Fast	Large driving force	Small driving force
k_L^0	i[a]	$\propto k_L^0$	i	$\propto k_L^0$	$\propto k_L^0$
A	i	$\propto A$	$\propto A$	$\propto A$	$\propto A$
V	$\propto V$	i	i	i	i
$\lvert a_i - a_0 \rvert$	$\propto \lvert a_i - a_0 \rvert$	$\propto \lvert a_i - a_0 \rvert$	$\propto \lvert a_i - a_0 \rvert$	Almost i	$\propto \lvert a_i - a_0 \rvert$
Apparent activation energy for k_L	Energy of activation of reaction	Very small	$\frac{1}{2}$ energy of activation of reaction	Very small	Heat of reaction
	$k_L \ll k_L^0$	$k_L = k_L^0$	$k_L \gg k_L^0$	$k_L \gg k_L^0$	$k_L \gg k_L^0$

[a]Independent.

The value of the equivalent kinetic constant k_0 can, therefore, be extracted from the data provided the VLE behavior is known (so that p^* is known), as well as the value of H. The value of k_0 depends on the liquid-phase composition; in the nomenclature of Section 7.2, we are considering the likely case where Eq. 7.2.14 holds and, therefore,

$$k_0 = g(b_{10}, \ldots, b_{N0}) \tag{13.5.2}$$

If a thermodynamic model is available, the values of the bulk-liquid concentrations $b_{10}, \ldots, b_{N0}$ can be calculated as functions of the molarity m and the degree of saturation y. Therefore, the experimentally determined dependency of k_0 on m and y should yield the form of the $g(\cdot)$ function, from which the chemical mechanism and the true kinetic constant can be obtained. An example of this procedure was discussed in Section 5.4; the results are shown in Figure 5.4.1. It is interesting to note that values of a true kinetic constant in excess of 10^6 liter/gmol sec have been determined with this procedure; it would be extremely difficult to measure such large kinetic constants by any direct kinetic technique.

Next consider the case where the data fall in the instantaneous reaction regime. The equation for $\overline{V}$ becomes in this case

$$\overline{V} = vI_\infty k_L^0 A \frac{p_i - p^*}{\mathrm{H}} \tag{13.5.3}$$

from which values of I_∞ can be calculated and compared with predictions of models such as those discussed in Chapter 6.

It is often useful to make the tentative assumption that the diffusivities of all solutes are equal to each other. If that is the case, Eq. 3.5.19 can be used for I_∞ and Eq. 13.5.3 reduces to

$$\overline{V} = \upsilon k_L^0 A(\alpha_i^* - \alpha_0) \tag{13.5.4}$$

The value of α_i^* can be simply read off a VLE curve as the value of α corresponding to the interface partial pressure of the volatile component at the bulk-liquid molarity. Therefore, a quantitative check of the validity of Eq. 13.5.4 can be performed.

Often in actual fact, the diffusivities are not all equal and, therefore, the measured value of $\overline{V}$ will not be equal to the right-hand side of Eq. 13.5.4. However, although $\overline{V}$ itself will depend strongly on the composition of both gas and liquid phases, its ratio to the right-hand side of Eq. 13.5.4 will be almost constant and close to unity; values between 0.7 and 1.3 are typical. This ratio can be taken as an empirically determined correction factor for non-equal diffusivities.

Finally, consider the case where the data appear to fall in the transition region between fast and instantaneous reaction behavior. In this case, it is useful to perform experiments with a stirred cell, so that different values of k_L^0 (and hence of ϕ) can be obtained by changing the stirrer speed. This allows a region of ϕ values at a given liquid-phase composition to be scanned. The experimentally determined I versus ϕ curve can then be compared with a theoretical model of the type discussed in Chapter 7 and an empirical value of n in Eq. 7.4.1 can be determined.

By appropriate choice of the laboratory units to be used, it is, in general, possible to obtain a reliable model for the prediction of I values for any given system.

14 Process Selection

It is very easy but perhaps impudent for me, an academic, an engineer on paper, to criticize those who have successfully created a large-scale commercial reactor but I ask you to ponder on this point. This is a simple reaction between two molecules at 360°C. Surely the chemical rate is not terribly slow, and at 15 cm/sec in a 10 m deep bed the reactants are in contact for more than a minute. Surely you would be surprised if the reaction was not nearly complete in such a deep bed.

P. N. Rowe

In contrast with process analysis which is (or should be) analytic and quantitative, process selection is largely an art, and the choice of any given process is often determined not uniquely by technical reasons. It is, therefore, difficult to develop a methodology for process selection without risking the vagueness of an essentially descriptive procedure.

In this chapter we focus attention on the problem of process selection for removal of CO_2 and H_2S; flue gas desulfurization is discussed briefly in Section 14.4. The subject material is organized as follows. Section 14.1 is a brief description of the available alternatives; the discussion of aqueous chemical solvents is brief, since these have been the subject of most of the balance of this book; a somewhat more detailed discussion is given of physical and hybrid solvents. There is considerable overlap in this section with material discussed in Chapter 1 and in Sections 9.1, 10.1, and 11.1; we have tried to make this chapter one that can be read on its own.

Section 14.2 tries to present a methodology of process selection, to the extent that such a goal can be reached—and it is not, unfortunately, a satisfactory extent. Again, some overlap exists with material discussed in other chapters. We also include a few examples of process selection.

Section 14.3 is dedicated to energy considerations in CO_2 removal and Section 14.5 to a review of future trends. Shifts in relative cost of energy, air quality standards, types of raw materials, and so on, may cause shifts in process selection; the boundary conditions that today make one process more attractive may, in the future, be changed enough to suggest a different process selection.

No industrial gas treating process is optimum for all applications. Indeed, there are large numbers of processes available for the removal of acidic impurities such as carbon dioxide, hydrogen sulfide, and sulfur dioxide. Inasmuch as gas treating is generally integrated with at least one other process, the selection of the optimum gas treating process is always a complex problem involving both technical and economic considerations. Our attention in this chapter will be focused mainly on developing a methodology for selection of the solvent since the critical differences between gas treating processes are so highly dependent on the solvent.

Selection of a solvent will ultimately be based upon a thorough economic analysis of the gas treating process integrated with the associated process units and the process(es) chosen to prepare sulfur from the recovered acid gases. For example, the Claus process is the conventional and usually the most economical process used to prepare sulfur from H_2S-containing acid gases. However, a Claus unit generally requires an H_2S concentration in its feed greater than about 10% for effective operation. When the H_2S concentration is above 50% H_2S, a Claus unit operates most efficiently and with a minimum of difficulty. Liquid-phase oxidation of acid gases can be used at H_2S concentrations in the acid gas of less than 5%. However, liquid-phase oxidation units are significantly more expensive from both a capital investment and operating cost point of view than the comparable Claus unit. As a generalization, it is preferable to process as much of the acid gas as possible through a Claus unit. Therefore, significant expenditures can be made for a gas treating unit that provides an acceptable feed to a Claus unit.

14.1. AVAILABLE PROCESSES

CO_2 and H_2S, as well as other impurities such as ammonia and hydrochloric acid, have been removed from gas streams by the use of water and basic materials such as lime, alkali hydroxides, and carbonates since the start of the industrial revolution. Water and lime were used in the early nineteenth century in both coal gas and the LeBlanc soda process. Acid gas scrubbing using sodium and potassium carbonate solutions is a very old process on which sufficient information had been compiled by the mid-1920s to permit rational design; CO_2 was also recovered from flue gas by this technique to produce liquid and solid CO_2.

However, modern industrial gas treating began with the discovery in the late 1920s by Robert Roger Bottoms (of the Girdler Corporation) that aqueous solutions of aminoalcohols such as mono-, di-, and tri-ethanolamine, and later diisopropanolamine, could rapidly absorb CO_2 and H_2S from gases and could easily be regenerated. Selective absorption of H_2S from CO_2-containing gas was pioneered by the Koppers Company, which in 1938 employed potassium and sodium carbonate solutions to purify coke oven gas in the vacuum carbonate process. The high-temperature high-concentration carbonate process was pioneered by Benson, Field, and coworkers at the U.S. Bureau of Mines in 1950; research in Europe and in the United States improved this process by addition of small amounts of promoters (weak inorganic acids, alkanolamines, and aminoacids) that increase both the capacity and the rate of absorption.

The impurities present in the feed gas to a treating unit may be conveniently divided into the following classes:

Acid gases—carbon dioxide and hydrogen sulfide.

Organic sulfur compounds—carbonyl sulfide, mercaptans, thiophenes, etc.

Other impurities—water, sulfur dioxide, sulfur trioxide, ammonia, cyanide, etc.

Hydrocarbons and tars.

Trace inorganics—selenium, arsenic, mercury.

The acid gases are generally present in the feed in significant quantities; their concentration, which can be as high as 50%, is one of the most important considerations in process selection. However, the presence of organic sulfur compounds and other impurities can be critical in solvent selection either to meet certain treated gas specifications or because of their potential for causing solvent degradation.

Industrial gas treating processes for CO_2 and H_2S removal can be divided into the following classes:

Physical solvents.

Chemically reactive solvents such as aqueous aminoalcohol and promoted alkaline salt solutions.

Hybrid solvents, i.e., solutions of a chemically-reactive base in a physical solvent.

Direct conversion of H_2S into elemental sulfur.

Dry bed processes that use a fixed bed of solid material such as iron oxide or molecular sieves (Zeolites) to remove CO_2, H_2S, and other impurities.

Cryogenic distillation.

Of these, the first three (i.e., the ones that make use of a solvent) are generally used for bulk removal of acid gas impurities, whereas the others are used for removal of trace quantities, possibly in tandem with a bulk-removal solvent process. Our review will focus primarily on solvent processes; the principal ones in commercial use are listed in Table 14.1.1.

Physical Solvents

The main physical solvents of interest, and some of their physico-chemical properties, are listed in Table 1.2.4. Desirable characteristics of physical solvents include high acid gas solubility, low hydrocarbon solubility, low vapor pressure, low viscosity, absence of side reactions, minimum corrosion of carbon steel and other common materials of construction, and ready availability at low cost.

TABLE 14.1.1. Processes Used Commercially for CO_2 and H_2S Removal from Gases

Type	Solvent	Trade name	Typical composition
Physical absorption systems	Propylene carbonate	Fluor solvent	–
	Polyethylene glycol dimethyl ether	Selexol	–
	N-methyl-2-pyrrolidone	Purisol	–
	Methanol	Rectisol	–
Aqueous chemically reactive systems	MEA	–	20–35% MEA
	DEA	–	30% DEA
	DGA	Econamine	60% DGA
	DIPA	ADIP	–
	MDEA	ADIP	–
	Promoted K_2CO_3	Benfield; Catacarb; Giammarco Vetrocoke	25–30% K_2CO_3: 5% promoters
Hybrid systems	DIPA-Sulfolane-water	Sulfinol	40%–40%–20%
	MDEA-Sulfolane-water	–	–

Inasmuch as the physical solubility of acid gases decreases with increasing temperature, low temperature absorption is more effective and some processes require refrigeration. Regeneration of the solvent is accomplished by pressure reduction with only moderate heat input, and/or gas stripping. Representative physical solvent flowsheets are shown in Figures 1.3.5, 9.1.2, and 9.1.3. Generally, heat requirements for physical solvent processes are lower than those for chemical solvents.

The reason for the lower heat requirement of physical solvent processes is related, though not trivially, to the fact that the heat of dissolution in a physical solvent is less than the heat of reaction in a chemical solvent. However, it should be borne in mind that, for the same reason, the temperature swing required between the absorption and desorption steps is smaller for chemical solvents (see discussions in Sections 1.7 and 15.2). Since the heat requirement for a process increases with increasing temperature swing, the advantage of physical solvents over chemical ones in this regard needs to be checked for each specific case.

At high acid gas partial pressures in the feed gas (say, above 50 psi), physical solvents may be more economical than aqueous chemical solvents. The relative attractiveness of the physical solvent is likely to be highest when the acid gas specification on the treated gas is not too low and the feed gas does not contain appreciable quantities of ethane and heavier hydrocarbons. The reason is that the loading of acid gas in the liquid may be higher in a physical solvent, as discussed below.

Within the approximation of the validity of Henry's Law, the highest load possible in a physical solvent α_{max} is related to the acid gas partial pressure in the feed gas p_B by the following equation:

$$\alpha_{max} = \frac{p_B}{H_p} \tag{14.1.1}$$

where H_p is Henry's Law constant in the physical solvent. Typical values of H for industrially important physical solvents are given in Table 1.2.4. In an aqueous chemical solvent at high acid gas partial pressures, the reactive solvent is chemically combined with the acid gas almost completely. Therefore, the amount of chemically combined acid gas is of the order of the molarity m of the solution. Some additional acid gas is physically dissolved and, therefore, the value of α_{max} is, as a first approximation, given by

$$\alpha_{max} = m + \frac{p_B}{H_c} \tag{14.1.2}$$

where H_c is Henry's Law constant in the chemical solvent. If the latter is an aqueous one, the value of H_c is between one and two orders of magnitude larger than H_p.

Now consider, for example, absorption of H_2S in sulfolane and in an aqueous DIPA solution. The values of H_p and H_c at room temperature are 1.06 and 13.3

atm 1/g mol, respectively. Suppose the DIPA solution is 3 molar. At a p_B value of 20 psia, Eq. 14.1.1 will give α_{max} = 0.64 gmol/1, whereas Eq. 14.1.2 will give α_{max} = 3.1 gmol/1; i.e., the equilibrium loading in the chemical solvent is almost five times that in the physical solvent. However, at a p_B value of 200 psia, Eq. 14.1.1 will give α_{max} = 6.4 gmol/1, and Eq. 14.1.2 will give α_{max} = 4.0 gmol/1; i.e., the equilibrium loading in the physical solvent is appreciably larger. As a rough rule of thumb, a physical solvent becomes attractive at that acid gas partial pressure at which the physical solubility exceeds the highest molarity that can be used in an aqueous chemical solvent. Calculations of the exact equilibrium loadings in chemical solvents are discussed in Chapter 2.

The common physical solvents all possess a thermodynamic (equilibrium) selectivity for H_2S relative to CO_2. H_2S-to-CO_2 solubility ratios are plotted in Figure 2.7.1. Chapter 11 gives a discussion of selective H_2S removal from CO_2-containing gases with physical solvents. One process is based on selective absorption. Another is based on simultaneous absorption combined with selective regeneration.

A disadvantage of physical organic solvents is their tendency to co-absorb hydrocarbons heavier than methane from the gas stream. Although the selectivity for H_2S directionally helps in producing a better Claus sulfur plant feed, this is often offset by an unacceptable level of hydrocarbon co-absorption. A certain amount of hydrocarbon co-absorption may be tolerated if the acid gas from the solvent process regenerator is processed through an activated carbon bed. This will minimize problems in sulfur recovery. Other disadvantages of physical solvents are their relatively high cost and the difficulty of meeting tight treated gas purity specifications.

Propylene carbonate was one of the first of the physical solvents to be applied commercially. Originally it was used for bulk removal of large concentrations of CO_2 (up to 50% and above) from high-pressure natural gas; the CO_2 concentration in the treated gas was 1-2%. H_2S pipeline specifications on a high H_2S feed gas can be achieved only by a process modification such as addition of a cleanup amine treatment. By 1970, the process had been applied in five plants for removing acid gases from natural gas, in one hydrogen plant and in one ammonia plant. Carbonyl sulfide, carbon disulfide, and mercaptans are removed to the extent of about 50% without measurable degradation. Carbon steel is generally used for all process equipment.

The dimethylether of polyethylene glycol (DMPEG) has been used primarily for bulk removal of CO_2 and H_2S and for selective H_2S removal. Commercial applications have been primarily in natural gas treating. DMPEG removes about 50% of COS and CS_2. Mercaptan removal is poor. Carbon steel equipment can be used throughout the plant. The solvent can simultaneously dehydrate natural gas if provisions are made for removing water in the regenerator.

N-methyl-2-pyrrolidone has been utilized largely for bulk removal of acid gases. The H_2S content in the treated gas can be reduced to as low as 0.25 gr/100 SCF (i.e., pipeline specification). It will dehydrate feed gases if a steam-heated solvent drying column is provided. All carbon steel construction is used throughout except the upper section of the slipstream drying column. Commercial applications have all been in the 600–1000 psig range. The exit gases are generally water washed to minimize solvent losses.

Cold methanol, operating at temperatures below 273K, has been applied for the treatment of synthesis gas made from coal and for treating gas from steam reforming and partial oxidation. Cold methanol is used in the SASOL oil-from-coal complex. The process will simultaneously remove CO_2, H_2S, and COS in one step. In a two-stage selective treating flowsheet, it is possible to recover a hydrogen sulfide fraction containing 25–30 mole% H_2S if the feed gas contains 0.2–0.6 mole% H_2S. A CO_2 fraction containing only 5 ppm H_2S or less can be obtained.

For CO_2 removal, cold methanol will reduce CO_2 from high concentrations (e.g., 35 mole% down to a few parts per million). Other impurities such as hydrocyanic acid, ammonia, and hydrocarbons are removed automatically. The gas can be dehydrated to a low dew point if a dehydration still is used. Methanol is relatively cheap compared to the other physical solvents but it is quite volatile. The process is often combined with a nitrogen-wash unit for residual carbon oxides removal.

Aqueous Chemical Solvents

Acid gas removal processes based on aqueous chemical solvents represent the majority of such processes in industrial operation (see Table 1.2.2). Analysis of the performance of chemical solvents constitutes the bulk of the contents of this book. The principal advantages of aqueous chemical solvents are:

Ability to reduce acid gas partial pressures in the treated gas to very low values; this is related to the favorable equilibrium behavior at low partial pressures (see Chapter 2).

High equilibrium loadings and hence low liquid circulation rates, which can be achieved up to moderately high acid gas partial pressures in the feed gas.

Low hydrocarbon solubility, from the aqueous nature of the solvent.

In addition to these advantages, aqueous chemical solvents are of interest in selective removal of H_2S from CO_2-containing gases. This will be discussed in some detail later.

There are basically two categories of aqueous chemical solvents: aminoalcohol solutions and carbonate solutions. Aminoalcohol solutions have as their greatest advantage the ability to meet the H_2S specification for pipeline quality (typically 0.25 gr/100 SCF) that carbonate solutions cannot. However, chemical degradation from side reactions of the alkanolamines with CO_2, CS_2, and COS occurs in all cases, often quite extensively, and thermal reclaiming of the undegraded alkanolamine is usually needed; potassium carbonate solutions are chemically stable in this regard. Removal of trace sulfur contaminants is partial with aqueous alkanolamines, fairly complete with hot potassium carbonate.

As was discussed earlier in this section, as well as in Chapter 1, it is desirable to have the highest possible cyclic capacity in a chemical solvent which can be achieved by working with the largest possible molarity and highest possible conversion in the rich solution—say, the absorber unit needs to be designed so that the rich solvent is almost in equilibrium with the feed gas. In the case of carbonate solutions, the upper limit is set by bicarbonate precipitation; the bicarbonate is more soluble at higher temperature. Therefore, a very large cyclic capacity is obtained in the hot carbonate process. In the case of aqueous alkanolamines, the upper limit to acid gas loading in the solvent is set by corrosion considerations. Historically, aqueous alkanolamines have a lower cyclic capacity than hot carbonate solutions, but new corrosion control technology has allowed higher loads comparable to those obtained with the hot carbonate process.

Aqueous primary and secondary alkanolamines are the most commonly employed chemical solvents for the simultaneous removal of CO_2 and H_2S from sour gases in petroleum refineries and natural gas fields. The most commonly used alkanolamines are listed in Table 1.2.3. The main characteristics of the common aqueous alkanolamine gas treating processes are compared in Table 14.1.2. The table also includes corresponding information for a physical solvent, a potassium carbonate solution, and a chemical-physical hybrid solvent. Hybrid solvents generally use an alkanolamine dissolved in an organic solvent containing a minor amount of water.

The range of maximum rich loadings attainable in practice is quite broad because alkanolamine corrosivity depends strongly on amine concentration, the ratio of CO_2 to H_2S in the feed gas, and the H_2S level. CO_2 is more corrosive than H_2S in an amine solution since H_2S can, if present in sufficient quantity, react with ferrous materials of construction to form a protective sulfide film. Representative operational capacities of alkanolamine solutions are summarized in Table 14.1.3. When very high amine concentrations are used (such as with DGA), stainless steel is required in critical plant areas. Substantially higher MEA concentrations can be used when corrosion inhibitors are present. The corrosion inhibitors used are multicomponent proprietary systems; royalties must be paid for their use. However, the higher amine concentrations result in substantial energy reductions.

Aqueous sodium and potassium carbonate solutions have been used to absorb acid gases for many years. Acid gas scrubbing using aqueous K_2CO_3 solutions is by no means a modern innovation; it has been used for more than 50 years to recover CO_2 from flue gas to produce liquid and solid carbon dioxide. Absorption of CO_2 was usually carried out at a temperature below 110°F and solution strength was limited to about 12 wt% K_2CO_3 to avoid precipitation. Concentrated CO_2 was recovered by steam stripping.

Current hot carbonate processes generally use solutions to which promoters have been added; the influence of the promoters on equilibrium and mass transfer has been discussed in Section 9.5. Hot carbonate solutions without promoters are currently used only if the product gas cannot be contaminated with trace quantities of the promoters. Hot carbonate processes require less sensible heat than conventional alkanolamine processes, and the size of heat exchange equipment is lower since absorption and regeneration are carried out at almost the same termperature.

TABLE 14.1.2. Characteristics of Gas Treating Processes

Characteristic	MEA	DEA	DGA	Physical solvent	Potassium carbonate	Hybrid solvent
Typical H_2S content in product gas (gr/100 SCF)[a]	0.1–0.25	0.25–0.5	0.25	0.25	1.0	0.25
Chemical degradation under COS, CS_2, CO_2	Yes	Some	Yes	No	No	Some
COS removed	Partial	Partial	Most	Yes	Yes	Yes
CS_2 removed	Yes	Yes	Yes	Yes	Yes	Yes
Mercaptans removed	Partial	Partial	Most	Yes	Some	Yes
Compatible with rich gas stream	Yes	Yes	Usually	No	Yes	Usually
Selective H_2S removal	No	No	No	Some	Yes	Some (in certain instances)

[a]0.25 gr/100 SCF ≈ 4 ppm v at 1000 psig.

TABLE 14.1.3. Operational Capacities of Alkanolamine Solutions for CO_2 and H_2S Removal from Gases

Alkanolamine	Amine concentration range (wt%)	Net acid gas loading range (mole/mole amine)	Solution capacity (SCF/gal)
MEA	10–20	0.15–0.35	0.8–3.7
DGA	50–65	0.15–0.35	2.2–7.0
DEA	20–40	0.50–0.85	3–100
DIPA	20–40	0.50–0.85	2.4–8.1

One of the main advantages of hot carbonate processes is the fact that they usually can be heat integrated with other processes. Exhaust steam often can be used for a large portion of the total heat requirement. In modern ammonia plants and hydrogen reforming units, the waste heat available in the gas from the shift converter is well matched with the process heat requirement. The absorption and regeneration pressures are ordinarily 300–400 psig and 0–10 psig, respectively, although 100–1000 psig absorption pressures and 35 psig regeneration pressures have been used in some plants. To achieve the optimum combination of capital investment and operating costs, several variations of the process can be used. These are discussed fully in Chapter 9 (Figure 9.5.8).

Hot carbonate processes are uniquely suited for bulk CO_2 removal at moderate partial pressures when waste heat is available from an adjacent unit. Their use can offer significant economic advantages compared to alternative gas removal systems when the acid gas partial pressures are in the range of 15–200 psi. However, hot carbonate processes cannot remove H_2S in the absence of CO_2 (see Section 11.3). Moreover, pipeline gas specifications for H_2S cannot usually be met. Thiophenes, COS, and CS_2 are absorbed although somewhat more slowly than CO_2. Aqueous potassium carbonate is degraded by CO, SO_2, and SO_3.

Before concluding this brief review of aqueous chemical solvents, it is useful to discuss their ability to be used for selective removal of H_2S from CO_2-containing gases. Inasmuch as it is often necessary to process gases containing too low an H_2S/CO_2 ratio to provide an acceptable feed to a Claus unit, selective H_2S removal processes will provide both economical and ecological advantages.

In absorbers, aqueous alkanolamines either exhibit a very slight selectivity for H_2S or, in fact, a strong selectivity for CO_2. The latter is typical of primary and secondary amines that form stable carbamates. Equilibrium selectivity is discussed in Section 2.7 and in Chapter 11. However, H_2S is absorbed at a significantly faster rate than CO_2 and, therefore, by appropriate choice of the operating conditions, it is possible to obtain a good non-equilibrium or kinetic selectivity for H_2S, particularly with weak secondary and with tertiary alkanolamines. Kinetic selectivity is discussed in Chapters 8 and 11. Carbonate solutions exhibit a good selectivity in absorbers for H_2S, though not as good as some physical solvents.

However, again a kinetic selectivity is obtainable with carbonate solutions, particularly at low temperature and without rate promoters.

Existing industrial processes for selective H_2S removal are based on aqueous tertiary alkanolamine solutions such as triethanolamine and methyldiethanolamine. These processes are not yet widely used, but they are becoming of increasing importance owing to the projected abundance of gases of high CO_2 content from coal and heavy oil gasification.

Hybrid Processes

Hybrid processes represent an attempt to combine the desirable features of chemically reactive solvent processes, particularly their ability to meet pipeline specifications, with those of physical solvents. Chemically reactive solvents rely mainly on acid-base chemistry, which is typical of aqueous solutions; however, water exhibits a low physical solubility for acid gases. A solution of base in a polar organic solvent will, in general, not exhibit the appropriate acid-base chemistry. Relatively minor amounts of water, however, are sufficient to allow the chemical reactivity of the base to be efficient so as to reduce residual acid gas in the treated gas to very low values. The bulk of the organic solvent provides physical solubility at high acid gas partial pressures.

As a rough first-order approximation, Eq. 14.1.2 holds at high acid gas partial pressures for hybrid solvents with H_c, however, having a much lower value than in an aqueous chemical solvent. Therefore, a large value of α_{max} is obtained. At low acid gas partial pressures, the vapor-liquid equilibrium is governed by the acid-base chemical reactions and comparatively high loadings are possible at low partial pressures (or, equivalently, very low equilibrium partial pressures are obtained at moderately high loadings). The methods discussed in Chapter 2 can be used to model the vapor-liquid behavior of hybrid solvents, keeping in mind that $\bar{y}$ and y are *not* almost equal as they are in aqueous solvents. It is surprising that, even with minor amounts of water in the hybrid solvent, the chemical behavior is often essentially the same as in aqueous solvents.

An early hybrid process used MEA in diethyleneglycol. Another hybrid process employs a Sulfolane solution of diisopropanolamine. This process, developed about 1964 by Deal and coworkers at Shell, is used on a significant commercial scale today. Other hybrid solvent systems such as DIPA in DMPEG have been tested in laboratory and pilot plant operations.

In hybrid systems, the physical solvent allows very high acid gas loadings at high acid gas partial pressures, giving it bulk removal capacity. The chemical component reduces residual acid gases in the treated gas to very low values. A typical solvent composition is 40% Sulfolane, 40% DIPA, and 20% water. However, considerable variation in solvent composition is possible depending on specific removal requirements, particularly with respect to COS and mercaptan removal as well as operating pressure.

The high solvency of the physical solvent component for acid gases at high acid gas partial pressures results in significantly reduced circulation rates. This is largely the basis for the economic advantage of the hybrid process over aqueous amine systems. However, this advantage disappears at acid gas partial pressures below 100 psia. Low foaming tendency permits high solvent and amine concentrations to be used with a close approach to the equilibrium acid gas solubilities. Regeneration and utility requirements are lower than those for conventional aqueous amines because of the lower solution circulation rate.

The advantages of DIPA-Sulfolane-water relative to aqueous amines is greatest when the H_2S/CO_2 ratio in the feed gas is 1 : 1 or greater. The solvent circulation rate is lowest when the feed gas contains all H_2S; and highest when it contains all CO_2. The solution can also be used to remove mercaptans, carbonyl sulfide, and carbon disulfide, as well as hydrogen sulfide and carbon dioxide, from a feed gas. In most instances, all of these components can be successfully removed to meet pipeline specifications. If COS, CS_2, and mercaptan removal are not required, selective H_2S removal is possible at high CO_2-to-H_2S ratios.

COS and CO_2 gradually degrade DIPA to various cyclic derivatives (see Figure 15.8.2); the extent of degradation can be quite significant at high CO_2 concentrations. This degradation product can readily be removed from the system by use of a thermal reclaimer. The degradation product may be allowed to accumulate up to a content of 10 wt% in the solvent without significant influence on operation of the unit.

Corrosion has not been a problem where both hydrogen sulfide and carbon dioxide are present in the feed gas. Often carbon steel can be used throughout the unit with long equipment life. In the absence of hydrogen sulfide, however, stainless steel must be used in those areas of the unit where high acid gas concentrations exist. Low vapor pressures of Sulfolane and DIPA keep vapor losses of chemicals low. Sulfinol generally does not dehydrate the treated gas. The gas will be almost saturated with water (with a 5-10°F dewpoint depression). Although certain mixtures of the Sulfinol solvent can be used for gas dehydration, it is generally not utilized for drying.

Sulfolane has a great affinity for aromatics and, to a lesser extent, for heavy paraffin hydrocarbons. Pretreatment facilities must be installed when hydrocarbons are present in the feed or charcoal treatment of the acid off-gases before the Claus sulfur plant will be required. Although solvent vapor losses are low, the cost of the hybrid solution is high, several-fold higher than that for an aqueous alkanolamine or hot potassium carbonate system.

14.2. SELECTION OF TREATING SYSTEMS FOR CO_2 AND H_2S REMOVAL

The general objective of an acid-gas removal system is purification of a gas stream prior to its subsequent processing or use. At a minimum, the acid gas impurities to be removed are hydrogen sulfide and carbon dioxide. Generally, the CO_2 concentration in the treated gas will be less than 1-2% although there are exceptions. At high residual concentrations, CO_2 will act as a diluent possibly impeding downstream processes and ultimately will have to be purged. Sulfur compounds must generally be removed to very low concentrations—often less than 1 ppm—because they poison downstream catalysts or cause corrosion or pollution.

Not all purification processes may be used alone. For example, cryogenic distillation, liquid nitrogen washes, and sulfur guard chambers are used in ammonia plants in addition to acid gas removal systems. These systems can remove the residual quantities of carbon dioxide and sulfur compounds that have escaped from the acid gas removal system. However, a solvent acid gas removal system is the lowest cost approach for the removal of the bulk of the acidic impurities.

The simplest approach for acid gas removal is a non-selective process, as shown in Figure 14.2.1. The light hydrocarbon oils may be removed from the process gas stream in any of three locations, depending on the specific acid gas removal system used. The bulk of the acid gases (carbon dioxide and all the sulfur compounds) are removed from the process gas stream in a single contacting vessel. Therefore, the H_2S/CO_2 ratio in the regenerator off-gas will be almost identical to that in the feed gas.

Some acid gas removal systems, as discussed previously, can have a selectivity for H_2S relative to carbon dioxide. With these systems, the processing conditions can be chosen so that one of the off-gases will contain most of the sulfur, thus generating a sulfur-rich gas stream. Figure 14.2.2 illustrates how a selective process can be

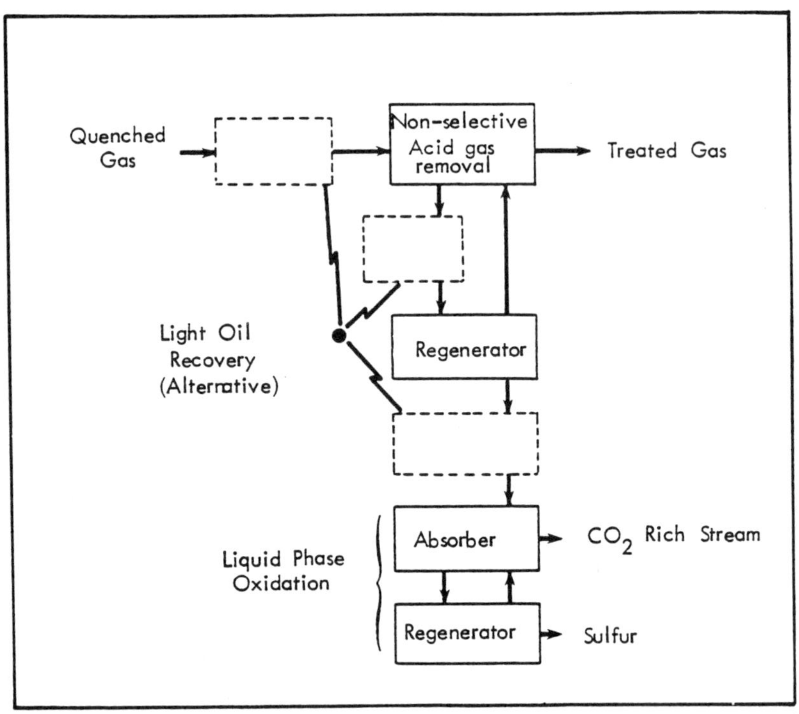

FIGURE 14.2.1. Non-selective acid gas removal.

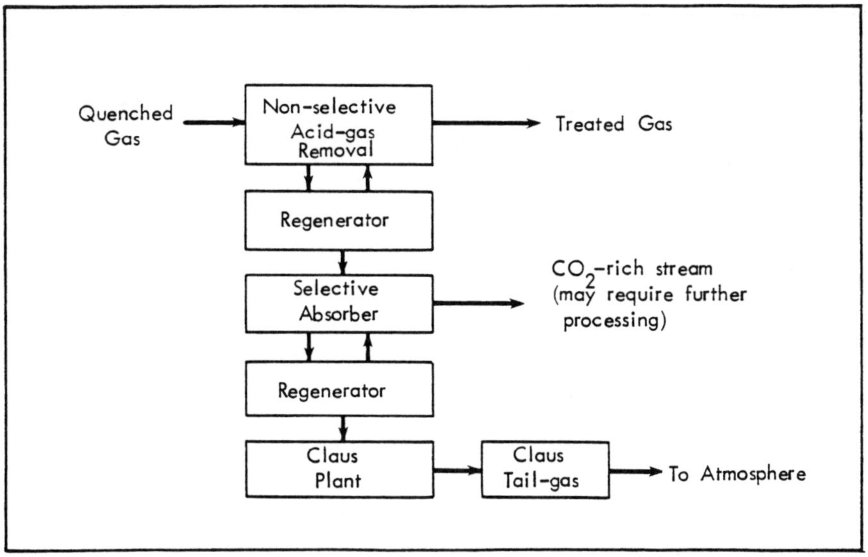

FIGURE 14.2.2. Non-selective, selective two-stage acid gas removal.

used. (Hydrocarbon recovery is not shown but it would be similar to that in Figure 14.2.1) Conditions are chosen so that most of the sulfur will be recovered in a stream sufficiently concentrated to feed a Claus unit. If the secondary acid gas system is sufficiently selective (which is rare), the CO_2-rich gas would be sulfur-free and could be vented to the atmosphere directly. At lower selectivities, a liquid-phase oxidation, some other direct conversion process, or incineration would be needed to further purify the CO_2 stream before venting. However, most of the sulfur-containing gases would be processed in a Claus unit. The economy of operation of a Claus unit in most cases more than offsets the cost of a second acid gas removal and sulfur disposal system.

An alternative approach to selective acid gas removal is indicated in Figure 14.2.3. Here, the selective unit is used to treat the raw process gas stream; its off-gas is the feed for the Claus unit. The CO_2-rich gas from the non-selective gas system is either vented or treated further. Figures 14.2.2 and 14.2.3 are simple representations of the commercial processes that are normally used. The number of towers may be greater than the four implied by the number of boxes, and a variety of feed and recycle loops between the towers are often incorporated into a commercial unit to improve selectivity.

Although the selectivity of some acid gas removal systems for H_2S over CO_2 can be high enough that the CO_2-rich gas is essentially H_2S-free, the fate of carbonyl sulfide is not well defined. Most acid gas systems will remove this species from the feed gas sufficiently so that downstream catalyst beds will not be poisoned. However, its precise disposition is not completely certain in all acid gas removal systems. Moreover, carbonyl sulfide is not removed significantly by most liquid-phase oxidation systems.

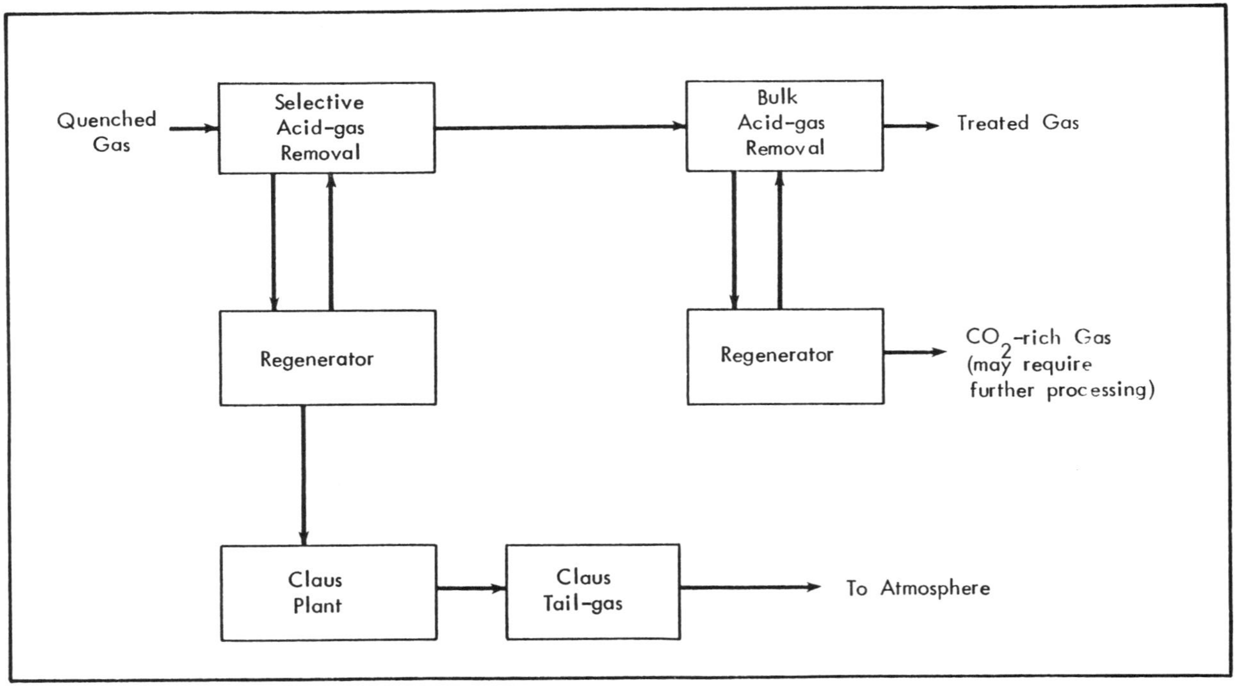

FIGURE 14.2.3. Selective, non-selective two-stage acid gas removal.

The separation between carbon dioxide and hydrocarbon species is also critical. Some acid gas removal systems do not have a high selectivity between carbon dioxide and C_2^+ hydrocarbons. Therefore, when hydrocarbons are present, they will be lost in the acid off-gas. This is particularly so for the CO_2-rich gas from a selective system. Hydrocarbons reduce the overall Claus plant efficiency through an increase in the quantity of undesirable side reaction products such as COS and CS_2. These are difficult to convert to elemental sulfur.

Variations in hydrocarbon content require changes in the quantity of air needed for combustion. If the air is not properly controlled, the H_2S and SO_2 in the reaction gas can be "off ratio." Either too much or too little air will reduce the conversion efficiency and can cause serious operating problems. Hydrocarbons will also produce large quantities of CO_2 and water vapor diluents, fouling the Claus catalyst with coke.

Selection Methodology

The essential elements of the methodology for acid gas removal processes are sketched in Figure 14.2.4. Starting from the top, the *definition* of a process is provided by two elements: the available feed gas characteristics (composition, pressure, temperature, etc.), and the treated gas specifications (i.e., the process requirements). These two elements provide a preliminary evaluation of the solvent working capacity which may, however, be influenced by several other elements, of which the most important ones are sketched in Figure 14.2.1. Analysis of these elements constitutes the body of this section.

The controlling factors in selecting an acid gas treating system are:

Raw feed gas pressure.
Raw feed acid gas composition and content.
Required treated gas purity.

To a first approximation, these factors can be used to establish the working capacity of possible solvents with the methods discussed in Section 1.5 and in Chapter 2. The working capacity establishes the solvent circulation rate (expressed often as a solvent/acid gas ratio) which has a major impact on both plant investment and operating cost of acid gas removal systems. It has a direct impact on the size of the absorber tower, the piping, the circulation pumps, and the size of the regenerator facilities. In a solvent acid gas removal system, 50–70% of the plant investment is directly associated with the magnitude of the solvent circulation rate. Obviously, there are advantages in employing the highest possible concentration of chemical or physical solvent as limited by solubility, precipitation, and corrosion considerations. Although there are exceptions for a chemically reactive solvent process to be economically attractive, the concentration of reagents must be greater than about 3 molar.

The partial pressure of acid gas in the feed to a first approximation determines the solvent circulation rate. In a countercurrent column, the feed gas contacts the

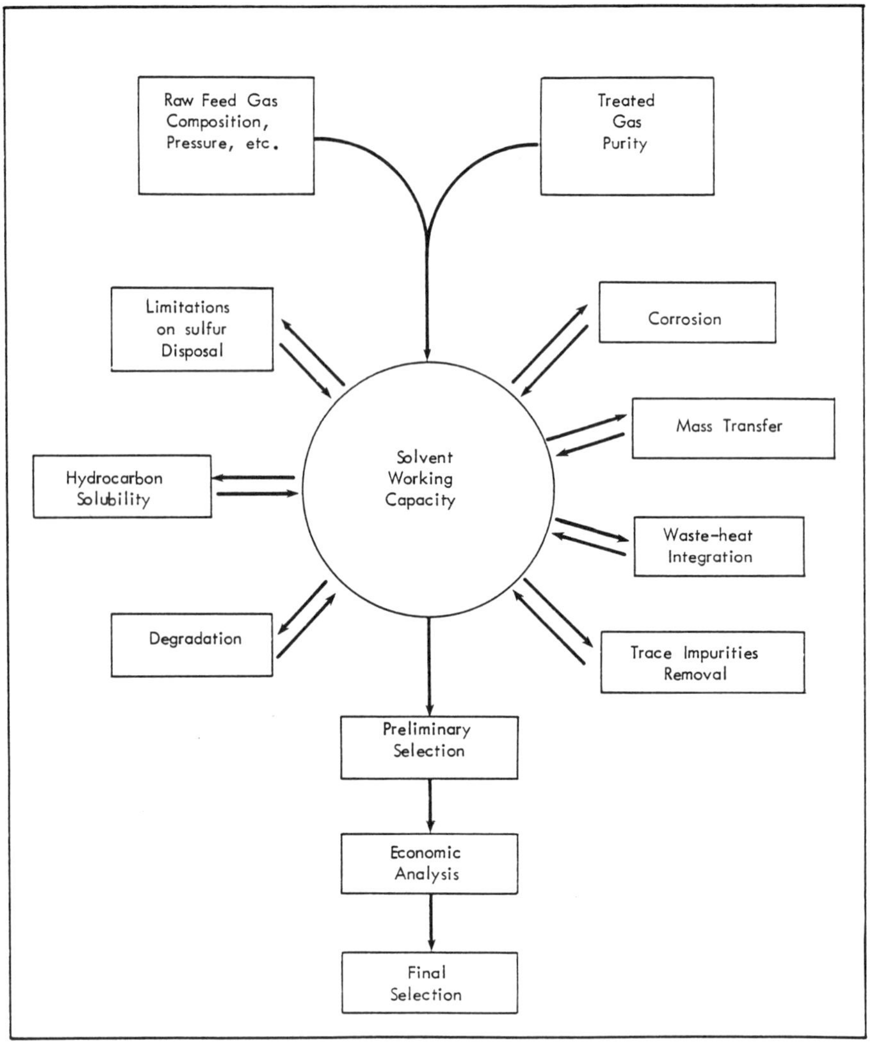

FIGURE 14.2.4. Selection methodology for acid gas removal solvent.

fully loaded rich solvent at the bottom of the column. At a minimum, therefore, there must be sufficient solvent to absorb substantially all of the acid gas in the feed and provide a driving force for absorption at the bottom of the column. Therefore, the partial pressure of acid gas in the feed is controlling.

The partial pressure of acid gas in the product determines the degree of regeneration required. This is because the partial pressure of each acid gas in the product must be higher than the equilibrium partial pressure of that same component in the regenerated solvent. Otherwise, there will not be a positive driving force for absorption at the top of the absorber.

TABLE 14.2.1. Operating Cost Breakdown for Acid Gas Removal System

	Percent of variable cost
Steam	70
Chemicals makeup	20
Electricity	10
Cooling water	7

Therefore, two most important cost factors—circulation rate and regeneration energy duty—are established to a significant extent by the acid gas partial pressures in the feed and the product. Ten to twenty percent of the investment of a chemically reactive acid gas removal system is dependent on the regeneration energy duty or the steam/acid gas ratio. The importance of energy in the variable operating costs of an acid gas removal system is shown in Table 14.2.1. We may expect that energy costs will increase in importance in future years.

Over the years, criteria have been developed for narrowing the choice of available processes based solely upon the partial pressure of acid gas in the feed and product. For example, if the partial pressure of acid gas in the *feed* is low (below about 50-100 psi), all processes using physical solvents can be eliminated. If the partial pressure of acid gas in the *product* is very low, physical solvents can generally be eliminated unless the total pressure is quite high.

The selection criterion can be reduced to sets of graphs with the two pertinent acid gas feed and product pressures as the axes. The areas show which process(es) are generally applicable. These graphs can be used with a high degree of confidence for choosing among the various available acid gas removal processes. Graphs of this type are presented in Figures 9.1.4, 10.1.4, 11.1.8, and 11.1.9. However, it should be borne in mind that these graphs reflect existing industrial process selection, and not necessarily an optimal one for a specific situation. Elements other than feed gas composition and pressure, and treated feed gas purity, enter into the optimal selection procedure, as shown in Figure 14.2.1. Where several processes are shown as possible, additional studies will be required. Initially, the presence of impurities other than CO_2 and H_2S will be ignored; these are discussed separately.

Carbon Dioxide Removal

Carbon dioxide removal from a gas containing essentially no H_2S is encountered, for example, in purification of ammonia synthesis gas made by steam reforming of natural gas. Feed gas compositions can vary from about 3-4% CO_2 to as high as 65%. Absorber operating pressure ranges from inches of water (gage) to over 1000 psi. As a result, the inlet CO_2 partial pressure may vary from 1-650 psi or higher. Product gas specifications generally are in the range of 0.1-2% CO_2.

For treating gas with a low inlet CO_2 partial pressure (i.e., up to about 10-15 psi), aqueous amines are the preferred process as shown in Figure 9.1.4. At higher CO_2 partial pressures but below 75-100 psi, promoted hot carbonate solutions should be considered. Above 100 psi, physical solvents may be advantageous.

This assumes that the process under consideration is indeed capable of producing the required CO_2 content in the product gas. Aqueous amines will reduce the CO_2 content of a gas stream to about 0.005 psi partial pressure (15 ppm at 300 psi total pressure). Promoted carbonate processes are capable of reducing the CO_2 partial pressure in the product to 0.2-0.25 psi, whereas physical solvents are generally economical for producing gases with CO_2 partial pressures of about 1-3 psi. Therefore, when treating a gas having a high inlet partial pressure of CO_2 to produce a gas with very low CO_2 content, it can be economical to combine a physical solvent unit with an aqueous amine process.

Promoted carbonate processes are sometimes used when physical solvents would appear to be preferable (see Figure 9.1.4). Physical solvents remove some of the C_3^+ hydrocarbons from the feed gas, exhausting them with the waste CO_2. The resulting complications in the unit handling the acid off-gases and the loss of potential hydrocarbon fuel can be avoided by using an aqueous process.

H_2S Removal

H_2S removal from a gas containing essentially no CO_2 is the situation found in the purification of recycle hydrogen used for oil desulfurization, refinery gas, and natural gas. Feed gas H_2S composition for these situations varies from about 1-2 gr/100 SCF to as high as 35%. Absorber operating pressures range from inches of water gage to over 1000 psig. Product gas specifications usually lie in the range of 0.1-1.0 gr/100 SCF. In the preparation of Figure 10.1.4, it has been assumed that if small quantities of CO_2 are present, the concentration is so small that the choice of H_2S removal process is not affected.

H_2S removal cannot be considered without also deciding the disposition of the waste H_2S. Normally it is converted to elemental sulfur, although production of aqueous H_2SO_4 is also a possibility.

Simultaneous Removal of Carbon Dioxide and Hydrogen Sulfide

The simultaneous removal of both CO_2 and H_2S from a gas stream is the most common treating problem. The simplest case to be considered is the one where the following conditions are met:

1. The treated gas specifications are such that essentially all of both CO_2 and H_2S need to be removed.
2. The H_2S/CO_2 ratio in the feed gas is an acceptable one for a feed to a Claus plant (H_2S is at least about 10% of total acid gas).

When these conditions are met, no problem of selective removal of H_2S arises, and process selection is again essentially governed by the pressure and composition of the raw gas, and the required purity of the treated gas (see Figure 11.1.8).

At low acid gas partial pressures, aqueous amine or amine-organic solvent processes are preferred, as shown in Figure 11.1.8. In the middle pressure range, the choice of possible processes widens to include promoted hot carbonate processes. Physical solvents are preferred, as in the case involving either CO_2 or H_2S removal, when the partial pressure of the acid gases in the feed exceeds about 60 psi.

The indicated boundaries between processes in Figure 11.1.8 do depend to some extent on the ratio of CO_2/H_2S in the feed. At high CO_2/H_2S ratios, for example, promoted hot carbonate will often be selected for use rather than amine-water or amine-organic solvent processes. At low CO_2/H_2S, aqueous amine or organic solvent-amine processes will be preferred. This results from the increasing difficulty of regenerating promoted hot carbonate solutions containing low concentrations of HCO_3^- and high concentrations of HS^- (see Section 11.3).

The liquid-phase oxidative H_2S removal processes directly convert H_2S to elemental sulfur. However, all the available liquid-phase direct oxidation processes have limitations. One solvent is an aqueous solution of sodium carbonate, sodium vanadate, and anthraquinone disulfonic acid. This process becomes less efficient at high CO_2 levels in the feed gas. Another selective process uses arsenic, a reagent that poses toxicity problems.

At low H_2S concentrations, the liquid-phase oxidative processes would be preferred. Between H_2S partial pressures of 3 and 10 psi, the aqueous amine and hybrid processes become of interest for H_2S removal. Since a catalytic sulfur recovery unit will be required with the aqueous amine and hybrid units, the capital investment required is increased. The economic breakpoint of an amine-Claus unit with a liquid-phase oxidative direct H_2S conversion process is not well established. This is, in part, because the impact of total unit pressure on the relative economics of H_2S removal from gas streams at low H_2S concentrations has not been accurately determined.

Plant size is also a critical parameter affecting relative treating costs for H_2S removal. Direct oxidation processes, because of their relatively low solution capacity, require very large circulation volumes and plot space in large sizes. Therefore, for world size units, aqueous amine–Claus and hybrid process–Claus combinations would have lower investments.

Above 60 psi H_2S inlet partial pressure, physical solvents such as chilled methanol or the dimethyl ether of polyethylene glycol show sufficient selectivity for H_2S over CO_2 to be considered. Some physical solvents can also have equilibrium selectivity towards H_2S. In the intermediate H_2S partial pressure range, kinetically selective H_2S absorption in a chemical absorbent may be economic (see Section 14.2). Certain physical solvent processes *cannot* easily reduce the H_2S concentration in the product gas to meet the tight $\frac{1}{4}$ gr/100 SCF specifications. An aqueous amine "clean-up" process may be needed following the physical solvent unit. The heavy glycol and chilled methanol processes are exceptions under some conditions to this "rule of thumb."

Process selection is also influenced by the presence of C_2^+ hydrocarbons. Although physical solvents have low energy requirements, the solubility losses of C_2^+ hydrocarbon can exceed the fuel requirement of a promoted carbonate process. In addition to being a fuel loss, C_2^+ hydrocarbons have a significant and serious impact on the operation of Claus units. Therefore, promoted hot carbonate processes are often used in such situations.

Selective H_2S Removal

When the raw gas to be treated contains both CO_2 and neither condition 1 nor 2 above is met, a problem of selective removal of H_2S arises. In selective removal, the H_2S partial pressure in the feed generally determines the selective treating process that is most economical. Kinetically selective H_2S absorption processes must compete with liquid-phase direct oxidation schemes for treating gases with low H_2S partial pressures, and with physical solvent processes when the H_2S partial pressure is high. Figure 11.1.9 shows those processes that are usually economical in different zones of the plane of H_2S feed pressure against H_2S product pressure. Aqueous diisopropanolamine and aqueous *N*-methyl diethanolamine are the most attractive kinetically selective H_2S absorption solvents available today.

When H_2S in the feed gas is at low partial pressure and the H_2S level in the product gas must be extremely low, the liquid-phase direct oxidation approach to produce elemental sulfur directly in the regeneration step appears best. In these processes, H_2S reacts with a reagent which is thereby reduced chemically. The sulfide-rich solution from the absorber flows to an oxidizer-regenerator in which it is contacted with air. The reagent is regenerated and the sulfur is released in elemental form (see Figure 10.1.3).

Where a selective removal step is required, two possible flowsheet configurations are available, as shown in Figures 14.2.2 and 14.2.3. The latter, without the bulk acid-gas removal step, is to be used when CO_2 need not be removed from the gas, as in the case of low heat content fuel gases. Selective removal of H_2S is discussed in detail in Chapter 11.

Impact of Impurities

The applicability of a given acid gas removal process must often be revised because of the presence of impurities such as NH_3, HCN, COS, CS_2, SO_2, SO_3, and mercaptans. In certain cases, the presence of hydrocarbons and tars must also be considered. Even water can be an impurity in certain instances.

All chemically reactive gas treating systems undergo degradation reactions in addition to the desired reversible reaction involving the acidic impurities. However, all aqueous aminoalcohols are quite stable in the presence of just hydrogen sulfide.

For MEA, a number of impurities (namely, COS, CS_2, SO_2, and SO_3) can react with and partially deactivate the solution (see Figure 15.8.1). However, during regeneration a portion of the degradation products is freed from the MEA. The portion of MEA combined with SO_2 and SO_3 along with that combined with CO_2, COS, and CS_2 degradation products can be recovered by treating with sodium carbonate or caustic soda. Permanent MEA loss is caused only by a small portion of the CO_2, COS, and CS_2 present in the gas to be treated.

The decision as to whether MEA should be used and whether a reclaimer is required will depend on the quantities of these impurities present and the specifications of the treated gas. When the feed gas contains appreciable quantities of both COS and CS_2, DEA may not be the most economic choice since some irreversible reactions can occur. In the promoted hot carbonate processes, COS and CS_2 are not harmful. They are partially hydrolyzed to CO_2 and H_2S and then absorbed. The portions that have not been hydrolyzed pass through with the product stream.

Ammonia and hydrogen cyanide are readily removed in acid gas systems other than MEA and discharged with the acid off-gases. Ammonia can cause problems in the downstream sulfur plant because of chemical reactions within that system. Cyanide problems result primarily from the formation of an iron complex (Prussian Blue) that fouls the acid gas recovery system.

Primary aminoalcohols and some secondary aminoalcohols are degraded irreversibly by CO_2. Secondary aminoalcohols are generally more stable than the primary ones. However, the reaction paths are not well understood and there are subtle effects of amine structure on stability. Carbon dioxide concentration and temperature are particularly critical variables.

Although diethanolamine and diisopropanolamine are used widely in gas treating, they are not particularly stable towards carbon dioxide. Tertiary aminoalcohols such as methyldiethanolamine are quite stable. Degradation in the presence of carbonyl sulfide and carbon disulfide is quite similar to the degradation behavior with carbon dioxide; secondary aminoalcohols being much more stable than primary aminoalcohols, and tertiary aminoalcohols being extremely stable.

Components in the gas stream such as carbon monoxide, sulfur dioxide, and sulfur trioxide also interact with a number of solvents other than MEA. Carbon monoxide can lead to the production of formate ion in alkaline scrubbing solutions. At high carbon monoxide partial pressures and high temperature, irreversible formation of formate and other organic compounds occurs. Oxygen concentration may be a factor in this reaction.

Sulfur dioxide and sulfur trioxide, since they are much stronger acids than hydrogen sulfide, react irreversibly with most basic reagents just as with MEA. If present in any significant quantities, the costs involved in prewashing them from the gas must be balanced against the solution loss that will otherwise occur.

Special challenges arise when large quantities of hydrogen sulfide and carbon dioxide must be removed in the presence of carbonyl sulfide, carbon disulfide, or mercaptans. Carbonyl sulfide can be removed with primary amines but only at the expense of some amine degradation. The degradation products unfortunately can be only partially recovered by reclaiming the solvent at high temperature with the possibility of other amine degradation reactions occurring. With secondary amines, reversible thiocarbamate formation can occur but the rate of reaction is about one-hundredth that of the corresponding reaction with carbon dioxide. Complete carbonyl sulfide removal with secondary amines is difficult.

The principal approach to complete carbonyl sulfide removal is catalytic hydrolysis over solid or liquid catalyst. Aminoalcohol-physical solvent systems can completely remove carbonyl sulfide. However, this requires a large number of contacting stages at high solvent rates. The low solubility of carbonyl sulfide in all aqueous systems results in an inherent limitation on attainable removal.

Carbon disulfide and mercaptans are at least as difficult to remove as carbonyl sulfide. Low aqueous solubility is again the limiting factor. Although carbon disulfide does react with secondary amines reversibly, the rate is extremely slow compared to carbon dioxide. Similarly, hydrolysis to carbonyl sulfide and hydrogen sulfide does occur in basic solution but only at very low rates. Mercaptans such as methyl and ethyl mercaptan are absorbed only slightly by all chemically reactive solvent systems. Only caustic is a strong enough base to result in significant mercaptan removal.

Trace metals such as selenium, arsenic, and mercury are often found in natural gas and coal and shale-derived gases. Selenium will probably go with the sulfur-rich off-gas. Arsenic compounds would probably pass through an amine or hot carbonate system, but would probably be recovered with the sulfur compounds in a physical solvent system. The presence of these compounds must be checked because the recovered sulfur might otherwise prove unsalable.

For physical solvents, the quantity of mercury in the treated gas is a function of the operating temperature of the system. For example, in a chilled methanol acid gas process operating at $-40°F$, 99% of the mercury would be found in H_2S off-gas. However, a system operating at $+40°F$ would only condense out about half of the mercury. At higher temperatures, most of the mercury would remain in the treated gas.

Hydrocarbon components in the feed gas will pass through both aqueous amine and hot carbonate systems except in high pressure treaters. Physical solvent systems, on the other hand, may remove a significant portion of the fuel gases from the process gas stream. In particular, C_2^+ hydrocarbons will go with the sulfur-rich gas with resulting problems in the sulfur plant. Some C_2 or C_3 hydrocarbons can be lost to the CO_2-rich gas, perhaps causing environmental concern as well as loss of valuable by-products. An extremely cold methanol process ($-100°F$) could be used for separation of methane from the hydrogen product gas, but the problem of C_2^+ hydrocarbon remains. Specialized systems, including solvent washing and selective solvent regeneration have been developed.

Oils, naphtha, or aromatics should be separately removed prior to amine or hot carbonate systems to avoid operating difficulties. In a physical solvent system, these oils are removed either prior to H_2S absorption or from the H_2S-rich gas.

Examples of Process Selection

The concepts discussed in Section 14.2 are perhaps best understood by considering some practical examples. Information about three gases is summarized in Table 14.2.2. The natural gas stream is a high-volume low-pressure gas with significant quantities of ethane and heavier components. The process gases are relatively high-pressure lean gases. Since removal of CO_2 is *not* desired for Process Gas *A*, it probably is going to be blended off with some other gas stream after the acid gas removal step.

The large quantities of natural gas that must be handled eliminate dry bed and direct conversion units. The large quantities of C_2^+ hydrocarbons in the gas make physical solvents a poor choice. Moreover, the acid gas partial pressure of 22 psig

TABLE 14.2.2. Feed Gas Conditions and Specifications Used in Examples of Process Selection

	Natural gas	Process gas *A*	Process gas *B*
Inlet gas composition			
C_2^+	36%	0.3	0.3
H_2S, %	3	10	5
CO_2, %	11	10	30
COS	–	–	0.1
HCN	–	–	0.1
Pressure, psig	160	1100	500
Volume, MMSCF/day	1400	70	100
Treated gas specification			
H_2S	4 ppm	2 ppm	1 ppm
CO_2	300 ppm	Maximum possible	300 ppm

would not result in effective treating with a physical solvent. Promoted hot carbonate would not attain the required H_2S specification for the treated gas without an aqueous amine wash. Therefore, we should first consider aqueous amines and organic solvent-amine processes.

Aqueous MEA and DEA are probably not economical compared to aqueous DGA and DIPA-Sulfolane-water. If the gas contains significant quantities of mercaptans and COS, aqueous DIPA would be a poor choice because it does not remove these impurities to a great degree. A caustic wash would be needed if DIPA were to be used. Aqueous DGA and DIPA-Sulfolane-water would merit further study.

For Process Gas *A*, a physical solvent process is the ideal choice as it would leave over 75% of the CO_2 in the treated gas. Its energy requirements are lower than for an aqueous amine. Of course, a physical solvent can be used because the gas is available at high pressure and it does not contain heavy hydrocarbons. If high COS removal were needed, a hybrid solvent would probably be the better choice.

For Process Gas *B*, considering only CO_2 and H_2S removal, promoted hot carbonate or aqueous amines would seem suitable. If large quantities of COS were present, chemical consumption would be excessive for aqueous MEA, and aqueous DEA would be preferred. However, the process gas also contains HCN. If aqueous DEA were used, a reclaiming step would be required. Reclaiming is not needed for promoted hot carbonate, since COS and HCN do not degrade the solvent. The presence of these impurities and any hydrocarbons favors promoted hot carbonate.

14.3. ENERGY REQUIREMENTS AND PROCESS INTEGRATION IN CO_2 REMOVAL

Energy costs are an important factor in gas treating. Indeed, for a large number of gas treating processes, energy can account for 70–80% of the variable operating costs, or over 50% of the total operating costs. Therefore, a critical factor in the selection of an acid gas treating process is its energy requirement and whether that requirement can be obtained from an associated process unit. Moreover, with the high energy costs being encountered today and projected for the future, there is an interest in refitting existing gas treating processes to improve their energy economy. Indeed, in many cases, modifications to promoted hot carbonate and MEA systems built in the 1960s or early 1970s can be justified.

Even in the case of units treating natural gas, which often stand alone, a portion of the energy required for the acid gas treating unit can be obtained through its integration with the Claus sulfur recovery unit. A Claus unit can generate steam at a pressure of 250 psig. When there is very little CO_2 in the feed gas, the amount of steam generated in a Claus unit can be approximately one-half of the energy requirement for an aqueous MEA gas treating unit. Large Claus plants can, in principle, generate more steam than required by some acid gas removal units.

The energy requirements for acid gas removal processes are dependent, not only on the solvent system used, but also on the specific process configuration employed. In Figure 14.3.1, the heat requirements for a number of gas treating processes have been summarized. Most promoted hot carbonate process schemes in operation today have reboiler heat duties of 90–120 kJ/gm mole of CO_2 (40–50,000 BTU/lb mole of CO_2). This heat requirement can be reduced to almost 30 k BTU/lb mol CO_2 if several stages of regeneration with intermediate flashing are employed. However, the additional cost of the flash tank, injectors, and pumps must be balanced against the energy savings.

For aqueous MEA gas treating processes, the development of corrosion inhibitors allows using higher strength MEA solution which reduces the heat consumption to 140 kJ/gmol CO_2 (60,000 BTU/lb mol). Lower energy consumptions can be achieved with single stage regeneration if the lean solution is first flashed. However, this requires using steam driver injectors as with hot potassium carbonate.

Physical absorption systems usually operate at reduced temperatures to improve the CO_2 absorption rate and also reduce solvent vapor pressure. Physical absorption systems relative to chemical solvent systems can require recycle compressors, air blowers, refrigeration, feed/product cold exchanger, and a solvent/water distillation unit. In a partial oxidation unit, the refrigeration and stripping gas requirement is available from the air liquefaction unit supplying the oxygen. The absorption requirements for a methanol system are of the order of 40–50 kJ/gm mol CO_2 (19,000 BTU/lb mol). However, the power requirements for physical absorption systems are very dependent on the system pressure, increasing at higher pressures. When physical solvent systems are operated above 400 psig, their energy advantage relative to promoted hot carbonate units is reduced.

Steam Reforming and NH_3

Efficient utilization of energy is particularly important for the manufacture of ammonia. A review of the development of steam reforming and ammonia, therefore, will show how attitudes towards the utilization of energy have changed over time.

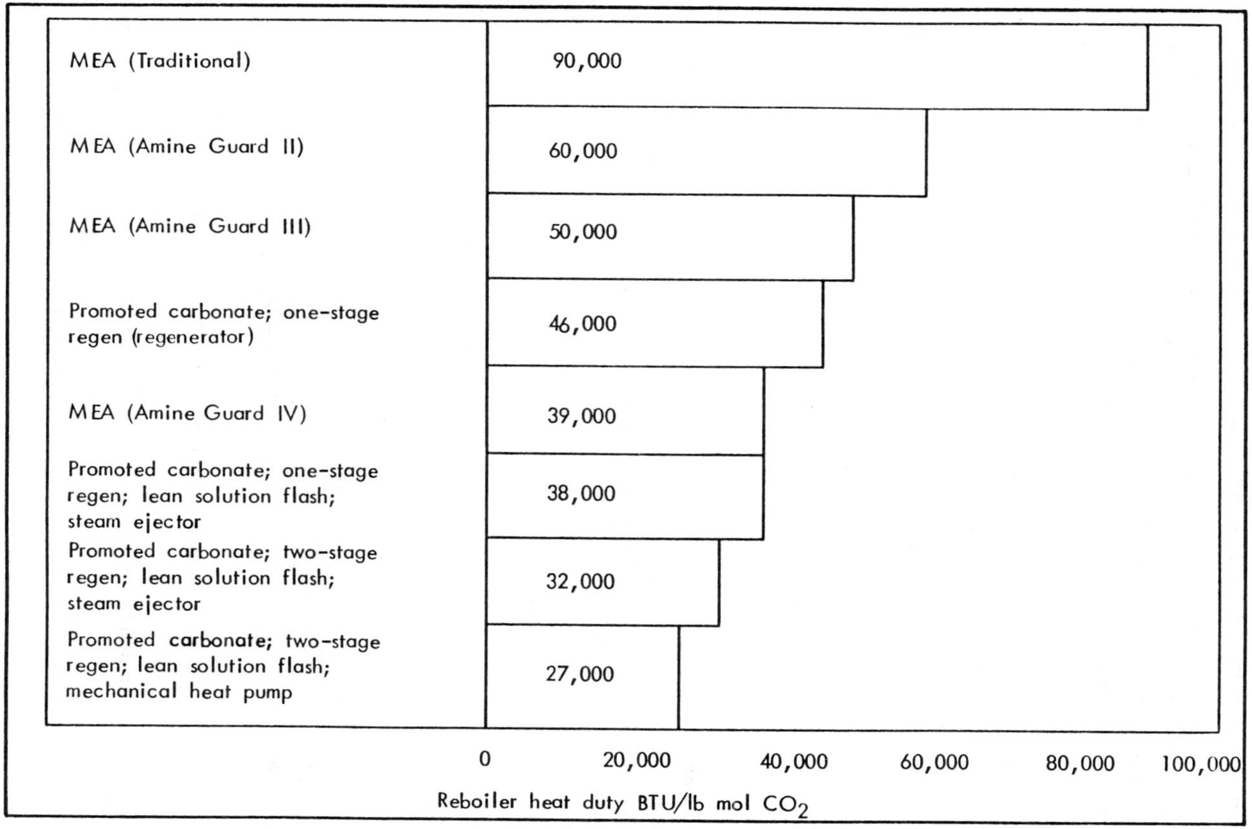

FIGURE 14.3.1. Energy requirements for CO_2 removal processes.

Commercial scale steam reforming of natural gas for the manufacture of hydrogen dates back to the early 1930s when the process was applied commercially for the first time. These plants were small—one or two MSCF/day—and were limited to feedstock such as natural gas. In addition, they operated at low pressure, delivering the product hydrogen at pressures only slightly above atmospheric.

Beginning in the early 1940s, steam reforming was adapted to produce synthesis gas for production of ammonia. Plant capacities increased to an average of about 10 MMSCF/day of hydrogen corresponding to an ammonia plant of about 150 tons/day, a size typical for that period.

Pressure reforming was introduced in the early 1950s with hydrogen being produced at pressures of 60–75 psi. Further increases in operating pressure quickly followed; first to 150 psig and then to 275 psig by the early 1960s. Simultaneously, plant capacities increased to the point where single-train plants were producing a minimum of 20 M SCF/day of hydrogen equivalent to about 300 tons/day of ammonia.

During the period from the early 1940s to the 1960s, a number of developments contributed significantly to a reduction in hydrogen production costs. Developments included improved design of process and mechanical equipment, improved materials of construction, catalysts with higher activity, greater durability, and the capability of being applied to a broader range of feedstocks. Feedstocks also began to change in this period from natural gas to light naphtha.

Improved acid gas removal systems were also introduced. Then, in the 1960s, the hydrogen production unit was totally integrated with the acid gas removal process. Operating pressures increased to over 500 psig and efficient waste heat recovery systems were developed. Capacity of hydrogen units increased to 100 M SCF/day or more, equivalent to 1500 tons/day of ammonia.

Flow schemes for steam reforming to produce ammonia synthesis gas or hydrogen are quite similar although the former has a secondary air oxidation section. The major reactions in steam reforming are:

reforming $$H_2O + CH_4 \longrightarrow 3H_2 + CO \qquad (14.3.1)$$

shift conversion $$H_2O + CO \longrightarrow H_2 + CO_2 \qquad (14.3.2)$$

methanation $$CO + 3H_2 \longrightarrow CH_4 + H_2O \qquad (14.3.3)$$

Often the first step is the desulfurization of the gaseous feedstock by activated carbon, molecular sieves of hot zinc oxide to protect against catalyst poisoning. The gas is then mixed with process steam and reformed at high temperatures (1073K) over a nickel-based catalyst, to yield an equilibrium mixture of hydrogen, carbon monoxide, carbon dioxide, methane, and water. The highly endothermic reforming reaction is propagated by the reformer furnace burners. The critical parameters for methane conversion and energy consumption are the reaction temperature and the steam-to-carbon ratio (defined as mols of steam/atom of carbon).

After reforming, the effluent is cooled to approximately 646K, before the exothermic high-temperature shift reaction. Here, the bulk of the carbon monoxide is converted to supplemental hydrogen and carbon dioxide. The discharge gases are cooled to about 498K and sent to the low-temperature shift converter for residual carbon monoxide conversion. Ultimately, a carbon monoxide concentration of about 0.2 vol% is attained.

The low-temperature shift gases are cooled to about 343K, prior to being sent to the carbon dioxide removal system where the CO_2 is reduced from the inlet level of approximately 20 vol% to 0.01 vol%. Residual carbon oxides still present in the synthesis gas after the shift conversion and carbon dioxide removal are methanized exothermically at 598K to levels less than 5 ppm. The product hydrogen is then cooled and sent to the downstream units. Hydrogen typically has a purity of 95–97% with virtually the only impurity being methane, a result of unconverted reformer feed and carbon oxide methanation.

Optimum operating conditions and the specifics of the process design are dependent both on the feedstock and the required gas composition. The design and the operating parameters for a hydrogen plant depend on:

Permissible methane content of the product hydrogen.
Pressures at which the hydrogen is produced and used.
Quantity of hydrogen to be produced.

The impact of these variables upon hydrogen production economics is considerable.

Optimization of a unit for the production of hydrogen essentially revolves around the proper utilization of the large quantities of heat associated with the unit. Heat at extremely high temperatures must be supplied to sustain the reforming reaction. This results in considerable heat being available in the furnace flue gases and the process gas stream from the reforming step. The flue gas energy can be converted to high pressure superheated steam in a waste boiler. The steam can be utilized through pressure letdown in a hydrogen compressor turbine. The exhaust steam from the compressor turbine can then be used as process steam in the reformer.

Similarly, the energy in the reformer process gas effluent can be used to generate additional steam as well as preheat reactants. In addition, the steam in the reformer process gas is used in the regeneration step of the carbon dioxide treating system. Heat integration, while desirable, results in an increase in investment.

The quantity of steam provided to the process determines the completeness of the reforming and of the shift reactions, and the effectiveness of the CO_2 removal step. As a result, the quantity of steam determines the overall hydrogen product purity. Once the hydrogen purity has been established, it is possible to develop an optimal combination of reforming temperature, steam-to-carbon ratio and carbon dioxide treating system to produce the most efficient heat balance for the process (see Figure 14.3.2). The goal is to have an overall system that is virtually self-sustaining and in steam balance—that is, one that requires no additional fuel firing for extra steam production, and does not produce copious quantities of excess low pressure steam that cannot be used within the hydrogen product unit. However, the recovery and purity of the hydrogen and carbon dioxide products cannot be sacrificed.

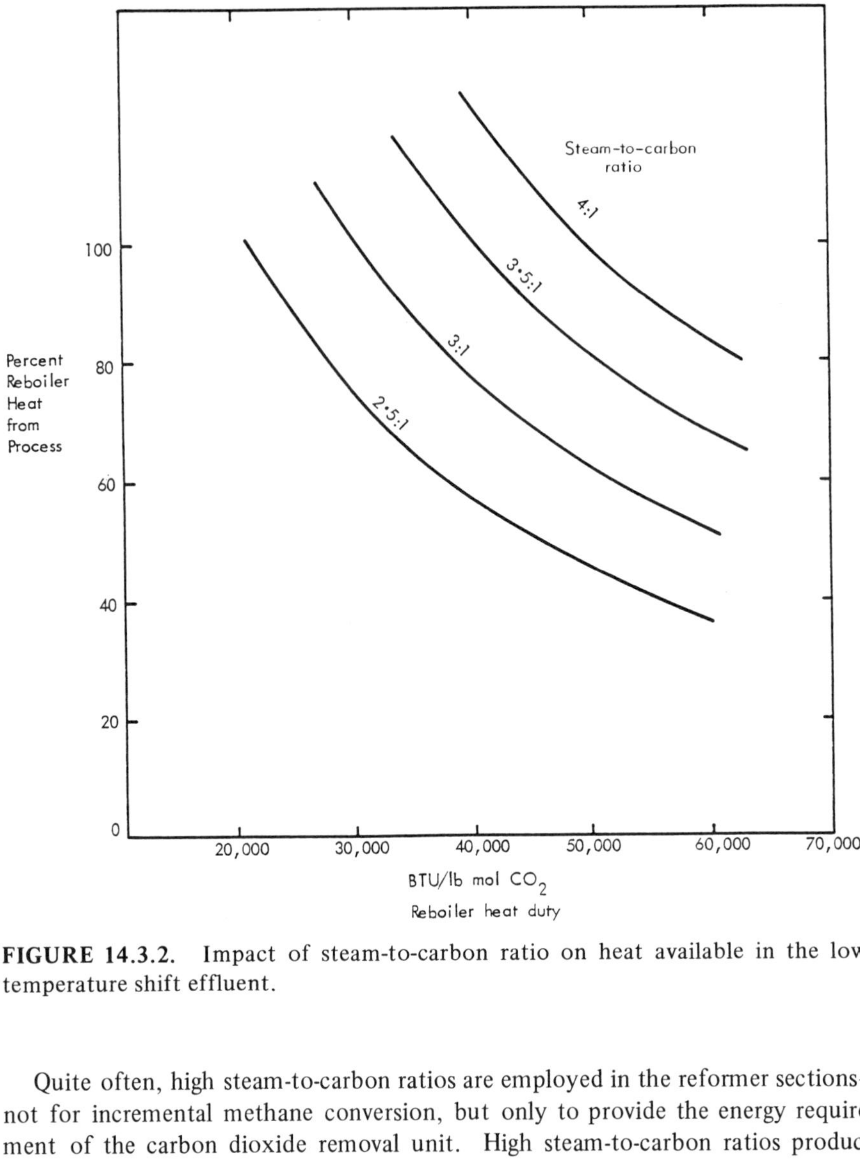

FIGURE 14.3.2. Impact of steam-to-carbon ratio on heat available in the low-temperature shift effluent.

Quite often, high steam-to-carbon ratios are employed in the reformer sections—not for incremental methane conversion, but only to provide the energy requirement of the carbon dioxide removal unit. High steam-to-carbon ratios produce only a marginal increase in hydrogen purity at higher overall reforming costs. This is because the regeneration requirements are supplied by the larger quantities of steam in the process synthesis gas exiting the low-temperature shift unit.

The steam-to-carbon ratio can be sufficiently high that a low pressure steam reboiler is not required and the reboiler heat duty is provided completely by the hot shift gas. Although this was acceptable when energy was inexpensive, today it represents a severe economic debit for a hydrogen plant.

Reductions in reboiler heat duty will permit significant reductions in reformer steam/carbon ratio inasmuch as the synthesis gas low-temperature shift gas heat requirement has been reduced. Obviously, the extent to which the regeneration reboiler duty should be reduced is dependent not only on the heat available in the synthesis gas, but also on the potential export value of steam to other units.

In Figure 14.3.2 is shown the impact of the steam-to-carbon ratio on the heat available in the shift gas stream after the low-temperature shift step in a conventional methane reforming plant. Typical conditions were used to prepare Figure 14.3.2.

A plant with a steam-to-carbon ratio of 3.5-to-1 and a reboiler heat duty of 60,000 BTU/lb mol of CO_2 could obtain 70% of the heat duty from the shift gas. However, new plants designed for reboiler duties of 40,000 BTU/lb mol CO_2 or less, and steam-to-carbon ratios of 3.5-to-1 do not need any supplemental steam, since all the reboiler heat can be provided by the synthesis gas. Hydrogen units with physical absorption CO_2 removal systems utilize the heat in the shift gas stream to operate absorption refrigeration systems or to generate electric power.

14.4. SELECTION OF FLUE GAS DESULFURIZATION PROCESS

Process selection for flue gas desulfurization is somewhat simpler than the selection of processes for CO_2 and H_2S removal from gases. There are two main reasons for this. First, fewer processes are available for SO_2 removal and, second, the SO_2 removal problem is essentially the same wherever it arises (i.e., the inlet SO_2 concentration and treated gas specification are nearly always the same).

The choice is basically between the throwaway types based on scrubbing with water slurries of lime or limestone, and regenerable types in which the SO_2 is recovered as SO_2, sulfuric acid, or sulfur. Electric utilities have tended to favor the former as they have little experience with plants for sulfur conversion or the marketing of sulfur by-products.

Commercial flue gas desulfurization processes for SO_2 removal are listed in Table 14.4.1. Many potential second generation processes are under serious development and these are listed in Figure 14.4.1.

The types based on water slurries of lime or limestone are non-regenerable. The sulfur from SO_2 is formed into a solid product intended for discharge as a solid waste. These systems generate large volumes of waste sludge and in some locations there is a shortage of landfill area required for disposing of the solid waste. Another

TABLE 14.4.1. Commercial Flue Gas Desulfurization Process

Type	System	Trade name
Throwaway	Limestone ($CaCO_3$) water slurry	–
	Lime (CaO) water slurry	–
	Double alkali	–
Regenerable, salable product	Magnesium oxide slurry	–
	Aqueous sodium sulphite	Wellman–Lord

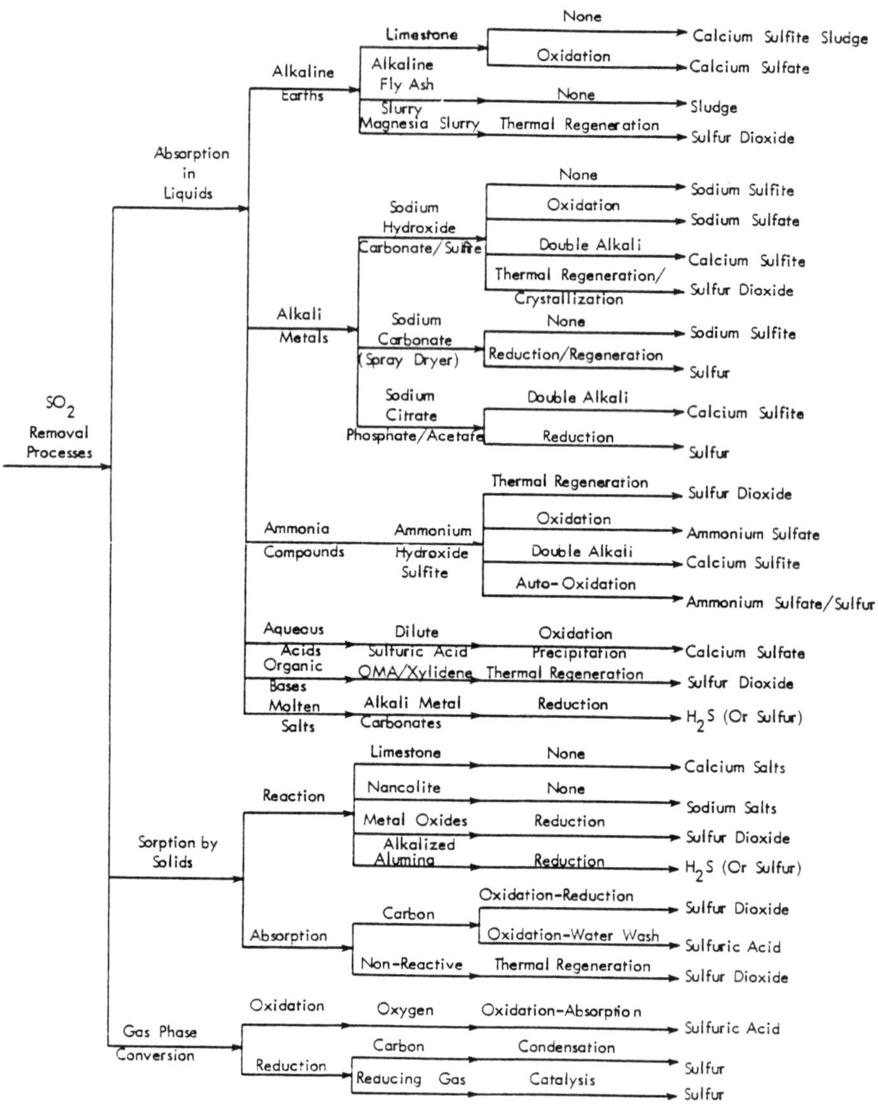

FIGURE 14.4.1. Flue gas desulfurization processes under study.

disadvantage is that despite considerable commercial experience with these units, operating problems continue to occur with the chemistry, controls, and materials of construction. Chapter 12 includes a description of throwaway flue gas desulfurization processes.

Regenerable solvent flue gas desulfurization processes are of considerable interest as they offer an alternative to lime or limestone slurry scrubbing. The most widely used regenerable process employs a homogeneous aqueous solution of sodium sulfite. The sulfur from SO_2 in the flue gas may be recovered in a useful form such as SO_2, sulfuric acid, or elemental sulfur. Another thermally regenerable process is based on an aqueous sodium citrate solution. The regenerable sulfite and citrate processes are described in Chapter 12.

14.5. FUTURE TRENDS IN GAS TREATING

The major traditional uses of gas treating for CO_2 and H_2S removal in hydrogen manufacture, ammonia production, petroleum refining, and natural gas purification are expected to experience significant growth in the coming decades. The importance of gas treating will grow faster than industrial activity as a whole, owing to the growing use of heavier more sulfurous petroleum, coal, shale, and tar sands as feedstocks and fuels. Large quantities of acid gases will have to be removed from these new energy sources. Moreover, the complexity of gas treating will increase since the acid gases will contain large quantities of CO_2, significant amounts of H_2S, and troublesome levels of other sulfur contaminants such as COS. These changes will require improved energy-efficient treating technology for both simultaneous and selective gas treating.

In the case of natural gas, new gas finds are expected to be at greater depths and in more remote locations than was the case in the past. New natural gas is expected to be more highly contaminated with H_2S and CO_2. These gases will require extensive treating to make them suitable for pipelining or conversion to LNG, methanol, or gasoline. Remote production of gas (e.g., offshore or on permafrost) will require improved technology with emphasis on small size units of light weight and high reliability.

Partial oxidation of heavy residual oils will become a major process alongside steam reforming for the generation of synthetic gases from which to make H_2, NH_3, and chemical synthesis gas (CO/H_2). Traditionally, steam reforming of sulfur-free natural gas or naphtha was used and only CO_2 removal from the gas was needed. With partial oxidation, in addition to CO_2, significant amounts of H_2S and COS will have to be removed.

For bulk removal of carbon dioxide from gases, organic promoted hot potassium carbonate processes and inhibited high-concentration monoethanolamine scrubbing will be favored. However, as sulfur-free feedstocks for synthesis gas manufacture become scarce, the need arises for treating processes that can remove CO_2, H_2S, and COS.

For simultaneous removal of H_2S and CO_2, a range of chemical absorption processes will continue to be important depending on acid gas partial pressure, H_2S to CO_2 ratio, hydrocarbon content of the gas, and degree of acid gas removal required. Processes based on aqueous aminoalcohols, solvent-aminoalcohol-water, promoted hot potassium carbonate, and physical solvents will continue to be used. Preferred will be processes showing versatility for handling H_2S, CO_2, and trace sulfur compounds.

Selective separation of hydrogen sulfide from gases containing both H_2S and CO_2 will become important for treating low H_2S–high CO_2 natural gas and gases produced in the gasification of coal and heavy oils. Processes based on kinetic selectivity for H_2S at short contact time using solutions of tertiary aminoalcohols and alkali metal salt solutions will receive much attention. There is an incentive to develop a versatile selective H_2S removal process that will convert the H_2S to sulfur directly in the regeneration step.

The emerging synthetic fuels industry based on coal and shale will create large new demands for gas treating. Coal and lignite gasification, coal liquefaction (both direct and indirect), and shale pyrolysis will all have very large requirements for acid gas removal.

A lower cost combination of processes is needed for removing acid gases from synthesis gas made by gasification of coal and heavy oils. This combination must perform the multiple functions of selective H_2S removal, bulk CO_2 removal, and trace organic sulfur compound removal while handling heavy hydrocarbons. Many possible combinations exist, including those based on physical solvents, hot carbonate, and amine-solvent-water systems. Gasifier pressure and type will be an important variable in deciding whether physical or chemical solvents are most economical.

15 Design and Scaleup

Computation of the number of trays required (for chemical absorption) is an exercise in futility.

R. N. MADDOX

This chapter covers the principles and methodology of design for industrial processes of acid gas absorption/desorption in chemically reactive solvents. The emphasis is on the differences between design procedures for physical solvents and for chemical solvents; the procedures for physical solvents are reviewed briefly, whereas those for chemical solvents are discussed in some detail.

Sections 15.1 and 15.2 are concerned with those aspects of design that are related mainly to the thermodynamic behavior of the system (i.e., the calculation of the minimum liquid rate and of the minimum steam rate). Section 15.3 is concerned with sizing the height of absorption units, which is related to the mass transfer behavior of the system. Pilot plant operation is discussed in Section 15.4, and desorption in Section 15.5. The last three sections are devoted to the selection of contactor type, flowsheet design, and technological problems such as corrosion, foaming, and solvent degradation.

We advance the thesis that a perfectly well understood process can be designed on the industrial scale using only data obtained on the laboratory scale. This is the ultimate goal of a scaleup procedure, and in this chapter we illustrate the methodology for reaching such a goal, based on the theoretical analysis discussed in earlier chapters. When this ultimate goal of scaleup methodology is attained, the need for pilot plant work may be eliminated.

15.1. MINIMUM LIQUID RATE

The calculation of the minimum liquid rate is based only on thermodynamics in the case of an *isothermal* absorption tower. Consider, in particular, the case where only one acid gas is to be removed; let G be the molar flowrate of inert gas per unit cross-sectional area, and let L be the volumetric liquid flowrate per unit cross-sectional area. A mass balance between the top of the countercurrent absorber and a generic cross section where the gas-phase mol fraction of the acid gas is Y, and its total concentration in the liquid phase is α, yields

$$G\left[\frac{Y}{1-Y} - \frac{Y_T}{1-Y_T}\right] = L(\alpha - \alpha_T) \qquad (15.1.1)$$

In the $Y - \alpha$ plane (see Figure 15.1.1) the operating line, i.e., the curve representing Eq. 15.1.1, is concave down since the local slope

$$\frac{dY}{d\alpha} = \frac{L(1-Y)}{G} \qquad (15.1.2)$$

decreases with increasing Y.

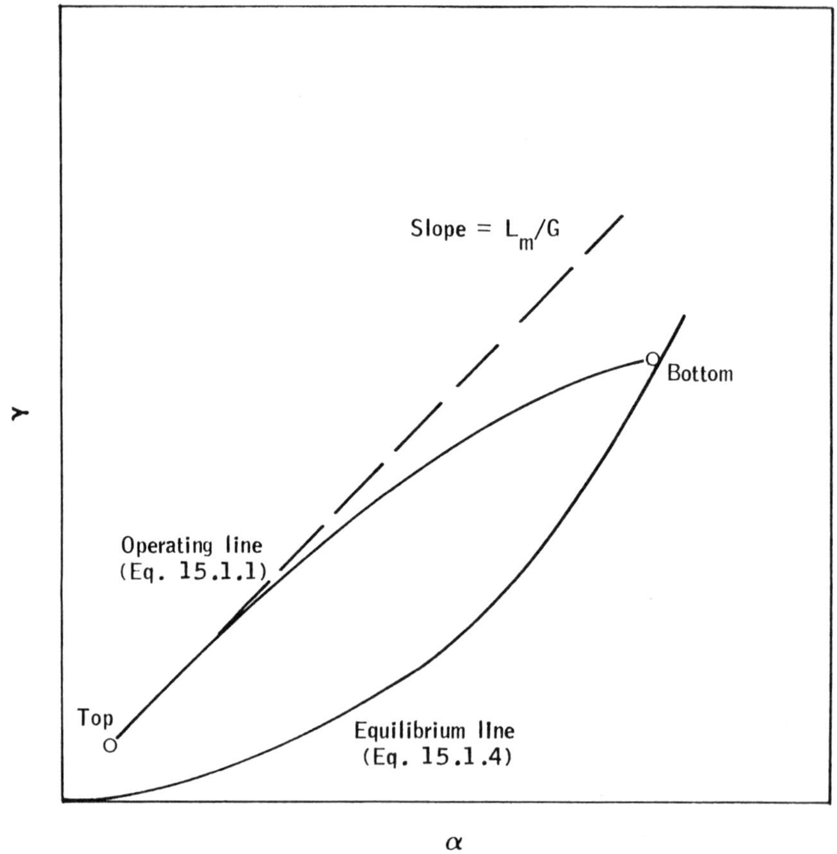

FIGURE 15.1.1. Determination of minimum liquid rate for isothermal countercurrent absorption.

Let $f(\alpha, T)$ be the equilibrium function, say

$$p^* = f(\alpha, T) \tag{15.1.3}$$

(it is useful to recall that, for chemical solvents, α can be approximated with *my* at all but extremely large values of p^*; see the discussion in Section 2.3). The equilibrium line in the $Y - \alpha$ plot is

$$Y^* = \frac{p^*}{\pi} = \frac{f(\alpha, T)}{\pi} \tag{15.1.4}$$

where π is the total pressure. For an isothermal chemical absorber (T = constant), the equilibrium curve is concave upwards since $\partial^2 f/\partial\alpha^2$ is always positive, as was discussed in Section 2.9.

Since the operating line is concave down and the equilibrium line is concave upwards, the minimum liquid flowrate L_m corresponds to an operating line that just touches the equilibrium line at the bottom of the absorber. Therefore, the value of L_m can be calculated on the basis of only a mass balance (Eq. 15.1.1) and of knowledge of the equilibrium behavior of the system considered (Eq. 15.1.3).

Industrial absorption towers where chemical solvents are used usually show significant deviations from isothermal behavior. The heat of absorption of the acid gas (which is large since a chemical reaction is involved) causes the liquid temperature to increase. Due to the high capacity of chemical solvents, the heat of absorption is released in a liquid that flows at a comparatively low rate and, therefore, the temperature rise may be very significant. The temperature rise may be mitigated by solvent evaporation and by heat transfer to the gas phase (industrial size absorption towers can be regarded as essentially adiabatic).

Inasmuch as the equilibrium function $f(\alpha, T)$ has values that strongly increase with temperature, the calculation of L_m based on Eq. 15.1.4 with T the liquid *inlet* temperature may result in a gross underestimate of L_m. An alternate approach is to use Eq. 15.1.4 with T the largest possible temperature of the liquid (i.e., the one that would result if all the heat of absorption were released in the liquid phase and none transferred back to the gas phase by heat transfer or solvent evaporation). Again, this largest possible liquid temperature can be calculated based on only thermodynamic information. However, the calculation of L_m based on the maximum possible liquid temperature is likely to lead to a significant overestimate of L_m since solvent evaporation and heat transfer to the gas phase take place significantly. Indeed, the maximum liquid temperature does not necessarily occur at the bottom of the absorption tower, but at some intermediate point. When such is the case, the pinch point that determines L_m may, in fact, be located at a section other than the rich end.

A calculation of L_m that takes into account the actual temperature profile in the absorber requires an analysis based on the *rates* of heat of absorption release, heat transfer to the gas phase, and solvent evaporation. It is, therefore, a calculation that cannot be performed only on the basis of thermodynamic information but requires understanding and mathematical modeling of the gas absorption rate behavior of the system.

The methodology for the calculation of L_m for nonisothermal absorbers is not completely developed even for physical solvents. A procedure that has been suggested in the literature is briefly reviewed below for the case of ammonia absorption in water. The procedure is based on the assumption that the resistances to both mass and heat transfer are entirely in the gas phase; such an assumption is presumably valid only for some special cases of chemical absorption.

Consider a packing segment of differential height dZ, over which three transport phenomena take place: ammonia absorption, water evaporation, and heat transfer. It is useful to define the corresponding incremental number of transfer units, dN, dN', and dN_H, as follows:

$$dZ = \frac{\tilde{G}}{\text{k}_G \tilde{a} \pi} dN = \frac{\tilde{G}}{\text{k}'_G \tilde{a} \pi} dN' = \frac{C_p \tilde{G}}{h \tilde{a}} dN_\text{H} \tag{15.1.5}$$

where $\tilde{a}$ is the interfacial area/unit volume, $\tilde{G}$ is the gas molar flowrate/unit cross-sectional area, k_G and k'_G are the mass transfer coefficients for ammonia and water, respectively, h is the heat transfer coefficient, and C_p is the gas-phase specific heat. With the definitions in Eq. 15.1.5, the transport equations take particularly simple forms; when these are combined with the mass balances for ammonia and water, and the heat balance, we obtain

$$-dY = (Y - Y^*)\, dN \tag{15.1.6}$$

$$-dY' = (Y' - Y'^*)\, dN' \tag{15.1.7}$$

$$-dT_G = (T_G - T_L)\, dN_\text{H} \tag{15.1.8}$$

where T_G and T_L are the local gas-phase and liquid-phase temperatures, and Y, Y' are the gas-phase mol fractions of ammonia and water, respectively. Equations 15.1.6-15.1.8 have been obtained for the dilute gas ($Y + Y' \ll 1$) case.

The three incremental numbers of transfer units dN, dN', and dN_H are not independent of each other, as can be seen from Eq. 15.1.5. Using classical transport theory for the evaluation of the transfer coefficients and taking the appropriate ratios leads to

$$\frac{dN'}{dN} = \beta \tag{15.1.9}$$

$$\frac{dN_\text{H}}{dN} = \beta_\text{H} \tag{15.1.10}$$

where β and β_H are approximately constant and, in fact, rather close to unity. Therefore, Eqs. 15.1.6-15.1.8 become

$$-dN = \frac{dY}{Y - Y^*} = \frac{1}{\beta}\frac{dY'}{Y - Y'} = \frac{1}{\beta_H}\frac{dT_G}{T_G - T_L} \tag{15.1.11}$$

Finally, the liquid temperature T_L and the liquid-phase concentration α are related to Y, Y', and T_G through a mass balance and a heat balance:

$$L\,d\alpha = G\,dY \tag{15.1.12}$$

$$\rho_L L C_L\,dT_L = (C_p\,dT_G + Q\,dY + Q'\,dY')\,G \tag{15.1.13}$$

where ρ_L, C_L are the liquid phase density and specific heat, and Q, Q' are the heats of solution of ammonia and water.

Equations 15.1.11-15.1.13 can be integrated starting from one extreme section of the tower (say, the top one) for different values of L/G, and a value of L_m can thus be calculated as the value that corresponds to $N \rightarrow \infty$ (i.e., to an infinitely high tower). The calculation also provides temperature and concentration profiles as well as the number of transfer units required for any assigned value of L/G; therefore, the *calculation of L_m is coupled with the sizing of the tower height.*

The results of the calculation for the ammonia-water system are given in Figures 15.1.2 to 15.1.4. The temperature of the entering water is 61°F. At the highest liquid rate, the temperature rise is about 20°F, with the maximum temperature occurring close to the base of the countercurrent absorber. As the water rate is reduced, the temperature rise increases quite sharply and the peak of the temperature maximum (bulge) moves up the column. The corresponding calculated composition profiles are shown in Figure 15.1.3, whereas Figure 15.1.4 gives the operating and equilibrium lines. The operating lines have slightly decreasing slopes as G/L increases; the equilibrium lines move towards the operating lines because of the temperature rise. At $G/L = 0.4515$, the operating and equilibrium lines nearly touch. It is clear that basing the minimum liquid rate on the assumption of phase equilibrium at the *bottom* of the absorber leads to a liquid rate that is too low.

Equation 15.1.13 can be used to estimate the safe liquid rate above the minimum to prevent a maximum internal temperature (i.e., the value of G/L at which dT_L/dN is zero at the base of the column). This will produce a better estimate of L_m than the one based on the maximum possible liquid temperature, but clearly still leads to an overestimation of L_m.

The procedure discussed above can, in principle, be extended to chemical absorption. This is, however, not an easy task; it involves, of course, also the sizing of the required packing height, which, as will be discussed in Section 15.3, is not a straightforward procedure even in the isothermal case.

One general observation concerning heat effects is that the solvent content of the inlet gas can have an important influence on the performance of the absorber.

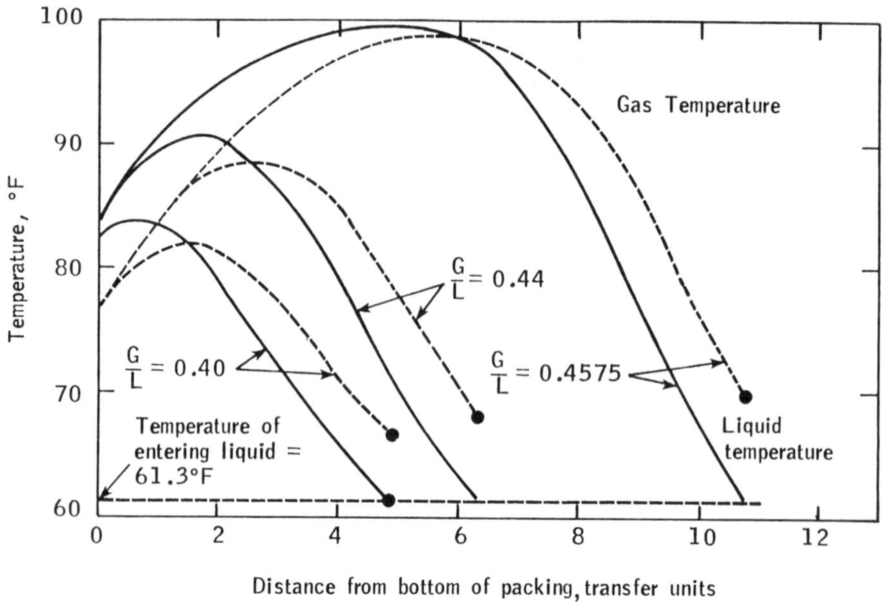

FIGURE 15.1.2. Calculated temperature profiles for adiabatic absorption of ammonia in water.

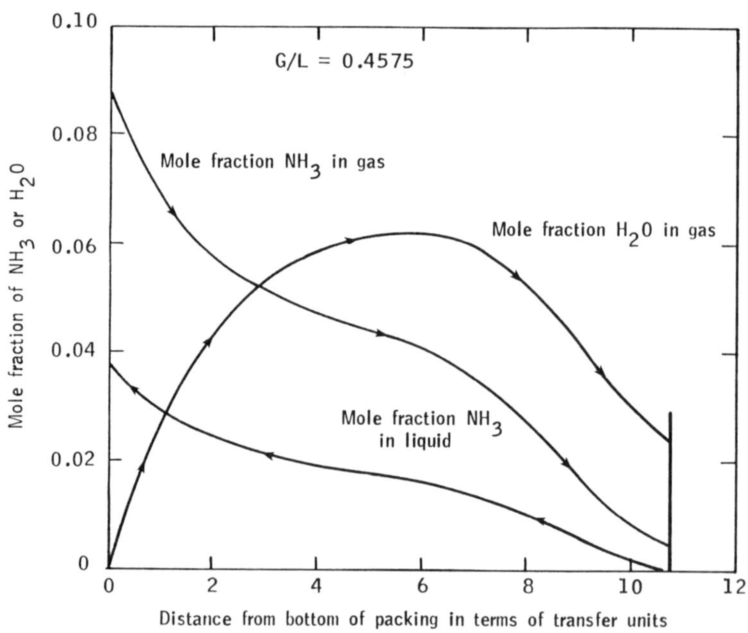

FIGURE 15.1.3. Calculated composition profiles in adiabatic ammonia absorber.

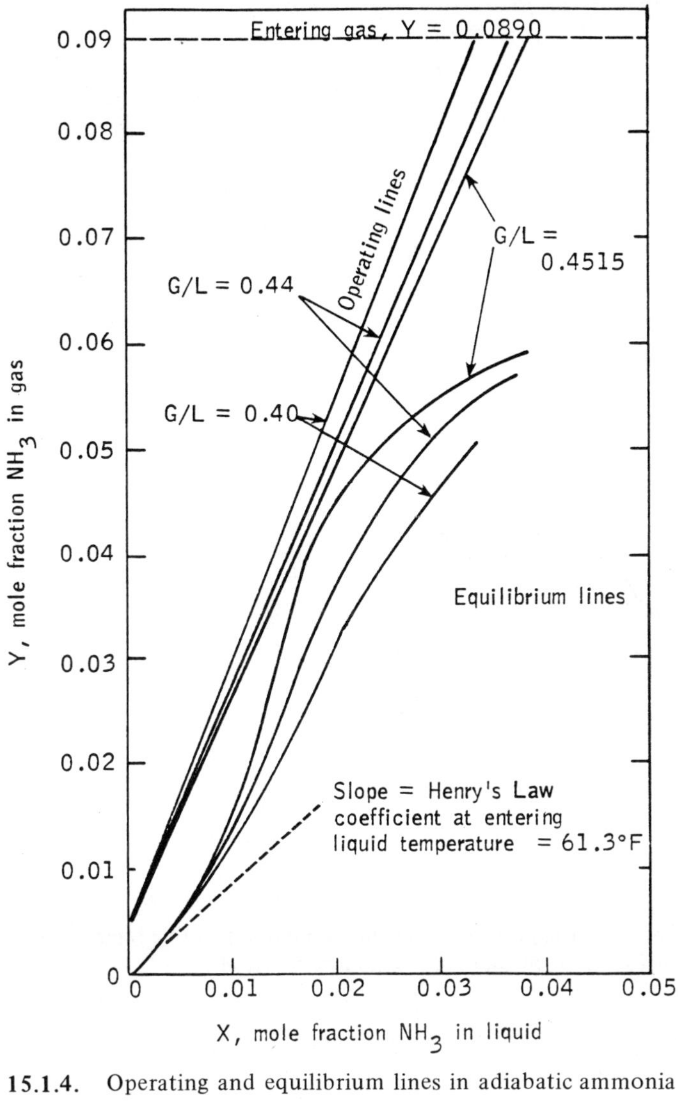

FIGURE 15.1.4. Operating and equilibrium lines in adiabatic ammonia absorber.

A low inlet solvent concentration will permit substantial adiabatic cooling of the liquid and generally improve the degree of absorption. In the case of water or aqueous solutions as the solvent, it may be worthwhile to dehumidify the inlet gas to permit greater evaporation of water within the tower.

The maximum temperature rise across the absorber depends mainly on the following factors: the heat of reaction, the molarity of the chemical solvent, and the specific heat of the solvent. This is well illustrated by the data reported in Figure 15.1.5. The most dramatic effects are, of course, observed at very high

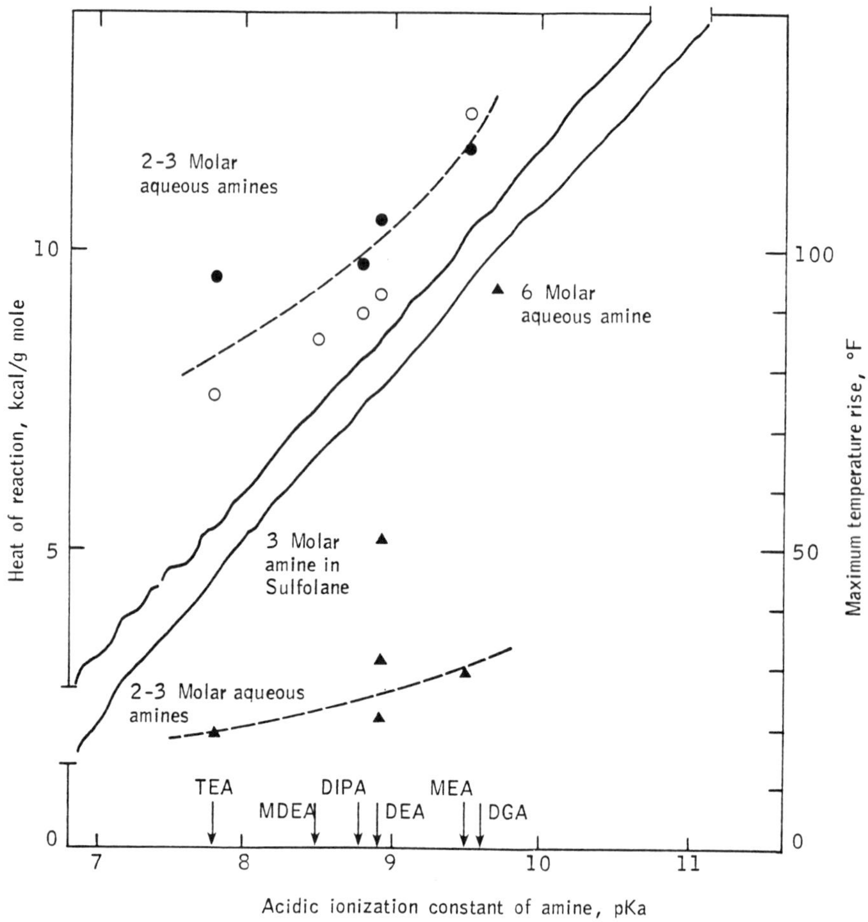

FIGURE 15.1.5. Heats of reaction and maximum possible temperature rise across absorber for H_2S absorption in aminoalcohol solutions.

molarity (see the six-molar DGA point) and at low specific heat (see the Sulfolane point).

15.2. MINIMUM STEAM RATE

The steam injected into the regeneration unit serves two purposes: it provides the sensible and latent heat required for the desorption operation, and it represents the diluent gas needed to keep the partial pressure of acid gas in the gas phase low enough to allow stripping to take place. Consequently, the required steam rate may be dictated either by the heat balance or by the stripping operation. The minimum steam rate needs to be calculated for both requirements, and the actual minimum is the larger one of the two.

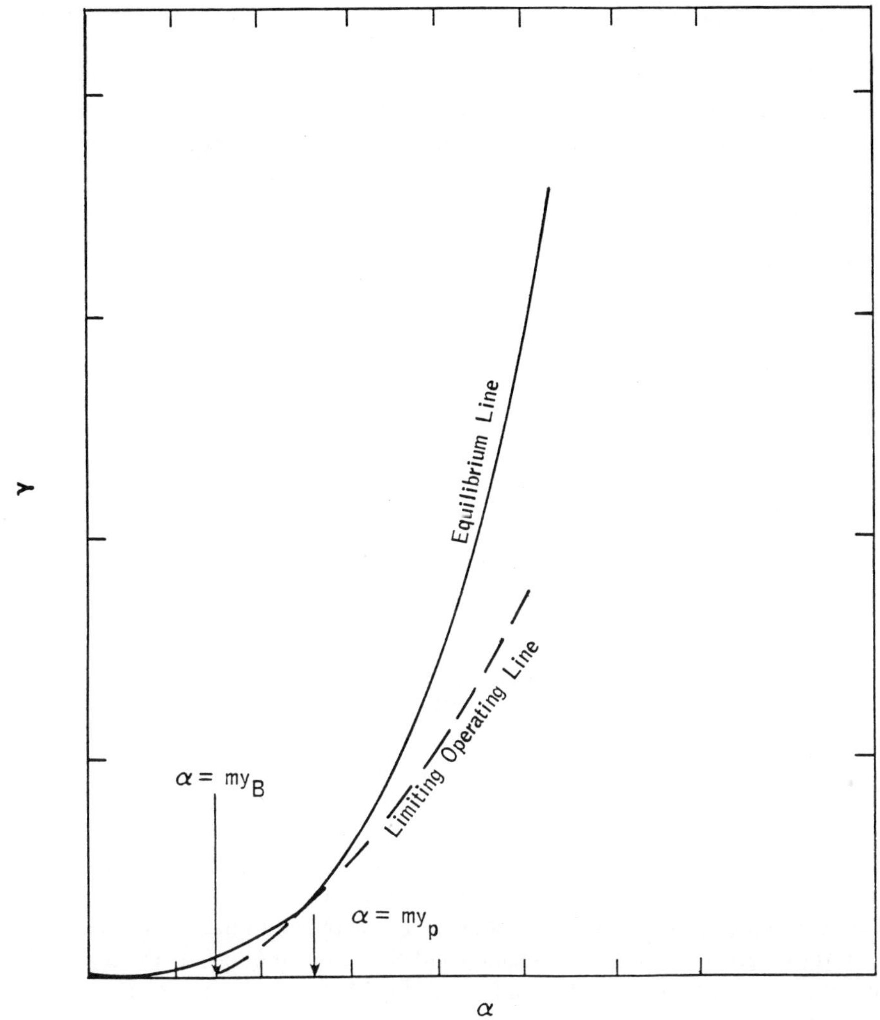

FIGURE 15.2.1. Equilibrium and limiting operating lines at bottom regenerator conditions.

Consider the stripping requirement. In a $Y - \alpha$ plane, the equilibrium line is strongly curved upwards (see Figure 15.2.1). The operating line also has a positive curvature since steam condenses along the regeneration tower to supply sensible heating as well as the heat of desorption. However, the curvature of the operating line, particularly at low α values, is much less than that of the equilibrium line, and the minimum steam rate generally corresponds to an internal pinch point at which the equilibrium and operating lines are tangent to each other.

The pinch point is likely to occur at a very low value of Y and, therefore, the analysis can be based on the simplified forms that the equilibrium equation takes in the limit $Y \to 0$, i.e., (see Section 2.9)

$$p^* = \mathrm{K}_0(T)\, m^q F_0(y) \tag{15.2.1}$$

Furthermore, since the pinch point occurs at a low value of Y, the temperature at the bottom of the regeneration unit T_1 may be used in Eq. 15.2.1; in other words, the problem of non-isothermal behavior is not an important one as far as the determination of the minimum steam rate is concerned.

The equation of the operating line is, if S is the molar steam rate,

$$p = \pi Y = \frac{Lm\pi}{S}(y - y_B) \tag{15.2.2}$$

(with $\alpha = my$ being certainly a very good approximation near the lean end). Therefore, the condition that the equilibrium and operating lines are tangent to each other can be expressed by the following two equations:

$$y_p - y_B = \frac{S_m \mathrm{K}_0(T_1)\, m^{q-1}}{L\pi} F_0(y_p) \tag{15.2.3}$$

$$1 = \frac{S_m \mathrm{K}_0(T_1)\, m^{\nu-1}}{L\pi} F_0'(y_p) \tag{15.2.4}$$

Equations 15.2.3 and 15.2.4 require the equilibrium and operating lines to have the same value and the same tangent at the pinch point $y = y_p$. The function $F_0'(y)$ is the first derivative of the function $F_0(y)$.

Elimination of y_p between Eqs. 15.2.3 and 15.2.4 allows the calculation of the minimum steam rate S_m. This will, of course, depend on the value of y_B, the fractional saturation of the lean solution. The maximum value of the latter can, in turn, be calculated from the requirement that a pinch occurs at the top of the absorber, i.e., by setting the left-hand side of Eq. 15.2.1 equal to the partial pressure in the exit gas from the absorber, and solving for y.

The equations simplify considerably whenever $F_0(y)$ is expressible as

$$F_0(y) = y^2 \tag{15.2.5}$$

which, as was discussed in Section 2.9, is very likely to be the case. When Eq. 15.2.5 applies, the solution to Eqs. 15.2.3-15.2.4 is

$$y_p = 2y_B \tag{15.2.6}$$

$$\frac{S_m}{L} = \frac{\pi}{4y_B \mathrm{K}_0(T_1)\, m^{q-1}} \tag{15.2.7}$$

Let the allowable partial pressure in the clean gas be p_T. Then the pinch at the lean end of the absorber condition ($p^* = p_T$ in Eq. 15.2.1) yields

$$y_B = \sqrt{\frac{p_T}{\mathrm{K}_0(T_2)\, m^q}} \qquad (15.2.8)$$

where T_2 is the temperature at the top of the absorber. Elimination of y_B between Eqs. 15.2.7 and 15.2.8 yields

$$\frac{S_m}{L} = \frac{m^{1-q/2}}{4\sqrt{p_T \mathrm{K}_0(T_2)}} \exp\left[-\frac{Q_0}{R}\left(\frac{1}{T_2} - \frac{1}{T_1}\right)\right] \qquad (15.2.9)$$

where use has been made of the identity

$$\frac{\mathrm{K}_0(T_2)}{\mathrm{K}_0(T_1)} = \exp\left[-\frac{Q_0}{R}\left(\frac{1}{T_2} - \frac{1}{T_1}\right)\right] \qquad (15.2.10)$$

with Q_0 the heat of solution at $y \to 0$.

Equation 15.2.9 is very important and should be discussed in some detail. It shows that strong absorbents, characterized by a very low value of $\mathrm{K}_0(T_2)$, require a large steam rate for stripping. Notice, however, that S_m increases only proportionally to the square root of $1/\mathrm{K}_0(T_2)$. The equation also reflects the beneficial effect of a large temperature swing between regenerator bottom and absorber top that can be achieved by deeper solvent cooling or by the use of solvents with high boiling point.

Calculated values of S_m for H_2S desorption from amine solutions based on a simplified form of Eq. 15.2.9 for $T_2 = 40°C$ (313K) and $\pi = 1.8$ bar are reported in Figure 15.2.2. For these systems, $\mathrm{K}_0(T_2)$ is simply proportional to the protonation constant of the amine (see Eq. 10.2.6); indeed, Figure 15.2.2 shows that the minimum steam rate increases with increasing amine basicity.

We now turn attention to the heat balance limitation. The minimum steam rate required for the heat balance is

$$\frac{S_m}{L} = \frac{\rho_L L C_L (T_1 - T_3) + m\bar{Q}\Delta y}{Q_s} + \left(\frac{S}{L}\right)_R \qquad (15.2.11)$$

where T_3 is the temperature of the liquid leaving the rich-lean heat exchanger, $\bar{Q}$ is the average heat of absorption over the absorber, Δy is the change of fractional saturation over the absorber, Q_s is the latent heat of steam, and $(S/L)_R$ is the steam rate required at the regenerator top. Equation 15.2.11 states that the heat supplied to the reboiler needs to: (1) supply sensible heat to raise the liquid temperature from T_3 to T_1; (2) supply the heat of desorption of acid gases; and (3) supply a residual steam rate at the regenerator top such that the water vapor pressure over

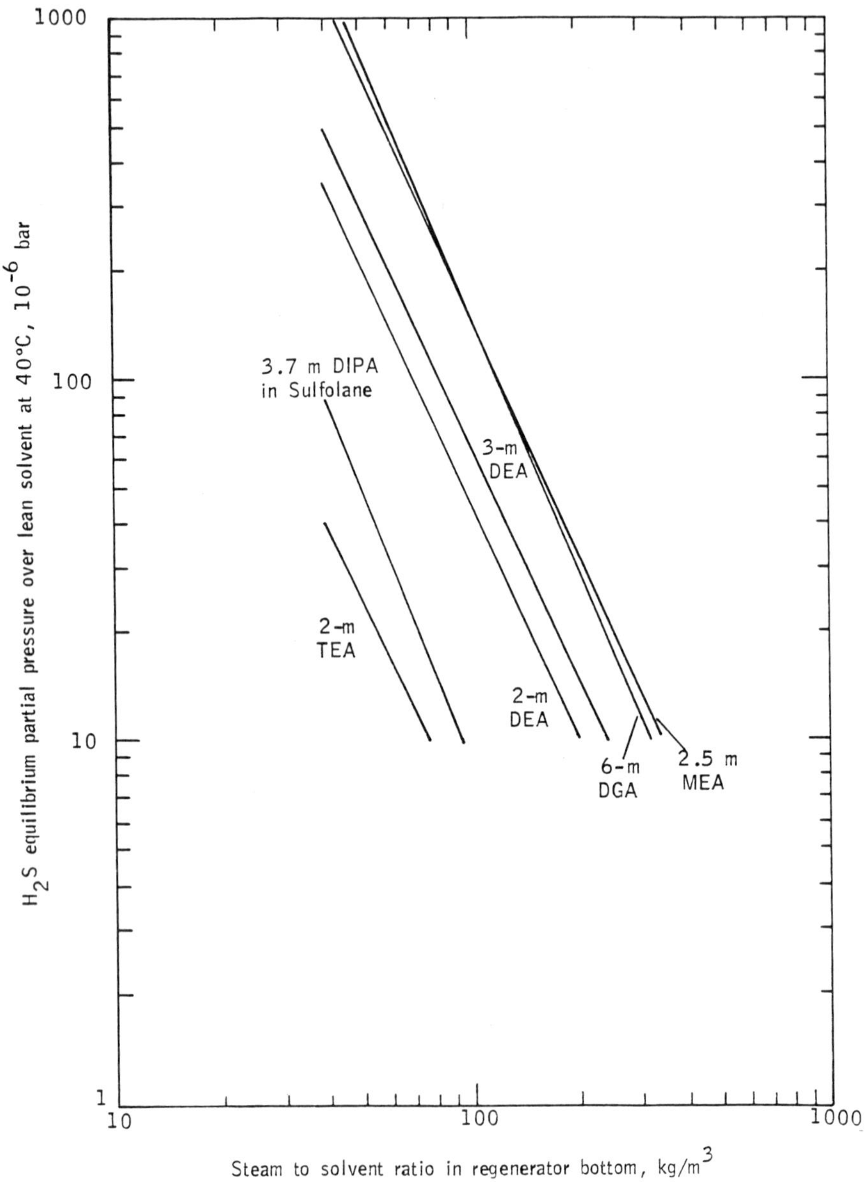

FIGURE 15.2.2. Stripping-limited minimum steam ratio as a function of equilibrium H_2S pressure of lean solvent.

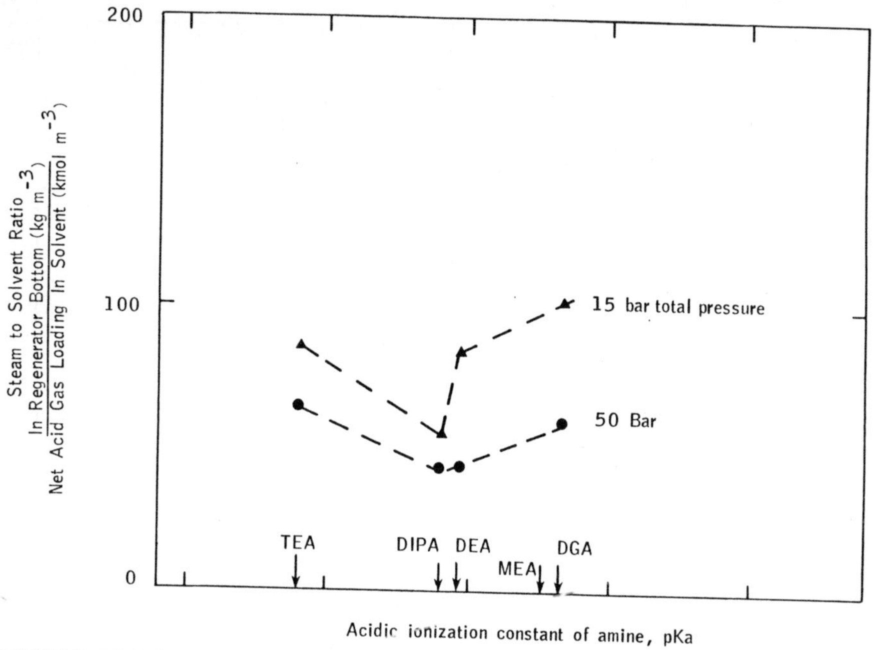

FIGURE 15.2.3. Heat balance-limited steam requirement for H_2S regeneration from aqueous aminoalcohols.

the solution entering the regenerator is matched by the partial pressure of steam in the vapor upflow. Figure 15.2.3 illustrates some steam rate values calculated from Eq. 15.2.11 for H_2S removal with amine solutions.

15.3. TOWER SIZING

The diameter of a tower is set by the flowrates of the gas and liquid phases flowing through it. The tower height is determined by the rate of mass transfer. Packings or other tower internals are designed to promote a large interfacial area of contact between the phases, and a high degree of mixing in both phases, with a minimum resistance to flow of the two phases (large void volume).

For estimating the diameter of packed columns with random packings, the generalized pressure drop correlation for dumped (random) packings shown in Figure 15.3.1 should be used. The packing factor F should be taken from Table 15.3.1 for the appropriate size of packing. Packings smaller than 1 in. size are intended for towers one ft or smaller in diameter; packings 1 in or 1.5 in. in size for towers over one ft and up to 3 ft in diameter; and 2-3.5 in. packings are used for towers 3 ft or more in diameter. The packing factor F is an experimentally determined quantity, obtained for the air–water system in 16 in. and 30 in. columns. In

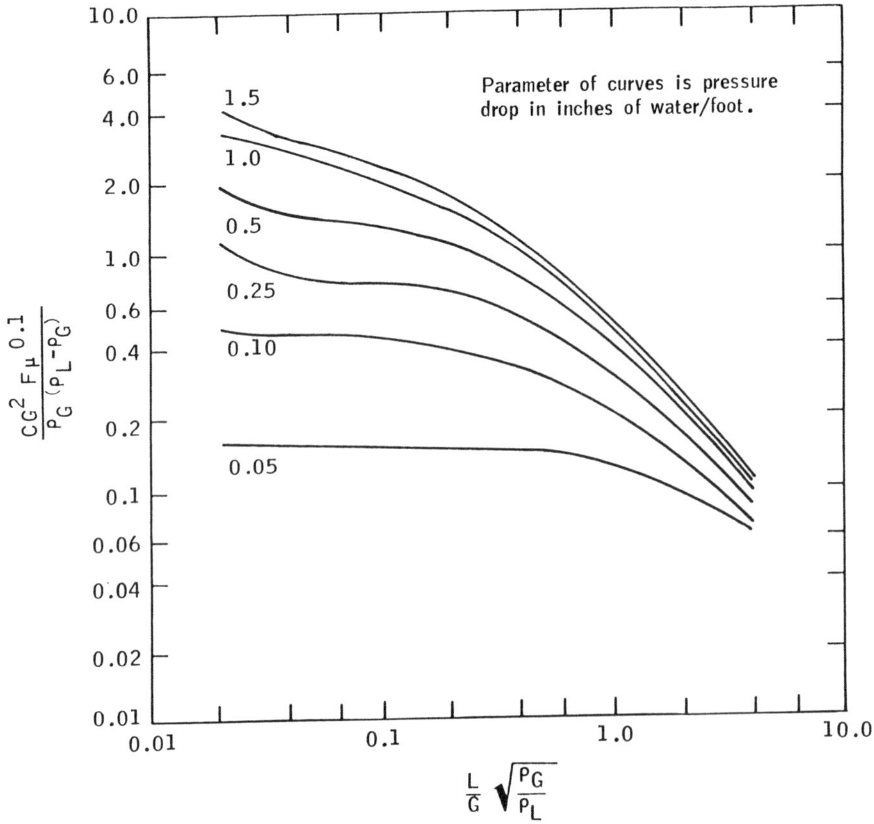

FIGURE 15.3.1. Generalized pressure drop correlation for dumped packings.

general, the largest size packing pieces that meet the packing-size-to-tower-diameter ratio criteria given above should be selected for initial designs. This will minimize tower diameter. If tower height is excessive, smaller packing pieces may provide a more optimum design, especially for the absorber and regenerator cleanup sections. Based on the packing factor data in Table 15.3.1, it appears that Pall rings and Intalox saddles may have certain advantages over other types of random packings. The cost of packing is also a factor in its selection.

A different but similar pressure drop correlation can be used to determine the diameter of trayed towers.

Aqueous amine and potassium carbonate solutions are generally designed to operate at 50–70% of the flooding limit (as compared to 70–90% of flood for physical solvents) owing to their tendency to foam. As far as tower diameter is concerned, the principal difference between physical and chemical solvents is that physical absorption processes operate at higher L/G ratios owing to the lower acid gas capacity of physical solvents. With chemical solvents, lower L/G ratios are

TABLE 15.3.1. **Packing Factors for Dumped Packings (determined with air–water system in 30 in. I.D. tower)[a]**

		Nominal packing size (inches)										
Packing type	Material	$\frac{1}{4}$	$\frac{3}{8}$	$\frac{1}{2}$	$\frac{5}{8}$	$\frac{3}{4}$	1	$1\frac{1}{4}$	$1\frac{1}{2}$	2	3	$3\frac{1}{2}$
Hy-Pak™	Metal						43		26	18		15
Super Intalox saddles	Ceramic						60			30		
Super Intalox saddles	Plastic						33			21		16
Pall rings	Plastic				97		52		40	24		16
Pall rings	Metal				70		48		33	20		16
Intalox saddles	Ceramic	725	330	200		145	92		52	40	22	
Raschig rings	Ceramic	1600	1000	580	380	255	155	125	95	65	37	
Raschig rings	$\frac{1}{32}$ in. metal	700	390	300	170	155	115					
Raschig rings	$\frac{1}{18}$ in. metal			410	290	220	137	110	83	57	32	
Berl saddles	Ceramic	900		240		170	110		65	45		
Tellerettes	Plastic						38			19		
Maspac	Plastic									32		20
Partition rings	Ceramic										80	

[a]Packing factors are essentially independent of liquid flowrate in the range 5,000 to 50,000 lb/hr ft^2.

used and tower diameter may be determined either by the vapor flow or by the liquid flow.

We now turn attention to the problem of sizing the required height of packing for a process of chemical absorption. The procedure for such a calculation is a rather involved one and, in fact, often the problem is circumvented by simply resorting to deliberate overdesign. However, overdesign is never an economical solution, and it is not even a technically acceptable one in cases where kinetic selectivity is required.

We only discuss the *methodology* of the design procedure and attention is, therefore, limited to the case of an *isothermal* absorber. The basic ideas to be developed if non-isothermal behavior needs to be taken into account have been introduced in Section 15.1.

We first briefly review the methodology of design for the case of physical solvents; furthermore, we restrict attention to the case where Henry's Law applies (i.e., the equilibrium relationship between the interface partial pressure and the interface liquid-side concentration is linear). Therefore, the equation for the mass transfer flux N takes the form of Eq. 1.6.6. The incremental rate of absorption per unit cross-sectional area over a length dZ of packing can thus be written as

$$N\tilde{a}\, dZ = \mathrm{K}_G^0\tilde{a}\pi(Y - Y^*)\, dZ \tag{15.3.1}$$

where $\mathrm{K}_G^0\tilde{a}$ is an overall gas-phase mass transfer coefficient per unit volume:

$$\mathrm{K}_G^0\tilde{a} = \frac{1}{1/\mathrm{k}_G\tilde{a} + \mathrm{H}/\mathrm{k}_L^0\tilde{a}} \tag{15.3.2}$$

The important consideration is that $\mathrm{K}_G^0\tilde{a}$ is, to a first approximation, constant along the tower axis. Its value depends on the efficiency of the packing being used and on the *equilibrium* behavior of the system (through Henry's constant H).

In analogy to Eq. 15.1.5, an overall incremental number of transfer units dN^0 can be defined as follows:

$$dZ = \frac{\tilde{G}}{\mathrm{K}_G^0\tilde{a}\pi}\, dN^0 \tag{15.3.3}$$

so that the transport equation takes the following simple form:

$$Na\, dZ = (Y - Y^*)\, \tilde{G}\, dN^0 \tag{15.3.4}$$

When this is combined with the local mass balance equation, one obtains

$$-d\left(\frac{Y}{1 - Y}\right) = (Y - Y^*)\, dN^0 \tag{15.3.5}$$

Equation 15.3.5 can be integrated directly to yield

$$Z_T = \frac{\tilde{G}}{K_G^0 \tilde{a} \pi} N^0 = \left(\frac{\tilde{G}}{K_G^0 \tilde{a} \pi} \right) \int_T^B \frac{dY}{(1 - Y)^2 (Y - Y^*)} \tag{15.3.6}$$

where T and B identify top and bottom conditions in the absorber. The integral on the right-hand side of Eq. 15.3.6 represents the total number of transfer units required (and is thus a measure of the difficulty of the separation process to be carried out), whereas the term in round brackets represents the height of packing required for one transfer unit, or the "height of a transfer unit" *HTU*. The *HTU* is an inverse measure of the efficiency of the packing ; the lower its value, the more efficient the packing. As a *rough* rule of thumb, the number of transfer units is also a good approximation of the number of theoretical trays required for the same process if carried out in a trayed column.

In the special case where the gas is *dilute* ($Y \ll 1$), a considerable simplification arises. The denominator of the integrand in Eq. 15.3.6 simply becomes the driving force $Y - Y^*$; the latter, in turn, is a linear function of Y, inasmuch as both the operating and the equilibrium lines are linear. Straightforward algebra leads to the following result:

$$Z_T = \frac{M}{K_G^0 \tilde{a} \pi} (Y - Y^*)_{\text{lm}} \tag{15.3.7}$$

where M is the total absorption rate per unit cross-sectional area,

$$M = \tilde{G}(Y_B - Y_T) \tag{15.3.8}$$

and $(Y - Y^*)_{\text{lm}}$ is the log mean driving force,

$$(Y - Y^*)_{\text{lm}} = \frac{(Y - Y^*)_B - (Y - Y^*)_T}{\ln [(Y - Y^*)_B/(Y - Y^*)_T]} \tag{15.3.9}$$

Now consider the case of a *chemical* absorption process. The transport equation for the gas phase is still written in the form of Eq. 1.6.3, but the transport equation for the liquid phase needs to be written in the form of Eq. 3.2.1, with k_L related to k_L^0 by Eq. 3.2.2. Thus one has

$$N\tilde{a}\, dZ = k_G \tilde{a} \pi (Y - Y_i)\, dZ = I k_L^0 \tilde{a} \frac{\pi}{H} (Y_i - Y^*) \tag{15.3.10}$$

where the enhancement factor I is a complex function of the bulk-liquid composition, interface partial pressure, and liquid-side mass transfer coefficient:

$$I = I(Y^*, Y_i, k_L^0) \tag{15.3.11}$$

For a non-isothermal absorber, I should be regarded as also depending on temperature. In Chapters 3-8 the theory required for the calculation of the functional dependency of I on Y^*, Y_i, and k_L^0 was developed.

If the right-hand side of Eq. 15.3.11 is known (i.e., if the *rate* behavior of the system considered has been modeled satisfactorily), Y_i can be eliminated between Eqs. 15.3.10 and 15.3.11. This elimination will never lead to any simple mathematical expression and, in fact, often the resulting equation will need to be solved numerically. Formally, the result is as follows:

$$N\tilde{a}\, dZ = K_g^0 \tilde{a} \pi (Y - Y^*)\, \xi\, dZ \tag{15.3.12}$$

where

$$\xi = \frac{1 + P}{1 + P/I} \quad \text{and} \quad P = \frac{k_G a H}{k_L^0 a} \tag{15.3.13}$$

The quantity P is the ratio of the effectiveness of ***physical*** mass transfer in the gas phase to that in the liquid phase. Since the physical solubility of acid gases is low (say, the value of H is large), in chemical absorption processes, P is likely to be a large number, say,

$$P \gg 1 \tag{15.3.14}$$

(A "typical" value of $k_G a / k_L^0 a$ is 10 gmol/liter atm. Therefore, with H values of order 10^2 atm liter/gmol, values of P are of the order of 10^3.)

The crucial term in Eq. 15.3.12 is ξ. A brief discussion of the values that ξ may take is given in the following.

Consider the case where the enhancement factor I is not extremely large; say, the following condition is fulfilled.

$$I \ll P \tag{15.3.15}$$

(Condition 15.3.15 is likely to occur at the rich end of the absorber.) In this case, in view also of Eq. 15.3.14, one has

$$\xi \simeq I \tag{15.3.16}$$

Furthermore, from Eqs. 15.3.2 and 15.3.14 one obtains

$$K_G^0 a \simeq k_L^0 a \tag{15.3.17}$$

and, therefore, Eq. 15.3.12 reduces to

$$N\tilde{a}\, dZ = k_L^0 \tilde{a} I \pi (Y - Y^*)\, dZ \tag{15.3.18}$$

The condition discussed above is the one where the enhancement factor is small enough to make the liquid-side resistance to mass transfer the rate-determining step. Therefore, $Y_i \simeq Y$ and Eq. 15.3.11 can be used to estimate the value of I by substituting Y for Y_i.

Conversely, consider the case where the enhancement factor is very large indeed, say,

$$I \gg P \tag{15.3.19}$$

(Condition 15.3.19 is likely to occur at the lean end of the absorber.) In this case, one has

$$\xi = 1 + P \tag{15.3.20}$$

and Eq. 15.3.12 reduces to

$$N\tilde{a}\, dZ = k_G a\pi (Y - Y^*)\, dZ \tag{15.3.21}$$

This situation is the one where the enhancement factor is so large that the liquid-side resistance to mass transfer has essentially been eliminated; the local rate of mass transfer is governed by the gas-phase resistance.

Combination of Eq. 15.3.12 with the local mass balance and integration yields the following design equation:

$$N^0 = \frac{K_G^0 \tilde{a} \pi}{\tilde{G}} Z_T = \int_T^B \frac{dY}{\xi (1 - Y)^2 (Y - Y^*)} \tag{15.3.22}$$

Although Eq. 15.3.22 bears a superficial analogy to the physical absorption Eq. 15.3.6, there is a crucial difference. The value of the integral on the right-hand side of Eq. 15.3.22 (i.e., the required number of transfer units) depends not only on the physico-chemical properties of the system considered and on the operation to be carried out, but also on the individual mass transfer coefficients $k_G a$ and $k_L^0 a$. Furthermore, even if the gas is dilute (say, $Y \ll 1$), the quantity $\xi(Y - Y^*)$ is a highly non-linear function of Y and, therefore, no analogue of the simple log mean rule is ever likely to occur.

The overall mass balance between the top section and a generic one yields

$$\frac{Y}{1 - Y} = \frac{Y_T}{1 - Y_T} + \frac{L}{G}(\alpha - \alpha_T) \tag{15.3.23}$$

which can be solved for $\alpha(Y)$. If a reliable thermodynamic model is available, one can then calculate $Y^*(\alpha)$ and, therefore, $Y^*(Y)$. Simultaneous solution of Eqs. 15.3.10 and 15.3.11 then yields $I(Y)$ and, therefore, $\xi(Y)$. With the values of Y^* and ξ available at any value of Y, the integral in Eq. 15.3.22 can be calculated. This exhausts the design procedure.

It is important to point out that the design procedure outlined above requires both a reliable thermodynamic model and a knowledge of the function on the right-hand side of Eq. 15.3.11. If, as is likely to be the case, the absorber is nearly pinched at the lean end, this information is needed particularly in the range of very low values of both Y and α, the range where both thermodynamic and rate data are very difficult to obtain. Therefore, the thermodynamic model and the rate model need to have firm theoretical ground because they both are crucially used in a range of Y and α values where they are essentially extrapolated from the range where laboratory data have been collected.

The analysis discussed above refers to the general case. Considerable simplifications arise when the mass transfer process in the liquid phase is known to take place in a specific regime. When the regime is known to be the instantaneous reaction one, the transport equation in the liquid phase is written in the form of Eq. 3.6.20, and one obtains

$$N\tilde{a}\, dZ = k_G\tilde{a}\pi(Y - Y_i)\, dZ = k_L^0\tilde{a}(\alpha_i^* - \alpha)\, dZ \tag{15.3.24}$$

where α_i^* is the liquid load that equilibrates Y_i. Therefore, the thermodynamic model furnishes the $Y_i(\alpha_i^*)$ relationship, and Eq. 15.3.23 the $Y(\alpha)$ relationship. Equation 15.3.24 can be combined with the local mass balance and integrated to yield Z_T.

When the regime is known to be the fast reaction one, the transport equation is written in the form of Eq. 3.5.11, and after algebraic rearrangement one obtains

$$N\tilde{a}\, dZ = \frac{\pi(Y - Y^*)\, dZ}{1/k_G\tilde{a} + (H/\tilde{a}\sqrt{D/t_r}\,)} \tag{15.3.25}$$

which can be combined with the local mass balance and integrated to yield

$$Z_T = \frac{\tilde{G}}{\pi}\int_{Y_T}^{Y_B} \frac{1/k_G\tilde{a} + (H/\tilde{a}\sqrt{D/t_r}\,)\, dY}{(1 - Y)^2\,(Y - Y^*)} \tag{15.3.26}$$

The integral on the right-hand side of Eq. 15.3.26 can be calculated by obtaining the $Y - \alpha$ relationship from Eq. 15.3.23, the $Y^* - \alpha$ relationship from the thermodynamic model, and the $t_r - \alpha$ relationship from the rate model.

It is important to point out that, in order to apply the procedures discussed above, one needs to know the individual mass transfer coefficients k_G and k_L^0 and the interface area $\tilde{a}$. This is in contrast with the case of physical solvents, where only the products $k_G\tilde{a}$ and $k_L^0\tilde{a}$ need to be known.

Available correlations for k_G, k_L^0, and $\tilde{a}$ are far from adequate. No correlations are available in the literature for the newer high-efficiency random packings such

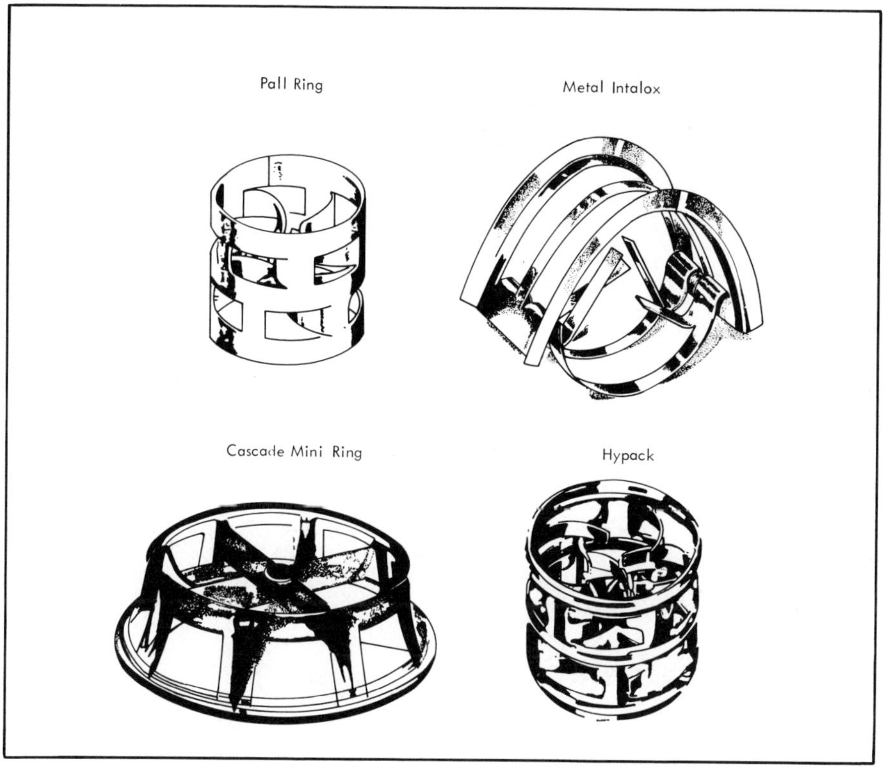

FIGURE 15.3.2. Random tower packings.

as those shown in Figure 15.3.2. For any particular packing type, quantitative correlations must be developed by pilot or mockup studies using the methods to be discussed in Section 15.5. Information in the literature for some of the older packings (e.g., Raschig rings) can be used to suggest the form that $k_G\tilde{a}$ and $k_L^0\tilde{a}$ correlations may take. The following discussion covers available information which may be a useful guideline for studies with new packings.

When operating in the trickle flow regime below the gas loading point, the gas-liquid interfacial area $\tilde{a}$ is always less than the total dry area of the packing and it depends on liquid flowrate L, as shown in Figure 15.3.3. It depends also on the physical properties of the liquid. We recommend the following general correlation which relates the gas-liquid interfacial area $\tilde{a}$ to the surface area a_t of the dry packing by the equation

$$\frac{\tilde{a}}{a_t} = 1 - \exp\left\{-1.45\left(\frac{\sigma_C}{\sigma}\right)^{0.75}\left(\frac{\tilde{L}}{a_t\mu_L}\right)^{0.1}\left(\frac{L^2 a_t}{\rho_L^2 g}\right)^{-0.05}\left(\frac{\tilde{L}^2}{\rho_L \sigma a_t}\right)^{0.2}\right\} \tag{15.3.27}$$

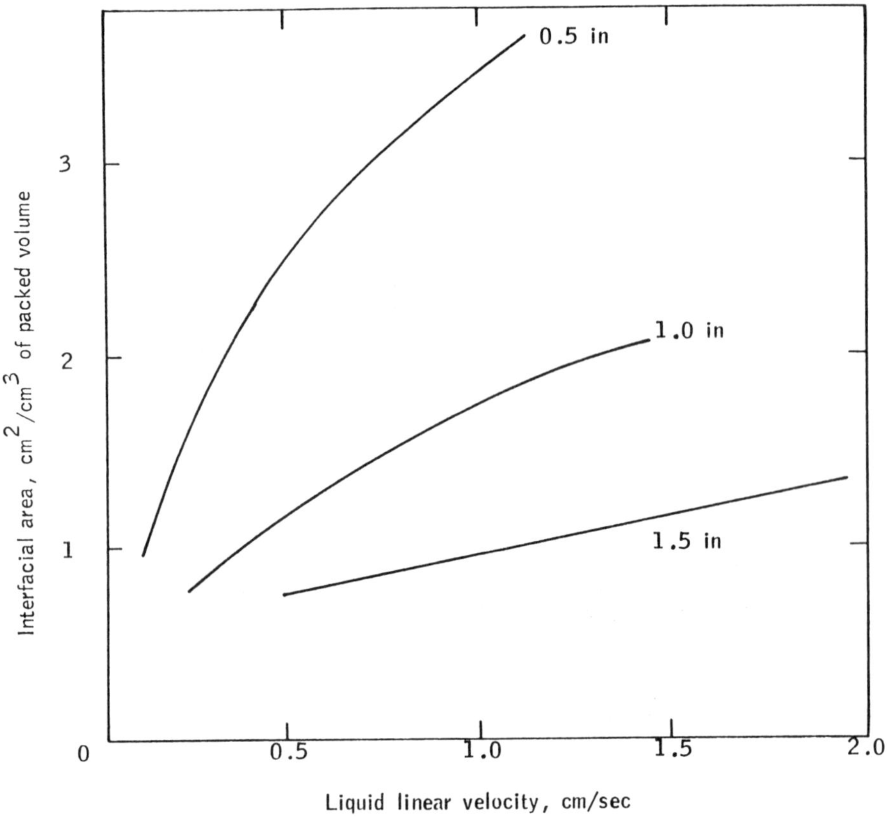

FIGURE 15.3.3. Interfacial areas in countercurrent packed beds of Ceramic Intalox saddles.

where σ_C = critical surface tension of the liquid for the particular packing material, σ = surface tension of the liquid, $\tilde{L}$ = superficial liquid mass flow rate, μ_L = viscosity of liquid, ρ_L = density of liquid.

Equation 15.3.27 correlates experimental results for a wide range of random packings having a continuous-flow configuration (e.g., uninterrupted rings, spheres, saddles). For Pall rings, which have an interrupted surface configuration, the interfacial area may be underestimated by as much as 50%.

Correlations for the product $k_L^0\tilde{a}$ have been obtained by means of experiments of absorption and desorption of sparingly soluble gases such as O_2 and CO_2 in water. For these systems, the liquid-phase resistance to mass transfer is the controlling one (say, H is so large that Eq. 1.6.9 applies). Very short bed heights (of the order of 20 cm) need to be used to ensure measurable terminal driving forces and, therefore, extensive precautions have to be taken to eliminate end effects (see also the discussion in Section 15.5).

Several correlations are available in the literature; we recommend the following, which is consistent with Eq. 15.3.27 in the sense that it has been obtained by subtracting out the effect of $\tilde{a}$ from available $k_L^0\tilde{a}$ data:

$$k_L^0\left(\frac{\rho_L}{\mu_L g}\right)^{1/3} = 0.0051\left(\frac{\tilde{L}}{\tilde{a}\mu_L}\right)^{2/3}\left(\frac{\mu_L}{\rho_L D}\right)^{-1/2}(a_t\,d_p)^{0.4} \qquad (15.3.28)$$

where d_p is the nominal packing size. The exponent $-\frac{1}{2}$ on the Schmidt number confirms the penetration theory predictions (see Eqs. 3.2.22 and 3.2.28). The correlation appears to work well for several types and sizes of packing and with both aqueous and organic liquids.

Correlations for gas-side mass transfer coefficients have been less successful than those for liquid-side mass transfer. The reasons for this are not well understood, but experimental problems associated with end effects in short columns may be a contributing factor; see also the discussion in Section 15.5. We recommend the following correlation, which again is consistent with Eq. 15.3.27:

$$\frac{k_G RT}{a_t D_G} = C\left(\frac{\tilde{G}}{a_t\mu_G}\right)^{0.7}\left(\frac{\mu_G}{\rho_G D_G}\right)^{1/3}(a_t\,d_p)^{-0.2} \qquad (15.3.29)$$

where the subscript G refers to properties in the gas phase. The dimensionless constant C is a function of the size of the packing; see Figure 15.3.4.

We now turn attention to the sizing of tray towers (i.e., to the calculation of the actual number of stages required).

Three efficiencies are commonly defined. The *overall* plate efficiency is the ratio of the number of equilibrium stages required for a given separation to the number of actual trays required. The overall efficiency is a composite index of separation that takes into account globally the rates of mass transfer, cross-flow effects on actual plates, and plate-to-plate entrainment effects.

The overall efficiency of plate towers for chemical absorption is, in general, considerably less than it is for operations carried out in the absence of chemical reactions. This is related to the fact that the effectiveness of rate enhancement is, in general, considerably less than unity; see the discussion in Section 3.7. Therefore, although chemical absorption processes usually require fewer equilibrium stages than comparable physical absorption processes (see the discussion in Section 1.8), the saving in the number of actual plates required may be considerably less.

The Murphree efficiency of a single plate E_M is defined as

$$E_M = \frac{Y_N - Y_{N+1}}{Y_N - Y^*} \qquad (15.3.30)$$

where plates are numbered bottom to top, and Y^* is the gas-phase mole fraction in equilibrium with the liquid-phase composition exiting from plate N. The Murphree efficiency in the absence of chemical reaction can be calculated if the assumption is made that the liquid on a given tray is perfectly well mixed, so that Y^* is constant

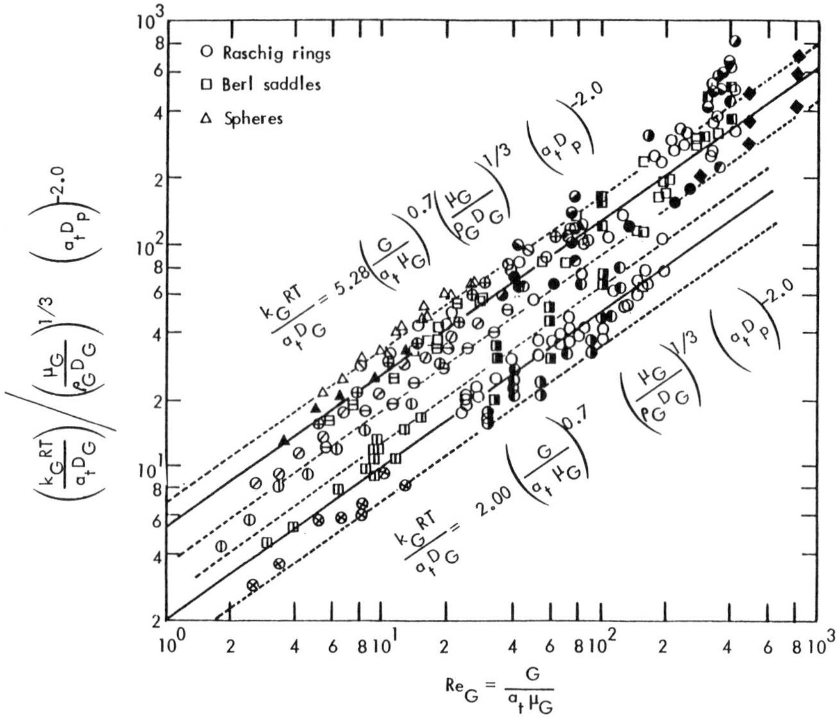

FIGURE 15.3.4. Correlation of gas-phase mass transfer.

throughout the plate. Equation 15.3.5 applies, and can be integrated to yield

$$\int_{Y_{N+1}}^{Y_N} \frac{dY}{(1 - Y)^2 (Y - Y^*)} = N^0 \tag{15.3.31}$$

Equation 15.3.1 can be solved for Y_{N+1} if N^0 is known, and the Murphree efficiency can thus be calculated. N^0 is related to the liquid height on the plate Z_P by Eq. 13.3.6, say,

$$N^0 = \frac{Z_P \pi K_G^0 \tilde{a}}{\tilde{G}} \tag{15.3.32}$$

In the special dilute case ($Y \ll 1$), solution of Eq. 15.3.31 and substitution into Eq. 15.3.30 yields

$$E_M = 1 - e^{-N^0} \tag{15.3.33}$$

The same approach can be taken in the case of chemical absorption; instead of Eq. 15.3.31, one should write

$$N^0 = \int_{Y_{N+1}}^{Y_N} \frac{dY}{\xi(1-Y)^2\,(Y-Y^*)} \tag{15.3.34}$$

However, several complications arise in the use of Eq. 15.3.34. In fact, even if the liquid is perfectly well mixed on the tray, so that Y^* is a constant, ξ may not be a constant, since it may depend on the gas-phase composition. Two important subcases can be considered and are briefly reviewed in the following.

1. Mass transfer in the liquid phase takes place in the fast reaction regime. In this case, ξ depends only on the liquid-phase composition and can, therefore, be regarded as constant. Integration of 15.3.34 can be carried out just as integration of 15.3.31; if the gas is dilute (i.e., at the lean end of the absorber where fast reaction conditions are, in fact, likely), one obtains

$$E_M = 1 - e^{-\xi N^0} \tag{15.3.35}$$

Since ξ is always larger than unity (see Eq. 15.3.13), it would appear that the Murphree efficiency is larger in a chemical absorption process than in a comparable physical absorption one. However, it should be borne in mind that the G/L ratio in a chemical process is much larger than in a physical process, due to the much larger capacity of the liquid; hence N^0 tends to be considerably smaller in a chemical process.

2. Equation 15.3.19 applies (i.e., the process is gas-phase controlled). If such is the case, the Murphree efficiency of a plate is, of course, just the same as it would be for a physical process but, of course, again N^0 tends to be smaller.

Although the assumption that the liquid phase is well mixed is a reasonable one in the vertical direction, in the horizontal direction along which the liquid flows it is a poor one and the effect of cross-flow may be very important. Therefore, it is usual to define a *point* efficiency through Eq. 15.3.30 applied at one particular point along the liquid flow path; the point efficiency can then be calculated from Eq. 15.3.31 for a physical solvent, or equivalently from Eq. 15.3.34 for a chemical one. However, now the question arises of modeling the cross-flow effect. Although this has been dealt with in some detail for the case where no chemical reactions occur, we do not know of any published analysis of the same problem for chemical absorption.

15.4. DESORPTION

The loaded solvent from the absorber is usually regenerated by desorbing the acid gases from the liquid into a gas or vapor stream. Conceptually, desorption comprises two steps in sequence: pressure reduction (flashing), followed by stripping with heating.

In practice, it is not uncommon to flash into the top of the desorption column, making the separation of the two steps difficult. For purposes of process design and analysis, it is desirable to think of flashing and stripping as two separate sequential steps with the flashed liquid flowing to the top of the regenerator packing. In pilot plant work it is desirable to provide a separate flash drum with a separate collection and measuring system for the flashed liquid and the acid gases that are flashed off. If this is not done, analysis of the desorption operation becomes very complicated because conditions at the top of the stripping section are poorly defined.

It is generally desirable to maximize the amount of acid gases released by flashing. The reason for this is that acid gases that are flashed off require no direct energy (steam) input for desorption. Flashing does involve an indirect energy input in that flashing, which is essentially an adiabatic operation, leads to a drop in liquid temperature that must be made up by supplying steam to the reboiler.

The result of a flashing operation can be calculated if the assumption is made that the flash drum behaves like an equilibrium stage. The calculation is based on writing the following equations:

1-2. Mass balances on the solvent (water for aqueous solutions) and on the acid gas. These are the two components that may be present in both the flashed solution and the flashed gas.

3-4. Equilibrium conditions on the solvent and on the acid gas. These relate the partial pressures of the two components in the flashed gas to the flashed solution's composition and temperature.

5. The condition that the sum of the two partial pressures equals the total pressure in the flash drum.

6. The expression of Dalton's Law; i.e., the partial pressure of the acid gas equals the total pressure times the acid gas mole fraction.

7. An enthalpy balance.

In these seven equations, the amounts of solvent and of acid gas in the liquid feed to the flash drum are given, as well as the pressure. The unknowns are:

1-4. The amounts of acid gas and of solvent in the flashed gas and in the flashed solution.

5-6. The two partial pressures in the flashed gas.

7. Temperature.

It follows from the argument above that the temperature in the flash drum is not an independent variable, but is determined by the operating pressure. The solution of the system of seven equations discussed above generally involves a trial-and-error procedure. In principle, allowance should be made for the fact that departures from equilibrium may take place in an actual flash operation. However, very little is known on the kinetics of flashing. Inasmuch as flashing occurs when the total equilibrium vapor pressure (i.e., the sum of the acid gas and solvent equilibrium partial pressures) exceeds the total pressure on the system, it should be regarded essentially as a boiling operation, the kinetics of which are governed mainly by nucleation, though mass transfer to already nucleated bubbles may also play a role.

From the viewpoint of comparison of different chemical solvents, it is evident that the larger amount of flashing will be exhibited by that solvent for which a given pressure difference corresponds to the larger difference in liquid loads. Therefore, the slope of the equilibrium curve in the usual range of pressures used for flashing should be as small as possible. Figure 15.4.1 shows that potassium carbonate solutions are likely to flash off more CO_2 than do aqueous aminoalcohol solutions.

Other possibilities for maximizing the amount of acid gas that can be released by flashing are:

1. A high temperature in the liquid entering the flash section (this can be attained by operating in such a way that the temperature of the rich solution leaving the bottom of the absorber is high and/or through use of a rich/lean heat exchanger).

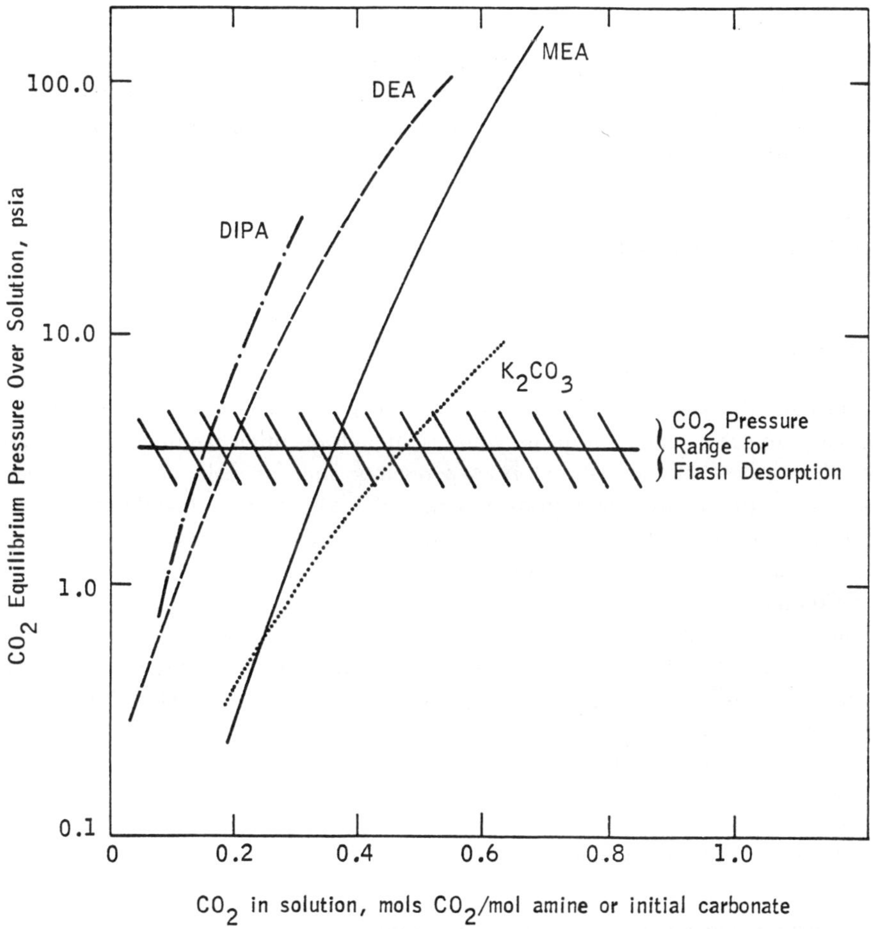

FIGURE 15.4.1. Equilibrium pressures of CO_2 over aqueous aminoalcohol and potassium carbonate solutions at flash regeneration conditions (100°C).

2. Use of a solvent of low volatility. With aqueous chemical solvents, a significant part of the liquid cooling is due to water evaporation. Use of a high boiling organic solvent in place of some or all of the water increases the amount of acid gas that can be recovered by flashing. For example, solvent-amine processes have an advantage over aqueous amine processes in this regard.

We now turn attention to stripping desorption, which takes place when the equilibrium partial pressure of the acid gas is larger than its partial pressure in the gas phase, but the total equilibrium vapor pressure is less than the total pressure on the system. Stripping desorption is clearly a mass transfer dominated phenomenon, and its theoretical modeling has been discussed in Chapters 2–7.

In principle, the design procedure for a desorption tower should be based on the same methodology as that for an absorption tower. In practice, heat effects are very important in desorption, so that a non-isothermal procedure should be developed, with all the difficulties that this entails.

Regimes of mass transfer that are likely to apply in different desorption units have been discussed in Chapters 9 and 10. Owing to the curvature of the operating line for desorption, which is related to steam condensation (see Figure 15.2.1), a desorber unit may work under conditions where it is moderately pinched throughout, so that the small driving force asymptote discussed in Section 6.4 may be particularly relevant.

15.5. PILOT PLANT DATA

The thesis was advanced at the beginning of this chapter that, if the physicochemical behavior of a particular gas–liquid system is perfectly well understood, it should be possible to design an industrial size process from information obtained on the laboratory scale. However, such a perfect understanding is not usually available and pilot plant work is, therefore, often required as an intermediate stage between laboratory experiments and full-scale design. Certain aspects of the actual operation of processes such as corrosion, foaming tendency, utilization losses, and the like, are often hard to predict on the basis of laboratory scale data and pilot plant operation is, therefore, needed.

Consider a pilot plant with a flowsheet such as in Figure 15.5.1. Two liquid feed points are available on the absorber. When the lean solution is fed at the top point, absorption takes place over a large packing height (a typical value may be 5 m); when it is fed at the low point, absorption takes place over a small packing height (a typical value is 1 m).

The operating variables of the pilot plant (i.e., the parameters that may be set at some preassigned value) are the following:

Steam rate to the desorber
Liquid circulation rate
Feed gas rate, composition, and temperature
Pressures in the absorber and in the desorber

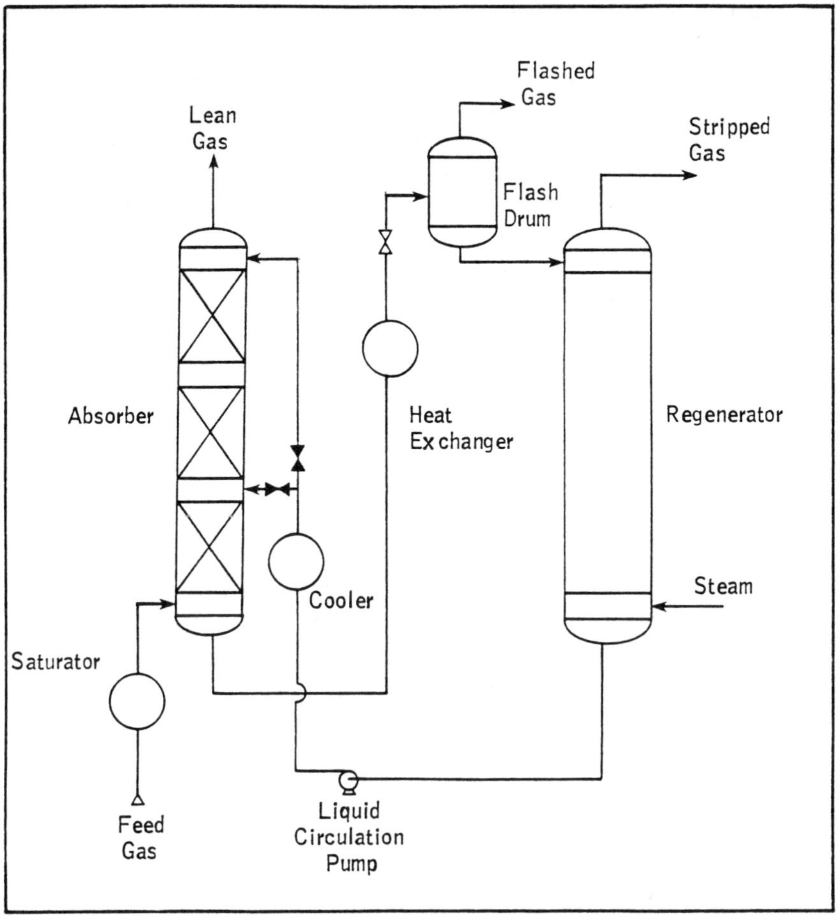

FIGURE 15.5.1. Flowsheet for a gas treating pilot plant.

The variables that are *usually* measured are:

Composition of the lean and rich solutions and of the flashed solution

Composition of the gas exiting from the absorber

Temperatures at the lean and rich end of both absorber and desorber

In addition to these, it would, in general, be desirable to also measure temperature, liquid composition, and gas composition at intermediate points along the packing in both the absorber and desorber. This is comparatively easy as far as temperature is concerned, but sampling of the liquid and gas phases at intermediate points presents some difficulties.

The pilot plant can be run in three different modes, each of which provides useful information. The modes will be termed rich end pinch (REP), lean end pinch (LEP), and unpinched (UP). In the REP and LEP modes, the lean solution is fed at the top of the absorber; in the UP mode, it is fed at the low feed point. The total packed height of the absorber should be large enough to ensure that a pinch *will* occur in the REP and LEP modes; conversely, the packed height below the low feed point should be small enough to ensure that no pinch will occur in the UP mode.

When the pilot plant is run in the REP mode, a high steam rate is maintained at the desorber bottom; ideally, the steam rate to liquid circulation rate should be held constant. If the steam rate is sufficiently high, the solution fed to the absorber is always very lean. The liquid flow rate is gradually decreased, and the composition in the exit gas corresponding to each liquid flowrate is recorded (each liquid flowrate needs to be kept constant long enough to ensure steady-state operation).

At low liquid flowrates, the liquid cannot pick up most of the acid gas because that would result in such a large fractional saturation in the rich liquid that the equilibrium vapor pressure would be larger than the partial pressure in the gas phase. Consequently, the leak rate is high; say, the partial pressure of the acid gas in the exit gas stream is high. Since the liquid feed is lean, the top of the absorber is unpinched. It follows that the absorption tower is pinched at the rich end. The rich solution exiting from the absorber is almost at equilibrium with the feed gas; hence the REP name (this can be checked, inasmuch as the rich solution composition should be almost in equilibrium with the feed gas). However, as the liquid flowrate is increased, a value is reached where the solution can pick up most of the acid gas without pinching at the bottom. As this condition is reached, the absorber will be pinched at the top; the leak rate will be very low, since the exit gas is now almost at equilibrium with the (very lean) feed liquid. A typical plot of leak rate

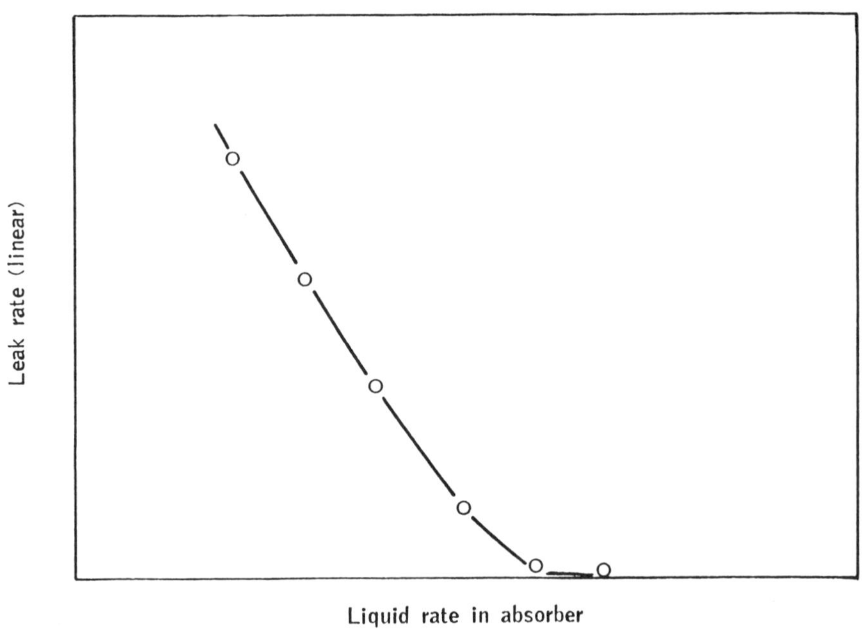

FIGURE 15.5.2. REP mode plot, linear scale.

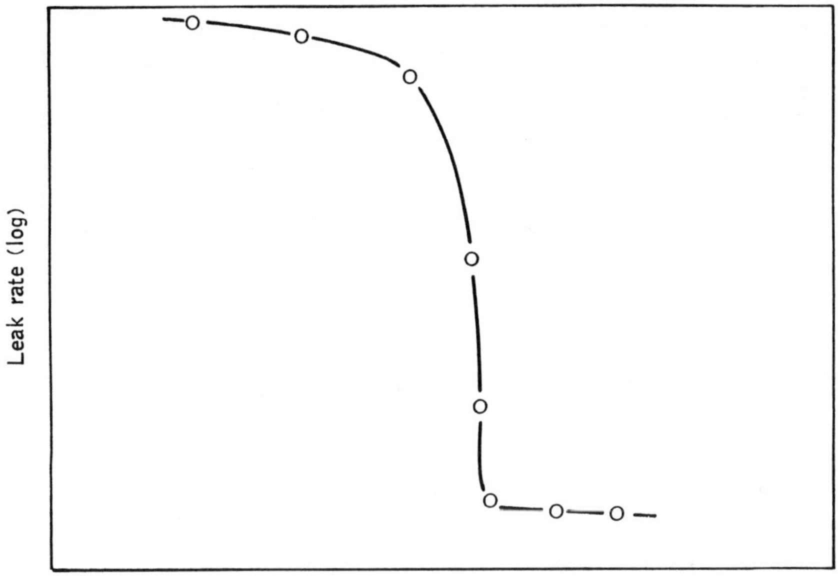

FIGURE 15.5.3. REP mode plot, logarithmic scale.

versus liquid flowrate is reported in Figure 15.5.2. If plotted in log-log coordinates, such as exemplified in Figure 15.5.3, the leak rate versus liquid flowrate curve defines the minimum liquid flowrate very sharply. The leveling off of the leak rate at very high liquid rates simply corresponds to the condition where the exit gas is in equilibrium with the liquid feed; often these last points cannot be obtained accurately because the leak rate is so small as to be unmeasurable. Indeed, if it can be measured, and if the fractional saturation of the lean solution can also be measured, one obtains useful VLE data corresponding to very low partial pressures.

Operation in the REP mode should be carried out in the range of L values corresponding to 30–75% of flooding; less than 30% may result in insufficient wetting of the packing, and more than 75% in excessive liquid entrainment in the gas phase.

The plot in Figure 15.5.3 exhibits a sharp transition because the operating-equilibrium curves have, for chemical solvents, the slope sketched in Figure 15.1.1. With physical solvents, should the equilibrium and operating curves be almost linear, the transition would be less sharp since a range of liquid flowrates exists where the absorber is moderately pinched throughout its length. In fact, a sharp transition is likely to be observed also in the temperature profiles. At flowrates less than L_m, the absorber is pinched at the rich end, and hence most of the absorption takes place in the top section; conversely, at liquid flowrates larger than L_m, the absorber is pinched at the top and most of the absorption takes place in the bottom section. The corresponding temperature profiles are expected to be of the shape sketched in Figure 15.5.4, with the high slope region corresponding to the unpinched part of the absorber where most of the absorption takes place.

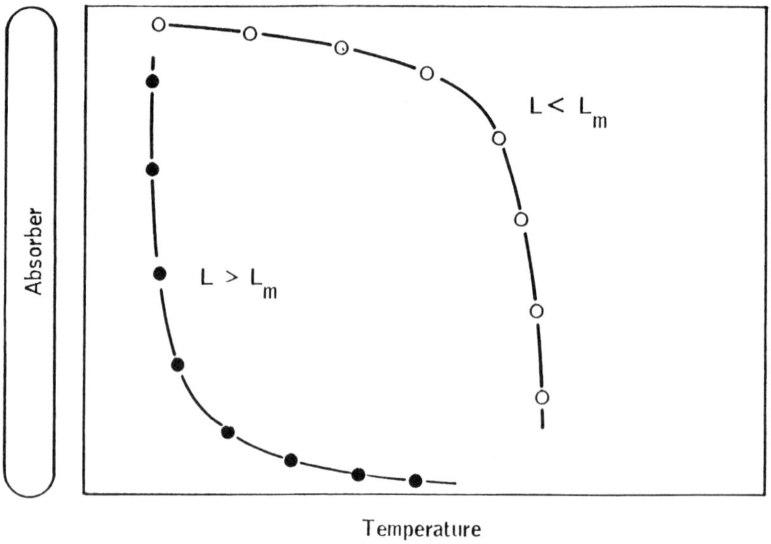

FIGURE 15.5.4. REP mode temperature profiles.

When operated in the LEP mode, the liquid flowrate is kept constant at some value exceeding the minimum rate L_m; this ensures that the absorber operates with a pinch at the lean end. The steam rate to the reboiler is progressively decreased. As the steam rate decreases, so does the regeneration efficiency, and the lean solution fed to the absorber become less and less lean. Correspondingly, the leak rate increases; a typical leak rate versus steam rate plot is sketched in Figure 15.5.5. From such data, the minimum steam rate required for obtaining an acceptable leak rate is determined. Again, these data can be used as VLE data at very low partial pressures.

When the pilot plant is operated in either the REP mode or the LEP mode, it is pinched at one of the extremes (or, if temperature variations are sufficiently large, somewhere in between). It follows that the driving force at the pinched end cannot be measured with sufficient accuracy to allow an accurate estimate of the effectiveness of mass transfer. Therefore, from LEP and REP mode data, the height of packing required for a given process cannot be calculated.

Meaningful mass transfer pilot plant data can only be obtained by operating in the UP mode (i.e., with a sufficiently small packed height). However, it should be borne in mind that operation in the UP mode *does not* represent actual industrial operation and, therefore, the UP mode data need to be well understood if one wants to use them to size the required packed height in the industrial unit. This is a particularly important point in the cases where a *kinetic* selectivity is required, since the usual rule of thumb of overdesigning the packed height just to be on the safe side will backfire in such cases.

When the pilot plant is operated in the UP mode, the data collected are usually

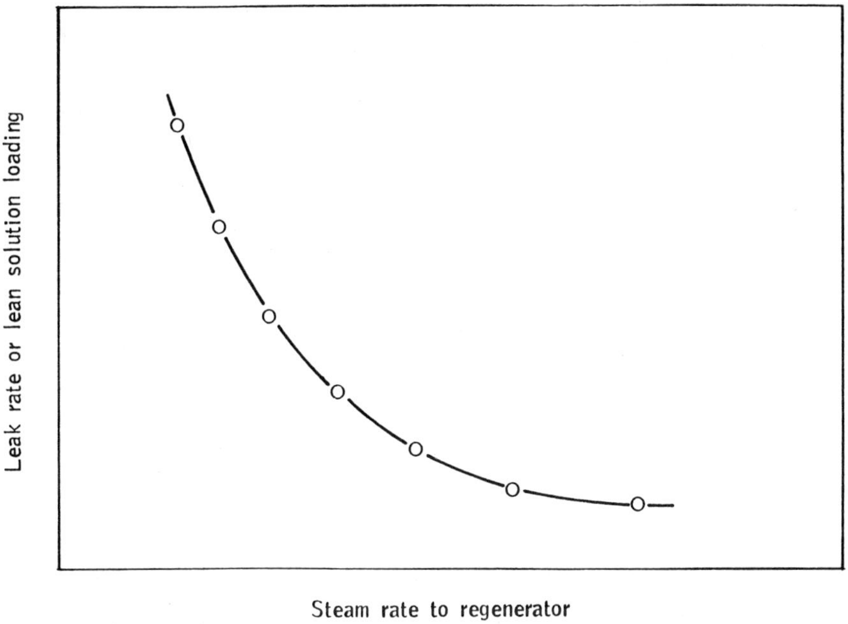

FIGURE 15.5.5. Leak rate plot obtained in LEP mode.

elaborated to the form of *apparent* $K_g\tilde{a}$ data, say, on the basis of a log mean driving force:

$$(K_g\tilde{a})_{app} = \frac{M}{Z_1\pi(Y - Y^*)_{lm}} \tag{15.5.1}$$

Equation 15.5.1 is nothing else but the solution for $K_g\tilde{a}$ of Eq. 15.3.7 and, therefore, it yields a physically significant value of $K_g\tilde{a}$ only if all the assumptions leading to Eq. 15.3.7 are justified. However, these assumptions are certainly *not* justified in the case of chemical absorption. It follows that values of $(K_g\tilde{a})_{app}$ obtained in this way will strongly depend on all operating parameters in rather unpredictable ways and cannot be extrapolated to conditions different from the ones for which they have been obtained.

An example may be illuminating. Figure 15.5.6 (lower graph) reports $(K_g\tilde{a})_{app}$ data obtained in a 6 in. diameter packed tower, with a packed height of 41 in. of $\frac{1}{2}$ in. Raschig rings for CO_2 absorption in 40% unpromoted K_2CO_3 (circles). It may be seen that the data are scattered over a range of one order of magnitude, and appear to be very strongly dependent on the partial pressure of CO_2 in the inlet gas. The data also appear to depend on liquid-phase composition, temperature, and liquid flowrate. This behavior would not be plausible for a true K_g^0a, as defined in Eq. 15.3.2; in particular, a true K_g^0a would be independent of the CO_2 partial pressure.

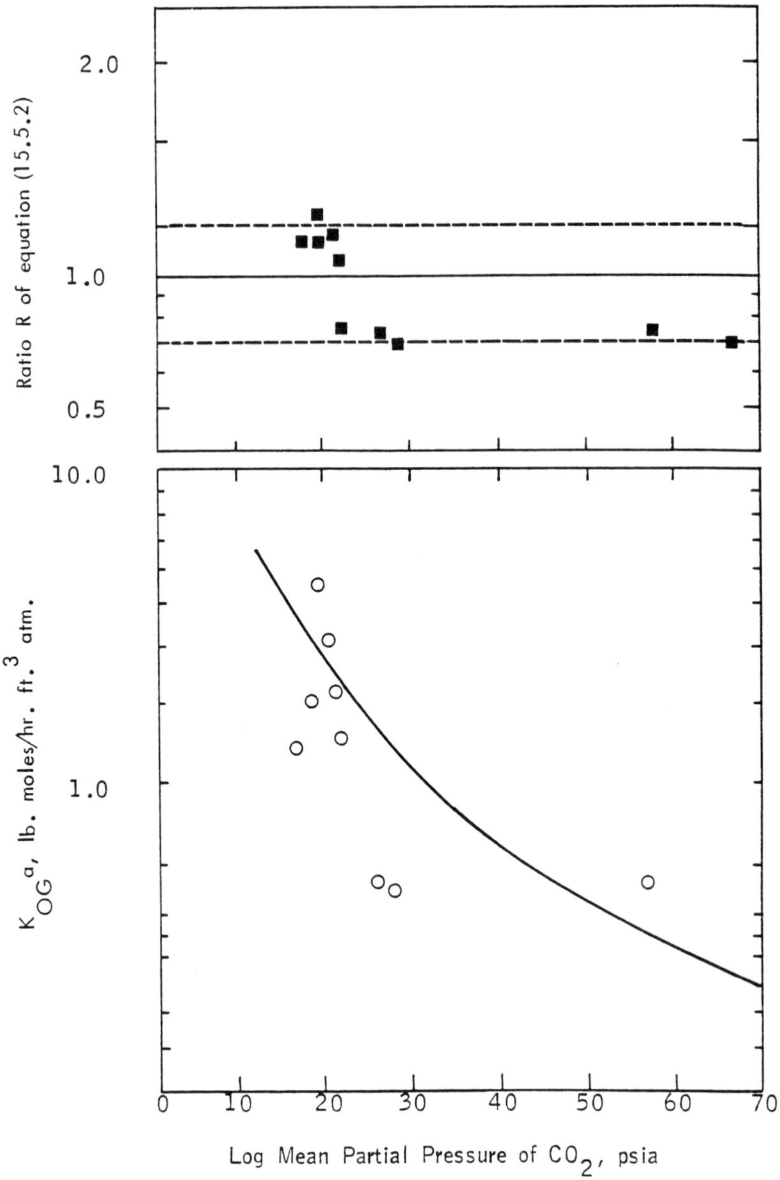

FIGURE 15.5.6. Test of integrated rate equation (15.5.2) for predicting pilot plant absorber performance.

The analysis of the pilot plant should be based on a correct integrated rate equation, say, on Eq. 15.3.22. The latter can be rewritten in the following form:

$$R = \frac{\tilde{G}}{K_g^0 \tilde{a} \pi Z_T} \Bigg/ \int_T^B \frac{dY}{\xi (1 - Y)^2 (Y - Y^*)} = 1 \qquad (15.5.2)$$

If a thermodynamic model is available for calculating Y^*, and a rate model for calculating ξ, the left-hand side of Eq. 15.5.2 can be calculated for each pilot plant UP mode run. This, of course, requires an *independent* estimate of the physical mass transfer coefficients $k_G\tilde{a}$ and $k_L^0\tilde{a}$, as well as $\tilde{a}$ itself.

For the CO_2-K_2CO_3 system, both a thermodynamic and a rate model are available (see Sections 2.4, 5.4, and 9.5). Thus it is possible to reanalyze the data reported in Figure 15.5.6 (lower graph) on the basis of Eq. 15.5.2, making use of the correlations in Eqs. 15.3.27–15.3.29. When this is done, the results are as shown in Figure 15.5.6 (upper graph). It can be seen that the calculated value of the left-hand side of Eq. 15.5.2 is, for all the runs, within 30% of the expected value of unity, and does not show any specific trend with varying CO_2 partial pressure.

Once UP mode pilot plant data have been analyzed satisfactorily on the basis of Eq. 15.5.2, a complete check of both the thermodynamic and the rate models has been obtained, and design of the industrial unit's required height of packing can be done with confidence on the basis of Eq. 15.5.2.

However, for such a program to be successful, values of $k_G\tilde{a}$, $k_L^0\tilde{a}$, and $\tilde{a}$ for the packings used in both the pilot plant and the industrial unit need to be known. Techniques for measuring these quantities are discussed in the following.

We begin by considering the experimental determination of the liquid-phase physical mass transfer coefficient k_L^0a. A model system can be used, inasmuch as corrections for minor differences in physical properties between the model system and the one of actual interest can be calculated from available correlations. The model system should be one where the gas-phase resistance to mass transfer is effectively zero, so that the measurable overall mass transfer resistance equals the liquid-side resistance.

If one chooses a physical system, condition 1.6.9 should be satisfied; a moderately soluble gas should be used (e.g., CO_2, with water as the liquid phase). However, even if for such a system condition 1.6.9 is likely to be satisfied, some gas-phase resistance to mass transfer may be present, at least over some part of the packed height.

The gas-phase resistance can be eliminated entirely if the gas phase is pure CO_2 saturated with water vapor; in this case, no partial pressure differences can develop in the gas phase, and mass transfer is entirely liquid-phase controlled. The equilibrium-operating plot will be of the form shown in Figure 15.5.7.

The difficulty arises here that, due to the shape of the equilibrium-operating plot, the liquid phase will become nearly saturated over a very short packed height, of the order of one height of a transfer unit. Therefore, experiments should be carried out with very short packed heights but this, in turn, introduces the difficulty that end effects become very important, and the measured values of $k_L^0\tilde{a}$ may not be representative of the true values after end effects have died out. Experiments of this type have been carried out, but great care is needed in order to eliminate the end effect problem.

A more efficient technique is to choose a *chemical* system, where mass transfer in the liquid phase is known to take place in the slow reaction regime. CO_2 absorption in dilute aqueous tertiary aminoalcohol solutions at low temperature is possibly

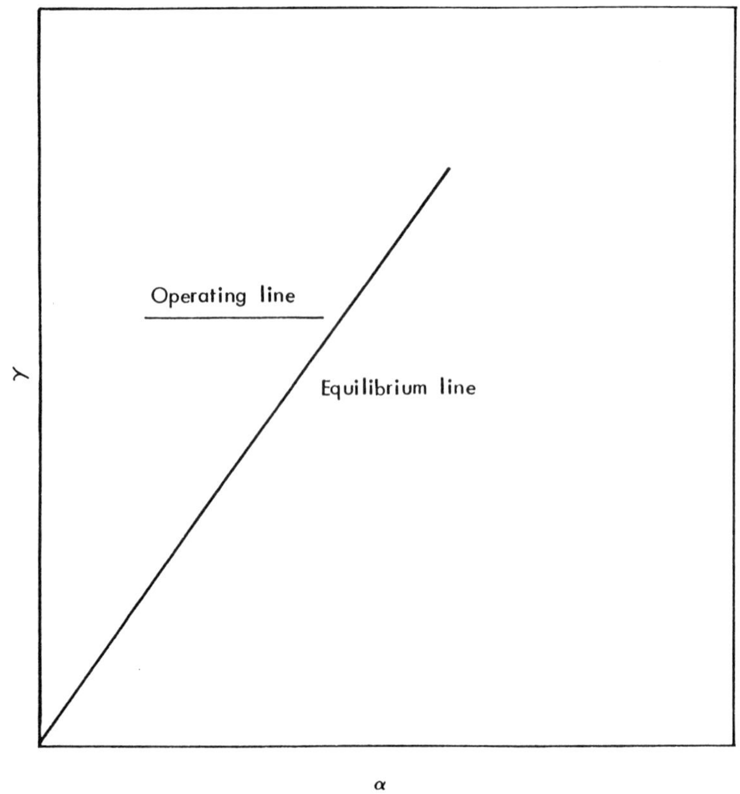

FIGURE 15.5.7. Equilibrium-operating plot for physical absorption of pure CO_2 in water.

the *best* system. The advantage over a physical system is that, due to the much higher capacity of the chemical solvent, the liquid *does not* get saturated even on rather large heights of packing.

In fact, with aminoalcohol solutions at low temperature in contact with pure CO_2, the equilibrium vapor pressure of CO_2 is likely to be negligible as compared to the partial pressure of CO_2 in the gas phase (which is almost equal to the total pressure) at all points of the packing height. If such is the case, the mass transfer equation reduces to

$$\frac{k_L^0 a p}{H} \, dZ = L \, d\alpha \tag{15.5.3}$$

which integrates to

$$k_L^0 a = \frac{H}{p Z_T} \, \Delta\alpha \tag{15.5.4}$$

Therefore, measurement of the change in total CO_2 concentration in the liquid phase $\Delta\alpha$ produces directly the value of $k_L^0 a$.

Measurement of $k_G a$ presents essentially the same type of problems as measurement of $k_L^0 a$. If one chooses a physical system, condition 1.6.7 should be satisfied; again, the best procedure is to have a pure component in the phase where the mass transfer resistance needs to be eliminated, say, in this case, the liquid phase. An attractive system is water–air, with the evaporation of water being the mass transfer phenomenon. However, the problem arises again that the gas phase will become saturated over a very short packed height; experiments should be carried out with very short packings, and elimination of end effects becomes very difficult.

Again, a chemical system produces a more efficient procedure. One needs to choose a system where the enhancement factor in the liquid phase is so large that the liquid-side resistance is effectively eliminated. Mass transfer in the liquid phase should proceed in the instantaneous reaction regime; an analysis following the lines discussed in Chapter 6 should be performed in advance in order to make sure that, indeed, the resistance to mass transfer in the liquid phase is negligible. In particular, it is possible to select a system where the equilibrium vapor pressure of the absorbing gas is negligible as compared to its partial pressure in the gas phase. In this case, the transport equation becomes

$$k_G a \pi Y \, dZ = G \, d\left(\frac{Y}{1 - Y}\right) \tag{15.5.5}$$

which integrates to

$$k_G a = \frac{G}{\pi Z_T} \int_T^B \frac{dY}{Y(1 - Y)^2} \tag{15.5.6}$$

Therefore, measurement of Y at the top and bottom of the packed height yields directly the value of $k_G a$. If the gas is dilute (i.e., $Y \ll 1$ throughout the packed height), the integral on the right-hand side of Eq. 15.5.6 reduces simply to

$$\int_T^B \frac{dY}{Y(1 - Y)^2} \cong \ln \frac{Y_B}{Y_T} \tag{15.5.7}$$

Finally, it should be mentioned that, in order to evaluate the parameter ξ appearing in Eq. 15.3.22, the enhancement factor I needs to be known; the latter, in turn, is given by Eq. 15.3.11 and depends on the value of k_L^0. Therefore, knowledge of the product $k_L^0 \tilde{a}$ is not by itself sufficient, unless a reasonably good estimate of the value of $\tilde{a}$ is available.

Values of the interfacial area per unit volume $\tilde{a}$ can be measured by running an absorption experiment where the following conditions are fulfilled:

1. Resistance to mass transfer in the gas phase is negligible.
2. Mass transfer in the liquid phase takes place in the fast reaction regime.
3. Kinetics of the reaction in the liquid phase are well known.

When these conditions are satisfied, the transport equation takes the following form:

$$\sqrt{\frac{D}{t_r}}\ \frac{\tilde{a}}{\mathrm{H}}\ \pi(Y - Y^*) = G\pi\ \frac{dY}{(1 - Y)^2} \tag{15.5.8}$$

which integrates to

$$\tilde{a} = \frac{G\mathrm{H}}{\sqrt{D}} \int_T^B \frac{\sqrt{t_r}\, dY}{(1 - Y)^2\,(Y - Y^*)} \tag{15.5.9}$$

Equation 15.4.23 yields $\alpha(Y)$; once α is known, Y^* can be calculated from a thermodynamic model, and t_r from the known kinetics of the reaction in the liquid phase; therefore, the integral on the right-hand side of Eq. 15.5.9 can be calculated. Notice that the quantity $\mathrm{H}\sqrt{t_r/D}$ (which appears in Eq. 15.5.9) can be measured on a laboratory scale absorber; see the discussion in Section 5.4.

Possibly the best system to be used is CO_2 absorption in hydroxide solutions, though other ones have also been used. The equations to be used have been discussed in Section 9.3. The equilibrium vapor pressure πY^* is given by Eq. 9.3.4 and is, in practice, negligible. The reaction time is given by Eq. 9.3.6. If the gas is dilute everywhere ($Y \ll 1$), combination of Eqs. 9.3.6, 15.3.23, and 15.5.9, and rearrangement gives

$$\tilde{a} = \mathrm{H}L\ \sqrt{\frac{m}{Dk}} \int_T^B \frac{dy}{(\sqrt{1 - 2y})\,[Y_T + (L/G)\,m(y - y_T)]} \tag{15.5.10}$$

When the data reported by the Norton Company are analyzed on the basis of Eq. 15.5.10, calculated values for different packings agree well with the correlations in Eq. 15.3.27.

15.6. SELECTION OF CONTACTOR TYPE

Although many different types of gas–liquid contactors have been developed, only three or four types have found significant industrial use in gas treating. The principal types available are shown in Figure 15.6.1.

The majority of industrial contactors are either packed or trayed towers, with packed towers taking an increasing share of the market in recent years. This trend is due largely to the development of high-capacity high-efficiency random packings that are available in a range of materials including steel and polypropylene. For a given pair of gas and liquid flowrates, a packed tower will usually be smaller in diameter than a tray tower. Packed towers cause less gas-side pressure drop than plate towers and they are more easily designed for resisting corrosion. In addition, packed towers are usually better for solvents that have a tendency to become foamy. This may be an important consideration with some chemical solvents.

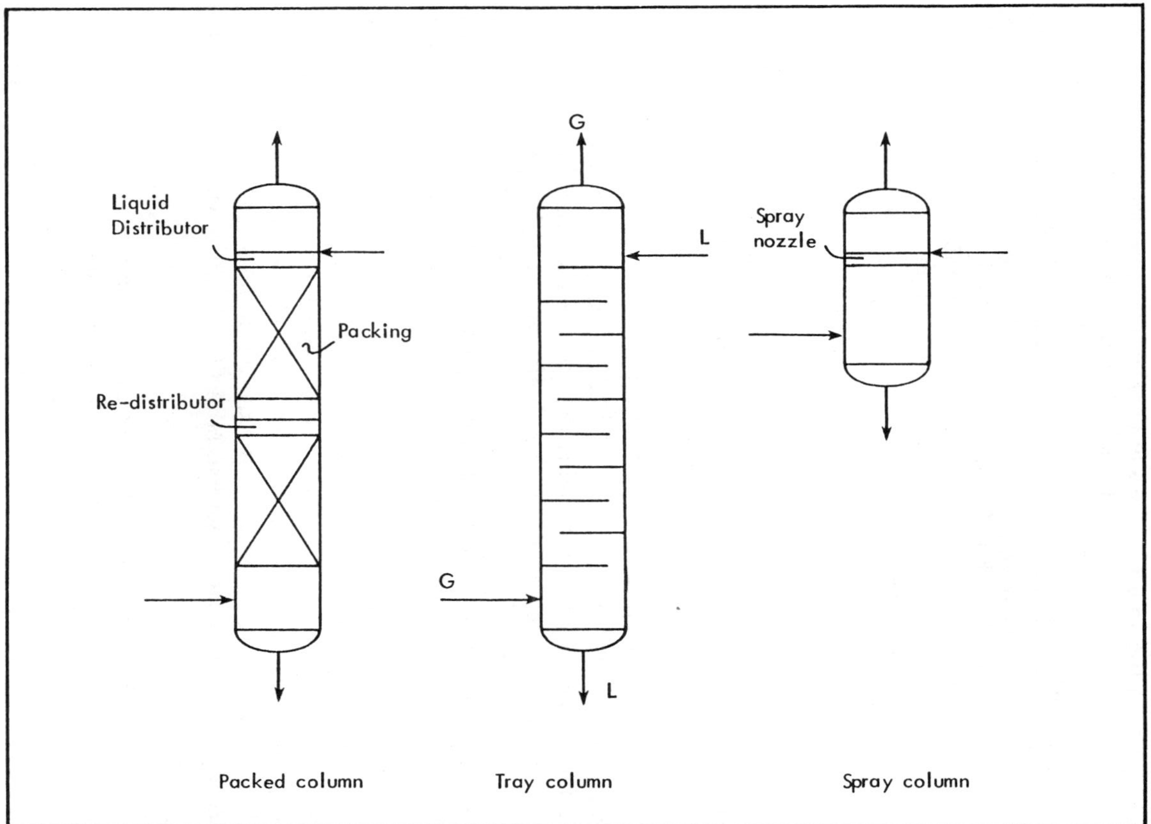

FIGURE 15.6.1. Principal types of gas–liquid contactors for reactive absorption.

TABLE 15.6.1. Types of Contactor Internals for Differential-Contact Columns

Random packings	Ordered packings
Saddles (Intalox)	Flexipac
Slotted rings (Pall, Hypak)	Goodlee
Tellerettes	Hyperfil
Maspac	Sulzer
Partition rings (Lessing)	Glitsch grid
Mini rings	Leva film trays

Packed towers do have a narrower range of vapor and liquid loads for efficient operation, but this is not a severe disadvantage for steady loadings that are common in many industrial processes. It can be a disadvantage for plants with fluctuating loads.

Packed contactor internals may be divided into random (dumped) packings and ordered packings (see Tables 15.6.1 and 15.6.2) Random packings (see Figure 15.3.2 for sketches of some common types) have been discussed in Section 15.3. Ordered packings, which combine very high efficiency with low gas-side pressure drop, are relatively new and have not received much commercial use to date in gas treating. Two types of ordered packing are shown in Figure 15.6.2. Ordered packings are relatively expensive compared to random packings. Figure 15.6.3 compares a random packing, an ordered packing, and a sieve tray contactor.

Plate towers are generally preferred over packed towers where fluctuating vapor or liquid loads are expected. Figure 15.6.4 shows representative tray efficiency profiles for various tray styles. Plate towers may also be specified for systems involving slow chemical reactions since trays with high weirs can be used to provide long liquid residence times. This requirement arises rarely in gas treating. One example is the rich end of aqueous MEA CO_2 absorbers where it is desired to

TABLE 15.6.2. Types of Contactor Internals for Staged, Tray Columns

Common types	Proprietary types
Sieve	Angle
Valve	Unitlux
Bubble cap	Montz
	Londe
	Thoman
	Jet
Pseudo equilibrium stages (downcomerless trays)	
Perforated (sieve)	
Turbogrid	
Ripple	

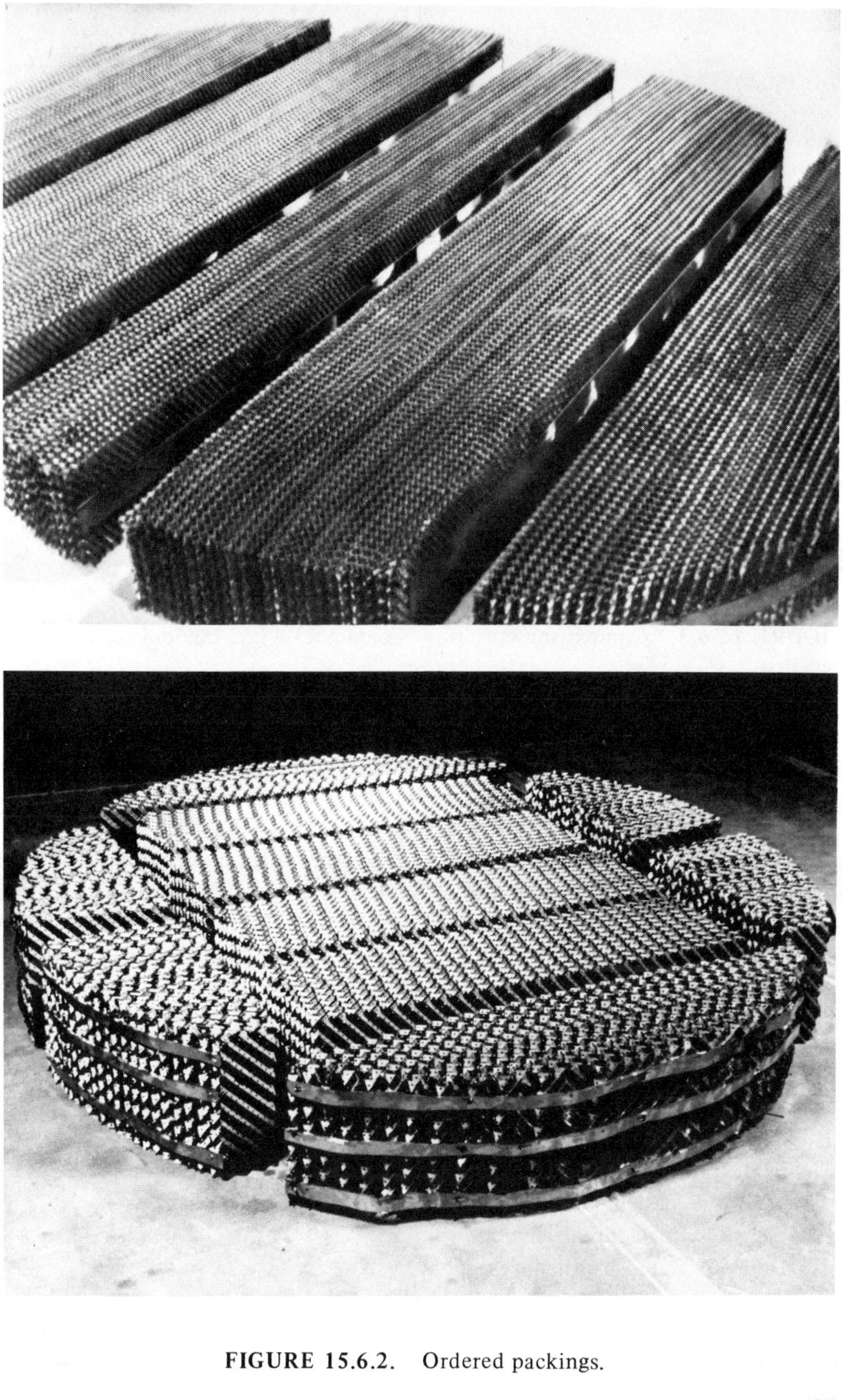

FIGURE 15.6.2. Ordered packings.

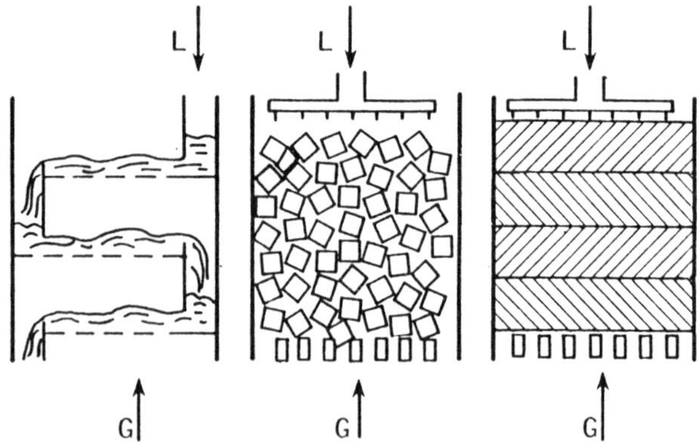

Example	Tray column Sieve trays	Column with random packing Pall rings 2"	Column with ordered packing Koch Sulzer BX packing
F_s factor $\left[\frac{ft}{S}\left(\frac{Lb}{ft^2}\right)^{\frac{1}{2}}\right]$	2	2	2
HETP (inches)	25	40-20	8-6.5
ΔP/TP (inches W.G.)	1.5-4	0.3-1.6	0.04-0.2

FIGURE 15.6.3. Comparison of tray contactor with random and ordered packings.

achieve rich loadings in excess of 0.5 moles CO_2/mole MEA. When liquid rates are low or when solid matter may be deposited, plate columns are indicated, since packed columns do not handle these situations well (a minimum liquid rate of about 2 GPM/ft^2 of packed tower section is required with aqueous systems for satisfactory liquid distribution). Plate columns are much more convenient when cooling is necessary throughout the column. Plate towers may also be preferred when future increase in plant capacity is anticipated. The plate tower can be designed to operate satisfactorily at low throughputs yet ultimately the tower can be expanded, by substituting packing for plates, to much higher throughputs as production demands increase.

Several common and proprietary types of staged tray columns are available. Sieve (perforated) trays are usually the first choice owing to their low cost and remarkably good flexibility. Sieve trays offer about 3-to-1 operating flexibility. If greater flexibility is needed, valve trays may be used. Figure 15.6.5 shows the essential features of sieve and valve trays.

Special tray designs are available for special contacting duties. Jet trays, for example, offer high liquid handling capacity, whereas downcomerless sieve trays are useful with liquids containing suspended solids.

Table 15.6.3 offers a qualitative selection guide for column internals. Nearly all types of gas–liquid contactors require the use of mist eliminators (de-entrainment devices) at the top of the columns to prevent liquid carryover with the treated gas

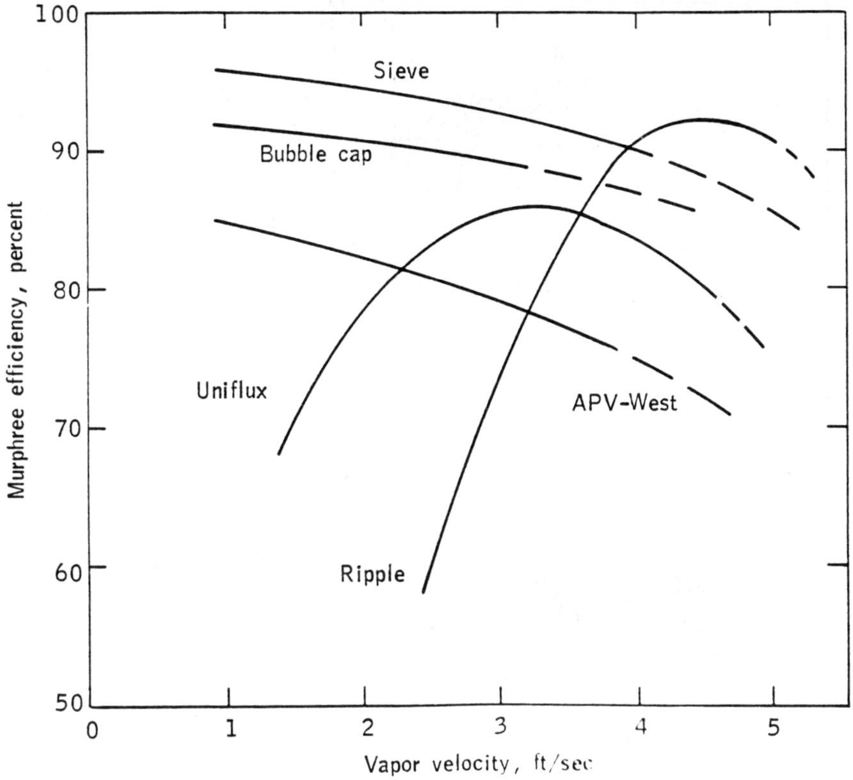

FIGURE 15.6.4. Performance profiles for tray contactors.

as fine droplets. This is especially true of liquids that have a tendency to become foamy. The common types of mist eliminators are shown in Figure 15.6.6.

15.7. FLOWSHEET DESIGN

Many flowsheet improvements have been made to reduce energy consumption in chemical absorption-desorption gas treating processes. Minimum energy designs for potassium carbonate and aqueous aminoalcohol processes have evolved along different paths, but the end result is two remarkably similar flowsheets.

In the past, three flowsheets were used for promoted carbonate CO_2 removal. These designs—single stage, split flow, and two-stage—were discussed under CO_2 removal, Chapter 10. The standard two-stage system is shown in Figure 9.5.9. In attempts to save energy, there has been a general trend towards a greater severity of operation. The specific flowsheet modifications include:

1. Increase of solution concentration (molarity of potassium carbonate) and of the acid gas-rich loading of the solution in order to decrease the solution circulation rate.

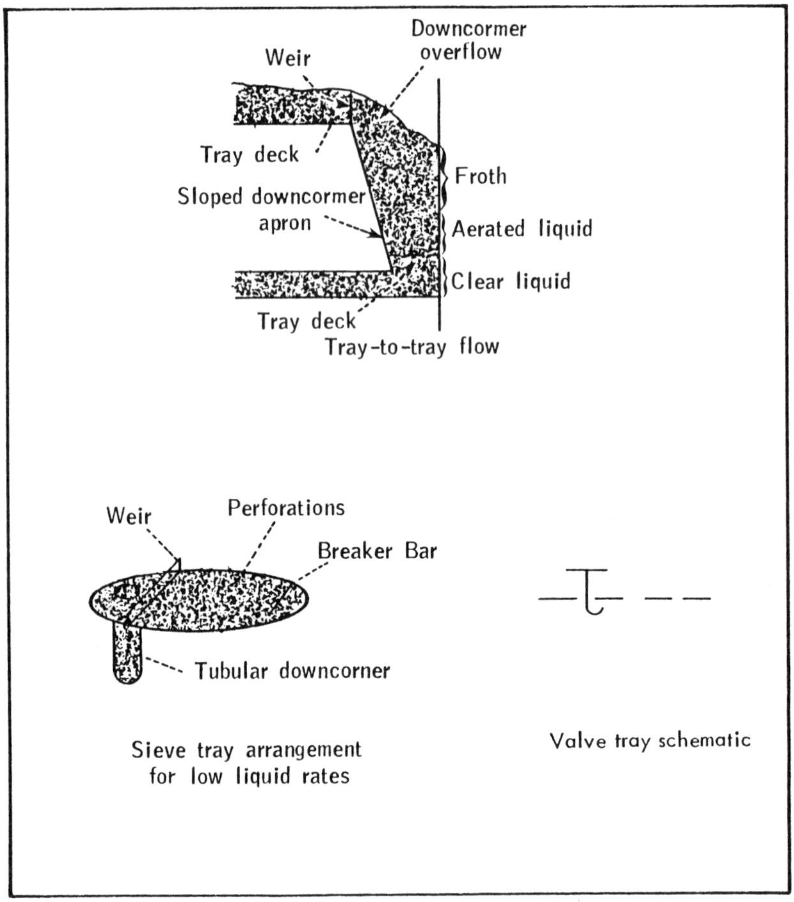

FIGURE 15.6.5. Sieve and valve tray layout.

2. Increase of the solution temperature at the base of the absorber by raising the gas inlet temperature of the absorber. High rich solution temperature improves the extent of flash desorption.
3. Increase of the pressure (and thus the temperature) of the first regeneration section. This reduces the flash at the top of the first stripper and makes possible the production of stripping steam for the low-pressure regenerator by flashing the lean and semi-lean solutions. This arrangement is shown in Figure 15.7.1. Flash cooling of the lean solution eliminates the need for the conventional lean solution cooler.

The changes to greater severity of operation described above tend to put stress on the corrosion inhibition system. Most promoted potassium carbonate processes use an organic activator (either an aminoalcohol or an amino acid) and a vanadate corrosion inhibitor. The normally slow chemical interactions between vanadate ion

TABLE 15.6.3. Selection Guide for Column Internals[a]

	Differential columns		Staged columns		Pseudo-equilibrium	
	Random packings	Ordered packings	Perforated or valve trays	Bubble cap or tunnel trays	Downcomerless	Disk and doughnut
Low pressure	2	3	2	1	0	1
Moderate pressure	2	1	3	2	1	1
High pressure	2	0	3	2	2	0
High turndown ratio	1	2	2	3	0	1
Low liquid rate	1	2	1	3	0	0
Foaming system	3	0	2	1	2	1
Internal tower cooling	1	0	2	3	1	0
Solids present	1	0	2	1	3	1
Dirty solution	1	0	2	1	3	2
Multiple feeds/side streams	1	0	3	3	2	1
High liquid rates	3	0	2	1	3	2
Small dia columns	3	2	1	1	1	1
Columns 3–10 ft dia	2	2	3	2	2	1
Large dia columns	2	1	3	1	2	1
Corrosive fluids	3	1	2	1	2	2
Viscous fluids	3	0	2	1	1	0
Low ΔP (% no concern)	2	2	1	0	0	3
Expanded column capacity	2	3	2	0	2	0
Low cost (performance no concern)	2	1	2	1	3	3
Available design procedures	2	1	3	2	1	1

[a]*Key*: 0 = don't use, 1 = evaluate carefully, 2 = usually applicable, 3 = best selection.

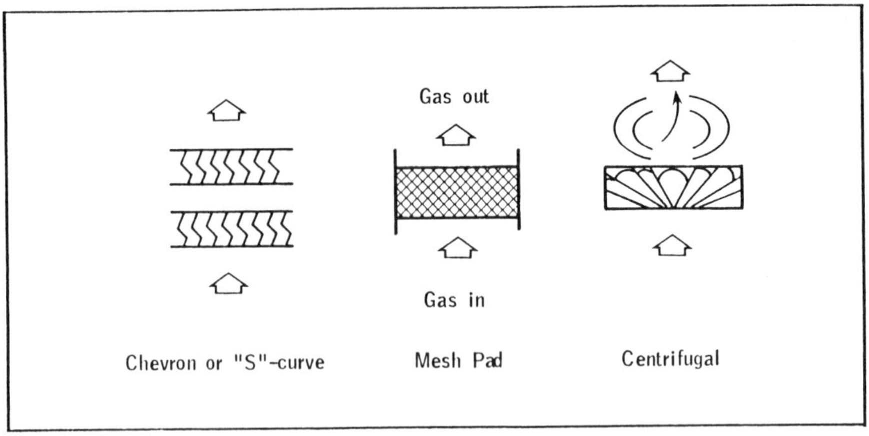

FIGURE 15.6.6. Mist eliminators.

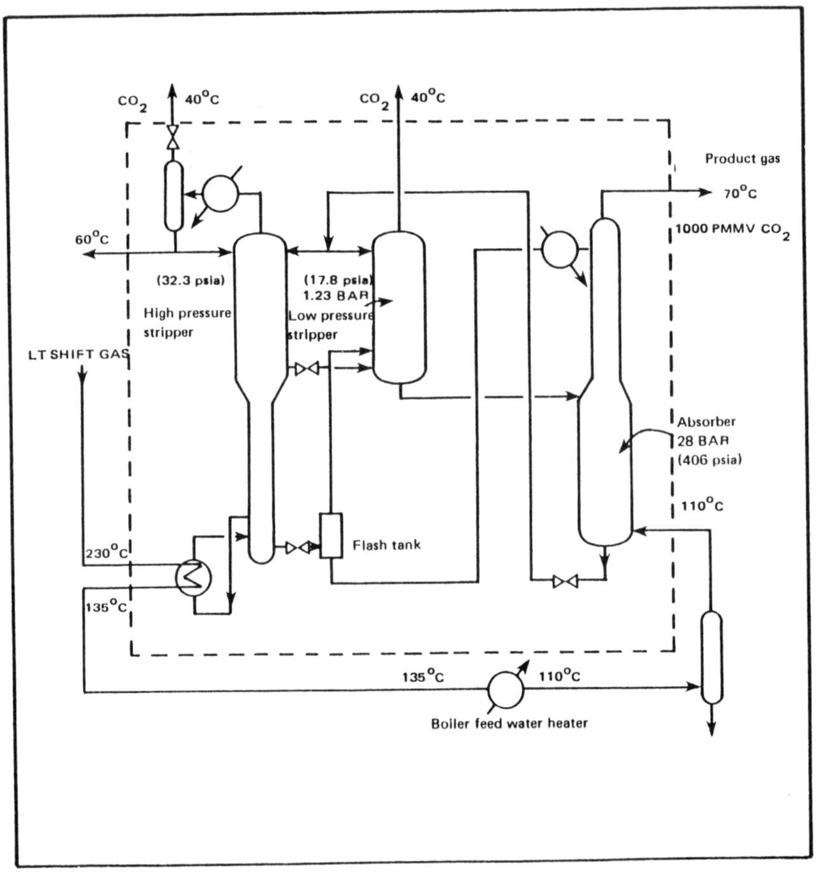

FIGURE 15.7.1. Improved two-stage hot carbonate CO_2 removal with dual pressure regeneration and lean solution flash.

and the organic compounds, which gradually reduce the concentration of pentavalent vanadium resulting in a decrease of corrosion inhibiting action, are accelerated by higher loadings and higher carbonate molarity. Thus considerations of corrosion place a practical upper limit on the degree to which the energy reduction modification can be implemented.

Another improvement concerns the optimum utilization of hot carbonate technology for treating gases for simultaneous removal of CO_2 and H_2S, in situations where H_2S must be reduced to a very low level in the treated gas (e.g., <4 ppm). The improvement, shown in Figure 15.7.2, involves integration of promoted hot carbonate for bulk removal of CO_2 and H_2S, with an aqueous aminoalcohol process for final removal of trace H_2S. The process incorporates a common overhead condenser for the two separate regeneration towers.

Another improvement with hot carbonate treating is the use of a power recovery turbine on the rich carbonate stream in place of the flash valve. Figure 11.1.1 shows a process for treating natural gas for simultaneous removal of CO_2 and H_2S to pipeline specification that incorporates all of the improvements mentioned above—integrated hot carbonate/aqueous amine with common overhead condenser, use of

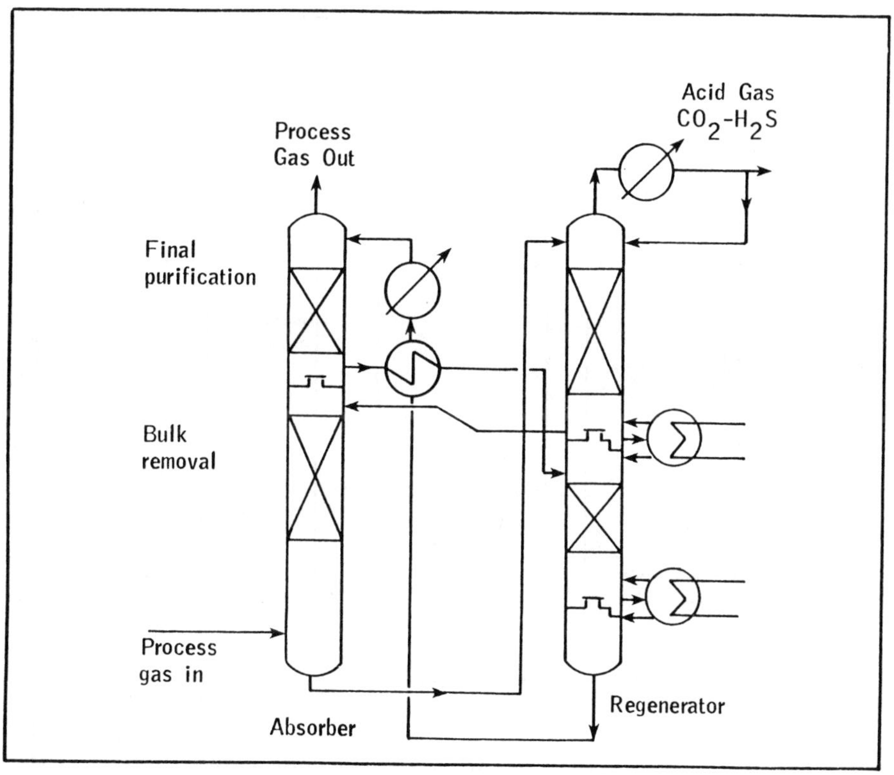

FIGURE 15.7.2. Integrated bulk removal/final purification acid gas removal process.

flash cooling of the lean carbonate solution, and a power recovery turbine on the rich carbonate stream.

Historically, the aqueous monoethanolamine CO_2 removal process was limited by corrosion considerations to an amine molarity of about $3M$ and rich loadings not exceeding about 0.4 moles CO_2/mole MEA. A rich-lean heat exchanger was used, as shown in Figure 9.4.5. The recent sharp rise in energy costs promoted a complete re-examination of the chemistry and chemical engineering of the aqueous MEA-CO_2 removal system. As a result, energy requirements have been reduced by 50% or more, as shown in Figure 14.2.4. Four principal improvements, some of them adapted from hot carbonate technology, have been made:

1. Discovery of new corrosion inhibitors allows MEA molarity to be increased to 4-6 M.
2. Operation with a high rich solution temperature allows elimination of the rich/lean heat exchanger. This is achieved by bringing in the feed gas hot.
3. The flash evaporation-thermocompressor ejector scheme has been introduced using extraneous low pressure steam; see Figure 15.7.3. Although this

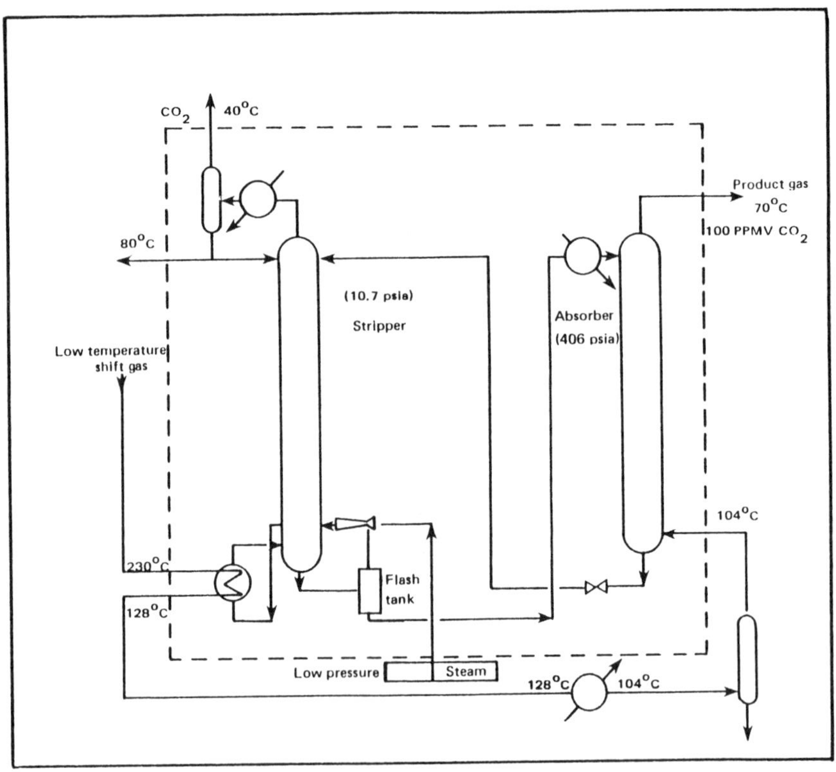

FIGURE 15.7.3. Improved aqueous MEA–CO_2 removal system without rich-lean heat exchanger and incorporating lean solution flash.

scheme reduces energy, it suffers from the disadvantage that water must be discarded from the system, and this water purge may be contaminated with up to 5 ppm MEA. This constitutes a possible environmental problem.

4. A high approach to gas–liquid equilibrium has been attained at the absorber rich end by using hot feed and a submersible gas inlet.

Recent improvements have led to a significant lowering of heat duties for CO_2 removal systems. The improvements can be applied with similar results to other acid gas removal situations.

15.8. TECHNOLOGICAL PROBLEMS

The corrosion of carbon steel by chemical gas treating solvents is a very significant issue with all types of solvents. Corrosion has an impact on the operating reliability and the expected life of the plant. The great majority of the equipment is carbon steel and no competing material is economical. The types of metallurgical problems that can arise fall into three categories:

General corrosion, including erosion–corrosion
Stress corrosion cracking
Hydrogen embrittlement (when treating H_2-rich gases)

A few generalized comments can be made about general corrosion with chemical solvents:

1. Organic solvent solutions of aminoalcohols tend to be less corrosive than aqueous aminoalcohol solutions.
2. Aqueous potassium carbonate solutions, with or without promoters, tend to be more corrosive than aqueous aminoalcohols.
3. In CO_2-removal service, corrosion inhibitors must be used with potassium carbonate solutions and with high-concentration aqueous aminoalcohol solutions. Organic solvent solutions of aminoalcohols may not need inhibitors.
4. The rate of corrosion with CO_2 present is generally greater in regions of the plant where CO_2 is flashing or boiling out of solution.
5. Corrosion is greatest at high K_2CO_3 or aminoalcohol molarity, high acid gas loading and temperature, and high solution turbulence.
6. Solutions loaded with CO_2 are much more corrosive than solutions loaded with H_2S. Corrosion with mixed acid gases increases as the ratio of CO_2 to H_2S in solution increases. Conversely, corrosion is generally reduced as the H_2S to CO_2 ratio in the solution increases.
7. Inhibitors may not be needed if H_2S is the only acid gas present, as passivation of the steel is achieved by the sulfide layer.

Corrosion in sour gas treating contactors using amine-organic solvents [e.g., Sulfolane solutions of diisopropanolamine (DIPA)], has been discussed. DIPA is apparently used up to about 3.5 *M* concentration without corrosion inhibitors. DIPA-Sulfolane solutions rich in CO_2 and H_2S are inherently not corrosive to carbon steel because films tend to form that protect the metal from the aggressive liquid. Corrosion can occur under certain circumstances when there is flashing or boiling, especially of CO_2. This can occur in the bottom (rich end) of the absorber tower under conditions where CO_2 is being displaced from the descending liquid by a high H_2S concentration in the feed gas. The probability of corrosion occurring for a given design is decreased by decreasing column temperature by cooling the feed gas or the lean solution. Alternatively, the problem can be handled by installing Type 304 stainless steel cladding in the critical lower part of the contactor.

In aminoalcohol-water scrubbing solutions, the allowable amine molarity/loading depends strongly on the nature of the amine. The reason is that it is the amine degradation products rather than the amine itself that promote severe corrosion, and amines differ as to their chemical stability and the nature of their degradation products. For example, MEA cannot be used in CO_2 removal above about 3 *M* concentration without a corrosion inhibitor, whereas DGA (which is also a primary amine) can be used up to 6 *M*. Both aminoalcohols are degraded extensively by CO_2, yet DGA solutions are much less aggressive than MEA solutions to carbon steel.

Owing to the strong relationship between amine degradation and corrosion, practices such as solution filtration and redistillation (reclaiming), which remove degradation products, help to alleviate corrosion. In redistillation, the degradation products are "fixed" by adding sodium carbonate so that the undegraded amine can be distilled off and recovered.

Two proprietary corrosion control schemes are available for amine plant users. These offer advice and additives for control of corrosion, sludge, and foam on a licensing basis. If good corrosion control is utilized, high amine concentrations can be used in H_2S and CO_2 mixed gas service, leading to significant energy savings. For mixed acid gas service with MEA, one corrosion inhibitor comprises a transition metal and a selective oxidizing agent.

Equipment and procedures are available for measuring corrosion rates in amine units. The use of both metal coupons and metal-resistance corrosion probes is feasible, and plant results correlate well with laboratory measurements. Corrosometer probes for plant use have been described.

For CO_2 removal with aqueous MEA, one corrosion inhibitor consists of 4–6 parts by weight antimony compounds and 6–4 parts by weight vanadium compounds.

Corrosion inhibitors are always needed when using hot potassium carbonate solutions for CO_2 removal. For simultaneous CO_2/H_2S removal, the hydrosulfide ion acts to passivate the carbon steel, provided the H_2S concentration is not too low. Without inhibitors, potassium carbonate solutions present a generalized attack on carbon steel at rates of several hundred mils/year.

Probably the most effective corrosion inhibitor for use with potassium carbonate solutions is arsenite ion. Unfortunately, its use is severely restricted by health and environmental considerations. Also some technical problems (e.g., high iron content in solution) tend to limit its effectiveness. Furthermore, arsenite is not compatible with certain CO_2 promoters (e.g., amino acids).

Vanadates are the most practical corrosion inhibitors for use with aminoalcohol-promoted hot carbonate solutions. The active form for corrosion inhibition is potassium metavanadate (KVO_3) in which V is in the pentavalent form. Vanadium is unusual in exhibiting five different valences. Metavanadate may be reduced in solution to the 4-valent form which is less effective, or ineffective, as a corrosion inhibitor. In practice, the V^{5+}/V^{4+} ratio must be maintained in a specified range for good corrosion control. Favorable experience with the addition of sodium nitrite ($NaNO_2$) to preserve the desired ratio has been described. The literature also contains a discussion of operating experience with promoted hot carbonate plants in which V reduction has been controlled by air injection.

The successful application of corrosion inhibitors usually requires the use of special equipment cleaning and pre-passivating techniques. The importance of solution maintenance—filtering, carbon treating, reclaiming— cannot be underestimated.

Stress corrosion cracking in arsenite-promoted carbonate units in CO_2 removal service has been discussed. It was shown that cracking only took place when the corrosion potential exceeded certain values depending on the carbonation index.

All of the common primary and secondary aminoalcohols—MEA, DGA, DEA, and DIPA—are subject to chemical degradation in the presence of CO_2 and COS. Tertiary aminoalcohols such as TEA and MDEA are relatively stable in the presence of CO_2. There is no chemical degradation when H_2S is the only acid gas present, and H_2S does not appear to influence the deactivation caused by CO_2 and COS.

Degradation is detrimental in that it leads to loss of active aminoalcohol; it can also lead to operational problems such as corrosion, foaminess, increase in solution viscosity, phase separation, and so on. Means for removing degradation products are essential for smooth plant operation. Proven techniques include solution filtration, carbon treating, and thermal reclaiming.

The chemical degradation reaction pathways are shown in Figure 15.8.1 for primary aminoalcohols and Figure 15.8.2 for secondary aminoalcohols. In all cases, the desired reaction that causes CO_2 or COS removal from the gas phase is formation of the carbamate or thiocarbamate ion (this chemistry was discussed in Chapter 9). Under normal regeneration conditions (393K), most of the carbamate and thiocarbamate will revert back to aminoalcohol and CO_2, which is stripped out. However, a small fraction may react further to form one or more aminoalcohol degradation products. These further reactions, which involve both the -NH and -OH groups in the aminoalcohols, may be reversible, partly reversible, or irreversible. High temperature treatment (thermal reclaiming) at 448K may allow recovery of undegraded aminoalcohol in certain cases.

A summary of the severity of aminoalcohol degradation is given in Table 15.8.1. There are remarkable differences in extent of degradation and ease of degradation product reversion to aminoalcohol among the four rather similar aminoalcohols. Some references report that DEA does not degrade chemically under CO_2. This statement does not appear to be true.

Potassium carbonate solutions react reversibly with CO_2 without any side reac-

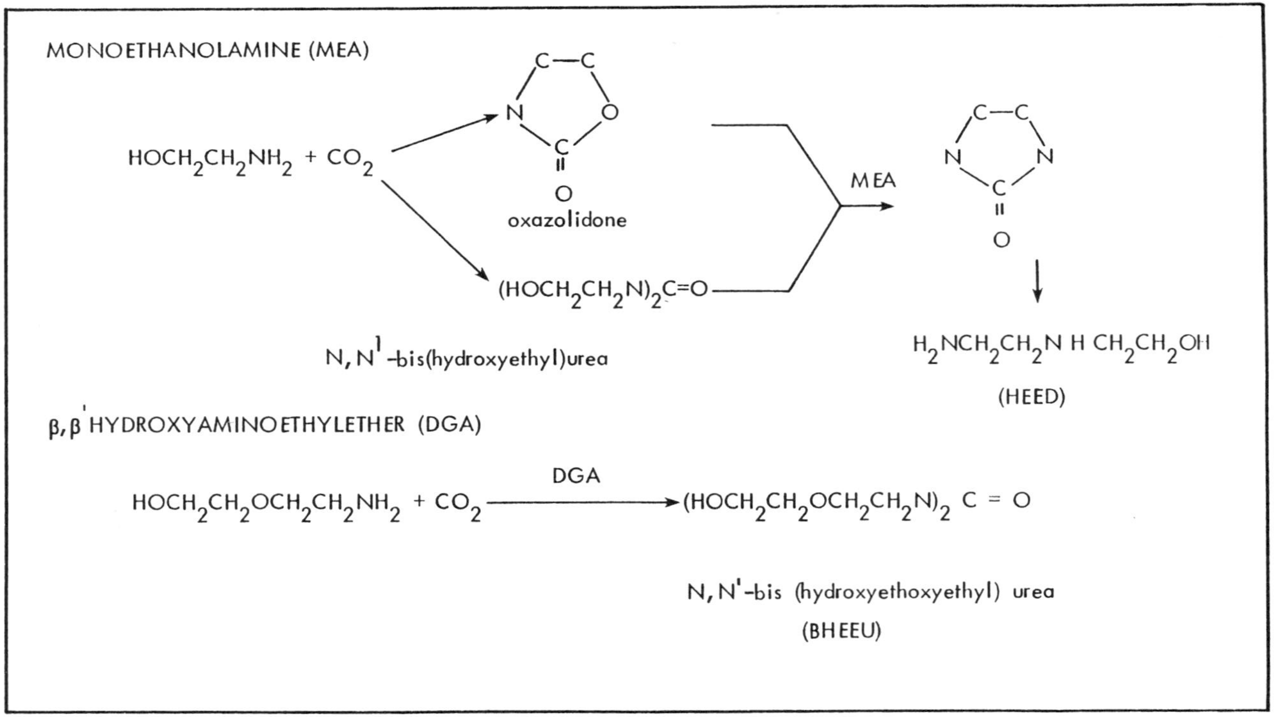

FIGURE 15.8.1. Primary amine degradation in presence of CO_2.

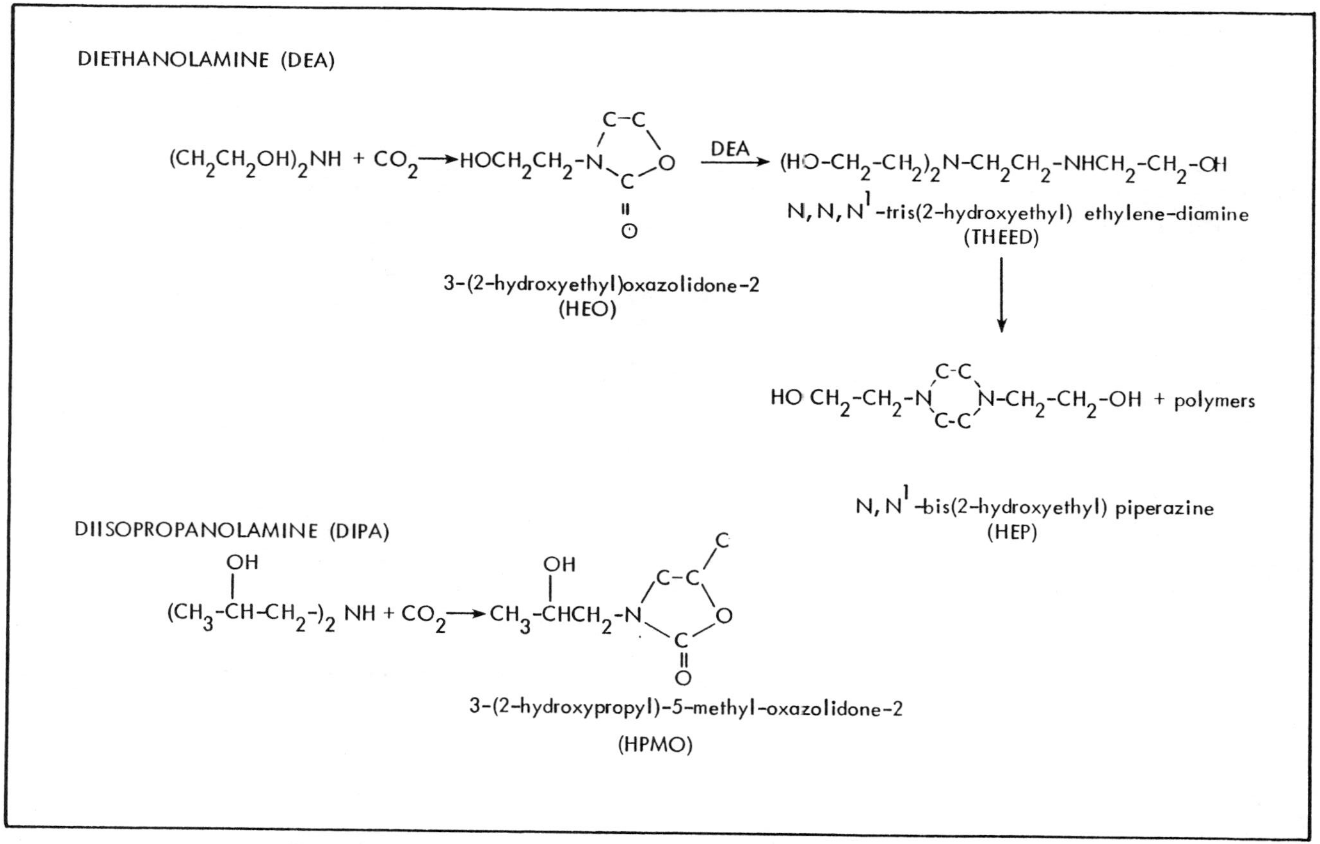

FIGURE 15.8.2. Secondary amine degradation in presence of CO_2.

TABLE 15.8.1. Chemical Degradation and Reclaiming Conditions for Aminoalcohols

		Primary amines		Secondary amines	
		MEA	DGA	DEA	DIPA
Chemical degradation	H_2S	No effect			
	CO_2	Some irreversible degradation (reclaiming required)		No effect	Severe irreversible degradation (reclaiming required)
	COS	Irreversible degradation (20% of COS feed)	Partially reversible degradation	Minor degradation	Severe irreversible degradation
Reclaiming		290–300°F Atmospheric reclaimer	370–380°F w/steam sparge	None	Vacuum distillation 50–100 MMHq @ 350°F

tions taking place. Potassium carbonate is degraded by H_2S in the absence of CO_2, the solution being converted to potassium bisulfide. Another degradation path for K_2CO_3 solutions occurs when treating gases that contain carbon monoxide, CO. Hydrogen and ammonia synthesis gases contain low concentrations of CO, whereas chemical synthesis gases contain high CO concentrations, usually at high pressure. K_2CO_3 reacts with CO to produce formate ion and possibly oxalate ion:

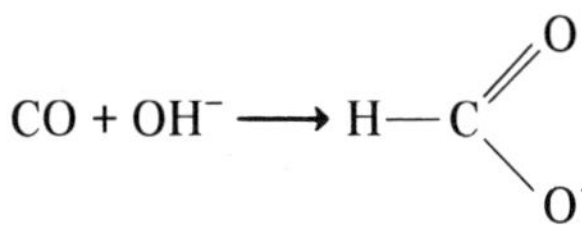

The common organic solvents used in gas treating–Sulfolane, alcohols, glycols, propylene carbonate, pyrrolidones–are relatively stable liquids in the presence of CO_2, H_2S, COS, CS_2, and RSH.

Gas treating solution foaming can be a persistent and troublesome problem, especially in the processing of natural gas. Foaming can increase operating costs owing to treating solution losses, reduced gas throughput rates and off-specification product gas. In nearly all circumstances, foaming can be controlled or eliminated efficiently and economically. Foam is usually brought under control through the use of antifoam agents (defoamers), solution filtration, carbon treating, or a combination of all three.

Foaming may be caused by trace dissolved impurities (extraneous materials or degradation or corrosion products of the treating solution), fine solids (e.g., corrosion products), or immiscible liquids (e.g., condensed hydrocarbons). The theory of foam formation and destabilization has been discussed in the literature.

A foam can be considered as a type of emulsion in which the inner phase is the gas. It seems necessary to have some surface active component present to stabilize the foam. Pure liquids do not foam: all liquid films tend to contract unless that tendency is counteracted by the presence of a surface active component. The type of foam encountered in chemical gas treating solvents contains mostly gas phase surrounded by thin films of solution. The mechanism of foam collapse is film drainage by a process whose rate is governed by what happens at the film boundaries. Foam lifetime depends on drainage rate. The properties of the liquid that increase foam stability are high elasticity, high resilience, and high surface viscosity. Surface tension of the solution, other than its value as an indicator of presence or absence of surfactant material, is not a particularly important parameter.

The action of antifoam agents in preventing foaming can be analyzed in terms of the above parameters. First, the agent should be able to displace the foam-producing surfactant from the interface. This enables it to insert its own poor foam stabilizing characteristics into the system. It has been suggested that effective antifoam agents must be liquids that are insoluble in the foamy solution and have the ability to spread spontaneously over the surface of the foamy liquid. This purely mechanical action of the spreading liquid ruptures the foam film. For inhibiting foam in aqueous solutions, the desired qualities in an antifoam include insolubility in water, lower density than water, relatively low surface tension, and relatively low interfacial tension against the aqueous solution.

In plants, foam makes itself recognizable in several ways. One of the best detection methods is use of differential pressure cells across the absorber and stripper columns. Foaming is characterized by higher than normal or erratic gas-side pressure drop, accompanied by lowered liquid levels in the reboiler and the absorber sump. Carryover of treating solution may occur if intense foaminess develops. A short-term remedy is to reduce gas throughput, but this leads to loss of production.

Foaminess can also be detected by withdrawing solution samples and subjecting them to a standard foam test in the laboratory. A relatively simple foam test apparatus has been described. The method involves sparging gas through the liquid under standard conditions, and measuring both the steady-state foam volume, and the foam collapse time after the gas flow has been stopped.

An important cause of foaming in natural gas treating is entrainment of liquid hydrocarbons and corrosion inhibitors from the upstream gathering system into the gas treating plant. Provision of an inlet knockout drum can help to prevent these potential foam producing chemicals from entering the gas treating plant. In new plants, various oils, inhibitors, and welding fluxes adhering to metal surfaces may promote foam. Thorough cleaning of new equipment and packings is essential to achieving a smooth startup. Certain types of filter elements may cause foam.

When a foaming problem develops, immediate results can usually be obtained by introducing an antifoam (that has shown effectiveness in the laboratory test). Antifoams that have been found to be effective with aqueous chemical solvents include hydrocarbon types such as polyethers, and silicone types such as polydimethylsiloxane containing fine particles of silica. Antifoam addition must usually be continued for an extended period because most antifoams have a limited life, either because they are degraded or because they are removed from solution by filters or carbon treaters.

Activated carbon adsorption has also proven to be highly effective in controlling foam, often in conjunction with antifoam addition. Many plants ensure a high service factor by continuous slipstream carbon treating plus continuous or regular additions of antifoam. If foaming persists, the cause of the problem must be identified and steps taken to eliminate it.

Environmental considerations can be classified under the following headings:

Solvent constituents and properties
Potential for loss and contamination
Objectional community visibility
Noise

The major and minor constituents of some common CO_2 scrubbing processes are listed in Table 15.8.2. All, except arsenic, are considered to be environmentally acceptable and controllable under normal plant operations. Certain metal ions used for corrosion inhibition (e.g., V, Sb) may come under future scrutiny.

TABLE 15.8.2. CO_2 Scrubbing Processes Used in Ammonia Production and Their Major Constituents

Classification	Type	Constituents	
		Major	Minor
Chemical	Aminoalcohol (alkanolamine)	MEA	–
		MEA DEA	Vanadium/antimony
	Alkaline salt	K_2CO_3	Organic/vanadium
		K_2CO_3	Organic borate/ vanadium
		K_2CO_3/arsenite	–
		K_2CO_3	Glycine/vanadium
Physical	–	Polyethylene glycol	–
		Propylene carbonate	–
		Methanol	–
Physical-chemical	–	DIPA/sulfolane	–

The main chemicals lost from acid gas scrubbing units are generally controllable and contamination of the surroundings can be minimized by good design and operation. The principal causes are tower carryover and system leaks. Tower carryover may be prevented by monitoring tower pressure drop, avoiding sudden pressure changes, preventing foaming, avoiding solution contamination, maintaining clean solution, proper use of antifoam, and avoiding operation too close to flooding. System leaks can be prevented by proper pump seal maintenance. Good practice includes the use of water-flushed seals and piping to collect frequent leaks into a recovery system. Other factors that lead to minimizing solution loss include proper backwash of filters, proper disposal of filtration media and activated carbon, and proper disposal of reclaimer residues. Monitoring of corrosion by the use of coupons, sampling, and analysis for dissolved metals and annual inspections can help to avoid unexpected leaks.

Special attention must be paid to system water balance to eliminate or minimize effluent streams leaving the system. Common liquid effluent streams associated with acid gas treating systems include (1) the process condensate separated from the absorber inlet, and (2) other system surplus water usually associated with the regenerator overhead reflux system. A "zero water balance" flowsheet for a hot potassium carbonate plant has been described.

Noise from the CO_2 vent can be prevented by the use of silencers. A visible plume problem associated with the CO_2 vent can be avoided by cooling the gas before venting and installation of a demister.

References and Notes

CHAPTER 1

There are two classical books on gas treating that cover in considerable detail the general background material discussed in this chapter:

R. N. Maddox, *Gas and Liquid Sweetening*, Campbell Petroleum Series, Norman, OK, 1977.

A. L. Kohl and F. C. Reisenfeld, *Gas Purification*, 3rd Ed., Gulf Publishing Co., Houston, 1979.

Table 1.1.1 was developed from information in R. W. Tennyson and R. P. Schaaf, *Oil and Gas J.*, **75**(2), 78 (1977).

Table 1.2.2 is based on information in *Hydrocarbon Processing*, **58**(4), 99 (1979).

Gas Treating in Ammonia and Hydrogen Manufacture:

O. J. Quartulli, W. Turner, and K. W. Padgett, "CO_2 Removal in Ammonia Plants," in J. J. McKetta (Ed.), *Encyclopedia of Chemical Processing and Design*, Vol. 3, Marcel Dekker, New York, 1977, p. 273 (Figure 1.3.1 was taken from this reference).

D. J. Kubek and K. F. Butwell, "Amine Guard Systems in Hydrogen Production," presented at the AIChE 86th National Meeting, Advances in Refining and Petrochemical Technology, Houston, TX, April 1-5, 1979.

A. V. Slack and G. R. James (Eds.), *Ammonia, Part II*, Marcel Dekker, New York, 1974.

Sour Natural Gas Treating Technology:

J. W. Estep and E. W. Plum, "Economics of Sour Gas Industry," presented at the 19th Canadian Chemical Engineering Conference, Edmonton, Alberta, October 19-22, 1969.

S. P. White and D. J. Morgan, *Oil and Gas J.*, **77**, 78 (1979) (Figure 1.3.3 and part of Table 1.3.2 were developed from information in this reference).

B. G. Goar, "Today's Gas Treating Processes," Paper 21, Selected Proceedings from Gas Conditioning Conferences, Vol. III, 1952-1974, University of Oklahoma, 1975.

G. L. Fawar, *Oil and Gas J.*, **70**, 95 (1972) (part of Table 1.3.2 was taken from this reference).

J. W. Sweny and J. P. Valentine, *Chem. Eng.*, **77**, 54 (1970) (Figure 1.3.5 was taken from this article).

Flue Gas Desulfurization Technology:

A. V. Slack, *Chem. Eng. Prog.*, **74,** 71 (1978) (Figure 1.3.6 was taken from this article).

Proceedings 3rd International Conference on Control of Sulfur and Other Gaseous Emissions, University of Salford, England, April 24-26, 1979.

Symposium Papers from Advances in Coal Utilization Technology, Symposium sponsored by Institute of Gas Technology, Louisville, Kentucky, May 14-18, 1979, Institute of Gas Technology, Chicago, 1979.

A. L. Kohl and F. Reisenfeld, *Gas Purification*, 3rd Ed., Gulf Publishing, Houston, 1979 (Table 1.3.3 was taken from this source).

T. W. DeVitt, B. A. Laseke, and N. Kaplan, *Chem. Eng. Prog.*, **76,** 45 (1980).

K. S. Murthy, H. S. Rosenburg, and R. B. Engdahl, *J. Air Pollution Control Assoc.*, **26**(9), 851 (1976).

A. B. Welty, *Hydrocarbon Processing*, **50,** 104 (1971).

F. T. Princiotta, *Chem. Eng. Prog.*, **74,** 58 (1978).

G. T. Rochelle and C. J. King, *Chem. Eng. Prog.*, **74,** 65 (1978).

Selective H_2S Removal from High-CO_2 Gases:

B. G. Goar, *Oil and Gas J.*, **78,** 239 (1980).

R. R. Johnson and G. R. Say, "Treating Low Joule Gas Produced in the Flexicoking Process," presented at 3rd International Conference on Control of Sulfur and Other Gaseous Emissions, University of Salford, England, April 24-26, 1979.

A. L. Kohl and F. Reisenfeld, *Gas Purification*, 3rd Ed., Gulf Publishing, Houston, 1979 (Figure 1.3.9 and Table 1.3.5 were taken from this reference).

R. A. Busch, J. J. Kociscin, H. F. Schroeder, and G. N. Shah, *Hydrocarbon Processing*, **58,** 136 (1979) (Figure 1.3.8 was taken from this reference).

Economic Factors in Gas Treating:

L. V. Kunkel, "The Fundamentals of Gas Sweetening," Paper I, Proceedings Gas Conditioning Conference, Norman, OK, 1977 (Table 1.4.1 was derived from information in this reference).

C. Ouwerkerk, "Process Selection for the Treating of Natural Gas," Paper I-36, Natural Gas Processing and Utilization Conference, Irish Group of the Institution of Chemical Engineers, April 1976.

Mass Transfer and Thermodynamics:

The theory of mass transfer is discussed in detail in a number of textbooks; possibly the best one is:

T. K. Sherwood, R. L. Pigford, and C. R. Wilke, *Mass Transfer*, McGraw-Hill, New York, 1975.

The material in Sections 1.7 and 1.8 is discussed in detail in all good textbooks on chemical engineering thermodynamics. The problem of the evaluation of p^0 in Equation 1.7.3 when the temperature is above the critical temperature of the soluble gas is discussed in:

J. M. Prausnitz, *Molecular Thermodynamics of Fluid Phase Equilibria*, Prentice-Hall, Englewood Cliffs, NJ, 1969.

Acid Gas and Light Hydrocarbon Solubility in Liquids:

E. W. Washburn (Ed.), *International Critical Tables*, McGraw-Hill, New York, 1926.

R. H. Perry and C. H. Chilton (Eds.), *Chemical Engineer's Handbook*, 5th Ed., McGraw-Hill, New York, 1973.

Solubility Data for Water:

E. Wilhelm, R. Battino, and R. J. Wilcock, *Chemical Reviews*, **77**(2), 219 (1977) (Figure 1.7.2 and Table 1.7.1 are based on information taken from this reference).

Solubility Data for Methanol:

J. K. Ferrall, R. W. Rousseau, and D. G. Bass, "The Solubility of Acid Gases in Methanol," U.S. Environmental Protection Agency Report, EPA-600/7-79-097, April, 1979 (Figure 1.7.1 was taken from this paper). This report gives references to earlier studies of CO_2 and H_2S solubility in methanol.

Solubility Data for *N*-Methylpyrrolidone:

GAF Corporation, *M-Pyrol^R N-Methyl-2-Pyrrolidone Handbook*, GAF Corp., New York, 1972 (Table 1.7.2 and Figure 1.7.3 are based in part on information from this reference).

Acid Gas Solubility Data in Propylene Carbonate:

O. R. Rivas and J. M. Prausnitz, *Ind. Eng. Chem. Fund.*, **18**(3), 289 (1979).

E. E. Isaacs, F. D. Otto, and A. E. Mather, *Can. J. Chem. Eng.*, **55**, 751 (1977) (data from these references were used in preparing Table 1.7.2).

Acid Gas Solubility in Polyethyleneglycol Di-Isopropyl Ether:

W. Wolfer, E. Schwartz, W. Vodrazka, and K. Volkamer, *Oil and Gas J.*, **78**, 57 (1980) (data from this reference were incorporated into Table 1.7.2).

Light Hydrocarbon Solubility Data in Physical Organic Solvents:

J. W. Sweny and J. P. Valentine, *Chem. Eng.*, **77**, 54 (1970).

B. B. Woertz, *J. Petr. Tech.*, **13**, 483 (1971).

K. Wehner, W. Burk, and W. Kisan, *Chem. Tech.*, **29**, 445 (1977) (data from these references were used in preparing Figure 1.7.3).

Figure 1.8.1 is taken from K. E. Zarker, "Sulfinol," in A. V. Slack and G. R. James (Eds.), *Ammonia Part II*, Marcel Dekker, New York, 1974.

Figure 1.8.2 is taken from K. G. Christensen and W. J. Stupin, *Hydrocarbon Processing*, **57**(2), 125 (1978).

Chemistry of Acid Gas Reactions:

There is an ample literature on the chemistry of acid gas reactions with chemical solvents. The following references are representative examples.

P. V. Danckwerts, *Chem. Eng. Sci.*, **34**, 443 (1979) (amino-alcohol reactions with CO_2).

D. W. Savage, G. Astarita, and S. V. Joshi, *Chem. Eng. Sci.*, **35**(7), 1513 (1980) (CO_2 reactions with potassium carbonate solutions).

G. Astarita, *Mass Transfer with Chemical Reaction*, Elsevier, New York, 1967 (H_2S reactions with amines).

J. R. Phillips and R. N. Fleck, "Reaction Kinetics in a Stretford-Process Absorber," presented at 72nd Annual AIChE Meeting, San Francisco, November 27, 1979 (H_2S oxidation chemistry).

G. Astarita, D. W. Savage, and J. M. Longo, *Chem. Eng. Sci.*, **36**, 581 (1981) (CO_2 rate promotion chemistry in carbonate buffers).

Unfinished Problems:

Figure 1.10.2 is taken from A. L. Kohl and F. C. Reisenfeld, *Gas Purification*, Gulf Publishing, Houston, 1979, p. 79.

Figure 1.10.3 was developed from ideas in A. L. Shrier and P. V. Danckwerts, *Ind. Eng. Chem. Fund.*, **8**, 415 (1969).

Figure 1.10.4 was taken from A. L. Kohl and C. Blohm, *Petr. Eng.*, **22**, C-37 (1950).

CHAPTER 2

There are several textbooks on classical thermodynamics of two-phase systems and of chemically reacting systems. A comparatively recent and excellent one is:

S. I. Sandler, *Chemical and Engineering Thermodynamics*, John Wiley, New York, 1977.

VLE data reported in Figure 2.4.1 are from:

J. S. Tosh, J. H. Field, H. E. Benson, and W. P. Haynes, U.S. Bureau of Mines, Report of Investigations No. 5484 (1959).

The model for the CO_2/MEA system as discussed in Section 2.4 is based on:

G. Astarita, G. Marrucci, and F. Gioia, *Rend. Acc. Lincei*, **VII**(34), 176 (1963).

A model for CO_2-H_2S-Aqueous Alkanolamine systems, which includes activity coefficient variations, has recently been published by:

R. D. Deshmukh and A. E. Mather, *Chem. Eng. Sci.*, **36**, 355 (1981). (This paper also refers to the comparatively abundant previous literature on the subject.)

The data in Figures 2.6.2 and 2.6.3 are from:

R. L. Kent and B. Eisenberg, "Equilibrium of H_2S and CO_2 with MEA and DEA Solutions," Paper E, Gas Conditioning Conference, University of Oklahoma, 1975.

Methods for the evaluation of the constant K_i in Eq. 2.8.1 are discussed by:

P. V. Danckwerts and M. M. Sharma, *Chem. Eng.*, No. 202, CE244, October, 1966.

CHAPTERS 3-7

There are two standard reference books for the theory of mass transfer with chemical reaction in gas-liquid systems:

G. Astarita, *Mass Transfer with Chemical Reaction*, Elsevier, Amsterdam, 1967.
P. V. Danckwerts, *Gas-Liquid Reactions*, McGraw-Hill, New York, 1970.

Most of the concepts in Chapters 3-7 are discussed at length in these two books. However, the books are more than 10 years old and some new concepts, as well as more compact viewpoints of old ones, have been developed in the last decade. In particular, the material in Section 3.5 and in Chapters 6 and 7 follows largely the approach recently presented in:

G. Astarita and D. W. Savage, *Chem. Eng. Sci.*, **35**, 659, 1755 (1980).

The first of these papers is addressed particularly to the problem of chemical desorption; an older one dealing with a special case of chemical desorption is:

K. E. Porter, A. D. C. Cantwell, and C. McDermott, *AIChE J.*, **17**, 536 (1971).

The analysis in Section 4.5 is based on:

G. Astarita and W. J. Beek, *Chem. Eng. Sci.*, **17**, 665 (1962).

The data used in Example 4.1 and Figure 5.4.1 are from:

D. W. Savage, G. Astarita, and S. Joshi, *Chem. Eng. Sci.*, **35**, 1513 (1980).

The problem of the dependence of effective interfacial areas on the regime of mass transfer is discussed by:

W. Bergbauer, B. Horner, and K. Dialer, *Ger. Chem. Eng.*, **2**, 153 (1979).
C. E. H. Joosten and P. V. Danckwerts, *Chem. Eng. Sci.*, **28**, 453 (1973).

The latter reference also introduces the concept of rate enhancement effectiveness.

The analysis in Section 6.1 is based on:

T. K. Sherwood and R. L. Pigford, *Absorption and Extraction*, McGraw-Hill, New York, 1951.

The case where Eq. 6.1.20 holds with $m = \frac{2}{3}$ has been presented by:

S. K. Friedlander and M. Litt, *Chem. Eng. Sci.*, **7**, 299 (1958).

The penetration theory equations, Eq. 6.2.1, can be solved for some specially simple chemical mechanism; see:

D. R. Olander, *AIChE J.*, **6**, 233 (1960).

There is an ample literature on the material in Chapter 7, mostly concerned with the numerical solution of the relevant equations. The simplified procedure discussed in Section 7.3 is based on an approximation introduced as early as 1948 by:

D. W. Van Krevelen and P. J. Hoftyzer, Chimi. Ind., XX1 Congr. of Indust. Chem., Bruxelles, p. 168 (1948).

A special form of Eq. 7.4.1 was presented by:

W. J. DeCoursey, *Chem. Eng. Sci.*, **29**, 1867 (1974).

whereas in its general form, it is credited to:

S. Joshi, Ph.D. Thesis, University of Delaware, 1980.

CHAPTER 8

There is an abundant literature on simultaneous absorption, but it is unfortunately often misleading in that either the occurrence of the shift reaction is neglected altogether or its effect is buried in the numerical solution of the relevant equations. The first explicit discussion of the shift reaction is very recent:

R. Cornelisse, A. A. C. M. Beenackers, F. P. M. Van Beckum, and W. P. M. Swaaij, *Chem. Eng. Sci.*, **35**, 1245 (1980).

The approach followed in Sections 8.1 to 8.5 is largely original and is based on the following two papers:

G. Astarita and D. W. Savage, *Chem. Eng. Sci.*, **37**, 677 (1982).

D. W. Savage, E. W. Funk, and G. Astarita, "Selective Absorption of H_2S and CO_2 into Aqueous Solutions of Methyldiethanolamine," submitted to *Ind. Eng. Chem. Fund.*, (1982).

The two special problems discussed in Section 8.6 are from:

G. Astarita and F. Gioia, *Ind. Eng. Chem. Fund.*, **4**, 316 (1965).

G. H. Roper, T. F. Hatch, and R. L. Pigford, *Ind. Eng. Chem. Fund.*, **1**, 144 (1962).

CHAPTER 9

Figure 9.1.1 is taken from:

H. Thirkell, "General Considerations of CO_2 Removal," in *Amonia Part II*, A. V. Slack and G. R. James (Eds.), Marcel Dekker, New York, 1974.

Figure 9.1.2 is taken from:

T. P. Cook, "Fluor CO_2 Removal," in *Ammonia Part II*, A. V. Slack and G. R. James (Eds.), Marcel Dekker, New York, 1974.

Figure 9.1.3 is taken from:

W. Linde, G. Ranke, and H. Jungfer, "Rectisol CO_2 Removal," in *Ammonia Part II*, A. V. Slack and G. R. James (Eds.), Marcel Dekker, New York, 1974.

Figure 9.1.4 is taken from:

R. W. Tennyson and R. P. Schaaf, *Oil and Gas J.*, **75**(2), 78 (1977).

Thermodynamics and Chemistry:

Figure 9.2.1 is based on data from:

G. Houghton, A. M. McLean, and P. D. Ritchie, *Chem. Eng. Sci.*, **6**, 132 (1957).

A. J. Ellis and R. M. Golding, *Amer. J. Sci.*, **261**, 47 (1963).

Figure 9.2.2 is taken from:

R. L. Kent and B. Eisenberg, "Equilibrium of H_2S and CO_2 with MEA and DEA Solutions," Paper E, Gas Conditioning Conference, University of Oklahoma, 1975.

Hydroxide Solutions:

There is an abundant literature on CO_2 absorption in caustic solutions; see:

G. Astarita, *Mass Transfer with Chemical Reaction*, Elsevier, New York, 1967.

Equation 9.2.4 and the data on which it is based are discussed by:

T. K. Sherwood, R. L. Pigford, and C. R. Wilke, *Mass Transfer*, McGraw-Hill, New York, 1975.

Amine Solutions:

Figures 9.4.1 and 9.4.2 were developed from data in:

J. I. Lee, F. D. Otto, and A. E. Mather, *Can. J. Chem. Eng.*, **52**(5), 803 (1974).

J. I. Lee, F. D. Otto, and A. E. Mather, *Can. J. Chem. Eng.*, **52**, 125 (1974).

E. E. Isaacs, F. D. Otto, and A. E. Mather, *J. Chem. Eng. Data*, **22**(1), 71 (1977).

J. S. Tosh, J. H. Field, H. E. Benson, and W. P. Haynes, "Equilibrium Study of the System Potassium Carbonate, Potassium Bicarbonate, Carbon Dioxide and Water," U.S. Bureau of Mines, Report of Investigation No. 5484 (1959).

Table 9.4.1 and the discussion of improvements to aqueous MEA is based on:

K. F. Butwell and D. J. Kubek, *Hydrocarbon Proc.*, **56**, 173 (1977).

Figures 9.4.3 and 9.4.4 were constructed from data in:

M. M. Sharma, *Trans. Farad. Soc.*, **61**, 681 (1965).

P. V. Danckwerts and M. M. Sharma, *Chem. Eng.*, No. 202, CE244 (1966).

H. Hikita, S. Asai, H. Ishikawa, and M. Honda, *Chem. Eng. J. (Lausanne)*, **14**, 27 (1977).

The literature on CO_2 reactions with aqueous alkanolamines is a vast one. The best discussion of the chemistry involved is perhaps presented by:

M. Caplow, *J. Am. Chem. Soc.*, **90,** 6795 (1968).

Reactions of CO_2 with MEA and the problem of the kinetics of carbamate formation with DEA have been discussed in the following recent papers, which are also a reference source to the previous literature:

P. V. Danckwerts, *Chem. Eng. Sci.*, **34,** 443 (1979).

C. Alvarez-Fuster, N. Midoux, A. Laurent, and J. C. Charpentier, *Chem. Eng. Sci.*, **33,** 1717 (1980).

H. Hikita, S. Asai, Y. Katsu, and S. Ikuno, *AIChE J.*, **25,** 793 (1979).

Reactions of CO_2 with ethylenediamine have been discussed by:

E. Sada, H. Kumazawa, and M. A. Butt, *Chem. Eng. J.*, **13,** 213 (1977).

Kinetics of CO_2 reactions with a large number of amines have been discussed by:

M. M. Sharma, *Trans. Farad. Soc.*, **61,** 681 (1965) (Figure 9.4.4 is based on data from this paper).

The state of the art on CO_2 reactions with amines up to 1966 is reviewed in:

P. V. Danckwerts and M. M. Sharma, *Chem. Eng.*, CE244, October, 1966.

Figure 9.4.5 is taken from:

J. M. Jackson, "Monoethanolamine (MEA) CO_2 Removal," in *Ammonia Part II*, A. V. Slack and G. R. James (Eds.), Marcel Dekker, New York, 1974.

Industrial Use of Arsenious Acid and Boric Acid as Rate Promoters:

G. Giammarco, "Giammarco-Vetrocoke," in *Ammonia Part II*, A. V. Slack and G. R. James (Eds.), Marcel Dekker, New York, 1974.

CO_2 Reactions in Unpromoted Carbonate Solutions at Temperatures of Industrial Interest:

D. W. Savage, G. Astarita, and S. Joshi, *Chem. Eng. Sci.*, **35,** 1513 (1980) (Eq. 9.2.6 is based on the data presented in this paper).

General Area of Rate Promotion in Carbonate Solutions:

G. Astarita, D. W. Savage, and J. M. Longo, *Chem. Eng. Sci.*, **36,** 581 (1981).

Recent papers dealing specifically with the catalytic effect of arsenite and of amines are, respectively:

C. N. S. McLachlan and P. V. Danckwerts, *Trans. Inst. Chem. Engrs.*, **50,** 300 (1977).

T. L. Donaldson and Y. N. Nguyen, *Ind. Eng. Chem. Fund.*, **19**(3), 260 (1980).

G. Sartori and D. W. Savage, "Carbon Dioxide Removal from Gases with Sterically-Hindered Amines," submitted to *Ind. Eng. Chem. Fund.* (1982).

Figures 9.5.8 and 9.5.10 are taken from:

J. H. Field, "Hot Carbonate CO_2 Removal," in *Ammonia Part II*, A. V. Slack and G. R. James (Eds.), Marcel Dekker, New York, 1974.

Figure 9.5.9 is taken from:

L. Crabs, R. Pouilliart, and F. Van Hecke, "Energy Savings for Carbon Dioxide Removal Systems," Paper 110b, presented at 72nd AIChE Meeting, San Francisco, November 25–29, 1979.

Table 9.5.1 is taken from:

R. W. Parrish and H. B. Neilson, "Synthesis Gas Purification Including Removal of Trace Contaminants by the Benfield Process," presented at 167th ACS Meeting, Los Angeles, March 31–April 5, 1974.

Generalized thermodynamic modeling of CO_2-amine promoted potassium carbonate and CO_2-aqueous amine systems has been carried out by M. T. Melchior, private communication, 1977.

CHAPTER 10

H_2S Removal from Natural Gas:

C. Ouwerkerk, "Process Selection for the Treating of Natural Gas," Paper I-36, Natural Gas Processing and Utilization Conference, Irish Group of Institution of Chemical Engineers, Dublin, April, 1976.

Claus Plants and the Integration of Tail Gas Units:

A. L. Kohl and F. Reisenfeld, *Gas Purification*, 3rd Ed., Gulf Publishing Co., Houston, 1979.

J. Waller, C. H. Golz, and E. S. E. Werner, Compendium 78/79 (DGMK) 26th Deutschen Gesellschaft Mineral Kohlchimie, October, 1978 (Figures 10.1.1 and 10.1.2 are taken from this paper).

Direct Conversion Oxidative H_2S Removal Processes:

A. J. Moyes and S. Vasan, *Oil and Gas J.*, **72**, 57 (1974) (Figure 10.1.3 was taken from this reference).

Process Selection:

Figure 10.1.4 was taken from:

R. N. Tennyson and R. P. Schaaf, *Oil and Gas J.*, **75**(2), 78 (1977).

Figure 10.1.5 on hydrodesulfurization is from:

M. F. Oxenreiter, C. G. Frye, G. B. Hockstra, and J. M. Sroka, paper presented at Jap. Pet. Instn. Meeting, Tokyo, 1972.

Thermodynamics and Chemistry:

Figure 10.2.1 is taken from:

R. L. Kent and B. Eisenberg, "Equilibrium of H_2S and CO_2 with MEA and DEA Solutions," Paper E, Gas Conditioning Conference, University of Oklahoma, 1975.

Eq. 10.2.8 has been proposed by:

C. Ouwerkerk, in "Process Selection for the Treating of Natural Gas," Paper I-36, Natural Gas Processing and Utilization Conference, Irish Group of Institution of Chemical Engineers, Dublin, April, 1976.

Figure 10.2.2 is based on data presented by:

J. I. Lee, F. D. Otto, and A. E. Mather, *J. Chem. Eng. Data*, **21**(2), 207 (1976).

J. I. Lee, F. D. Otto, and A. E. Mather, *J. Chem. Eng. Data*, **18**(1), 71 (1973).

J. I. Lee, F. D. Otto, and A. E. Mather, *J. Chem. Eng. Data*, **18**(4), 420 (1973).

E. E. Isaacs, F. D. Otto, and A. E. Mather, *Can. J. Chem. Eng.*, **55**(2), 210 (1977).

Kinetics:

An estimate of the order of magnitude of the kinetic constant for the proton-transfer reaction 10.2.1 is given by:

H. Eigen, *Soum. Kemistil.*, **A34**, 24 (1961).

Figure 10.3.1 is taken from:

F. Gioia and G. Astarita, *Ind. Eng. Chem. Fund.*, **6**, 370 (1967) (This paper also deals in general terms with the rate of mass transfer of H_2S in alkaline solutions).

Figures 10.3.2 and 10.3.3 are from:

G. Astarita and D. W. Savage, *Chem. Eng. Sci.*, **35**, 1755, (1980).

Diffusivities:

A recent critical review on the estimation of diffusion coefficients in liquids is given in:

R. C. Reid, J. M. Prausnitz, and T. K. Sherwood, *Properties of Gases and Liquids*, McGraw-Hill, New York, 1977.

Diffusive coupling of ions is discussed by:

J. S. Newman, *Advances in Electrochemistry and Electrochemical Engineering*, C. W. Tobias (Ed.), Interscience-Wiley, New York, 1967.

Estimation methods for solubilities and diffusivities of gases in liquids have recently been reviewed by:

B. I. Morsi and J. C. Charpentier, NATO ASI on Mass Transfer with Chemical Reaction in Multiphase Systems, CESME-IZMIR-TURKEY, August 10–21, 1981. Also, A. E. Rodrigues, J. M. Calo, and N. H. Sneed (Editors), Multiphase Chemical Reactors, Sijkhoff and Noordoff, The Netherlands (1981).

CHAPTER 11

Process Selection:

Natural gas treating to pipeline specification and the operation of split-flow Claus plants is discussed by:

S. P. White and D. J. Morgan, *Oil and Gas J.*, **77**, 78 (1979) (Figures 11.1.1 and 11.1.2 are taken from this article).

Chemicals syn gas treating is described by:

C. L. Reed and C. J. Kahre, *Hydrocarbon Proc.*, **58**, 191 (1979) (Table 11.1.1 and Figure 11.1.3 are from this reference).

A selective regeneration physical solvent process is discussed by:

R. Zeller, *Linde Rep. Sci. Tech.*, **30**, 3 (1979) (Figure 11.1.4 is from this source).

Tables 11.1.2 and 11.1.3 are from:

C. Chin, *Hydrocarbon Proc.*, **57**, 266 (1978).

Figure 11.1.5 is based on:

J. L. Kennedy, *Oil and Gas J.*, **76**(22), 51 (1978).

Figures 11.1.8 and 11.1.9 are from:

R. W. Tennyson and R. P. Schaaf, *Oil and Gas J.*, **75**(2), 78 (1977).

Selective H_2S Removal

F. C. Vidaurre and L. C. Kahre, *Hydrocarbon Proc.*, **56**, 333 (1977).

C. Ouwerkerk, *Hydrocarbon Proc.*, **57**(2), 89 (1978).

R. Cornelisse, A. A. C. M. Beenakers, F. P. H. Van Beckum, and W. P. M. Van Swaajj, *Chem. Eng. Sci.*, **35**, 1245 (1980).

A. E. Cornelissen, *Trans. Instn. Chem. Engrs. (London)*, **58**(4), 242 (1980).

R. R. Johnson and G. R. Say, "Treating Low Joule Gas Produced in the Flexicoking Process," 3rd International Conference on Control of Sulfur and Other Gaseous Emissions, University of Salford, England, April 24–26, 1979.

P. W. Sigmund, K. F. Butwell, and A. J. Wussler, *Hydrocarbon Proc.*, **60**(5), 118 (1981).

M. W. McEwan and A. Marmin, "Developments in LNG Treating," Appendix, Paper 3, Sixth International Conference on Liquefied Natural Gas, April 7–10, 1980, Kyoto, Japan.

Sponsored by International Gas Union, International Institute of Refrigeration, Institute of Gas Technology, papers published by Institute of Technology, Chicago.

Figure 11.1.7 is taken from:

D. W. Savage, E. W. Funk, and G. Astarita, "Selective Absorption of H_2S and CO_2 into Aqueous Solutions of Methyl-diethanolamine," submitted to *Ind. Eng. Chem. Fund.* (this paper also discusses selective removal with tertiary amines).

Figure 11.4.1 was constructed from data in:

M. W. Wainwright, G. C. Egleson, C. M. Brock, J. Fisher, and A. E. Sands, U.S. Bureau of Mines, Report of Investigations No. 4891, October 1952.

A large fraction of the material discussed in Sections 11.2, 11.3, and 11.4 is original.

Coke Oven Gas Treating:

M. J. Massey and R. W. Dunlap, *J. Air Poll. Con. Assoc.*, **25**(10), 1019 (1975).

Table 11.1.6 is based on information in:

W. B. Watson, "The Manufacture of SNG from Coal," Paper R, Proceedings Gas Conditioning Conference, University of Oklahoma, March, 1977.

CHAPTER 12

There is an ample, but not well systematized, literature on flue gas desulfurization technology. The following references are representative:

R. L. Pedroso and K. M. Press, "Sulfur Recovery from Glue Gas at Large Coal-Fired Power Plants," Paper F, I. Chem. E. Symp. Series, No. 57, 3rd International Symposium on Control of Sulfur and Other Gaseous Emissions, Salford, England, April, 1979.

F. A. Ayer (Compiler), Proceedings: Symposium on Flue Gas Desulfurization, Las Vegas, NV, March, 1979, EPA-600/7-79-167a, July, 1979.

I. Miller, *Chem. Eng.*, **86**, 52 (1979).

W. J. Raymond and A. G. Sliger, *Chem. Eng. Prog.*, **74**, 75 (1978).

J. D. Cunic, S. A. Diamond, P. E. Reeder, and L. M. Williams, *Hydrocarbon Proc.*, **57**, 141 (1978).

G. T. Rochelle, "Desulfurization of Stack Gas," prepared for publication in *The Encyclopedia of Chemical Processing and Design*, J. J. McKetta and W. A. Cunningham (Eds.).

Figure 12.1.2 is taken from:

J. C. Yarze and W. Beiersdorf, "SO_2 Removal Process Using Magnesium Promoted Limestone Slurry," Paper G, I. Chem. E. Symposium Series No. 57. The Control of Sulfur and Other Gaseous Emissions, 3rd International Symposium, Salford, England, April, 1979 (Institution of Chemical Engineers, NW Branch in association with University of Salford).

Figure 12.1.4 is from:

M. S. Rosenberg, *Hydrocarbon Proc.*, **57**, 132 (1978).

Figure 12.1.5 is from:

L. M. Lehman and M. W. Cooper, *Hydrocarbon Proc.*, **56,** 109 (1977).

Equilibrium Vapor Pressures over Sulfite-Bisulfite Solutions:

H. F. Johnstone, H. J. Read, and H. C. Blankmeyer, *Ind. Eng. Chem.*, **30,** 101 (1938) (Eq. 1 in this paper is the equivalent of Eq. 12.2.4).

Thermodynamic Data for SO_2 in Aqueous Solutions:

A. M. Abdulsattar, S. Sridhar, and L. A. Bromley, *AIChE J.*, **23,** 62 (1977).

Thermodynamics of SO_2 in Citrate Solutions:

O. Erga, *Chem. Eng. Sci.*, **35,** 162 (1980).

Equilibrium Constants for the Relevant Reactions in Lime-Limestone Slurries:

L. W. Nannen, R. E. West, and F. Kreith, *J. Air Poll. Cont. Assoc.*, **24**(1), 29 (1974).

Mass Transfer Rates:

Section 12.3 follows an approach based on a series of papers by G. T. Rochelle and coworkers. The analysis in the appendix is based on:

C. S. Chang and G. T. Rochelle, submitted to *AIChE J.*

This paper also gives references to other papers by the same authors where the application of the methodology to specific cases is discussed in detail. The effect of additives in slurry scrubbing is discussed in detail in an older paper by:

G. T. Rochelle and C. J. King, *Ind. Eng. Chem. Fund.*, **16,** 67 (1977).

The same authors discuss alternatives to throwaway processes in:

Chem. Eng. Prog., **74,** 65 (1978).

There is much literature on the analysis of mass transfer rates in slurries, where the effect of the solids themselves on the enhancement of mass transfer rates is emphasized. The most recent paper in this area at the time of writing is by:

E. Sada et al., *Chem. Eng. Sci.*, **36,** 149 (1981) (earlier papers can be traced back from this one).

A recent contribution to the analysis of SO_2 mass transfer in aqueous solutions is:

D. L. Roberts and S. K. Friedlander, *AIChE J.*, **26**(4), 593/602 (1981).

CHAPTER 13

Rocking Autoclave Cell Used for the System $CO_2/K_2CO_3/H_2O$:

J. S. Tosh, J. H. Field, H. E. Benson, and W. P. Haynes, U.S. Bureau of Mines, Report of Investigations No. 5484 (1959).

Rocking Autoclave Cell for the CO_2-H_2S/Aqueous MEA and DEA Systems:

J. D. Lawson and A. W. Garst, *J. Chem. Eng. Data*, **21**(1), 20 (1976).

Vapor Recirculation Cell:

Figure 13.1.2 is taken from:

J. I. Lee, F. D. Otto, and A. E. Mather, *J. Chem. Eng. Data*, **18**(1), 71 (1973).

Figure 13.1.3 is taken from:

D. Lal, E. E. Isaacs, A. E. Mather, and F. D. Otto, "Equilibrium Solubility of Acid Gases in Diethanolamine and Monethanolamine Solution at Low Particle Pressure," presented at Gas Conditioning Conference, University of Oklahoma, March, 1980.

Figures 13.1.4 and 13.1.5 are taken from

E. E. Isaacs, A. E. Mather, and F. D. Otto, "Solubility of CO_2 and H_2S in MEA, DIPA, Sulfinol, and Propylene Carbonate Absorbents," presented to the Canadian Natural Gas Processing Association, Calgary, Alberta, November 19, 1976.

Henry's Law Constant:

Solubility data for O_2, N_2, and H_2 leading to Eq. 13.2.2 are reported by:

H. Hikita, S. Asai, and Y. Azuma, *Can. J. Chem. Eng.*, **56**, 371 (1978).

The data on H values for CO_2 in K_2CO_3 solutions given in Figure 13.2.1 are from:

S. Joshi, Ph.D. Thesis, University of Delaware, 1980.

Isothermal Calorimeter:

Figure 13.3.1 is taken from:

J. Cone, L. E. S. Smith and W. A. Van Hook, *J. Chem. Thermodynamics*, **11**, 277 (1979).

Adiabatic Calorimeter:

A. Kahrim and A. E. Mather, *Can. J. Chem. Eng.*, **58**, 660 (1980).

Gas–Liquid Microcalorimeter Used for Biological Studies:

S. A. Rudolph, S. O. Boyle, C. F. Dresden, and S. J. Gill, *Biochemistry*, **11**, 1098 (1972).

Corrections to the Idealized Hydrodynamics of Mass Transfer Units:

L. E. Scriven and R. L. Pigford, *Am. Inst. Chem. Engrs. J.*, **5**, 397 (1959).
W. J. Beek, Dissertation, Delft, 1962.
J. O. Wilkes and R. M. Nedderman, *Chem. Eng. Sci.*, **17**, 177 (1962).
S. Portalski, *Ind. Eng. Chem. Fund.*, **3**, 49 (1964).
R. A. T. O. Nijsing, Dissertation, Delft, 1957.
D. Roberts and P. V. Danckwerts, *Chem, Eng. Sci.*, **17**, 961 (1964).

High-Temperature, High-Pressure One-Sphere Unit:

Figure 13.4.8 is taken from:

D. W. Savage, G. Astarita, and S. Joshi, *Chem. Eng. Sci.*, **35**, 1513 (1980).

Special Versions of the Stirred Cell Unit:

P. V. Danckwerts and E. Alper, *Trans. Inst. Chem. Engrs.*, **53**, 34 (1975).
O. Levenspiel and J. M. Godfrey, *Chem. Eng. Sci.*, **29**,1723 (1974).

Interfacial Turbulence During Chemical Gas Absorption:

P. L. T. Brian, J. E. Vivian, and D. C. Matiatos, *AIChE J.*, **13**, 28 (1967).

Units for Simultaneous Absorption:

The unit sketched in Figure 13.4.10 is described in detail by:

G. Astarita and F. Giola, *Chem. Eng. Sci.*, **19**, 963, 1964.

Criteria for the Rational Design of Laboratory Scale Mockups of Industrial Towers:

E. Alper and P. V. Danckwerts, *Chem. Eng. Sci.*, **31**, 599 (1976).

The batch absorber illustrated in Figure 13.4.11 was developed by:

G. Sartori (private communication).

CHAPTER 14

The literature on process selection methodology is scant. Furthermore, due to partial overlap with Chapters 1 and 9–12, many of the relevant references have already been quoted. The following is a short list of papers of general interest in the area.

D. J. Kubek and K. F. Butwell, "Amine Guard Systems in H_2 Production," presented at the AIChE 86th National Meeting, Advances in Refining and Petrochemical Technology, Houston, TX, April 1-5, 1979 (Figure 14.3.2 is taken from this reference).

K. F. Butwell, D. J. Kubek, and P. W. Sigmund, "Primary vs. Secondary Amines," Proceedings Gas Conditioning Conference, University of Oklahoma, 1980.

K. G. Christensen and W. J. Stupin, *Hydrocarbon Proc.*, **57**(2), 125 (1978).

R. W. Tennyson and R. P. Schaaf, *Oil and Gas J.*, **75**(2), 78 (1977).

Table 14.2.1 is from:

C. Ouwerkerk, "Process Selection for the Treating of Natural Gas," Paper I-36, Natural Gas Processing and Utilization Conference, Irish Group of the Institution of Chemical Engineers, Dublin, April, 1976.

Figures 14.2.1, 14.2.2, and 14.2.3 are from:

D. K. Fleming, "Acid Gas Removal Systems in Coal Gasification," presented at Symposium on Ammonia from Coal, sponsored by Tennessee Valley Authority, Muscle Shoals, May 8-10, 1979.

Figure 14.3.1 is from:

K. J. Stokes, "The Economics of CO_2 Removal in Ammonia Plants," Paper 110a, 72nd AIChE Meeting, San Francisco, November, 1979.

Figure 14.4.1 is taken from:

A. L. Kohl and F. C. Reisenfeld, *Gas Purification*, 3rd Ed., Gulf Publishing Co., Houston, 1979.

Cryogenic distillation has not been discussed in this book; an excellent reference is the recent paper of:

A. S. Holmes, *Hydrocarbon Proc.*, **61**, 131 (1982).

CHAPTER 15

Calculation of minimum liquid rates and packed heights for physical solvents are discussed in a number of textbooks; an excellent recent one is:

T. K. Sherwood, R. L. Pigford, and C. R. Wilke, *Mass Transfer*, McGraw-Hill, New York, 1975. The results in Figures 15.1.2-15.1.4 are from this book.

The results in Figures 15.2.2-15.2.3 are from:

C. Ouwerkerk, Paper I-36, Natural Gas Processing and Utilization Conference, Irish Group of the Institution of Chemical Engineers, Dublin, April, 1976.

The generalized pressure drop correlation in Figure 15.3.1 is from:

Norton Co., "Design Information for Packed Towers," Bulletin DC-11, 1977,

and is based on a correlation originally proposed by:

T. K. Sherwood, G. H. Shipley, and F. A. L. Holloway, *Ind. Eng. Chem.*, **30**, 768 (1938).

General Procedure for Sizing the Tower Height for Chemical Absorption:

S. Joshi, G. Astarita, and D. W. Savage, *Chem. Eng. Prog. Symp. Series No. 202*, **77**, 63 (1981).

Specific cases where one particular regime is known to apply are discussed by:

R. M. Secor and R. W. Southworth, *AIChE J.*, **7**, 705 (1961).
O. Levenspiel, *Chemical Reaction Engineering*, 2nd Ed., Wiley, New York, 1972.

The correlations in Eqs. 15.3.27-15.3.29 have been proposed by and Figure 15.3.4 is taken from:

K. Onda, M. Takeuchi, and Y. Okumoto, *J. Chem. Eng. Japan*, **1**, 56 (1968).

An alternate correlation for the interfacial area has been proposed by:

S. S. Puranik and A. Vogelpohl, *Chem. Eng. Sci.*, **29**, 501 (1974).

and a general discussion of interfacial areas summarizing a considerable amount of available information of given by:

A. Laurent and J. C. Charpentier, *Chem. Eng. J.*, **8**, 85 (1974) (Figure 15.3.3 is based on this paper).

Correlations for $k_L^0 \tilde{a}$ are given by:

T. K. Sherwood and F. A. L. Holloway, *Trans. AIChE*, **36**, 39 (1940).
D. M. Mohunta, A. S. Vaidyanathan, and G. S. Laddha, *Indian Chem. Engr.*, **11**, 73 (1969).

The most recent review of gas-liquid mass transfer parameters has been given by:

J. C. Charpentier and B. I. Morsi, NATO ASION Mass Transfer with Chemical Reaction in Multiphase Systems, CESME-IZMUR-TURKEY, August 10-21, 1981.

The packings in Fig. 15.3.2 have been discussed in:

Norton Chemical Process Products Division Bulletin HY-30 (Hypack™); Bulletin PR-16 (Pall Rings); Bulletin IM-83 (Intalox Metal).
I. E. Eastham, D. van den Brock, and G. N. Miertschin, *Chem. Eng. Prog.*, **74**, 61 (1978) (Cascade Mini-Rings).

Tray efficiency predictions for systems where no chemical reactions occur have been reviewed by:

T. K. Sherwood, R. L. Pigford, and C. R. Wilke, *Mass Transfer*, McGraw-Hill, New York, 1975.

The only papers we know of that deal with the kinetics of stripping desorption in chemical solvents are:

G. Astarita and D. W. Savage, *Chem. Eng. Sci.*, **35**, 649 (1980).
K. E. Porter, A. D. C. Cantwell, and C. McDermott, *AIChE J.*, **17**, 536 (1971).

Pilot Plants for Gas Treating with Chemical Solvents:

H. E. Benson, J. H. Field, and R. M. Jameson, *Chem. Eng. Prog.*, **50**, 356 (1954).
L. J. P. Evans and Q. M. Siddique, British Gas Corp. Communication 980 (1975).

A portable pilot plant for natural gases has been described by:

C. L. Dunn, E. R. Freitas, J. W. Goodenbour, M. T. Henderson, and M. N. Papadopoulos, *Oil and Gas J.*, **62**, 95 (1964).

The data reported in Figure 15.5.6 are reported by:

H. E. Benson, J. H. Field, and W. P. Haynes, *Chem. Eng. Prog.*, **52**, 433 (1956).

Their analysis on the basis of Eq. 15.5.2 is discussed in detail by:

S. Joshi, Ph.D. Thesis, University of Delaware, 1980.

Selection and Design of Packed Tower Support Plates, Liquid Distributors, and Redistributors:

H. F. Rase, *Chemical Reactor Design for Process Plants*, Vol. 1, Wiley, New York, 1977.

Tables 15.6.1-15.6.3 and Figure 15.6.5 are taken from:

O. Frank, *Chem. Eng.*, **84**(2), 119 (1977).

Figure 15.6.2 is from:

M. S. Kuk and C. S. Brown, "Selection and Interpretation of Key Design Variables in Packed Tower Pilot Plant Data," presented at AIChE 71st Annual Meeting, Miami, FL, November 12-16, 1978.

Figure 15.6.3 is from:

W. Meier, W. D. Stoecker, and B. Weinstein, *Chem. Eng. Prog.*, **73**, 71 (1977).

Figure 15.6.4 is from:

F. Kastanek and G. Standart, *Sepn. Sci.*, **2**, 439 (1967).

Figure 15.6.6 is from:

J. E. McCarthy, *Chem. Eng. Prog.*, **17**(58), (1980).

Flowsheet Design:

Figure 15.7.1 is from:

L. Crabs, R. Pouilliant, and F. Van Heck, "Energy Savings for Carbon Dioxide Removal Systems," Paper 110b, 72nd AIChE Meeting, San Francisco, November, 1979.

Figure 15.7.2 is from:

H. E. Benson and R. W. Parrish, *Hydrocarbon Proc.*, **53**, 81 (1974).

Figure 15.7.3:

K. J. Stokes, "The Economics of CO_2 Removal in Ammonia Plants," Paper 110a, 72nd AIChE Meeting, San Francisco, November, 1979.

Table 15.8.1 is taken from:

K. F. Butwell, D. J. Kubek, and P. W. Sigmund, *Chem. Eng. Prog.*, **75**(7), 75 (1979).

Corrosion Problems:

W. R. Schmeal, A. J. MacNab, and P. R. Rhodes, *Chem. Eng. Prog.*, **74**(3), 37 (1978).

Belgium Patent, BE-873-127 (1979) to Union Carbide Corp. (discloses corrosion inhibitors for alkanolamine solutions).

T. J. Handley, "Operating Experience in a Benfield Carbon Dioxide Removal System," Paper 97, presented at Safety in Ammonia Plants and Related Facilities, AIChE Meeting, Miami, FL, November 12–17, 1978 (discusses vanadate inhibitors for promoted carbonate scrubbing solutions).

C. W. Martin, R. G. Asperger, J. R. Davidson, and R. L. Pearce, Energy Processing /Canada, 52 (1977) (discusses corrosion inhibitors for high strength MEA solutions).

C. G. Arnold, *Chem. Eng. Prog.*, **74**, 43 (1978).

H. U. Schutt, "Materials Experience in Sulfinol Plants," Paper #10 presented at National Association of Corrosion Engineers, "Corrosion 1979" Meeting, Atlanta, GA, 3/12–16/79.

Amine Degradation Phenomena:

L. D. Polderman, C. P. Dillon, and A. B. Steele, *Oil and Gas J.*, **54**(2), 180 (1955).

K. F. Butwell, D. J. Kubek, and P. W. Sigmund, "Primary vs. Secondary Amines," Proceedings Gas Conditioning Conference, University of Oklahoma, 1980 (Table 15.8.1 was taken from this reference).

J. C. Dingman and T. F. Moore, Paper 18, Selected Proceedings from Gas Conditioning Conferences, Vol. III, University of Oklahoma, 1952 through 1974.

Foam Control:

K. E. Meusburger and E. W. Segebrecht, "Foam Depressants for Gas Processing Systems," Gas Conditioning Conference, University of Oklahoma, March 3, 1980.

S. Ross, *Chem. Eng. Prog.*, **63**(9), 41 (1967).

R. D. Kulkarni, E. D. Goddard, and B. Kanner, *Ind. Eng. Chem. Fund.*, **16**(4), 472 (1977).

C. F. Pratt and S. Y. Hobbs, *Chem. Eng.*, **82**, 112 (1975).

Environmental Considerations:

R. W. Parrish and R. O. Seidel, "Environmental Considerations in Acid Gas Removal Systems," Paper A4f, presented at Safety in Ammonia Plants and Related Facilities Symposium, AIChE Second Pacific Chemical Engineering Congress, Denver, CO, August 28–31, 1977 (Table 15.8.2 is taken from this paper).

Appendix Gas Treating Patents

Gas treating is an industrial art. An important source of information about industrial practices, both current and historical, is the patent art that has developed over a long period of time.

Patents issued by the U.S. Patent and Trademark Office are classified into over 300 classes and 50,000 subclasses. Each patent is classified into a specific subclass; it is often cross-referenced to one or more subclasses.

Subclasses 220–234 (and Class 423, Chemistry, Inorganic) would seem to be of greatest interest for gas treating as shown below:

Subclass No.	Title
220	Carbon dioxide or hydrogen sulfide component
221	Utilizing thionate or thiosulfate as reactant
222	Reacting mixture with sulfur dioxide, sulfite, or bisulfite
223	Utilizing reactant containing arsenic, phosphorus, or boron
224	By oxidizing or burning component
225	By suspension of metal oxide or hydroxide particles in liquid
226	Utilizing organic reactant
227	+ Phenolate or phenolic type
228	+ Amine
229	- Ethanolamine
230	Utilizing solid sorbent, catalyst, or reactant
231	+ Iron oxide or hydroxide

Subclass No.	Title
232	Utilizing carbonate as reactant
233	+ And regenerating reactant by incoming actifying gas
234	Utilizing ammonium or metal hydroxide solution

It is quite reasonable to believe that patents on gas treating may be found in other subclasses.

Published United States patents (through early 1981) have been classified as shown in Table A-1. The relationship of the classification to the chapters of the book is quite straightforward. Each patent has been entered only *once* into the classification. This analysis of the patents is undoubtedly not a complete one; it should *not* be used to decide the potential patentability of an invention.

TABLE A-1. Classification of Issued United States Patents on Gas Treating in Subclasses 220–234

I. Gas treating processes
   - A. Adsorption
   - B. Chemical reactions
   - C. Chemical solvents
   - D. Physical solvents
   - E. Hybrid systems
   - F. COS, CS_2, mercaptan removal
   - G. Selective H_2S removal
   - H. HCN removal
   - I. Conversion of H_2O to sulfur
   - J. Conversion of H_2S to SO_3
   - K. H_2S removal with chemically reactive solutions
   - L. H_2S removal with adsorbents and reactive solids
   - M. H_2S removal by condensation
   - N. Removal of specific impurities
   - O. Purification of special gases

II. CO_2 Recovery and purification

III. Deodorization

IV. Personal protection devices, controlled atmosphere rooms, submarines

V. Design of gas treating processes
   - A. Analysis
   - B. Contacting devices
   - C. Control systems
   - D. Diolefin removal
   - E. Power cycles and turbines
   - F. Regeneration of aqueous solutions
   - G. Removal of impurities from gas treating solutions
   - H. Waste water purification
   - I. Prevention of sulfur clumping in alkaline solutions

TABLE A-1. (*Continued*)

VI. Processes incorporating gas treating
- A. Fluoride removal–aluminum manufacture
- B. Ammonia
- C. Ammonia recovery
- D. Boron trifluoride purification
- E. Brine purification
- F. Calcium cyanamide
- G. Chlorine
- H. Coal desulfurization
- I. Coal gasification
- J. Hydrogen fluoride
- K. Hydrogen
- L. Hydrogen sulfide
- M. Incinerator
- N. Magnesia
- O. Methanol
- P. Integration with oil field production
- Q. Pulp and paper
- R. Rayon
- S. Solar furnace
- T. Tar sands and oil shale
- U. Sewer sludge
- V. Water gas and coke oven gases
- W. Flue gas
- X. Iron ore reduction
- Y. Geothermal steam

VII. Miscellaneous
- A. Amine scrubbing of hydrocarbon liquids
- B. H_2S emissions from slag quenching
- C. CO_2 removal–electrolytic
- D. Hydrocarbon recovery–hydrate formation

The relationship of this outline to the chapters of the book is obvious. However, it should be noted that each patent has been entered only *once* in the patent classification.

I. Gas treating processes
- A. Adsorption
  - 1. Apparatus and processes
    - 1,947,779
    - 2,384,311
    - 2,446,076
    - 2,551,905
    - 2,603,553
    - 2,632,528
    - 2,682,444
    - 2,685,343
    - 3,907,967
    - 3,928,532
    - 4,259,301
  - 2. Adsorbents
    - a) CO_2 and H_2S removal
      - 3,199,946
      - 3,274,125
      - 3,466,138
      - 3,556,716

b) CO_2 removal
1,821,195
1,852,497
1,855,355
1,912,877
2,332,887
3,117,844
3,511,595
3,627,478
3,659,400
3,660,023
3,865,924
3,906,945
3,909,206
4,005,708
4,039,620
4,231,760

c) CO_2 from NH_3
2,992,703

3. Coated solids
1,396,586
2,193,570
2,818,323
3,511,594
3,859,417
3,935,294

4. Membranes
2,077,298
3,515,513
3,819,806
3,911,080
4,089,653
4,115,514
4,117,079
4,147,754
4,174,374
4,187,086

5. Pyrolysis of hydrocarbons on adsorbents
1,514,889
2,647,042

B. Chemical reactions

1. Alkali metal hydrides
1,681,702

2. Catalytic hydrogenation of CO
1,990,708
2,747,970
2,892,685

3. Catalytic oxidation
2,998,303
3,540,867
3,833,477
3,867,313
3,909,452
3,941,868

4. Catalytic oxidation of ammonia
3,661,507
3,773,912
3,812,236
4,181,706

5. Combustion
1,150,337
1,154,171
1,154,174
2,007,586

6. Molten salts
1,403,391
1,955,328
2,032,925
3,552,912
3,668,120
3,671,185
3,917,797
3,919,390
3,996,335
4,053,554
4,086,323
4,207,298
4,244,930

C. Chemical solvents

1. Alkaline salt solutions

a) Borates for CO_2 removal
1,964,808
2,395,564
2,395,565
2,477,636

b) Alkali and alkaline metal hydroxides

(1) H_2S and CO_2 removal
242,546
402,787
1,133,087
1,212,456
1,239,516
1,523,845
1,523,872
1,580,452
1,717,761
1,920,626
1,930,875
2,205,122
2,479,781
2,684,231
2,740,693
2,761,755
2,955,901
3,262,753
3,471,249
3,839,548
3,962,410

(2) SO_2 removal
2,176,196
3,767,765
3,932,584
3,970,740

(3) Cl_2 and CO_2 removal
357,011

(4) H_2O removal (drying)
1,169,371

(5) Reclaiming of solutions
4,041,129

c) Alkali and alkaline metal hydroxides containing suspensions of metal salts
1,609,872
1,771,153
1,824,103
1,834,764
2,081,960
2,798,802
3,021,189
3,514,283
3,519,384
3,557,011
4,058,587
4,163,043

d) Sodium phenolates
1,968,275
1,971,798
2,002,357
2,028,124
2,163,169
2,196,281
2,219,713
2,242,323
2,244,731
2,275,294
2,326,122

e) Potassium and sodium phosphates
491,365
1,945,163
2,064,838
2,157,879
2,164,194
2,182,305
2,368,595
3,911,093

2. Ammonia

a) NH_3 (aqueous) scrubbing for CO_2 and H_2S removal
Re 15,314
222,715
467,605
1,038,315
1,360,734

1,425,577
1,673,877
1,875,920
1,915,723
2,018,863
2,028,336
2,070,620
2,162,838
2,866,679
2,878,099
2,884,304
2,889,197
3,337,297
3,556,721
3,718,734
3,789,105
3,819,816

b) NH_3 (liquid) scrubbing for CO_2 and H_2S removal
1,694,594
2,966,392
3,298,949

c) NH_3 (aqueous) scrubbing for SO_2
3,681,024
3,943,237

d) Integration of NH_3 scrubbing for CO_2 and H_2S removal with NH_3 manufacture
1,932,507
2,106,734
2,785,045
2,955,909
3,107,149
3,524,722
3,640,052
3,684,442
3,719,740

3. Copper complexing
1,962,525
2,965,438
3,014,973
3,403,196
3,960,910
4,192,854

4. Aqueous amines

a) Amines for CO_2 and H_2S removal
1,834,016
1,951,992
1,990,217
2,011,386
2,065,112
2,106,435
2,106,446
2,139,123
2,139,375
2,143,393
2,486,778
2,722,500
2,801,151
3,387,917
3,502,428
3,622,267
3,632,519
3,773,896
3,793,434
4,096,085
4,101,633
4,112,502
4,217,236

b) Amines for SO_2 removal in presence of CO_2
4,071,602

c) Salts of organic acids for CO_2 and H_2S removal
2,053,650
2,176,441
4,080,423

d) Amine–hydrocarbon (emulsions) mixtures
2,497,954

2,600,328
3,660,016
3,696,162
4,137,294

e) Amine processes
Re 18,958
2,607,657
2,613,132
2,638,405
3,245,752
3,271,110
3,420,633
3,565,573
3,642,430
3,775,507
3,823,222
3,958,943
3,972,692
4,079,117
4,093,701
4,138,230

f) Operating features
2,487,577
3,829,521

g) Enhancing solubility of S
3,545,916
3,826,811
4,033,410

h) Foam prevention/defoamers
1,944,122
2,390,899
2,544,564
3,100,680
3,943,227

i) Additives to reduce degradation
Re 29,428
1,956,113
2,559,580
3,137,654
3,372,981
3,829,494
3,864,449
3,891,742

j) Reclaiming of degradation products
2,701,750
2,761,818
2,797,188
3,535,260
3,658,462
4,113,837
4,138,468

k) Regeneration of "rich solutions"
2,890,931
3,554,691
4,184,855

l) Recovery of amines from regenerator overhead gases
2,487,576
2,592,762
2,608,461
2,608,462
2,768,058
3,228,874

m) Corrosion inhibitors
2,031,632
2,377,966
2,487,578
2,544,649
2,668,748
2,776,870
2,826,516
2,869,978
3,518,048
3,808,140
3,896,044
3,923,954
3,959,170
4,071,470
4,100,099

4,100,100
4,102,804
4,143,119
4,205,050

5. Aqueous sodium carbonate
   a) Promoters
      1,679,858
      1,724,909
      1,850,388
      2,718,454
   b) Solutions containing suspensions of metal salts
      1,290,244
      1,545,703
      1,741,113
      1,752,382
      1,792,097
      1,800,297
      1,806,370
      1,866,229
      1,932,812
      2,002,365
      3,079,223
      3,988,422
   c) Vacuum carbonate process
      1,802,580
      1,849,526
      1,864,068
      2,256,962
      2,379,076
      2,458,505
      2,464,805
      2,464,824
      2,490,799
      2,671,008
      2,675,296
      2,837,399
      2,842,423
      3,030,182
   d) Processes
      1,390,037
      1,942,054
      1,942,071
      1,961,255
      2,043,109
      2,386,359
      3,264,056
   e) Regeneration of "rich solutions"
      Re 21,358
      1,440,977
      1,478,543
      1,533,773
      1,598,895
      1,653,933
      1,654,782
      1,690,437
      1,690,438
      1,727,559
      1,743,479
      1,802,580
      1,817,777
      1,831,731
      1,918,153
      2,014,250
      2,094,070
      2,134,507
      2,134,518
      2,142,917
      2,183,324
      2,318,522
      3,846,539

6. Aqueous potassium carbonate
   a) Promoters
      2,840,450
      2,993,750
      3,041,135
      3,086,839
      3,304,151
      3,637,345
      3,848,057
      3,851,041
      3,856,921
      3,896,212
      3,907,969

3,932,582
4,094,957
4,096,085
4,112,050
4,217,237
4,217,238

b) Borate mixtures—absence of promotors
2,374,876

c) Process
2,886.405
3,685,960
3,773,895
4,146,569
4,160,810
4,180,548
4,251,494

d) Foam prevention/defoamers
4,183,903
4,208,385

e) Removal of K_2SO_4
2,997,366

f) Regeneration of "rich solutions"
2,178,968
3,344,050
3,563,695
3,563,696
3,725,529
4,258,019

g) Corrosion inhibitors
(1) Absence of promoters
3,087,778
3,181,929
3,275,403
3,721,526
(2) Promoters present
3,144,301
3,863,003

3,951,844
4,061,714

D. Physical solvents
1. Water
655,436
4,087,258
2. Hydrocarbons
1,422,183
2,781,863
3. Propylene carbonate
2,926,751
4. Glycerin
769,860
5. Sulfolane
3,039,251
6. ϵ-caprolactam
3,653,809
7. *N*-methyl-2-pyrrolidone
3,103,411
4,208,382

E. Hybrid systems
1. Sulfolane/amines
3,161,461
3,338,664
3,347,621
3,352,631
3,363,989
3,463,603
3,551,102
3,630,666
3,989,811
4,025,322
4,100,257
4,112,051
4,240,923
2. Hydrocarbons/amines
1,936,570
2,161,663
2,475,334
2,614,904
4,100,257
4,112,051
4,240,922
4,240,923

3. Alcohols/amines
   2,445,468
   3,653,810
4. Alcohols/metal salts
   4,144,314
5. ε-caprolactam/amines
   4,088,735
   4,145,192
6. Acetone/amines
   1,774,633
   3,607,003
7. *N*-methyl-2-pyrrolidone/amines
   3,642,431
   3,656,887
   3,767,766
8. Glycols ethers–polyhydric alcohols/amines
   2,177,068
   2,435,089
   2,437,288
   2,518,752
   2,547,278
   2,550,446
   2,712,973
   2,860,030
   2,946,652
   3,516,793
   3,681,015
   3,831,348
   4,044,100

F. COS, CS_2 mercaptan removal
1,976,806
2,149,039
2,311,342
2,383,416
2,490,840
2,557,643
2,581,135
2,598,034
3,061,421
3,116,970
3,490,865
3,794,710
3,812,200
3,961,015
3,965,244
3,966,875
4,041,130
4,100,256
4,153,674
4,233,141
4,254,094

G. Selective H_2S removal
1. Processes
   137,713
   2,028,125
   2,082,213
   2,747,962
   2,994,578
   3,139,324
   3,266,866
   3,934,012
   3,963,825
   4,020,144
   4,085,192
   4,107,270
   4,112,049
   4,144,039
   4,198,387
   4,198,388
   4,241,032
2. Aqueous K_2CO_3
   3,931,389
3. Methanol
   2,813,126

H. HCN removal
624,383
832,466
1,604,565
1,855,090
1,971,799
2,113,002
2,137,602
2,143,821
2,419,225
2,615,787
2,784,055
2,837,398
2,839,356

2,855,277
2,859,090
3,855,390
3,887,682
3,904,734
3,923,957
3,953,577
3,956,460
4,062,931

I. Conversion of H_2S to S
1. Claus
1,782,590
1,807,528
1,895,724
1,900,751
1,922,872
2,037,790
2,083,895
2,298,641
2,355,147
2,787,519
2,787,598
3,186,789
3,363,401
3,407,040
3,414,524
3,445,181
3,666,418
3,692,480
3,717,699
3,822,337
3,851,050
3,885,020
3,939,250
3,947,547
4,012,486
4,077,777
4,086,963
4,088,743
4,092,404
4,123,505
4,169,136
4,171,347
4,197,277
4,198,386
4,210,627
4,233,276
4,241,040
4,263,270

2. Stretford
3,928,535
3,959,452
3,965,243
3,966,876
3,972,989
3,975,508
4,002,727
4,049,776
4,060,594
4,083,945
4,098,886
4,118,467
4,206,194
4,251,492
4,251,493
4,260,590

3. SO_2 in a solvent
3,928,548
4,107,269

4. SO_2 in a solvent-process
1,832,448
1,904,424
1,925,198
3,672,837

5. O_2 in a solvent
2,881,407
3,725,537
3,914,399
3,956,473
4,177,250

6. O_2 in a solvent–soluble metal salts
Re 16,631
1,597,964
2,088,814
2,641,527
2,972,512
3,077,488
3,097,925

3,226,320
3,622,273
3,676,356
3,761,569
3,849,540
3,932,583
3,933,993
4,009,251
4,011,304
4,036,942
4,091,073
4,189,462

7. O_2 in a solvent–insoluble metal salts
Re 16,631
Re 17,162
315,953
1,160,375
1,160,836
1,578,560
1,629,396
1,629,536
1,656,881
1,700,698
1,724,909
1,732,905
1,733,321
1,734,304
1,734,307
1,741,113
1,752,382
1,771,153
1,774,368
1,812,099
1,817,778
1,819,833
1,824,104
1,841,419
1,847,795
1,852,160
1,854,491
1,858,919
1,947,983
2,019,468
2,503,528
2,641,526
2,689,164
2,733,979
3,068,065
3,761,570
3,950,492
4,123,506

8. O_2 in a solvent–organic redox systems
2,868,617
2,975,026
3,035,889
3,459,495
3,937,795
4,238,462

9. O_2 in a solvent–as salts
1,179,177
1,719,180
1,719,762
1,809,818
1,844,694
1,852,014
2,937,071
2,943,910

10. O_2 in a solvent-activated carbon slurry
4,049,399
4,196,183

11. Electrolytic
3,933,608
4,196,183

12. Hydrogen peroxide
4,163,044
4,250,151

13. Iodine
3,607,004
3,716,620
4,075,282
4,094,962
4,220,505

14. Lignins
3,989,810

15. Glowing wire/high-temperature arc
1,120,475
2,984,548
4,171,347

16. O_2 with oil shale solids
    4,082,146
    4,086,962
17. O_2 with lime
    402,767
    2,052,892
18. O_2 with pumice
    1,317,583
    1,396,586
19. O_2 with carbon solids
    1,412,954
    1,430,920
    1,479,852
    1,911,498
    1,984,971
    3,469,934
    3,598,521
    3,634,028
    3,637,352
    3,701,824
    3,790,659
    4,196,183
20. O_2 with molecular sieves
    3,864,452

J. Conversion of H_2S to SO_3
   4,139,596

K. H_2S removal with chemically reactive solutions
   1. Alkaline salt/alkaline earth hydroxide solutions
      480,312
      1,681,702
      1,685,733
      1,819,658
      1,878,682
      1,919,778
      2,444,930
      2,772,945
      2,845,382
      3,579,305
      4,151,260
      4,162,298
   2. Sodium carbonate solutions
      1,665,013
      1,690,438
      1,700,698
      1,937,196
      3,846,538
   3. Aqueous ammonia
      1,795,120
      1,848,516
      1,915,755
      1,979,934
      2,045,747
      3,715,426
      3,728,440
      3,728,441
      3,840,653
      a) With metal salts
         (1) iron
             1,733,321
             1,822,380
             1,957,253
             2,169,282
         (2) Arsenic
             1,957,262
         (3) Calcium
             1,349,750
   4. Aqueous polythionates
      1,542,971
      1,589,749
      1,660,741
      1,826,779
      1,832,325
      1,840,579
      1,841,419
      1,843,224
      1,861,268
      1,868,843
      1,932,820
      1,932,820
      1,944,978
      2,029,262
      2,152,454
      2,772,146
      3,859,414
      3,941,875
   5. Sulfite solutions
      1,580,451

3,446,595
3,833,714

6. Cadium sulfate in sulfuric acid
   2,624,694
7. Methyl saluglate solutions
   3,415,633
8. Chromate solutions
   2,044,116
   2,405,672
9. Phosphoric ester-amine
   3,495,933
10. Insoluble carbonates
    1,716,028

L. H_2S removal with adsorbents and reactive solids

1. Massive iron
   Re 4,998
   76,544
   100,309
   109,268
   292,341
   314,342
   853,402
   1,389,980
   1,390,037
   1,436,196
   1,453,754
   1,520,726
   1,571,624
   1,578,560
   1,624,625
   1,626,664
   1,757,826
   1,809,990
   1,815,933
   1,816,533
   1,822,293
   1,851,312
   1,856,301
   1,878,609
   1,934,478
   2,007,741
   2,024,393
   2,110,241
   2,181,433
   2,208,029
   2,213,615
   2,429,759
   2,979,384
   2,983,573
   3,579,293
   4,201,751
   4,238,463
   4,246,243
   4,246,244
2. Supported iron
   3,672,125
   4,089,809
3. Calcium salts
   1,523,872
   1,523,845
   1,580,452
   1,779,024
   3,966,431
   4,092,128
   4,191,538
   4,252,778
4. Aluminates
   2,037,789
   2,442,982
   3,943,226
5. Copper salts and alloys
   83,417
   564,922
   737,579
   1,770,694
   1,947,776
   2,288,749
   2,747,968
   4,008,174
   4,101,641
   4,251,495
6. Lead salts
   137,615
7. Cobalt and nickel salts
   1,760,522
   1,812,526
   1,815,846
   1,818,615
   1,841,421

2,175,190
2,220,849
4,039,619
4,082,834
4,196,174

8. Metal phosphates
   2,930,673
9. Supported rare earths
   3,974,256
   4,002,720
10. Tellurium dioxide
    4,053,574
11. Zinc salts
    1,759,408
    3,427,253
    3,441,370
    4,088,736
    4,147,212
12. Manganese oxide
    4,164,544
    4,180,549
    4,268,487
13. Coatings on activated carbons
    214,412
    507,331
    1,281,597
    2,403,319
    3,491,031
14. Silica gel
    1,335,348
    4,181,704

M. H_2S removal by condensation
2,901,326
3,108,855
3,498,087
4,169,133

N. Removal of specific impurities

1. Hydrogenation of impurities in benzene
   2,957,925
2. Acid scrubbing of calcium oxide
   2,692,815
   3,961,032
3. Adsorption of hydrogen chloride
   3,935,295
4. Nitrogen oxides
   1,976,704
   2,073,083
   3,674,423
   3,677,708
   3,961,018
   4,152,296
5. Oxygen
   a) High temperature
      1,787,795
   b) Adsorption
      3,852,406
      4,104,192
      4,199,472
   c) Reactive solids
      1,787,795
      4,153,576
6. Particulates (containing adsorbed gases)
   3,616,610
   3,782,905
   3,907,520
7. Sulfur dioxide
   3,933,994

O. Purification of special gases

1. Acetylene
   209,350
   625,479
   701,995
   2,908,546
   3,299,161
   3,974,085
2. Carbon monoxide
   1,597,345
   1,999,159
   2,890,930
   3,927,176
3. Engine exhaust gases
   1,716,481
   3,961,019
4. Inert gases
   2,284,662
   3,061,403

4,012,490
4,080,429

5. Nitrogen
923,846
965,393
1,036,788
2,314,827
2,448,719
3,202,482

6. Olefin purification
2,413,254
2,471,550
3,840,611

7. Oxygen
2,490,986

8. Ozone
2,700,648

9. Radon/radium derived gases
1,570,834

II. CO_2 recovery and purification
Re 3688
135,001
496,546
740,700
793,707
1,044,531
1,519,932
1,619,336
1,687,229
1,924,178
1,931,817
1,942,485
1,968,899
2,016,815
2,033,933
2,205,962
2,225,131
2,726,930
3,992,293

III. Deodorization
316,465
666,044
2,220,849
2,793,096
3,071,433
3,803,290
3,923,955
3,946,101
3,949,055
3,949,056
3,969,479
4,070,300
4,125,589
4,212,852
4,256,710
4,256,728
4,238,461

IV. Personal protection devices, controlled atmosphere rooms, submarines
87,155
216,083
632,400
680,028
693,638
795,679
928,978
1,111,055
1,379,221
1,792,939
1,860,298
1,922,187
2,033,698
2,270,025
2,322,206
2,389,309
2,401,484
2,430,861
2,848,309
2,889,210
3,148,034
3,173,849
3,316,163
3,410,191
3,489,693
3,497,312
3,502,429
3,532,637
3,544,264

3,619,130
3,655,346
3,771,959
3,847,837
3,883,637
3,920,803
3,957,948
4,000,246
4,047,894
4,102,803
4,107,268
4,112,048
4,238,464

V. Design of gas treating processes

A. Analysis
1,512,893
2,174,349
2,554,819
3,837,807
4,127,386
4,141,800
4,172,880
4,238,460

B. Contacting devices
1,668,013
1,916,551
1,945,002
2,165,490
2,877,101
2,912,301
2,941,872
3,442,603
3,489,506
4,024,028
4,125,589
4,156,712
4,225,560

C. Control systems
2,695,836
2,799,851
3,411,866
4,106,916
4,172,880
4,208,383

D. Diolefin removal
3,911,082
3,926,591

E. Power cycles/turbines
3,276,203
3,418,082
4,208,384

F. Regeneration of aqueous solutions
1,665,013
2,368,600
2,466,183
2,477,134
2,635,039
3,035,900
3,101,996
3,208,557
3,362,891
3,690,816
3,709,976
3,714,327
3,787,559
3,897,227
3,962,404
4,038,366
4,073,863
4,085,688
4,152,117
4,198,378

G. Removal of impurities from gas treating solutions

1. Foaming agents
4,002,721

2. Oxygen from alkaline solutions
Re 29,428
3,642,448

H. Waste water purification
3,867,950
4,087,513
4,168,299

I. Prevention of sulfur clumping in alkaline solutions
2,215,754
4,013,426

J. Coal gasification
2,672,402
3,506,590
3,770,399
4,080,424
4,123,502
4,175,928
4,233,275

K. Hydrogen fluoride
2,526,585
2,726,932

L. Hydrogen
1,904,592
2,487,981
2,513,022
2,594,269
2,879,139
3,069,250
3,116,116
3,413,214
3,552,924
3,650,697
3,676,516
3,694,344
3,719,749

VI. Processes incorporating gas treating

A. Fluoride removal in aluminum manufacture
3,792,571

B. Ammonia
3,947,551
4,259,300

C. Ammonia recovery
1,916,980
2,067,311
3,649,190
4,009,243

D. Boron trifluoride purification
2,381,027

E. Brine purification
3,676,067

F. Calcium cyanamide
3,147,069

G. Chlorine
3,881,893
3,896,213

H. Coal desulfurization
4,013,426

I. Coal gasification
2,672,402
3,506,590
3,770,399
4,080,424
4,123,502

J. Hydrogen fluoride
2,526,585
2,726,932

K. Hydrogen
1,904,592
2,487,981
2,513,022
2,594,269
2,879,139
3,069,250
3,116,116
3,413,214
3,552,924
3,650,697
3,676,516
3,694,344
3,719,749

L. Hydrogen sulfide
1,580,451

M. Incinerator
3,706,445

N. Magnesia
3,357,787

O. Methanol
2,425,625

P. Integration with oil field production
3,097,917
3,274,101
3,812,030
3,817,038

Q. Pulp and paper
1,854,428
3,028,295
3,183,145
3,293,113
3,366,535
3,376,102
3,431,165
3,525,666
3,532,595
3,554,859
3,574,556
3,647,363
3,752,875
3,826,710
3,841,961
3,842,160
3,857,902
4,098,639

R. Rayon
2,378,689
2,899,267
4,049,775

S. Solar furnace
4,000,733

T. Tar sands and oil shale
3,122,493

U. Sewer sludge
2,097,454

V. Water gas and coke oven gases
676,209
1,758,796
2,850,352
3,993,730

W. Flue gas
4,208,381

X. Iron ore reduction
4,201,571

Y. Geothermal steam
4,202,864
4,242,305
4,224,151

VII. Miscellaneous

A. Amine scrubbing of hydrocarbons
2,238,201

B. H_2S emissions from slag quenching
3,900,304

C. CO_2 removal–electrolytic
3,519,488
3,527,618

D. Hydrocarbon recovery–hydrate formation
2,375,559
2,375,560

Index